COLLEGE LIBRARY
SUFFOLK UNIVERSITY
BOSTON, MASS.

DISCARDED BY
SUFFOLK UNIVERSITY
SAWYER LIBRARY

Lichen Ecology

Lichen Ecology

Edited by

MARK R. D. SEAWARD

School of Environmental Science
University of Bradford

1977

ACADEMIC PRESS
London   New York   San Francisco
A Subsidiary of Harcourt Brace Jovanovich, Publishers

ACADEMIC PRESS INC. (LONDON) LTD.
24/28 Oval Road,
London NW1

United States Edition published by
ACADEMIC PRESS INC.
111 Fifth Avenue
New York, New York 10003

Copyright © 1977 by
ACADEMIC PRESS INC. (LONDON) LTD.

All Rights Reserved
No part of this book may be reproduced in any form by photostat, microfilm, or any other means, without written permission from the publishers

Library of Congress Catalog Card Number: 77 76676
ISBN: 0 12 634350 0

PRINTED IN GREAT BRITAIN BY
THE WHITEFRIARS PRESS LTD, LONDON AND TONBRIDGE

Contributors

AHTI, T., *Botanical Museum, University of Helsinki, Unioninkatu 44, SF-00170 Helsinki 17, Finland.*
BRIGHTMAN, F. H., *British Museum (Natural History), Cromwell Road, London SW7 5BD, England.*
GERSON, U., *Faculty of Agriculture, The Hebrew University of Jerusalem, P. O. Box 12, Rehovot 76–100, Israel.*
GILBERT, O. L., *Department of Landscape Architecture, The University, Sheffield S10 2TN, England.*
HAWKSWORTH, D. L., *Commonwealth Mycological Institute, Ferry Lane, Kew, Surrey TW9 3AF, England.*
JAMES, P. W., *British Museum (Natural History), Cromwell Road, London SW7 5BD, England.*
LINDSAY, D. C., *Department of Botany, The University of Alberta, Edmonton, Alberta T6G 2E9, Canada.*
RICHARDSON, D. H. S., *Department of Biology, Laurentian University, Sudbury, Ontario P3E 2C6, Canada.*
ROGERS, R. W., *Botany Department, University of Queensland, St. Lucia, Brisbane 4067, Australia.*
ROSE, F., *Department of Geography, King's College, University of London, Strand, London WC2R 2LS, England.*
SEAWARD, M. R. D., *School of Environmental Science, University of Bradford, Bradford BD7 1DP, England.*
TOPHAM, P. B., *Scottish Horticultural Research Institute, Invergowrie, Dundee DD2 5DA, Scotland.*
WEBER, W. A., *University of Colorado Museum, Boulder, Colorado 80302, U.S.A.*
YOUNG, C. M., *Department of Biology, Laurentian University, Sudbury, Ontario P3E 2C6, Canada.*

Preface

Lichens play varying roles in pioneer, transition and climax ecosystems: they make specific demands on the environment and respond critically to changes in it. This book essays a compilation of current knowledge on lichens in relation to the physical and biological components of their environment. As many aspects as possible of the subject are discussed without undue encroachment on material in recently published books and major papers, which this volume is designed to complement. Readers of this book will judge the timeliness of the compilation: there can be no doubt that collation of the ecological and geographical aspects of lichenology has been neglected. Although there has been undoubted development of information on the subject over the years, it is widely scattered, often in obscure or inaccessible literature sources, and analysis and interpretation have been lacking. The present volume provides a body of information which will I hope stimulate further research and promote the further development of lichen ecology and geography, but makes no claim to be comprehensive.

The contents are essentially for the purposes of reference, but the approach of the numerous authors differs considerably: for example, some accounts have a review bias, others are mainly interpretative, while some incorporate original data. Furthermore, the extensive list of references accompanying each chapter, the bibliographic guide to the lichen floras of the world, the glossary and indexes lend the work an encyclopaedic schema. Every attempt has been made to present the chapters in logical format: any lack of continuity results from diversity of authorship or deficiency of existing knowledge in one or more aspects of the subject. At one time more contributions were envisaged, but, for various reasons, the number had to be reduced.

Few lichenologists can have been more fortunate in their friends than myself; the extent of their help cannot be adequately expressed. I would also like to refer here to the invigorating role of the British Lichen Society, particularly in the provision of active help and of access to information on a scale hitherto impossible.

I am especially grateful to Dr D. L. Hawksworth for the help and advice he has given me throughout the lengthy, and often arduous, task of compiling this book; he has always been ready to deal *instanter* with the problem of the moment, and no trouble has ever been too great for him. I am also grateful to him for the compilation of the taxonomic index.

I am also grateful to Dr D. H. S. Richardson for his constructive criticism and his very considerable practical help with the final stages of manuscript revision and cross-reference. Further thanks are due to Mr P. W. James, Dr T. Ahti, Mr B. J. Coppins and Mr A. Henderson for reading some of the draft manuscripts and making valuable suggestions, to Miss V. A. Hinton who has been an indispensable auxiliary and has cheerfully borne a heavy burden of work, and to the authors for their patience and cooperation.

Finally, I wish to thank the staff of Academic Press for their courtesy and care in seeing the book through to publication.

Bradford, 1977   MARK R. D. SEAWARD

Contents

Contributors v

Preface vii

Chapter 1
Introduction
MARK R. D. SEAWARD 1

Chapter 2
Environmental Modification and Lichen Taxonomy
WILLIAM A. WEBER 9

Chapter 3
Colonization, Growth, Succession and Competition
PAULINE B. TOPHAM 31

Chapter 4
Lichen–Invertebrate Associations
URI GERSON and MARK R. D. SEAWARD 69

Chapter 5
Lichens and Vertebrates
DAVID H. S. RICHARDSON and COLIN M. YOUNG . . . 121

Chapter 6
Lichens of the Boreal Coniferous Zone
TEUVO AHTI 145

Chapter 7
Lichens of Cold Deserts
DENNIS C. LINDSAY 183

Chapter 8
Lichens of Hot Arid and Semi-arid Lands
RODERICK W. ROGERS 211

Chapter 9
Lichens of Man-made Substrates
FRANK H. BRIGHTMAN and MARK R. D. SEAWARD . . . 253

Chapter 10
Lichen Communities in the British Isles: A Preliminary Conspectus
PETER W. JAMES, DAVID L. HAWKSWORTH and FRANCIS ROSE . . 295

Chapter 11
Lichen Conservation in Britain
OLIVER L. GILBERT 415

Appendix A:
A Bibliographic Guide to the Lichen Floras of the World
DAVID L. HAWKSWORTH 437

Appendix B:
Selected Glossary
MARK R. D. SEAWARD and DAVID L. HAWKSWORTH . . . 503

Taxonomic Index 519

Subject Index 543

1 | *Introduction*

MARK R. D. SEAWARD

Lichens are a highly diverse group, universally distributed, and exhibit a distinctive symbiotic relationship. However, they have until recently received comparatively little scientific attention despite the foundation of a classification at the beginning of the nineteenth century (e.g. Acharius, 1810) and an interpretation of their dual algal/fungal character about 60 years later (de Bary, 1866; Schwendener, 1867). Indeed, lichens were referred to for many years in the British Isles as the "neglected plants", and even today they usually pass unregarded in school, college and university syllabuses, and general botanical textbooks have been apt to give them only scant treatment, usually within the framework of mycology.

Lichenology was largely kept alive through work on local and national floras, recorded in taxonomic monographs, until the recent upsurge of interest. The lack of attention afforded to lichens has been associated with difficulties in identification and their limited economic importance; for example, the extensive use of lichens for dyes was superseded by synthetic processes in the nineteenth century. Lichens have proved difficult to investigate taxonomically, cytologically and physiologically, but over the past two decades electron microscopy, the use of radioactive tracers and the development of physiological equipment capable of measuring slow metabolic processes, have been instrumental in providing a better understanding of lichen symbiosis. Furthermore, the use of microchemical tests, developed by Asahina (1936–1940), the compilation of the chemical constituents in lichens (e.g. Culberson, 1969, 1970) and the use of sophisticated chemotaxonomic techniques (reviewed by Nourish and Oliver, 1974) to corroborate morphological characters used in species delimitation, have advanced lichen taxonomic knowledge considerably. These more recent developments, coupled with a rise in output of specialized lichenological literature on a wide range of topics stemming from (1) works by Scandinavian and German authors, (2) a limited number of major reviews (e.g. Smith, 1921;

des Abbayes, 1951; Barkman, 1958), and (3) the three journals, *Revue Bryologique et Lichénologique*, *The Bryologist* and *The Lichenologist*, have placed lichenology on a firmer footing.

Throughout this period of lichenology, the ecological aspects of the subject were pursued mainly for the purpose of monograph and flora compilation, but all too often the ecological and phytogeographical notes accompanying lichen lists and taxonomic descriptions are a poor reflection of data assembled from fieldwork. Field observations on lichens are to be found in a variety of journals and books—often obscure, as revealed from a detailed survey for a bibliography on the British lichen flora (Hawksworth and Seaward, 1977)—devoted to such topics as topography, travel, horticulture and numerous aspects of botanical and zoological natural history. Such observations often reveal the interest in, but inadequate interpretation of, the interrelationships of lichens with other organisms, and their temporal and spatial role in a defined ecosystem. Furthermore, the inability of the zoologist to identify the associated plants, and the botanist's reluctance to determine, or have determined, those animals sheltering, grazing etc. in specific plants or plant communities, are all too obvious. The accounts of the diverse lichen–animal relationships provided in Chapters 4 and 5 are an attempt to synthesize the widely scattered and multifarious data.

Attention is given in this book to colonization, growth, succession and competition (see Chapter 3), to the exploitation of man-made substrates by lichens (see Chapter 9), and to a phytosociological analysis of lichen communities in the British Isles (see Chapter 10). Detailed treatments of specialized topics in lichen ecology recently published elsewhere and not covered specifically in this book include (1) the role of lichens in soil formation (Syers and Iskandar, 1974), (2) dispersal, germination and establishment of propagules (Pyatt, 1974; Bailey, 1976), and (3) substrate ecology (Brodo, 1974). Other aspects of lichen ecology for which detailed and/or synthesized knowledge is lacking are enumerated at the end of this chapter. A general review of lichen ecology by Smith (1921) has not been bettered; a comparative account of the relative paucity of knowledge over 40 years later is provided in Haynes (1964).

Major modern works on lichenology which contain an ecological component include (1) those connected with air pollution surveys (see Ferry *et al.*, 1973), (2) reviews of one or a limited number of aspects of the subject (e.g. Ahmadjian and Hale, 1974; Brown *et al.*, 1976), or (3) brief treatments of the current knowledge (e.g. Hale, 1974; Henssen and Jahns, 1973; Richardson, 1975). Lichen phytogeographical enquiry too has been unsatisfactorily served by published works, and regions have received taxonomic descriptions of collections rather than thorough field investigations.

Taxonomic works remain the cornerstone of lichenological fieldwork, for without them ecological and phytogeographical studies are impracticable. Unfortunately, many areas of the world are without suitable texts for comprehensive lichen identification: even continents such as North America only have detailed sources with keys based on surveys at a regional or local level, or studies of a limited number of genera often involving a taxonomic revision on a worldwide basis. A bibliographic guide to the lichen floras of the world is provided in Appendix A.

The rationale justifying this book included a desire to present a more comprehensive account of lichen ecology and geography than is available from air pollution studies. The bio-indicational importance of lichens in air pollution surveys is widely acclaimed and suitably documented (e.g. Ferry et al., 1973; Hawksworth and Rose, 1976). Natural habitats are still to be found throughout the world; on other habitats man's influences are diverse—air pollution presents only one of many problems affecting lichen distribution. It is unfortunate that many geographical settings have not, and will not, receive adequate lichen surveys prior to man's destructive practices. It is to be hoped that suitable base-line botanical surveys, taking due account of the lichen component, will be possible in many underdeveloped countries. It is appreciated, however, that such work will have a low priority; a cost–benefit analysis on the value of a lichen flora to a community is impossible to undertake in developed, let alone underdeveloped, countries in the light of present trends in population growth and economic demand. Aspects of lichen conservation in the British Isles are discussed in Chapter 11.

There can be little doubt that biogeographical aspects of lichenology have been neglected. Phytogeography presents large-scale temporal and spatial problems; consideration should be paid on the one hand to large geographical areas and, on the other, to those which are more circumscribed. Chorological generalizations in the past have often been based on the accumulation of data derived from local observations. Although a comprehensive synthesis of lichen geography is not yet possible, certain broadly based treatments can be prepared for major vegetation types of temperate (e.g. Trass, 1970; Coppins, 1976), boreal (see Chapter 6), cold desert (see Chapter 7), and hot desert (see Chapter 8) zones. Unfortunately our knowledge of the lichen floras of subtropical and tropical forests, savanna and prairie is fragmentary, and it has proved impracticable to cover these aspects in this book, even in a most rudimentary manner. Although more data on maritime lichens have been assembled recently (e.g. Fletcher, 1973a, b), there is only limited information available on freshwater lichens, namely on their zonation in lakes (Santesson, 1939) and on their eco-physiology (e.g. Ried, 1960).

Our knowledge of the general lichen ecology and geography of many countries is still in its infancy; even the factual material is imperfect and incomplete. A full count of the taxa is needed before the consequences of their interactions and their relationship to the ecosystem as a whole can be fully appreciated. In the past, taxonomic differentiation of species and other taxa has all too often been based on dubious criteria. More recently, taxonomic research has been more ecologically orientated to take due account of the many lichen taxa which exhibit a greater range of form when observed in the field than can be accommodated by the narrow concepts imposed by some taxonomists working on limited herbarium material. Future progress in lichen taxonomy as in that of other plant groups, may lie to a large extent in the hands of the cytologist and geneticist; such studies will help lichenologists to judge what is a stable genetic phenomenon and what is a phenotypic response to varying environmental conditions (Seaward, 1976). In this context, transplant experiments to investigate moulding of the lichen phenotype and the eco-physiological performance of lichens under new environmental regimes will be of paramount importance. Similarly, problems relating to morphogenesis will be resolved by these approaches (James and Henssen, 1976), coupled with culture techniques which are still largely undeveloped. The question of environmental modification and lichen taxonomy is discussed in more detail in Chapter 2.

Field lichenologists should be more aware of the part played by interacting environmental factors. Laboratory experiments which single out one or more factors for correlation studies to corroborate field observations are required, but the few studies completed to date have been disappointing. Some optimism may, however, be derived from the considerable advances in lichen physiology which show that these plants can be successfully studied in the laboratory. These advances have in the main been concerned with an understanding of symbiotic relationships and nitrogen fixation; these topics are reviewed in Ahmadjian and Hale (1974) and Brown *et al.* (1976). On site eco-physiological experiments on lichens remain in their infancy, but work so far performed by O. L. Lange (see Kappen, 1974) and others (e.g. Harris, 1971; Kershaw and Larson, 1974; Larson and Kershaw, 1975) provides strong evidence to suggest that this line of enquiry will be fruitful.

The preparation of lichen distribution maps for particular geographical areas is progressing, but lichenologists are often more competent in the preparation of maps than in their interpretation. The value of automatic data-processing methods in the preparation of maps is widely known; phytosociological classification and ordination, pattern and factor-dependent analyses, similar to those used to evaluate local distributions, can be derived from data on a grand scale too. The ecological and geographical factors which determine lichen distributions can be usefully derived from computer analyses of map data, since the human eye, through unconscious bias, may discover groupings even when none are present. Little progress on phytogeographical interpretation can be made, however, until a detailed knowledge of the distribution of taxa is achieved; from such information it may be possible to determine origins, dispersion patterns, factor tolerance and evolutionary relationships which have hitherto largely eluded the lichenologist. Such precision will depend upon fieldworkers competent to collect detailed distributional data which can be easily handled in mapping programmes.

The chapters which follow deal with a wide range of topics associated with lichen ecology and geography by synthesizing the available data to give a current account of their subject. In these chapters the authors invariably emphasize the deficiencies in knowledge, and in many cases indicate where further research is needed. In general there are many aspects which would repay enquiry by professional or amateur lichenologists, particularly if working in conjunction with specialists from other disciplines. Those areas in most need of attention are:

1. Detailed taxonomic monographs, keys, identification manuals, and distribution maps needed to provide the lichen ecologist with the basic tools for fieldwork. Unfortunately, lichen taxonomists are still relatively few in number, even on a world scale.

2. An expansion of regional and international cooperation between lichenologists to maximize effort and prevent duplication, to be effected, for example, by joint authorship of taxonomic, ecological and geographical publications and by the circulation of draft keys and distribution maps, and herbarium material.

3. Detailed ecological investigations of tropical, subtropical and aquatic environments, especially those where human interference is, or will shortly be, an influence on the ecosystem. These and studies of other ecosystems should take due account of the interactions between lichens and other organisms, and thus provide detailed data on such topics as phytosociology, food-webs, energy flow, nutrient flow, productivity, reproductive capacity and eco-physiology.

4. Evaluation of knowledge gained from lichen ecological research for applied fields such as bio-indication (including environmental monitoring), lichenometry and land reclamation.

References

des Abbayes, H. (1951). Traité de Lichénologie. *Encycl. Biol.* **41**, 1–217.
Acharius, E. (1810). "Lichenographia Universalis". Danckwerts, Göttingen.

Ahmadjian, V. and Hale, M. E., eds. (1974) ["1973"]. "The Lichens". Academic Press, New York and London.

Asahina, Y. (1936–1940). Mikrochemischer Nachweis der Flechtenstoffe I–XI. *J. Jap. Bot.* **12**, 516–525, 859–872; **13**, 529–536, 855–861; **14**, 39–44, 244–250, 318–323, 650–659, 767–773; **15**, 465–472; **16**, 185–193.

Bailey, R. H. (1976). Ecological aspects of dispersal and establishment in lichens. *In* "Lichenology: Progress and Problems" (D. H. Brown, D. L. Hawksworth and R. H. Bailey, eds), pp. 215–247. Academic Press, London and New York.

Barkman, J. J. (1958). "Phytosociology and Ecology of Cryptogamic Epiphytes". Van Gorcum, Assen.

de Bary, A. (1866). "Morphologie und Physiologie der Pilze, Flechten, und Myxomyceten". Engelmann, Leipzig.

Brodo, I. M. (1974) ["1973"]. Substrate ecology. *In* "The Lichens" (V. Ahmadjian and M. E. Hale, eds), pp. 401–441. Academic Press, New York and London.

Brown, D. H., Hawksworth, D. L. and Bailey, R. H., eds (1976). "Lichenology: Progress and Problems". Academic Press, London and New York.

Coppins, B. J. (1976). Distribution patterns shown by epiphytic lichens in the British Isles. *In* "Lichenology: Progress and Problems" (D. H. Brown, D. L. Hawksworth and R. H. Bailey, eds), pp. 249–278. Academic Press, London and New York.

Culberson, C. F. (1969). "Chemical and Botanical Guide to Lichen Products". University of North Carolina Press, Chapel Hill.

Culberson, C. F. (1970). Supplement to "Chemical and Botanical Guide to Lichen Products". *Bryologist* **73**, 177–377.

Ferry, B. W., Baddeley, M. S. and Hawksworth, D. L., eds. (1973). "Air Pollution and Lichens". Athlone Press, London.

Fletcher, A. (1973a). The ecology of marine (littoral) lichens on some rocky shores of Anglesey. *Lichenologist* **5**, 368–400.

Fletcher, A. (1973b). The ecology of maritime (supralittoral) lichens on some rocky shores of Anglesey. *Lichenologist* **5**, 401–422.

Hale, M. E. (1974). "The Biology of Lichens", 2nd ed. Arnold, London.

Harris, G. P. (1971). The ecology of corticolous lichens. II. The relationship between physiology and the environment. *J. Ecol.* **59**, 441–452.

Hawksworth, D. L. and Rose, F. (1976). "Lichens as Pollution Monitors". Arnold, London.

Hawksworth, D. L. and Seaward, M. R. D. (1977). "Lichenology in the British Isles 1568–1975". Richmond Publishing, Richmond.

Haynes, F. N. (1964). Lichens. *Viewpts Biol.* **3**, 64–115.

Henssen, A. and Jahns, H. M. (1973) ["1974"]. "Lichenes. Eine Einführung in die Flechtenkunde". Thieme, Stuttgart.

James, P. W. and Henssen, A. (1976). The morphological and taxonomic significance of cephalodia. *In* "Lichenology: Progress and Problems" (D. H. Brown, D. L. Hawksworth and R. H. Bailey, eds), pp. 27–77. Academic Press, London and New York.

Kappen, L. (1974) ["1973"]. Response to extreme environments. *In* "The Lichens" (V. Ahmadjian and M. E. Hale, eds), pp. 311–380. Academic Press, New York and London.

Kershaw, K. A. and Larson, D. W. (1974). Studies on lichen-dominated systems. IX. Topographic influences on microclimate and species distribution. *Can. J. Bot.* **52**, 1935–1945.

Larson, D. W. and Kershaw, K. A. (1975). Acclimation in arctic lichens. *Nature, Lond.* **254**, 421–423.

Nourish, R. and Oliver, R. W. A. (1974). Chemotaxonomic studies on British lichens I. *Cladonia* subgenus *Cladina*. *Lichenologist* **6**, 73–95.

Pyatt, F. B. (1974) ["1973"]. Lichen propagules. *In* "The Lichens" (V. Ahmadjian and M. E. Hale, eds), pp. 117–145. Academic Press, New York and London.

Richardson, D. H. S. (1975). "The Vanishing Lichens". David and Charles, Newton Abbot.

Ried, A. (1960). Stoffwechsel und Verbreitungsgrenzen von Flechten. II. Wasser- und Assimilationshaushalt, Entquellungs- und Submersionsresistenz von Krustenflechten benachbarter Standorte. *Flora (Jena)* **149**, 345–385.

Santesson, R. (1939). Über die Zonationsverhältnisse der lakustrinen Flechten einiger Seen im Anebodagebiet. *Medd. Lunds Limnol. Inst., Lund* **1**, 1–70.

Schwendener, S. (1867). Über den Bau des Flechtenthallus. *Verh. schweiz. naturf. Ges.* **51**, 88–89.

Seaward, M. R. D. (1976). Performance of *Lecanora muralis* in an urban environment. *In* "Lichenology: Progress and Problems" (D. H. Brown, D. L. Hawksworth and R. H. Bailey, eds), pp. 323–357. Academic Press, London and New York.

Smith, A. L. (1921). "Lichens". Cambridge University Press, Cambridge.

Syers, J. K. and Iskandar, I. K. (1974) ["1973"]. Pedogenetic significance of lichens. *In* "The Lichens" (V. Ahmadjian and M. E. Hale, eds), pp. 225–248. Academic Press, New York and London.

Trass, H. (1970). The elements and development of the lichen-flora of Estonia. *Trans. Tartu St. Univ.* **268**, *Pap. Bot.* **9**, 5–233.

2 | *Environmental Modification and Lichen Taxonomy*

WILLIAM A. WEBER

I. Introduction	9
II. Types of Environmental Effects	13
A. Morphology	14
B. Colour and Pruinosity	17
C. Anatomy	17
D. Chemistry	19
III. Terminology	19
IV. The Vagant Life-form	19
V. "Fossil Modifications"	24
VI. Biogeochemical Erosion	25
VII. Discussion	26
Acknowledgements	27
References	27

I. Introduction

Environmental modification is the production of responses that are non-transmissible—in other words, acquired characteristics. Growth, survival and reproduction of plants occur inside a wide range of acceptable, tolerable or non-lethal environmental parameters. Towards the extremes of these ranges, plants are visibly abnormal and their morphology may evoke questions as to their taxonomic identity. The genetic basis of response is now fairly well understood. In the higher plants, genetic theory and practice, together with experimental physiology, are so far advanced that there remain almost no taxonomic dilemmas attendant upon environmental modification. In the nineteenth century, modification was often mistaken for taxonomically relevant variation. With our present knowledge, however, we can extrapolate from an enormous wealth of experimental evidence (Evans, 1963; Levitt, 1972) and confidently interpret the behaviour of given individuals as representing sun or shade forms, indicating specific mineral deficiencies, viral or insect injuries or responses to other environmental pressures such as grazing and competition.

The lichen thallus, however, cannot be equated with an individual of a vascular plant species. The former is a complex assemblage of one fungus living symbiotically with one, or rarely two, algal genera. While it can be said that most lichenized fungi are algal-genus specific (but see p. 14), there is little firm evidence as to whether they will form lichen thalli with one or a range of species of that algal genus (see Hawksworth, 1976; Tschermak-Woess, 1976). The symbionts usually are not casual acquaintances but, in most instances, appear to have co-evolved. Neither partner may be identical to its free-living ancestor. The principal symbionts are often joined by others. Some lichens have a second alga entering into the symbiosis in the form of a discrete second algal layer or of encapsulated blue-green algae in cephalodia (see p. 14). Some lichens are able to grow on others and a few with their own thalli are apparently largely or exclusively restricted to specific hosts (e.g. *Lecidea insularis* on *Lecanora rupicola* and *Lecidea sulphurea*; see Hertel, 1970; Poelt and Steiner, 1972). Fruit bodies anatomically identical to independently living lichenized species but without any thalli also occur on the thalli of other lichens and may be host-specific (e.g. *Arthonia clemens* on *Lecanora rupicola*; see Hertel, 1969). These, and a considerable number of other fungi, evidently do not harm the host unduly, but presumably utilize carbohydrate manufactured by the alga in the host. These lichenicolous fungi not damaging the host are termed parasymbionts. In such cases two fungi are sharing one alga in contradistinction to cephalodia where one fungus can utilize two quite different algae. Lichens are also frequently inhabited by bacteria (Ahmadjian, 1974). Fungal saprophytes are generally rather rare on healthy lichen thalli. In all these cases two or more separate organisms are interacting. The substrate upon which the lichen grows (especially if it is a rock subject to intense variations in moisture and temperature) represents a milieu in which neither participant in the symbiosis could long survive on its own. Thus, in terms of the environment of the lichen, the substrate also must be taken into account as one of the interacting agents.

It is difficult to analyse the behaviour of any interacting biological system because of the large number of variables. In lichens it is particularly frustrating because, while the separate symbionts may be grown successfully in the laboratory, it is currently almost impossible to maintain even fast-growing whole lichens under controlled conditions on their natural substrates for any length of time. The prognosis for eventual success in handling the slower-growing saxicolous lichens, the main subject of this chapter, is still more pessimistic. When a lichenologist, lacking experimental controls, attempts to predict and to interpret environmental stress and the lichen's response to it under field conditions, his plight is equivalent to that of a general plant ecologist who nowadays is asked to predict "environmental impact" of the past or future activities of man upon an ecosystem for which he has virtually no experimental evidence concerning the behaviour of its components. Nevertheless, he must try to make as true an assessment as he can. Depending on how clever he is as an observer and how experienced he is in his subject, some of his conclusions are bound to be valid. For lichen taxonomy, the fundamental problem to be faced is this: can the lichenologist distinguish environmental modifications from genetic ecotypes in the absence of a body of experimental evidence? In other words, must the lichenologist regard all morphological manifestations of lichen variability as worthy of formal taxonomic ranks, or may he attribute some of this variability to environmental modification? I believe that he can and must, even if some of his interpretations later prove false. In some genera, such as *Aspicilia, Staurothele, Verrucaria* and *Acarospora*, the pyramid of names based on what I interpret as environmental modifications has become so massive that it becomes difficult to apply species concepts. In order to prevent undue increases in names it is necessary to try as best as we can, with the primitive tools available, to assess the role of environmental factors in the determination of thallus form and other characters.

One does not have to consider the lichen to be a dual organism in order to appreciate the complicated situation, morphologically, physiologically and evolutionarily, that the lichen symbiosis implies. A sensitive and continuing interaction exists between the lichen fungus, the phycobiont and the substrate. This interaction, in turn, is affected by short or long-term changes in the physical environment. Nevertheless, many taxonomists tend to oversimplify the situation, which in reality is a delicately balanced relationship. Although the precise nature of the lichen association is far from clearly understood, Loegering (1966) presented a concept of the host–parasite relationship called the "aegricorpus", which should be examined carefully by lichenologists because, while relating to the wheat rust disease, it seems to apply as well to lichens. The aegricorpus is asserted to exist where there is "the intimate association of a host and a pathogen in which specificity occurs, and the pathogen clearly derives its nutrients from living host cells". Loegering's concept builds on the gene-for-gene hypothesis (Flor, 1942). In the following quotation from Loegering, I insert lichenological terms in square brackets where they apply:

> Where specificity occurs, there is a gene for reaction [in the phycobiont] for every gene for pathogenicity [in the fungus] or *vice versa*. . . . The infection type [lichen character] conditioned by a set of corresponding gene pairs is always the same in identical environments and is a constant genetic character.
>
> Reaction is a character of the host, pathogenicity of the pathogen, and infection type of the disease. To use the word "disease" here in the same generic sense as "host" and "pathogen" does not seem admissible because "disease" [lichen] is commonly defined as a condition or process and often refers to a sequence of events in the sense of the disease spreading from plant to plant or from field to field. Thus, it is essential that a word be found to replace "disease", and "aegricorpus" is proposed. The word comes from the Latin and is compounded of aeger=sick and corpus=body in the sense of being made up of component parts. "Aegricorpus" [lichen] is defined as a single living manifestation of specific genetic interactions in and between host and pathogen.

Loegering goes on to develop the genetics of the aegricorpus concept together with one environmental factor, temperature. His discussion of the latter is of particular concern to us here:

> . . . It can be postulated that the temperature so affected the physiology of the host that it did or did not produce a toxin or an ailment; however, the same can be said with respect to the pathogen. In addition, it is possible that the development of the aegricorpus depends on a physiological interaction involving host and pathogen and that this interaction was affected by the temperature. Thus, it is not possible to determine whether the effect of temperature was on the host, the pathogen, or the aegricorpus.

He concludes:

> In epidemiology [lichenology] the concept of the aegricorpus and its genetic and environmental determination is most useful because equal emphasis is placed on genetic and population changes of the pathogen [mycobiont] and of the host [phycobiont], and variations in the environment, thus making possible a more rational approach to the dynamics of epidemic development of disease [lichen].

The aegricorpus concept may be particularly apt for the lichen symbiosis. This relationship inevitably involves the interactions between the following: (1) fungal and algal organisms which have somehow come in contact and have undergone considerable co-evolution; (2) the substrates upon which they grow; (3) the environment that surrounds them and which can be considerably altered over time; (4) the growth curves of the symbionts, which may or may not be harmonious; and (5) other epiphytes or parasites casually or intimately associated with them. The variables we have to be concerned about in environmental modification in lichens, therefore, are more numerous and more difficult to analyse than in comparable units of organization among individual autotrophic higher plants.

Poelt (1974) aptly points out:

> Lichens lack the capacity to develop special resting stages for survival during unfavourable periods. Thus, as long-lived organisms, they may be subjected to the effects of an extreme and hostile environment over a period of many years. This leads to considerable environmentally-induced modification, and the range of phenotypic variation, especially among many of the crustose lichens, is often much greater than the phenotypically expressed genotypic difference between species and even species groups. Phenotypes of a single species growing in different habitats can appear so unlike that it calls for a detailed analysis and long experience to recognize them as relatives of one and the same population.

The key to a rational concept of the species, then, depends upon our attention to sound sampling methods for herbarium material to ensure a comprehensive permanent record of variation, intelligent observation of populations in the field, and sensitivity to the nuances of the microhabitat.

In the past, classification of lichen specimens has been primarily a herbarium exercise. Specimens had to be taken at face value, for the herbarium taxonomist worked at so many disadvantages—lack of locality and site data, virtually no ecological information, no first-hand knowledge of the habitat, and most of all, the inadequacy of the specimens at hand. The ramifications of this problem are well known and need not be elaborated here. Fortunately, most active lichen systematists today are fully aware of these problems and revisionary work not carried out in conjunction with field observations is becoming frowned upon.

II. Types of Environmental Effects

Hawksworth (1973) reviewed in detail the entire literature on ecological factors and species delimitation in the lichens. The concept of extreme environments and environmental extremes in terms of lichen ecology has been discussed very well by Kappen (1974) who reviewed not only the physiological literature but some of the taxonomic papers. Brodo (1974) brings together research concerning the influence of the physical and chemical character of the rock substrate upon thresholds governing production of various types of thalli. His paper on substrate ecology is a valuable contribution to the literature indirectly bearing on environmental modification. Farrar (1973) reviews the progress of studies in lichen physiology and presents information of critical importance to the problem of environmental modification. He warns that "both the complexity of the interactions between variables, and the significance of environmental fluctuations to lichen thalli, have been seriously underestimated".

My own research on the subject is contained in three papers (Weber, 1962, 1967, 1968) in which I developed a classification of the kinds of environmental modification relevant to lichen taxonomy and some interpretations of troublesome groups, particularly the genera *Acarospora*, *Aspicilia*, *Lecidea* and *Dermatocarpon*. Rather than cover the same ground here, I intend to develop some new aspects of the problem in which experimentation is at present impossible but where field observations are of critical importance in developing satisfactory interpretations. As Hawksworth (1973) says:

> [experimental] methods clearly have potential application in taxonomic investigations but at the present time the lichenologist has to base his species concept on individuals and populations growing in the field. . . . Experimental taxonomy is largely foreign to the literature of lichenology, remaining an unrealized ideal.

A. Morphology

First of all, however, it may be of value to summarize briefly the main non-inheritable types of environmental effects on lichen morphology, colour and anatomy. Hawksworth's (1973) paper should be referred to for fuller information on most examples. Eco-physiological effects are not considered here as they fall outside the scope of this contribution.

1. Effect of the Alga

The investigation of cephalodiate taxa, i.e. "lichens" with different algal genera (blue-green *v.* green) in morphologically dissimilar parts of the same thallus (combined morphotypes) etc., convinced James and Henssen (1976) that a single fungal partner can form morphologically quite different structures with different genera of algae. In some cases these are very striking as in the foliose *Sticta filix* (green alga) and its fruticose counterpart, *Dendriscocaulon* (blue-green alga). James and Henssen also presented evidence that in some cases the nature of the environment determined whether a green algal morphotype or a blue-green algal morphotype was produced. These authors also found that, contrary to some earlier reports, the genus of algae present could affect the production of at least some lichen substances.

2. Effect of Climate

While the role of climate in lichen speciation now seems clear to judge from a few well studied examples, its effects require detailed study in many groups of lichen "species". At the local and microclimatic level examples are easier to detect by field observations, as in my own studies on *Acarospora* (Weber, 1962). In the case of the crustose lichens under harsh environmental conditions areolae may become widely separated (see also pp. 25–26) whilst in particularly humid situations the production of soredia can be enhanced. Fruticose lichens in particularly wind-swept situations may tend to be stunted and contorted adopting unusual growth habits, and the foliose *Xanthoria parietina* on the sea-shore becomes narrower-lobed (Richardson, 1967).

3. Effect of the Substrate

Substrate can affect thallus form in a wide variety of ways depending on the type of substrate and the growth form of the species involved. Thus on particularly wet peaty soils many *Cladonia* spp. produce grossly inflated thalli; fruticose species in wind-swept heaths may become dorsiventrally flattened (e.g. in some *Alectoria* and *Cetraria* spp.); and the arrangement of lirellae (e.g. in Graphidaceae) may be affected by the nature of the bark. On long-persistent substrates (e.g. persistent *v.* flaking bark, hard *v.* soft rocks), crustose thalli may be well developed whilst on less persistent substrates they are poorly developed. Perithecia in *Arthopyrenia halodytes* are much larger when on soft chalky rocks (Swinscow, 1965). When a species spreads from rock on to soil even the habit can be changed from crustose to fruticose in at least one instance (see p. 22).

4. Vagant Life-forms

These particularly distinctive largely wind-produced modifications are discussed separately below (pp. 19–22).

5. Submergence

Several genera, such as *Verrucaria*, *Staurothele* and *Aspicilia*, include some species typical of dry sites, with verrucose or areolate thalli, and others, typical of irrigated sites, with continuous or merely rimose (irregularly cracked) thalli. These differences in thallus form are accorded considerable taxonomic importance, but it is likely that the respective thallus types may represent modifications. The occurrence of parallel modifications involving unrelated genera suggests a response to environmental stress, but the possibility of genotypic selection for a particular habitat should not be discounted. We have some evidence from field observations that transitions across the environmental gradient do in fact connect recognized taxa characterized by the extreme thallus types, but experimental evidence is needed for a satisfactory resolution of the problem. One approach might be to remove submerged rocks to sites alongside the stream that are not subject to the same degree of submergence and watch for changes in thallus form. Microhabitat boundaries involving water relations on rock surfaces, unfortunately, are usually sharp, and therefore the ecological gradients required to produce gradual modificational clines may be difficult to demonstrate despite their reality.

Wave action in maritime situations appears to affect the production of sterile raised black striae on the thalli of some *Verrucaria* spp. and conceivably might effect comparable structures in freshwater species of this genus.

6. Galls

Gall-like outgrowths on lichen thalli have been reviewed in some detail by Grummann (1960). In some cases these can arise due to the effects of lichenicolous fungi (e.g. *Pyrenopeziza lettaui* on *Evernia prunastri*, *Polycoccum galligenum* on *Physcia caesia*, *Polycoccum trypethelioides* on *Stereocaulon*, *Abrothallus parmeliarum* on *Usnea*). In many cases, the causal agents of lichen galls are obscure and there is scope for a great deal more work in this area. James and Henssen (1976) point out that it could be argued that lichens themselves are galls—the response of a fungus to an "invading" alga.

7. Other Modifications

Modifications which are not caused by galls include inhibition of isidia formation in the case of *Pertusaria coccodes* invaded by *Cyphelium sessile*, contortion of thalli caused by *Lichenoconium* sp. on *Cladonia arbuscula* and inducement of soralia formation in *Ochrolechia parella* invaded by *Leciographa parellaria*.

8. Damage

If a lichen is damaged by mechanical, cropping or grazing actions, regeneration can lead to abnormal modifications. Isidia have been shown, for example, to occur on the crushed edges of thalli in certain *Peltigera* spp.

(Thomson, 1948). Cropped fruticose lichens may regenerate by forming tufts of young branches which can also lead to abnormal branching patterns and habits.

B. Colour and Pruinosity

In particularly well-lit situations, pigmentation of lichen cortices becomes more intense due either to the deposition of greater amounts of pigmented lichen substances in or on the cortex or enhanced pigmentation in the walls of the fungal hyphae themselves. Intergrades occur on transects between well-lit and shaded sites and are readily seen in the field but the extremes can be very different (e.g. bright orange-red *v.* pale yellowish-grey); if sampling has been inadequate the importance of such colour differences may be exaggerated. The physiological basis for these differences requires investigation although it has been suggested that increased pigmentation shelters the algal layer from harmful amounts of light. Some very marked qualitative colour differences due to the presence *or* absence of single pigments may be genetically determined (see Hawksworth, 1976). The tendency for thalli to be somewhat more greenish in shaded sites may be enhanced by an increased proportion of algal cells in them.

The nature of the substrate itself can also influence the colour of the thallus in some crustose lichens. Thalli on iron-rich rocks, for example, may in certain species become somewhat rust-coloured; I have termed such thalli "oxydated" (Weber, 1962). Species from a wide range of genera, including macrolichens (Schade, 1970, 1975), may have deposits of calcium oxalates on their surfaces. If abundant this gives them a frosted, or "pruinose" appearance and is particularly marked in crustose and placodioid lichens growing on calcareous rocks (Weber, 1962). Some other effects related to oxalate pruinosity are discussed separately below (p. 24). (It should be emphasized that pruinosity in lichens can also arise from ultrastructural differences not due to oxalate deposition.)

C. Anatomy

1. Thallus Structure

The total thickness of the crustose lichen thallus, or of particular layers in it, is also influenced by environmental factors and thus should be treated with caution in taxonomic studies. In general, thicker medullary and algal layers tend to form under optimal growing conditions, whilst tougher cortical layers are a feature of species in harsher environmental situations. The phycobiont/mycobiont ratio varies significantly from one habitat to another (Hill and Woolhouse, 1966; Seaward, 1976). The degree of development of some other tissues, for example prothalline margins, may also be influenced by the nature of the substrate.

2. Algae

The algae within a lichen thallus tend to be somewhat modified by the process of lichenization. Filamentous species, for example, often appear almost unicellular in the thallus (see Tschermak-Woess, 1976) so that cultural studies are a prerequisite for the determination of the algal genus and species present.

The relevance of this to lichen systematics is evident from recent studies on the *Staurothele clopima* complex, a group of crustose lichens characterized by continuous or areolate thalli with embedded perithecia and brown muriform spores, and which are indicative of surfaces wetted by seasonally seeping or flowing water. A feature peculiar to *Staurothele* and a very few related genera is the presence of colonies of green algae submerged in the gelatinous interior of the perithecium. These algae are strikingly different in morphology from the thalline algae, being much smaller and in short chains of spheres or rods.

Classifications of *Staurothele* have been based to a great extent on the shapes of these hymenial algae. The close proximity of the hymenial algae to the spores themselves would suggest that this genus reproduces itself by simultaneous discharge of the spores with the appropriate algae. However, the endohymenial algae have traditionally been considered to belong to a genus different from that of the thalline algae. If this were true, the only possibility of reproduction for the lichen would be the dispersal of an entire fertile areole or of at least a fragment containing a portion of the vegetative thallus and the contents of a perithecium. Because the morphology of the hymenial algae loomed so large in the taxonomy of *Staurothele*, I asked Dr V. Ahmadjian to culture the thalline and hymenial algae. He found (Ahmadjian and Heikkilä, 1970) that the hymenial algae not only belonged to the same genus, *Stichococcus*, as the thalline algae, but that their small size was induced by the environment of the perithecial gelatine, and their shapes represented stages in development of the individual cells.

3. Ascospores

Differences in ascospores are accorded paramount importance in lichen systematics today, but at the species level one should perhaps be asking what effects the environment might have. Spore size, numbers and perhaps even the extent of septation suggest themselves as characters on which further studies should be carried out, but scant attention has been paid to these aspects in recent years.

D. Chemistry

As known environmentally related chemical differences have been reviewed recently elsewhere (Hawksworth, 1976), this aspect is not considered further here.

III. Terminology

The term "environmental modification" has been used in two senses: to denote the environmental *influence* upon organisms or to refer to the *organism* that has produced a response. This ambiguity evokes alternative terminology designed specifically to apply to the modified organism, such as "ecad" (e.g. Seaward, 1976) and "ecophene". Some lichenologists prefer to use the term "morphotype" when in doubt as to whether a given plant is a modification or a genetic ecotype. I prefer to use the term "modification", which can be abbreviated to "mod." and used along with the terms "subspecies", "varietas" and "forma"; thus, *Acarospora schleicheri* mod. *crassa squamulosa pruinosa saxicola* (see Weber, 1968). The category "mod." has no status under the Code, so these epithets are unofficial, therefore nomenclaturally null and void.

IV. The Vagant Life-form

"Vagant" or unattached lichens are characteristic inhabitants of windswept steppes and deserts all over the world. Smith (1921) reviewed the older literature concerning *Aspicilia esculenta,* which some authors believe was the "manna" of the Israelites. A group of species closely related to *A. esculenta* ranges throughout southwest Asia and North Africa, some of them forming dense pebble-like balls, others displaying looser branch systems, but all of them quite free from any connection with the substrate.

On the Namib Desert of southwest Africa, this role is played by *Parmelia (Omphalodium) convoluta,* a foliose lichen with an extremely tough horny texture, (see figures in Mattick, 1970); in South Australia by *Parmelia australiensis* and the monotypic genus *Chondropsis semiviridis;*

and on the desert-steppe of the Continental Divide in Wyoming by *Parmelia chlorochroa* and *Lecanora haydenii*. *Ramalina pulchribarbara* and *Teloschistes peruensis* are characteristic species of coastal fog deserts in western North and South America respectively (Rundel et al., 1972; Thomson and Iltis, 1968). In Alaska *Cetraria richardsonii* lies unattached in tundra depressions. Smith (1921) refers to some other examples, including *Parmelia revoluta* on the chalk downs in England.

The genus *Lecanora* contains a group of species that are characterized by having a peltate thallus attached to the rock by a central umbilicus. These are large lichens with orange, brown, blackish or pruinose apothecia contrasting in colour with the usually greenish or pale brownish thalli. The umbilicate thallus marks the group as unique in the genus though under certain substrate conditions the thalli are crowded and often almost crustose. Some lichenologists treat it as a distinct genus, *Omphalodina*, although it shares many characters with other lobate *Lecanora* spp. Two of these umbilicate species form unattached *Wanderflechte* that roll up into tight balls and lie loose on the ground. None are known to occur with apothecia. *Lecanora haydenii* is abundant on the windswept steppe-desert of Wyoming, and *L. baranowii* in similar habitats on the Altai steppe. Morphologically and chemically, these species are counterparts of the attached *L. chrysoleuca* and *L. melanophthalma*, respectively. *Lecanora chrysoleuca* tends to have its black prothalline tissue very weakly developed, while *L. melanophthalma* has a strong blackening along the edges and sides of the lobes, giving the lichen a variegated aspect. However, this character is variable. While sterile specimens lacking the hypothalline colour are more likely to be *L. chrysoleuca*, both species may develop marked blackening, so that the only way to distinguish them is by the bright orange disc of *L. chrysoleuca* v. the blue-grey pruinose disc of *L. melanophthalma*.

Dr G. Argus has collected vagant forms of *L. chrysoleuca* in the Lake Athabasca region of northern Saskatchewan, where it occurs on sandy till plains covered with ventifact gravel. The specimens consist of balled "pebbles", green with black lobule margins. At Little Gull Lake the specimens are tightly rolled with the marginal lobes meeting to form perfectly smooth contours. At William River they are looser, with roughly protruding lobe tips and a more open organization. Had the collector not noticed fertile *L. chrysoleuca* (a phase with prominent black hypothalline coloration) growing attached to boulders at the Gull Lake site, he might never have suspected that the two forms were related. But in the light of my own subsequent observations on another population, it may be inferred that the vagant lichen is a modification of the attached *L. chrysoleuca*.

The western summit of the Beartooth Plateau in northwestern Wyoming is a level alpine saddle at an altitude of 3300 m exposed to more or less continuous high-velocity winds. Here, typical *L. melanophthalma* is abundant on boulders, fixed outcrops and pebbles. Scattered here and there one finds small drifts of unattached thalli of "*L. baranowii*", particularly in wind-excavated depressions between tundra mats. In a few minutes one can scoop up hundreds of them. Closer observation reveals a series of transitions, ranging from specimens attached to pebbles and outgrowing them, to specimens only partly rolled up and still bearing a few apothecia. In fact, the attached and vagant specimens represent a continuum of variation displayed by a single taxon.

Three features of these specimens stand out. The first concerns alteration of thallus morphology: once they are free, the thalli lose their usual rotate form and tend to become more deeply dissected, with narrower marginal lobes and a concomitant loss of dorsiventrality. The lobes often become terete branches, prothalline blackening becomes less pronounced and more diffuse, the ends of the lobes lose their dorsiventrality and often terminate in rounded knobs. Thallus form becomes extremely variable, ranging from tight, smooth-surfaced balls to forms with divaricately spreading branches that eventually assume a more or less spherical outline. Second, once dissociated from the substratum, the apothecia disappear and thereafter the vagant thalli are sterile. Third, the lichens freed from a substrate apparently enjoy a greatly accelerated growth rate over that which they exhibited in their fixed state.

Although these field observations are not reinforced by experimental evidence, a circumstantial case can be made for modification. I submit the following hypothesis:

1. The substrate serving as an attachment for a lichen thallus exerts some form of control over the organizational capacity of the lichen. When the lichen is no longer attached, it is "without a rudder" and may display rapid, uncontrolled growth and bizarre morphology reminiscent of cells in tissue culture.

2. Once freed from the organizational control formerly effected by its attachment to the substrate, the lichen fungus loses its capacity to produce apothecia. This is as true for those lichens that have *evolved* to an unattached life style as it is for those that are environmental modifications.

Before we can learn why the vagant habit might be correlated with sterility, we should try to determine what happens to apothecia in the normal sequence of events. There seem to be no published observations on this question (cf. Pyatt, 1974). Yet apothecia emerge, reach maturity and presumably disappear without leaving scars of any kind. Are they pinched off, resorbed, or otherwise disposed of? In *Omphalodium arizoni-*

cum, a peltate monophyllous lichen attaining a diameter of 30 cm, bursts of new apothecial production appear on some sectors of the thallus while large old fruits dominate others. What is happening? To determine the fate of senescent apothecia, one might select a lichen of relatively rapid growth such as *Xanthoria polycarpa* and follow it photographically over a few years at monthly intervals.

Crustose and squamulose lichens also produce vagant modifications. On desert soil crusts, many lichens approach the vagant habit as the soil pedestals supporting them become smaller and more eroded basally under the influence of raindrop-splash erosion and eddying wind currents. For example, the dark blackish-brown thalli of *Endocarpon pusillum* are abundant along the margins of temporary rainwater pools on wind-excavated shallow depressions of sandstone ledges in Dinosaur National Monument, Colorado. Here the shallow soil supporting the lichen washes or blows away from the margin of the thallus and is undercut below the lichen itself. Subsequent wetting and drying causes alternate swelling and curling, shrinking and flattening. Eventually the lichen is detached and free to roll thereafter across the substrate as a vagant ball. (Collections of this population are distributed as Lich. Exsicc. COLO No. 429.)

In a paper on *Aspicilia* (Weber, 1967), I discussed the almost incredible behaviour of two species that are normally crustose and saxicolous, but which under certain conditions encroach on soil substrates where they assume fruticose life-forms. In one of these instances the fruticose phase was made the basis of a new genus termed *Agrestia* and referred to the Usneaceae. It is possible to find the lichen partly crustose on the stone, moving off it and assuming a fruticose form on the soil, then contacting the stone again, and becoming a crust. Rogers and Lange (1972) figured an identical variation from Australia. It is possible, on the Koonamore Vegetation Reserve in South Australia, to find numerous thalli showing the transition from crustose to fruticose morphotypes, even to the extent of finding crusts in which a single areole has developed into a tiny fruticose column a few millimetres tall. The extreme fragility of the transitional forms requires the utmost skill of the field botanist to preserve these as herbarium specimens.

In Asia certain species of *Aspicilia* behave quite differently in that they form vagant lichen balls. Whether these vagant forms are distinct taxa or are related to fixed crustose species is a problem that only future fieldwork can determine. Smith (1921, her p. 404), evidently paraphrasing Eversmann (1825) or Berkeley (1849), said of *A. esculenta:* "It grows abundantly . . . on the rocks or on soil. It is easily broken off and driven into heaps by the wind." This suggests the occurrence of attached counterparts in the same areas.

At first glance, the vagant habit might be interpreted as an adaptation providing mobility and enhanced dispersal for the lichen. On the contrary, the thalli accumulate in the lee of depressions relatively sheltered from the wind. Their often tight morphological conformation tends to streamline them rather than to expose "vanes" to the wind gusts. The exclusively vagant monotypic genus *Chondropsis semiviridis* affords an elegant instance of a lichen seemingly adapted for increased dispersal but not necessarily achieving it. *Chondropsis* spreads out quite flat when moist, with branches up to 3 mm wide, forming one of the most beautifully dichotomous thalli among lichens. As it dries, the lichen rolls into a ball in which the spaces between the branches become lacunae in a reticulum. As the thallus tumbles over the ground, it is impaled on bits of leaves and stems that penetrate these holes, tending to hold the lichen where it is rather than permitting it to roll further on. Wind-rows of *Chondropsis* do accumulate, however, where vegetational impediments are absent. Nevertheless, the fact that vagant life-forms among the lichens tend to be concentrated in areas of high and persistent winds suggests that wind is the prime mover in the production of vagant modifications, and that wind is also an important agent of natural selection in the evolution of vagant taxa.

The bizarre behaviour of vagant *Aspicilia* may have taxonomic implications for other species of the genus that, while being subjected to unusual stresses, never actually become separated from their substrates. Several montane and arctic *Aspicilia*, growing as crusts on rocks in areas of severe environmental stress, develop elongate, discrete, marginal ray-like extensions. These growth patterns may be modifications analogous to the fruticulose and creeping forms discussed earlier. Magnusson (1939) recognized some of these types as valid species (*Aspicilia disserpens*, *A. alboradiata* and *A. virginea*) and interpreted them as members of a section of the genus characterized by a marginally lobate thallus. The finger-like extensions, in these instances, cannot be interpreted as marginal lobes, but are probably reactions of non-lobate, effuse thalli to severe environmental stress. In a protected site, an effuse thallus tends to grow equiformally to produce a more or less circular thallus. Under severe stress of wind-scouring action, incipient areolae may be destroyed or inhibited differentially. Those areolae that survive may afford some protection for those distal to them, the end-product being a single row of areolae. Irregularities in the rock surface may also provide local protection for isolated chains of areolae. I have observed these forms in the field in Alaska, Sweden and in the Northwest Territories of Canada, and saw numerous transitions, convincing me that at least some of these irregularly radiant forms are of no taxonomic significance.

V. "Fossil Modifications"

One of the most puzzling phenomena observed in the field is the distinctly different appearance of the same lichen species either immediately adjacent to, or overlapping, each other on the same rock surface, even when there is no obvious change in rock texture, water availability or insulation. In a population of the coastal Californian *Dimelaena radiata* (Lich. Exsicc. COLO No. 189) I noted on the label: "Pruinose and epruinose forms grow together, the thalli of the epruinose form usually overlapping the pruinose one; both evidently are anatomically and chemically identical." Sheard (1974) confirms their conspecificity.

Another instance is a large plaque of the calcium oxalated phase of *Acarospora smaragdula* mod. *strigata* (*sensu* Weber, 1962) on a dry sandstone talus block in the Abajo Mountains of Utah (COLO 162950). This thallus forms an essentially unbroken areolate crust covering a diameter of about 5 cm. The oxalate pruinosity is evidently worn and impregnated with the reddish dust of the eroded stone. Scattered near and upon this old plaque are small fresh-looking squamulose rosettes of the same species which have evidently colonized small eroded loci on the old thallus where the weathered substrate was exposed following the removal of an areole. The vacant spot either accumulated a bit of loose detritus or eroded further, forming a small soil mound, upon which new thalli appeared. These obviously younger thalli are very white with oxalate and are more distinctly squamulose than the old thallus, which is crustose throughout. Internally both thallus types are identical, undoubtedly representing one and the same taxon, but their outward appearance is strikingly different.

Similar instances could be given, involving species of *Aspicilia*, *Lecanora*, *Lecidea*, *Buellia* and other genera. A certain burden of proof falls upon the observer who claims that the forms are phases of the same species, but in the instances chosen above, this hypothesis seems to be indisputable.

Lichenometric evidence (Webber and Andrews, 1973) supports the notion that saxicolous crustose lichens do not grow equally rapidly over their entire life-span, but that these lichens exhibit various growth phases beginning with a so-called "great period" of growth followed by a "long-term" growth rate. Probably the growth curve in most lichens is not linear but sigmoidal. Growth rates, according to the authors, are greatly affected by the relative oceanicity and continentality of the local climate (see also Chapter 3). Farrar (1973) develops a thesis that the slow growth of lichens is not the result of a slower activity of the metabolic process in the lichens than in the higher plants, but that carbon and nitrogen which move between the symbionts accumulate in soluble forms and are not routinely incorporated into the proteins and polysaccharides necessary for thallial growth. Hence the carbon remains in polyols and much is lost by respiration. Slow growth of lichens would then be primarily due both to the short percentage of time during which net carbon assimilation is possible, and to partitioning of fixed carbon, such that little is put into growth processes because most is needed for stress resistance. Slow growth would be a necessary concomitant of survival in adverse habitats, due to the small percentage of fixed carbon retained in the thallus for long periods. Thus, as lichens enter their senescent period of little or no growth, long-term effects of the adverse environment would accumulate as scars on the thallus, and regenerative activity might not erase the damage. This appears to be a reasonable explanation for the differences observed between young, rapidly growing individuals and older established thalli, even when they lie adjacent or overlapping on the same substrate.

VI. Biogeochemical Erosion

In my earlier work (Weber, 1962), I was unaware of the significance of another potent force in producing modifications. This is chemical erosion of the substrate underneath a thallus and the subsequent exfoliation of the lichen and its underlying substrate resulting from the alternate swelling and shrinking with wetting and drying of the lichen. A substantial amount of the erosion of large lichen plaques in the arid American Southwest occurs through the mass-wasting of thallus and substrate in this manner. Probably a combination of chemical weathering and sand-blasting is responsible for the myriad of degenerate and reconstituting thalli one encounters in these areas.

The extreme dryness of lichen thalli under desert conditions means that when dry, lichens are as readily subject to erosion as the substrate upon which they grow. They may actually be more so, because the lichen responds to slight changes in the humidity of the air. When the air is dry, the thallus is hard; when the air is humid, the thallus is soft. Each condition represents a different type of vulnerability to wind and chemical erosion. Apothecia consist of tightly packed vertical hyphae, often embedded in gelatinous or granular material. Their greater density makes them more resistant to erosion than the tissues of the vegetative thallus. Because of its contrasting characteristics, and its greater area, the thallus, on the other hand, is very likely to be more effective than the apothecium in causing biogeochemical erosion.

Many desert lichens may lose their entire thalli to erosion before any of the apothecia are affected. *Lecidea tessellata* is one of the most baffling of all desert lichen species because of this phenomenon. Where *L. tessellata* is abundant all manner of thalli may be found, ranging from perfectly intact examples in areas protected from sand-blasting, to fragmented, decorticated and otherwise degenerate examples in exposed sites. A leached area on the rock, peppered with isolated black apothecia, is often the only evidence remaining to show that the lichen once formed a continuous areolate pavement. Herbarium specimens are often too small to show this, but in the field the leached eroded area of the stone usually shows in ghostly outline the former extent of the thallus. Similarly, in Australia, erosion of *Diploschistes scruposus* results in wide variation in apothecial form. In more arid regions the exciple is worn away by continual sand-blasting so that the disc is quite superficial, instead of being more or less immersed in the thallus and partly enclosed by the exciple.

VII. Discussion

In drawing attention to these features of lichen variability and in suggesting that in many instances they may be explained in terms of environmental modification, I do not wish to imply that all of lichen taxonomy is a mare's nest of environmental plasticity. In the arid lands where crustose lichens predominate, certain genera obviously are problematical because of variation induced by the environment. These genera include predominantly *Aspicilia*, *Acarospora* and *Staurothele*, where the classification of almost all the taxa are affected, plus a few protean species in other genera, such as *Lecidea tessellata* and the *Omphalodina* section of *Lecanora*. Interpretation of taxa in these genera cannot be clarified without the aid of exhaustive herbarium documentation and critical field observations.

We urgently need to develop some understanding of the fundamental mechanisms of thallus organization and growth. Experimental culture of intact thalli needs to be encouraged and a suitable "laboratory animal" must be found soon. I think that several examples are awaiting inspection by a skilled technician. *Cetraria richardsonii*, *Chondropsis semiviridis*, *Parmelia chlorochroa* and similar vagant species are perfectly qualified subjects. They are abundant, easily handled and, lacking substrate, they can be moistened, dried, weighed, injured, marked, hung in growth chambers and subjected to intermittent or cyclic doses of a variety of conditions. When the right environmental regime is designed, one or more lichens may be grown routinely under it in an accelerated simulation of the natural environment. When we achieve this, we may expect to learn why and how lichen thalli are able to develop their unique forms and how these forms are modified by environmental extremes. Time-lapse photographic studies of rapid-growth lichens can be instituted immediately without special resort to laboratory subjects. Such studies will provide elementary information concerning the growth, periodicity, senescence and, especially in relation to the sterility of vagant lichens, the ultimate fate of apothecia.

Many lichenologists will read these pages and conclude that this writer sees a modification under every bush. I merely submit these observations for further testing, and in order to stimulate thought and discussion. To me it seems less damaging to the taxonomic system to suggest that taxa may be identical than to introduce into the formal nomenclatural structure hosts of new taxa for which there is little demonstrable biological justification, but on the other hand, taxonomically important variations should not be overlooked. A. H. Magnusson, the great Swedish lichenologist whom I have criticized for his overproduction of new species in *Acarospora*, nevertheless appreciated the problem when he wrote (Magnusson, 1932):

> What lichenology needs is not a name of each preserved specimen, the consequences of which are terrifying, but a survey of the variability of the species, founded if possible on studies in nature or at least on a large herbarium material.

Until this is accomplished, lichenologists should maintain a healthy scepticism of the enormous number of taxa in some lichen genera in which modification is clearly involved, plan field conferences in arid lands where the evidence can be discussed and notes compared, preserve herbarium material illustrating population parameters, and encourage experimental work, not only in the field of lichen synthesis, but in the culture of whole lichens.

Acknowledgements

I am grateful to Dr D. L. Hawksworth for his assistance in the preparation of this chapter, and to Mr B. J. Coppins, Dr D. H. S. Richardson, Dr R. W. Rogers and Dr M. R. D. Seaward for valuable criticism and additional data.

References

Ahmadjian, V. (1974) ["1973"]. Resynthesis of lichens. *In* "The Lichens" (V. Ahmadjian and M. E. Hale, eds), pp. 565–579. Academic Press, New York and London.
Ahmadjian, V. and Heikkilä, H. (1970). The culture and synthesis of *Endocarpon pusillum* and *Staurothele clopima*. Lichenologist **4**, 259–267.
Berkeley, M. J. (1849). Note on *Lecanora esculenta*. Gard. Chron. **1849**, 611.

Brodo, I. M. (1974) ["1973"]. Substrate ecology. *In* "The Lichens" (V. Ahmadjian and M. E. Hale, eds), pp. 401–441. Academic Press, New York and London.
Evans, L. T., ed. (1963). "Environmental Control of Plant Growth". Academic Press, New York and London.
Eversmann, E. (1825). In Lichenum esculentum Pallasci. *Nova Acta Leopoldina* **15**, 349–362.
Farrar, J. F. (1973). Lichen physiology: progress and pitfalls. *In* "Air Pollution and Lichens" (B. F. Ferry, M. S. Baddeley and D. L. Hawksworth, eds), pp. 238–282. Athlone Press, London.
Flor, H. H. (1942). Inheritance of pathogenicity in *Melampsora lini*. *Phytopathology* **32**, 653–669.
Grummann, V. J. (1960). Die Cecidien auf Lichenen. *Bot. Jb.* **80**, 101–144.
Hawksworth, D. L. (1973). Ecological factors and species delimitation in the lichens. *In* "Taxonomy and Ecology" (V. H. Heywood, ed.), pp. 31–69. Academic Press, London and New York.
Hawksworth, D. L. (1976). Lichen chemotaxonomy. *In* "Lichenology: Progress and Problems" (D. H. Brown, D. L. Hawksworth and R. H. Bailey, eds), pp. 139–184. Academic Press, London and New York.
Hertel, H. (1969). *Arthonia intexta* Almqu., ein vielfach verkannter fruchtkörperloser Flechtenparasit. *Ber. dt. bot. Ges.* **82**, 209–220.
Hertel, H. (1970). Parasitische lichenisierte Arten der Sammelgattung *Lecidea* in Europa. *Herzogia* **1**, 405–438.
Hill, D. J. and Woolhouse, H. W. (1966). Aspects of the autecology of *Xanthoria parietina* agg. *Lichenologist* **3**, 207–214.
James, P. W. and Henssen, A. (1976). The morphological and taxonomic significance of cephalodia. *In* "Lichenology: Progress and Problems" (D. H. Brown, D. L. Hawksworth and R. H. Bailey, eds), pp. 27–77. Academic Press, London and New York.
Kappen, L. (1974) ["1973"]. Response to extreme environments. *In* "The Lichens" (V. Ahmadjian and M. E. Hale, eds), pp. 311–380. Academic Press, New York and London.
Levitt, J. (1972). "Responses of Plants to Environmental Stresses". Academic Press, New York and London.
Loegering, W. Q. (1966). The relationship between host and pathogen in stem rust of wheat. Proc. 2nd Int. Wheat Genetics Symposium. *Hereditas* suppl. **2**, 167–177.
Magnusson, A. H. (1932). Lichens from western North America, mainly Washington and Alaska. *Annls cryptog. exot.* **5**, 16–38.
Magnusson, A. H. (1939). Studies in species of *Lecanora*, mainly the *Aspicilia gibbosa* group. *K. Svenska VetenskAkad. Handl.* (3), **17**(5), 1–182.
Mattick, F. (1970). Flechtenbestände der Nebelwüste und Wanderflechten der Namib. *Namib Meer* **1**, 35–44.
Poelt, J. (1974) ["1973"]. Systematic evaluation of morphological characters. *In* "The Lichens" (V. Ahmadjian and M. E. Hale, eds), pp. 91–115. Academic Press, New York and London.

Poelt, J. and Steiner, M. (1972). Über einige parasitische gelbe Arten der Flechtengattung *Acarospora* (Lecanorales, Acarosporaceae). *Annln Naturh. Mus. Wien* **75**, 163–172.

Pyatt, F. B. (1974) ["1973"]. Lichen propagules. *In* "The Lichens" (V. Ahmadjian and M. E. Hale, eds), pp. 117–145. Academic Press, New York and London.

Richardson, D. H. S. (1967). The transplantation of lichen thalli to solve some taxonomic problems in *Xanthoria parietina* (L.)Th.Fr. *Lichenologist* **3**, 386–391.

Rogers, R. W. and Lange, R. T. (1972). Soil surface lichens in arid and sub-arid southeastern Australia. I. Introduction and floristics. *Aust. J. Bot.* **20**, 197–213.

Rundel, P. W., Bowler, P. A. and Mulroy, T. W. (1972). A fog-induced lichen community in northwestern Baja California, with two new species of *Desmazieria*. *Bryologist* **75**, 501–508.

Schade, A. (1970). Über Herkunft und Vorkommen der Calciumoxalat-Exkrete in kortikolen Parmeliaceen. *Nova Hedwigia* **19**, 159–186.

Schade, A. (1975). Über das Vorkommen der Calciumoxalat-Exkrete bei den Usneaceen (Lichenes), nebst Bemerkungen über Höhlungen der Achse, gelegentliche Ol- und Großenverhältnisse bei den *Usnea*-Arten Afrikas und mit einem Nachtrag über die Usneen Japans. *Nova Hedwigia* **26**, 45–82.

Seaward, M. R. D. (1976). Performance of *Lecanora muralis* in an urban environment. *In* "Lichenology: Progress and Problems" (D. H. Brown, D. L. Hawksworth and R. H. Bailey, eds), pp. 323–357. Academic Press, London and New York.

Sheard, J. W. (1974). The genus *Dimelaena* in North America north of Mexico. *Bryologist* **77**, 128–141.

Smith, A. L. (1921). "Lichens". Cambridge University Press, Cambridge.

Swinscow, T. D. V. (1965). Pyrenocarpous lichens: 8. The marine species of *Arthopyrenia* in the British Isles. *Lichenologist* **3**, 55–64.

Thomson, J. W. (1948). Experiments upon the regeneration of certain species of *Peltigera*, and their relationship to the taxonomy of this genus. *Bull. Torrey bot. Club* **75**, 486–491.

Thomson, J. W. and Iltis, H. H. (1968). A fog-induced lichen community in the coastal desert of Southern Peru. *Bryologist* **71**, 31–34.

Tschermak-Woess, E. (1976). Algal taxonomy and the taxonomy of lichens: the phycobiont of *Verrucaria adriatica*. *In* "Lichenology: Progress and Problems" (D. H. Brown, D. L. Hawksworth and R. H. Bailey, eds), pp. 79–88. Academic Press, London and New York.

Webber, P. J. and Andrews, J. T. (1973). Lichenometry: a commentary. *Arct. alp. Res.* **5**, 295–302.

Weber, W. A. (1962). Environmental modification and the taxonomy of the crustose lichens. *Svensk Bot. Tidskr.* **56**, 293–333.

Weber, W. A. (1967). Environmental modification in crustose lichens. II. Fruticose growth forms in *Aspicilia*. *Aquilo*, ser.Bot. **6**, 43–51.

Weber, W. A. (1968). A taxonomic revision of *Acarospora*, subgenus *Xanthothallia*. *Lichenologist* **4**, 16–31.

3 | Colonization, Growth, Succession and Competition

PAULINE B. TOPHAM

I. Colonization	. . . .	32
II. Growth	. . . .	35
III. Succession	. . . .	46
IV. Competition	. . . .	55
References	. . . .	59

These little plants first clothe the naked rock; some of them almost too small for vision unless accumulated in myriads are yet the simple agents employed to change the rocky mountain into the waving forest, or the sandy desert into the fruitful province. So slow is the process at first that many years must elapse before the newly-erected edifice becomes even grey with age, or in other words puts on the first clothing of vegetation. After this the progress of vegetable life is more rapid. The Lichens in their decay leave a little earth where a green Moss soon springs up; this collects more earth, and a Fern or a Grass or some other plant succeeds, and thus a bed of flowers arises even upon the barren cliff and the mouldering wall; but not here only are the Lichens found, the earth abounds with various species, and on the trunks of trees they are still more numerous.

This account appears in an introduction to botany (Francis, 1842) dedicated to the "young ladies of England"; the themes of lichens as the first species colonizing rock, their slow growth and their role in succession form the basis of this chapter.

I. Colonization

What may be called the primary pedogenic succession, which changes the rocky mountain into the waving forest, has always attracted the interest of lichenologists, probably because northern Europe, Scandinavia and North America are all areas of immature landscapes left by the recession of the great Quaternary glaciations where the process may be observed in all its stages. The antiquity of lichens as pioneers on rock is shown by the cosmopolitan distribution of certain species including *Rhizocarpon geographicum* and some *Umbilicaria* spp.

The adaptations which permit certain species to "clothe the naked rock" and exploit an environment inimical to most other forms of life include resistance to desiccation and to extremes of temperature, longevity and a growth rate commensurate with the slow release of nutrients from unweathered rock. Air-dried thalli are remarkably resistant to drought—up to 62 weeks for *Lasallia pustulata* (Lange, 1953) which usually grows on rock surfaces with some degree of humidity. *Rhizocarpon geographicum* showed no permanent damage after 6 months desiccation although its metabolism was abnormal for 5 days after rewetting (Ried, 1960). Temperatures within lichen thalli may be 20–40°C above the surrounding air temperature (Lange, 1953); although doubts have been expressed (Smith, 1962) as to whether Lange's measurements of physiological activity following thermal insults of up to 101°C really indicated continued wellbeing, he was able to rank lichen species in order of resistance, and to show that this was of a remarkable order. These aspects of lichen adaptation are reviewed in Ahmadjian and Hale (1974).

Another adaptation to pioneer growth on bare rock is the ability to adhere, to penetrate and digest the substance of the rock. Little information is available as to the depth of penetration possible; Degelius (1962) quotes an early report of 19 mm for *Verrucaria marmorea*, and I have seen a weft of hyphae below *Baeomyces rufus* extending for about 1 cm along a cleavage in schist. A method for studying penetration published by Jones (1959) was used by Syers (1964) who reported depths of up to 16 mm for rhizinae in limestone. Endolithic thalli occur mainly among calcicoles, but occur also in occasional calcifuges such as *Lecidea auriculata*, *L. diducens*, *Sarcogyne privigna* and *S. simplex*, which often exploit lines of weakness in the granites or schists on which they grow.

The ability to digest minerals has been mainly studied from the pedogenic angle, but it is fundamental to our understanding of colonization. Earlier explanations of the accelerated weathering of rocks and other surfaces by lichens stressed the mechanical effect of penetrating hyphae alternately swelling when wet and contracting when dry (Mellor, 1921;

Fry, 1927). Chemical activity in etching glass and pitting limestone was attributed to the weakly acidic effects of carbon dioxide from respiration dissolved in moisture retained by the thallus (Mellor, 1922), but experimental results were inconclusive. Early speculations that oxalic acid was involved may be true for calcicolous species; Wade (1965) observed that the thallus of species of *Caloplaca* growing on calcareous substrates frequently has a pruinose appearance due to deposits of calcium oxalate, whilst obligate calcicoles such as *Caloplaca heppiana, Aspicilia calcarea* and *Rhizocarpon umbilicatum* are so much richer in calcium oxalate than non-obligate calcicoles from the same habitat (Syers *et al.*, 1967) that one is led to suggest that metabolic activity involving calcium oxalate is involved in their adaptation to growth on calcareous substrates. Soft calcareous rock may be protected from erosion by lichen thalli (Pentecost and Fletcher, 1974) at least for a time, but we are here concerned to focus attention on metabolic activity adapting lichens to their habitat, rather than to assess the amount of destruction to which this activity subjects their substrate. A small group of crustose lichens appears to be confined to iron-rich rocks and to metabolize and excrete ferric oxide; these include *Acarospora sinopica, Rhizocarpon oederi, Lecidea atrata* and *L. silacea*. They thus exploit an environment unfavourable to other lichens (Brodo, 1974). Silicates are much less chemically active than carbonates, and it was not until the Schatzes and their co-workers in the 1950s suggested that lichen acids play a role in pedogenesis by chelation that a plausible and experimentally demonstrable mechanism emerged (Schatz *et al.*, 1956). Iskandar and Syers (1971, 1972) showed that many lichen acids were water-soluble to a slight but measurable degree and that some complexed appreciable amounts of cations from powdered minerals and rocks. Both fumaroprotocetraric acid and the thallus of *Parmelia "conspersa"* were capable of chemically weathering granite (Syers, 1969). The ability of lichen extracts to effect chemical changes in siliceous minerals has been conclusively demonstrated by Ascaso *et al.* (1976) using modern crystallographic techniques. The importance of such evidence is twofold—by dissolving silicates hyphae can both penetrate the rock and gain nutrients.

Williams and Rudolph (1974) confirmed the activity of squamatic acid, but found that non-lichenized fungi were more active. It would be naïve to assert that lichens are the only organisms which hasten weathering. Degelius (1962) summarizes reports on endolithic blue-green algae, especially *Gloeocapsa* species; Williams and Rudolph (1974) report several microfungi whilst Webley *et al.* (1963) discuss the silicate-dissolving ability of fungi, myxomycetes and bacteria isolated from weathered rock surfaces. They found that the microflora increased and became more active as colonization by lichens increased. Their succession began with rock which had traces of crustose lichens, followed by those with a more continuous cover of crustose lichens, through foliose and fruticose species to the "raw soil" of rock crevices. We may thus accept lichens as the dominant, the largest and the most identifiable elements in a succession whose ability to alter its substrate steadily increases.

Colonization of bark by epiphytes seems less remarkable, since plant cells are growing on and penetrating other plant cells. Fry (1926) showed that fragments of bark are occluded by the lichen thallus and that many of the effects on bark of endophloeodal thalli could be duplicated by alternately drying and wetting a layer of gelatine attached to the bark. Rainwater has been shown to leach nutrients from the forest canopy (Ovington, 1965), so that we may conclude that trees offer lichens mere physical support. Bark usually contains resins and tannins which inhibit plant growth, but water relations, pH, light and nutrient status appear more important in determining which species are able to colonize a surface (Barkman, 1958; see also Chapter 9).

Soil surfaces, too, seem to demand fewer specialized abilities. Colonization by lichens is limited by their poor competitive ability relative to higher plants. Among the soil factors which favour lichen-rich vegetation are drought, high calcium content and low nutrient status. In the latter respect it is noteworthy that many pioneer soil species have the ability to fix atmospheric nitrogen, either because the phycobiont is itself capable, or because accessory phycobionts housed in cephalodia are active. Such species include *Stereocaulon*, *Peltigera*, *Solorina* and *Collema* spp. The phenomenon seems to be worldwide, occurring for instance in the Arctic (Schell and Alexander, 1973), Iceland (Crittenden, 1975), southern North America (Shields, 1957), Australia (Rogers et al., 1966) and the Antarctic (Fogg and Stewart, 1968; Horne, 1971). On other substrates this ability is not unknown but is much less common: for example, *Stereocaulon vulcani* is an important pioneer lichen on Hawaiian lava flows (Jackson, 1971) and *Placopsis gelida* occurs on the shingle banks left by braided glacial rivers in Iceland, whilst canopy lichens with blue-green algae are a nitrogen source in Columbian rain forest (Formann, 1975).

An essential feature for colonization on all these substrates is the ability to accumulate nutrients from rain or run-off, that is from extremely dilute solutions (Smith, 1960; Brodo, 1974). A figure of 0·07 p.p.m. ammonia-nitrogen has been quoted for rain in northern England (Raistrick and Gilbert, 1963). Rainwater also contains traces of potassium. Not only will such solutions be dilute, their contact with the lichen thallus will usually be of short duration.

The ecological aspects of dispersal and establishment have been treated by Bailey (1976) but two ecological aspects of colonization will be mentioned here, the effect of air pollution, and the colonization of unusual substrata. Air pollution may affect colonization rather than the survival of established plants. In London churchyards (Laundon, 1967), the percentage incidence of *Caloplaca heppiana* on limestone memorials declined from over 80% to 0 as the age decreased from 200 years old, whereas *Lecanora dispersa* was unaffected and its incidence was over 90% whatever the age of the monument.

It has been suggested that a phenomenon akin to inoculum potential may facilitate the establishment of species on abnormal substrata when the diaspore load is heavy (Lambinon, 1968). It is probable that similarities in the physical and chemical constitution of the unusual substratum to that of the usual one are more generally the cause, but critical investigation of particular instances is needed.

Quantitative studies of colonization are few and much further work is required. A mathematical study of the colonization of avenue trees by two fruticose lichens related the numbers of thalli to the distance of propagule dispersal (Tapper, 1976). The numbers of thalli colonizing appeared to decline exponentially rather than linearly with distance from the source and the estimated 50% dispersal distance for propagules ranged from 19 to 27 m, showing as much variation between dispersal along different rows of trees within each species as between species. Tentative conclusions may sometimes be drawn about the dynamics of colonization from studying the size–class structure of lichen populations. Since growth rate affects these conclusions the method will be discussed later.

II. Growth

Both the establishment and the competitive ability of new thalli are bound up with their growth rate, yet very little is known of lichen growth in the establishment phase. It is generally considered that a period of some years must pass before lichens establish themselves on new substrates; during the latter part of this latent period propagules must be passing from the microscopic to the macroscopic stage. Information about the onset of reproduction, changes in reproductive capacity with time, and the effect of such changes on growth are of vital importance to our understanding of lichens as members of plant communities. Little explicit information is available although Seaward's (1976) work on the autecology of *Lecanora muralis* in urban environments showed that the balance between growth and reproductive activity may vary in different environments.

Early observations, often of an anecdotal nature, of slow growth rates of mature thalli stimulated attempts to measure just how slow, and much effort has been devoted to their measurement (see the summaries by Hale, 1967, 1974). Apart from work simply establishing rates for individual species, growth studies have tended to demonstrate either the effect of environmental variables on growth or its internal regulation during the life cycle. Two main approaches have been used, direct measurements at intervals of individual thalli, and measurements of the size attained on dated substrates.

Information on the fruticose species is restricted. Frey (1959) measured the podetia of several species of *Cladonia* at intervals over periods of up to 35 years in various permanent quadrats in Switzerland. Andreev (1954) first demonstrated that the *Cladinae* produce a single whorl of branches each year, so that the age of the podetium may be calculated from the number of "internodes". Intercalary growth continues for about 10 years. He divided the life cycle into three periods of generation, renovation and decline: in the first, growth is continuous (5–25 years); in the second, the podetia in a mat decay from below whilst growth continues at the tips (80–100 years); in the final period (20 years), decay predominates. Ahti (1961) quotes the internode length of the longest living internode as an index of growth for different species. Advantage has been taken of this unusual and convenient growth habit to study growth rates and productivity of, for example, *Cladonia stellaris*, *C. mitis* and *C. rangiferina* in Canada (Scotter, 1963), and *C. arbuscula* and *C. impexa* in Scotland (Prince, 1974). The most detailed work has shown that younger parts of *C. stellaris* have a greater relative growth rate than older ones and that summer growth is correlated with rainfall (Kärenlampi, 1971a); as growth progresses in plants up to 12 years old, relationships between thallus length, diameter and weight are maintained (Kärenlampi, 1970).

The rates of growth of populations and of individual podetia in *C. stellaris* (Yarranton, 1975) differed with the successional maturity of the stand in which they occurred, rather than with the age of the population, suggesting that soil moisture was a factor. Population growth was logistic whilst the growth rates of individual podetia declined with age due to density-related factors.

Whilst fruticose species grow in three dimensions, foliose and crustose species are alike in that they form circular thalli in a plane, growing outward at the margin. The area of a thallus gives a good approximate representation of its size, though Fletcher (1972) showed the relationship was not exact for two *Verrucaria* spp. Frey (1959) and Rydzak (1967) list growth rates for many species showing great intra-specific variation without assignable cause. Clearly a statistical approach with adequate samples of carefully defined material is essential.

Growth varies over the geographical range of a species, and the size ratio of equal-aged thalli of two species may differ in different regions (Beschel, 1961).

For foliose species, within the U.S.A., the growth rates of *Parmelia baltimorensis* (sub *P. caperata*) (Hale, 1970) and *P. conspersa* (Hale, 1967) have been shown to vary with the geographic location. Beschel (1961) quotes "lichen factors" for West Greenland, i.e. the estimated diameter of 100-year-old thalli of *Rhizocarpon geographicum*, ranging from about 2 mm to over 40 mm, decreasing with increasing continentality assessed by altitude and distance from the sea. Similar variations in growth rates for this species in Europe are dependent on local climatic conditions (Jochimsen,

TABLE I

Growth rates of Rhizocarpon geographicum *aggr. on granite tombstones in three Scottish churchyards (P. B. Topham, unpublished data).*

	Kirkton of Durris, Kincardineshire	Old Struan, Perthshire	Crarae, Argyllshire
National Grid Reference	37/7796	27/8065	16/9897
Range of dates	1857–1933	1887–1964	1906–1949
Number of tombstones	10	12	12
Radial growth rate (mm year^{-1})	0·0949	0·1936	0·5025
Standard error	±0·0302	±0·0242	±0·0702
Lichen factor (mm)	20·06	38·1	87·3
Altitude (m)	30	152	8

1966). In Scotland also, variations occur from east to west of the country (Table I). Granite headstones cut by machinery were introduced in the middle of the nineteenth century and were used over the whole country alongside those of more local stone. The fronts were highly polished and were inscribed at intervals with family names. The date of the first recorded burial and that of the stone rarely coincide and piety frequently dictated that the family memorial be maintained in pristine splendour and unstained by lichen growth, but satisfactory series can be selected from the large numbers available. The unpolished vertical backs offer a uniform habitat to judge growth rates. The major factor in the four-fold increase from east to west is probably the increasing rainfall.

Rainfall is important for many species; over a 4-year period the mean annual growth rate of *Xanthoria elegans* was linearly related to the mean annual rainfall (Beschel, 1954), while Armstrong (1973) in Britain and Hale (1967) in America related monthly growth rates to rainfall for four species of *Parmelia* and one of *Physcia*.

Other climatic factors are important; Brodo (1965) found differences for three species in annual growth rates which he related to a particularly cold winter. *Parmelia caperata* has been reported to show frost damage after a hard winter (Laundon, 1966).

Alternations of light and darkness, humidity and dryness were essential for growth of *Parmelia sulcata* and *Hypogymnia physodes* (Harris and Kershaw, 1971). With such fluctuations, growth of small intact thalli was maintained for several months at about 15% dry weight increase for the first species and at about 5% for the second.

It has been generally observed that moderate nutrient enrichment improves growth; experiments (Hakulinen, 1966) have demonstrated the response. Pollution has various effects on growth, depending on the species. In Britain, *Lecanora muralis* showed a reduction in radial growth rate (Seaward, 1976), whilst in *Parmelia saxatilis* growth rate is maintained but cover is reduced and crescent-shaped colonies form, where portions of the thallus margin are locally more resistant (Gilbert, 1971).

Such variations from lobe to lobe of foliose species or from place to place of crustose thalli are the rule, reported by Frey (1959) for many species in Switzerland, by Andrews and Webber (1969) for *Pseudephebe minuscula* in the Canadian Arctic, by Hale (1959) for *Parmelia conspersa* and *Dimelaena oreina* in Connecticut, U.S.A., and by Rydzak (1961) for *Graphis scripta* and other species in Poland, to quote a few examples. The most precise study of lobe growth was carried out by Hale (1970) using *Parmelia baltimorensis* (sub *P. caperata*); extension growth was most active within a zone 1 mm behind the lobe tip and extended backwards for another 1 mm. Lobe division took place by generalized growth in a few of the lateral bulges, one of which ultimately took over as the tip of the new lobe. Daily growth in a mild humid period averaged 0·05 mm for a month. Radial growth may be calculated by averaging several lobes (Armstrong, 1973) or by measuring the area and computing the radius as that of a circle of that area (Hale, 1959).

Growth rates also depend on the substrate, sometimes in quite subtle ways. Non-vertical surfaces on the Scottish granite gravestones already referred to were clearly more favourable for growth than the vertical surfaces, probably because they were slower to dry out after rain. Those who have measured radial growth of lichens growing on tree trunks have commented on the degree of passive stretching due to the growth in diameter of the tree trunk and have allowed for it in their calculations of radial growth; presumably a degree of intercalary growth to compensate takes place over the whole lichen surface. Barkman (1958) suggested that the ratio of vertical to horizontal diameter for corticolous species could be the basis of a dating technique. Bioret (1921) showed that this ratio also depended on the tree species and particularly on the anatomy of the bark; *Graphis* and *Opegrapha* species grew faster with the grain of the bark and their thalli may be especially horizontally elongated on *Prunus avium*.

Alongside the accumulation of information on the average growth rates for different species and the demonstration that over both long and short time scales they are dependent on various aspects of the environment, has gone an interest in variations in growth with age, and attempts to establish whether lichen thalli show an S-shaped curve of cumulative growth, the so-called "grand period" of growth. Various sources of confusion must be borne in mind: it is important to distinguish between radial, absolute and relative growth rates (Woolhouse, 1968), i.e. growth expressed in units of cm year^{-1}, cm^2 year^{-1} or cm^2 cm^2 year^{-1}. There is a tendency to oversimplify by classifying thalli into young (small) thalli and old (large) thalli, sometimes without specifying the range; many studies have of necessity replaced the time axis by a size axis when the subject under discussion is the fact that the relationship between these two is not necessarily linear. The term, "grand (or great) period of growth" is frequently misused by lichenologists to refer solely to a period of rapid growth in young thalli. The phrase was coined by Sachs in the second half of the nineteenth century (e.g. Sachs, 1882) to refer to the whole course of growth of a typical plant part, beginning slowly, increasing in rate to a maximum and then declining, until growth ceases at maturity, and it is used in this sense by plant physiologists (Whaley, 1961). The cumulative growth curve which results will be sigmoid; neither the growth curve nor the resulting cumulative growth curve need be symmetrical (Richards, 1959) and the relative growth rate will usually also vary with time. The confusion in lichenological literature seems to have arisen because the radial growth of many slow-growing species, especially *Rhizocarpon geographicum* in a lichenometrical context, is often approximately constant with time, so that to establish that a period of more rapid growth preceded this phase is to make out a strong case for the existence of the whole of Sachs' "grand period".

Some evidence exists for growth rates increasing with age or size in very small thalli, as Beschel (1958) demonstrated using a *Parmelia* sp. on dated wooden grave crosses in Austria. By weighing foliose lichens from twigs of known age, Platt and Amsler (1955) showed that dry weight increased logarithmically for up to 13 years, which was the maximum age they studied. Phillips (1963) found that in general large thalli had larger radial growth rates. Andrews and Webber (1969) quote mean radial growth rates for six thalli of *Pseudephebe minuscula* of sizes ranging from 5 to 28 mm, from which a positive correlation of 0·855 between diameter and radial growth rate can be calculated. Increases in the radial growth rate with size for thalli less than 1·5 cm in diameter belonging to two species of *Parmelia* were considered by Armstrong (1974) to be consistent with an early logarithmic growth phase.

Evidence for constancy of growth in mature thalli requires careful formulation, since a constant radial growth increment implies an exponential decline in the relative growth rate (Armstrong, 1974). It also implies an absolute growth in area which is a time power function (Richards, 1969); various growth functions are tabulated for this function (Table II).

TABLE II

Growth functions of a circular thallus with radial growth constant in time (time power function).

Radial growth rate	b
Diameter at time t	$2bt$
Area at time t	$\Pi(bt)^2$
Absolute growth rate	$2\Pi b^2 t$
Relative growth rate:	
(a) at time t	$2/t$
(b) at area A	$2b(\Pi/A)^{\frac{1}{2}}$
(c) at diameter d	$4b/d$

Rydzak (1961) found that the relative growth rate (percentage increase in area) was greater in young thalli; his data were recalculated by Brodo (1965) and the radial growth rates, especially of *Pertusaria coccodes* and *Graphis scripta*, then showed the opposite trend. Brodo himself could not detect any differences in radial growth for two *Parmelia* spp. over sizes ranging from less than 2·8 to over 5·8 cm in diameter, though his samples were small. Hale (1967) found that in *Parmelia conspersa* radial growth was less in small thalli (under 1·5 cm in diameter), constant over a middle period, and declined again in large thalli (10–12 cm in diameter) when fragmentation of the centre occurred, so that the life-span was estimated to be about 39 years from observations over a 16-year period. Thirteen thalli named as *Caloplaca aurantiaca* were measured for 5 years in the Central Negev desert (Lange and Evanari, 1971). Percentage growth increase declined with thallus size, but radial growth was not affected.

Relative growth rates of carefully measured plants of three species of

Parmelia and one of *Physcia* growing under uniform conditions were reported by Armstrong (1973) to decline exponentially with size, consistent with constant radial growth rates.

We may then conclude that many lichens show a pattern of growth in which a preliminary exponential phase of growth increasing with size is succeeded by one in which radial growth is constant over a long period. Some values modelling growth of this type are presented in Table III, which owes much to Armstrong's ideas. All the evidence we have makes it appear that the growth of several *Parmelia* spp. and some other foliose and placoid species follows this pattern, except that the central area ultimately breaks down leaving fragmented portions of the margin. Some species appear to survive this period of disorganization, marginal growth continuing with little alteration as in *P. centrifuga* (Henssen and Jahns, 1973) or *Parmelia alpicola*. In others some individual lobes continue (Armstrong, 1974; Gilbert, 1971), or the growth rate slows (Hale, 1959, 1967). This points to some lack of coordination between the inner and outer zones of the thallus.

Another type of growth curve has been proposed for *Pseudephebe minuscula* (Miller, 1973), based on diameter increments for 27 thalli over two seasons and using the relationship between diameter and relative growth rate in diameter. Relative growth rate showed a negative exponential decline with size and yet the reconstructed cumulative growth curve, replacing the size axis with a time axis, does not show a linear phase, a warning against the tendency to interpret declining relative growth rate as always indicating constant radial growth.

Another major source of evidence comes from lichenometry, the use of lichen-size measurements to date surfaces. Details of the principles and techniques involved may be found in the papers by Beschel (1961) and Webber and Andrews (1973), but it is the growth curves which have been established which are relevant to this chapter. The most commonly used species is *Rhizocarpon geographicum* aggr. The difficulties of establishing dated reference points are formidable; the evidence from any single technique, such as radiocarbon dating of soils or artefacts, dendrochronology or historical records, requires expert interpretation, so that such points may be few. The absolute values may depend on the size of the area sampled (Matthews, 1974) but the shape of the curve should not.

Growth curves, based on radial increments, are reported that are linear with time (Stork, 1963; Reger and Péwé, 1969; Burrows and Lucas, 1967; Burrows and Orwin, 1971; Andersen and Sollid, 1971; Lindsay, 1973). Others are linear after an initial period of more rapid growth (Miller, 1969; Andrews and Webber, 1964; Benedict, 1967; Miller and Andrews, 1972; Denton and Karlén, 1973), whilst other workers have interpreted

Table III

Growth model for an idealised circular lichen thallus. A propagule with an area of 1 mm² grows exponentially for 10 years with a constant relative growth rate of 0·58 per year; thereafter there is a yearly radial increase of 1·4 mm.

	Time (years)	Area[a] (mm²)	Absolute growth[a] (mm² year⁻¹)	Relative growth[a] (%)	Diameter (cm)	Radial growth[a] (mm year⁻¹)	% error in age estimate[b]
Propagule	0	1	—	—	0·11	—	
Exponential growth phase	1	1·58	0·58	58	0·14	0·14	
	3	3·98	1·47	58	0·22	0·22	
	6	15·85	5·85	58	0·45	0·46	
	10	100·00	36·90	58	1·13	1·15	
Constant radial growth phase	15	498·5	104·6	21	2·5	1·4	40
	20	1206·3	166·2	14	4·0	1·4	30
	30	3544·9	289·3	8	6·7	1·4	20
	40	7114·5	412·3	6	9·5	1·4	15
	50	11914·9	535·4	5	12·3	1·4	12
	60	17946·2	658·5	4	15·2	1·4	10
	100	54380·3	1150·9	2	26·3	1·4	6

[a] Calculated on a yearly basis.
[b] If age is deduced from constant radial growth.

their curves as showing a gradual decline in growth without distinct phases (Mottershead and White, 1972; Matthews, 1974). It seems a reasonable generalization that the unqualified linear relationships between age and size are based on relatively few points; also that when a series of measurements of young thalli up to a century old exists, the resulting curve will show rapid early growth slowing with age. The evidence from southern Norway for a gradual decline which was summarized by Matthews (1974), where the curves for several glaciers obtained by different workers can be superimposed, seems most convincing and his data are well fitted by the theoretical curve (though with a three parameter model and four data points this is not entirely surprising); but with a longer time-span such as that of Benedict (1967) (from 500 B.C. instead of from A.D. 1750), the distinction between a linear phase and a slow exponential decline will be blurred.

It is perhaps appropriate to consider why growth curves are fitted. To say, for example, that the relationship between diameter and age is linear is purely descriptive, but it permits interpolation or comparison between areas. It also has certain purely mathematical implications in terms of absolute and relative growth rates (Table II), which we may compare with those of other possible growth curves (Richards, 1969).

But growth curves also have biological interpretations; they may be derived from biological considerations (Bertalanffy, 1957), or biological consequences may be derived from them and their plausibility examined. The S-shaped growth curve covering a "grand period of growth" and found in so many organisms has two implications: that absolute growth in the early stages is limited because the organism's own size denies it the necessary resources; and that its growth is finite, bounded by some combination of genetic, metabolic or environmental restrictions which gradually bring it to a halt. Numerous suitable curves have been suggested, many of which can be expressed as variations of a single function (Richards, 1959) and which differ in the shape of their growth and relative growth curves. But we have evidence that the radial growth of certain species is approximately constant over long periods and may infer that the conditions experienced by the outward growing hyphae are equally stable. Since the thallus is increasing in size, we conclude that the zone which acts as a basis for growth is constant in width, and probably quite narrow, since linear radial growth is established early. The size or time at which it is established is important because it may affect the competitive ability of plants in the early stages of colonization. It is also important when considering reproductive capacity, since the inner zone is then released for vegetative or sexual reproduction. On this consideration, linear radial growth should begin at about the same time as the first reproductive structures appear, and the non-reproductive margin is the "effective" zone, so that relative growth calculated on this basis will be approximately constant. This interpretation stresses growth as one aspect of the reproductive biology of the plant; it also provides an objective measure of maturity.

The use of size–frequency histograms of lichen populations to investigate lichen growth rates and succession was suggested by Farrar (1974). In any population, the frequency of thalli in a particular size category is proportional to the rate of colonization when that size class was established and inversely proportional to the growth rate in that class. The effects of death or fusion may be ignored in the well dispersed populations to which this type of analysis may be applied. In the same way, the size–class structure of forest trees is often used to explore their population dynamics. For example, conclusions may be drawn as to possible changes in death rate with age and on perturbations due to disease or changed land use (Johnson and Bell, 1975).

Benedict (1967) has already developed the use of size–frequency plots of populations of *Rhizocarpon geographicum* to make deductions about the past history of the population. He shows frequency distributions for two sites. The first was a mud-flow with a radiocarbon date for the underlying soil of 180 years B.P.; 1500 thalli were measured in 1-mm size classes from 1 to 18 mm. Because the distribution was smooth and unimodal he assumed that it represented a single population with a continuous history, so that no earlier plants had survived the mud-flow. At the other site, stones from sorted nets had been used to build a wall 970 years B.P. Thalli from undisturbed stones showed a negative log-linear relationship between frequency and diameter class, whilst thalli from the wall showed a break corresponding to the disturbance when the wall was built. It may be noted here that such a relationship, given a constant growth rate, implies an exponential increase in the numbers of thalli colonizing the surface.

Using the same technique, Andersen and Sollid (1971) found outlying points representing a small number of anomalously large lichens among over 1000 measured on a moraine in south Norway. These were interpreted as either fast-growing mutations or survivors of an earlier population, and the population was rejected for lichenometrical purposes. However, the estimate of size derived from the intercept in such a graph would seem the ideal (though not the most practical) approach to the problem of estimating the largest lichen in a population, since it uses all the evidence and is less subject to sampling fluctuation than the estimate using the largest specimen found.

Similarly in the mixed population studied by Benedict (1967) the intercept of the graph of the more recent population with the size axis, rather than the breakpoint between the two, is also an acceptable estimate of the size of the earliest thalli to colonize the surfaces provided when the wall was built. Lindsay (1973) also uses such size–frequency graphs, using the weight of a fruticose antarctic species, from two surfaces of different ages. None of these authors nor Denton and Karlén (1973) use population data to do more than decide on the homogeneity or otherwise of the population.

Assuming that colonization increases exponentially, the relationship between frequency and growth rate can be estimated by plotting a cumulative size–frequency graph (Fig. 1) with frequency on a log scale to remove the colonization effect. The same technique applied to Benedict's (1967) mud flow data yields a growth curve in close agreement with that deduced by Benedict from other evidence.

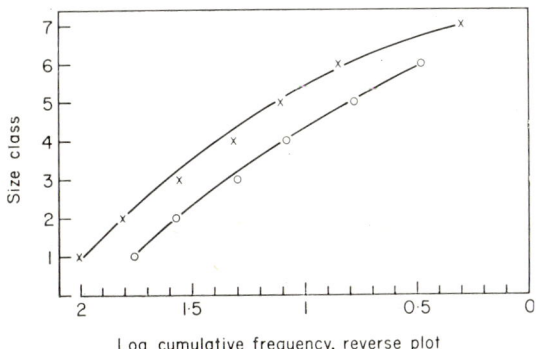

FIG. 1. Size–frequency graph, cumulative frequency on a logarithmic scale to show the potential use as a growth curve, assuming logarithmic increase in colonization (two species). Data from Lindsay (1973) by permission.

The assumption of exponential increase in colonization is not always justified. Farrar (1974) shows data for *Lecanora muralis* which imply a peak of colonization followed by a decline, for a population on an asbestos roof in Berkshire, England, whilst data, for the same species and similar habitat (Fig. 2), replotted from Brightman (1959) are probably best interpreted as indicating a constant rate of colonization. Brightman's population grew in the outskirts of London and was both sparser and slower-growing than that of Farrar; his figures for the distribution of thalli on individual tiles show close agreement with a Poisson distribution and suggest that distribution in space at least was random. A constant rate of colonization would imply no secondary spread from older plants and the spatial distribution of the population also supports this.

Growth may also be measured as productivity, that is by the accretion of biomass. Here again information on lichens is both less and more difficult to obtain than is the case for higher plants. In many instances recorded, lichens are of minor importance (Coppins and Shimwell, 1971; Wein and Bliss, 1974). Estimates of lichen biomass as a food resource in wildlife management are the main source of information (Edwards et al., 1960; Chapters 5 and 6) though work stimulated by the Tundra Biome section of the International Biological Project is also a source of information (Wielgolaski, 1972; Richardson and Finegan, 1977). The biomass of epilithic lichens may locally exceed that of higher plants in the Negev desert (Kappen et al., 1975).

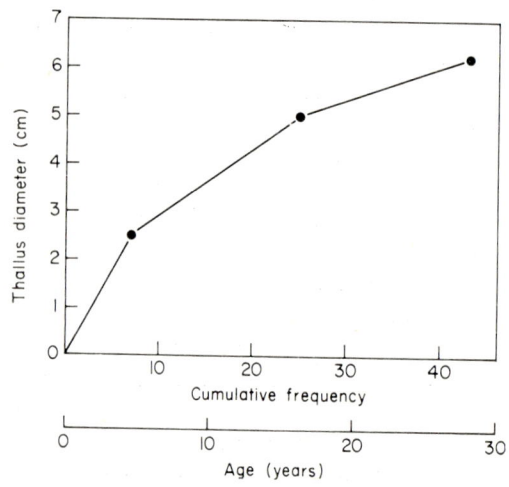

Fig. 2. Size–frequency graph: cumulative frequency on an arithmetic scale to show its potential use as a growth curve assuming constant colonization. Data from Brightman (1959) by permission.

An indication that the processes of growth and carbon assimilation have been well understood is the construction of a successful computer simulation combining physiological, climatic and environmental variables. Carbon assimilation in *Parmelia caperata* has been modelled as a function of light intensity and moisture availability (Harris, 1972) and the growth and decay of forage lichens as affected by lichen biomass, rainfall, tree cover and grazing has been plausibly simulated (Bunnell et al., 1973).

III. Succession

Because of the sheer volume of information available, and the widely differing treatment of the material, this section will deal firstly with a few problems, either general, or involving particular substrates, and then discuss how lichens follow the same trends as other organisms in developing ecosystems.

As in the study of lichen growth, consideration of successions involving lichens is made difficult by problems of dating substrates, so that similar techniques are used. These include: historical records of road cuttings, graves, buildings, quarries; tree ring counts and tree size; and the annual bud scars of deciduous trees. Dunes and raised beach systems are probably unique in providing habitats ordered in time, whilst radiocarbon dating, the rate of isostatic uplift and old land maps also provide evidence. For this reason, too, the early stages of succession on rock are much better documented than later ones. New Zealand work was exceptional in successional studies in examining rock surfaces up to 40 years old and moraines up to about 320 years old (Orwin, 1970). Schauer (1969) describes a 30-year-old surface of glacial gravel exposed near Munich, which was still immature.

In reverse, successional trends, including increase in percentage cover, in numbers of species per unit area, in thallus size of three species and in percentage frequency of eight species were used to characterize five lichen age zones to classify moraines on an Afghan glacier (Gilbert *et al.*, 1969), whilst Miller and Birkeland (1974) assessed the value of similar trends and of other techniques for the same purpose in Wyoming.

The need for caution in extrapolating to areas of differing climatic conditions is stressed by Orwin (1972) who found that New Zealand rock surfaces 320 and 40 years old had reached similar stages.

Another difficulty in general studies, though less apparent nowadays, is the inability of otherwise competent ecologists to recognize any lichens other than a few of the more conspicuous *Cladonia* or *Peltigera* spp. It is usually impossible to make deductions from the absence of lichen records, although successions may be found involving them on any substrate on which they grow. In many cases we are dealing with serrules, such as those on decaying logs and stumps (Barkman, 1958) or there may be lichen-rich phases in primoseres when climatic and edaphic conditions are favourable, but here the lichens are usually subordinate to the physiognomic dominants and can hardly be discussed without reference to those species. Little (1951) enumerated the species in the minor vegetation of pine barren swamps in New Jersey, U.S.A., of which, although not important elements in the vegetation, 22% or 27 species were lichens. Looman (1964) pointed out that their inclusion with other elements of the flora could contribute to more general ecological studies: he found that including lichen data reduced the similarity coefficients between stands and so increased his ability to discriminate between them.

Indeed, most lichenologists have been moved at some time or other to agree with Rydzak (1961) that "lichens could be an index of ecological conditions, if lichenologists succeeded in elaborating suitable coefficients of correlation". Studies using multivariate statistical techniques demonstrating sensitivity to ecological gradients, including successional ones, incorporate, for corticolous lichens, the construction of a phytosociological index showing a continuum from oak savannah to *Quercus–Acer* forest (Hale, 1955), and the demonstration of a complex relationship between the lichen flora, the age of the tree and exposure (expressed as distance from the nearest tree) (Yarranton, 1972). For terricolous lichens, one may quote a moisture gradient (Lambert and Maycock, 1968) and two axes of a Bray and Curtis ordination, representing respectively a range from xeric to mesic habitats and degree of shading, both of which were related to successional changes (Lechowicz and Adams, 1974) and the study of a raised beach system by principal component analysis (Kershaw and Rouse, 1973).

In dealing with changes in corticolous communities, it is not always clear that succession in the full sense of the word exists. Yarranton (1972) refers to succession among epiphytic lichens, but does not consider that there is a true succession. Kershaw (1964) found that zonation on young trees was similar to that on older trees but more compressed. So any site on a tree will experience directional change with time, but changes in the substrate are caused only to a minor extent by the corticolous community and mainly by the growth and development of the phorophyte and its congeners. There is evidence that the pH of bark is changed by the lichen cover (Kershaw, 1964) and that hyphae penetrating the bark may occlude fragments (Fry, 1926) and so change its water-holding capacity, but the influence of lichens on the trees which support them appears to be slight (Barkman, 1958). Furthermore, the longevity of communities on bark is limited by the life-span of the tree, so that whilst it is convenient to speak of succession, the species that comprise it have less power to influence their own environment than usual. Successions among corticolous communities also depend on the phorophyte. Thus the *Lobarion* in Europe seems to be associated with climax woodland (Rose and James, 1974). Within an ancient woodland area such as the New Forest in southern England, communities representing all stages in the probable succession exist, but whilst the earlier stages may be found in a range of woodland types, the later stages are almost diagnostic for climax woodland.

That the *Lobarion*, a European climax community, consists mainly of large foliose lichens, contradicts one of the shibboleths of lichen ecology, the existence of a succession of life-forms beginning with crustose, continuing through foliose and concluding with fruticose species. Like many such ideas, it is a useful oversimplification. It is not particularly useful when considering corticolous species. The large foliose species of the

Lobarion are adapted to compete with mosses by growing over the surface of the mat. Among fruticose genera, *Alectoria* and *Bryoria* spp. with haptera are similarly adapted, but *Ramalina* spp., despite their often considerable size, usually require anchorage on dry bark. Again, colonizing species may be foliose, for instance on tree bases (Kalgutkar and Bird, 1969) and on twigs (Degelius, 1964).

In successions on soil, there is considerable variation in the colonizing species. After heath fires, the peat exposed may be quickly covered by *Lecidea granulosa* and *L. uliginosa*. In the Arctic, species of foliose and fruticose genera such as *Stereocaulon*, *Cetraria* and *Peltigera* may predominate (Viereck, 1966; Crittenden, 1975). Beschel and Weideck (1973), discussing the foreland succession in West Greenland, found that silt was first colonized by about ten species of vascular plants before the first macrolichens appeared. The latter had reached about 2-cm diameter before the first crustose species could be identified. On the other hand, *Lecanora epibryon* is reported by Acock (1940) to figure fairly prominently in communities on calcareous beach deposits in Spitzbergen, whilst in northern Europe, dunes in the early open stages dominated by grasses are the habitat for some of the more interesting terricolous crustose species.

On rock, the generalization appears to have more justification (Faegri, 1933; Degelius, 1940), although Beschel and Weideck (1973), discussing the foreland succession in West Greenland, found that *Umbilicaria* spp. and members of the bird perch community preceded crustose species. The reversal in arctic environments of the order found with moderate consistency in temperate climates perhaps reflects an adaptation in the form of large diaspores, especially thallus fragments.

The importance of lichens in the lithosere, even the existence of the lithosere as a major successional entity, has been the subject of some debate; the quotation at the head of this chapter represents a simplified picture to which exception has been taken (Cooper and Rudolph, 1953). Winterringer and Vestal (1956) traced the theme back to Biberg, a pupil of Linnaeus, in 1749; Syers and Iskander (1974) to Linnaeus himself in 1762 (see also Smith, 1921). Despite the poetic hyperbole with which it is expressed, it must stand as a vision of the dynamics of plant succession remarkable for a period when ideas of the evolution of species and the occurrence of ice ages lay in the future. Many successions, especially in mesic or humid environments, do not involve lichens to any significant extent. Even in xeric environments, mosses are often ecologically equivalent to them, filling many of the same niches on trees and soil, collecting dust and humus on rock. In the temperate northern hemisphere, species of *Rhacomitrium* and *Andreaea* may be the most prominent colonizers of rock (Oosting and Anderson, 1937; Tallis, 1958). Polunin (1936), describing successions in Norwegian Lappland—in the type of terrain which must have inspired the original conception—divided the lithosere into eight stages of which the first two to three were lichen-dominated, non-essential and proseral. He noted the importance of chomophytic communities exploiting crevices and the wind-blown mineral and organic matter which accumulates in them. Similarly, several studies of granitic flat rocks, rock ledges and sloping cliffs in the southeastern United States (McVaugh, 1943; Keever *et al.*, 1951; Winterringer and Vestal, 1956) stress that the lichen communities do not develop further—it is mosses in general which initiate unstable mats and accumulate soil to be invaded by vascular plants along with *Cladonia* spp. Trees may appear in the older mats, but the system is fundamentally unstable and the mats eventually break loose and are eroded away. Studies of these habitats have stressed zonal or cyclic trends rather than succession; however, it is not unreasonable to consider that unfavourable conditions on these bald-faced crags have preserved early successional stages.

The role of lichens in pedogenesis, discussed earlier in relation to their ability to colonize rock surfaces, has been recently summarized by Syers and Iskandar (1973). It has been investigated largely by soil scientists, to whom soil is a complex produced by the action of living organisms on mineral matter. Lichens and mosses can extract nutrients from rocks that are unavailable to higher plants (Jacks, 1965). Increased weathering of lichen-covered areas together with changes in the concentration of certain metallic ions was reported from recent Hawaiian lava flows (Jackson and Keller, 1970). The depth of weathered rind was 0·142 mm beneath *Stereocaulon vulcani* and less than 0·002 mm on bare surfaces after about 60 years. The time-scale required to convert the rocky mountain into a waving forest is indeed long, but those involved in successions on dunes or moraines are also long. Over 1000 years are involved in a system of raised beaches (Kershaw and Rouse, 1973) and 2900 years for sand dunes in Ontario, Canada (Morrison and Yarranton, 1974). A climax surface 5000 years old and an immature surface 200–300 years old have been reported on glacial gravel outwash in Alaska (Viereck, 1966), whilst in Michigan, soil formation was still continuing in dune ridges 12000 years old (Olson, 1958). These are but a few examples; events on rock might be expected to proceed even more slowly. But in general rock surfaces are not homogeneous; the lithosere and the psammosere are two ends of a continuous spectrum of substrate-dependent development.

Because lichens are broadly speaking xerophytic and photophilous, they are frequently abundant in subclimax communities, as mentioned above, and in cyclic successions such as dwarf *Calluna* dominated communities (Watt, 1947; Gimingham, 1964) and the *Juncus trifidus* communities

(Ingram, 1958) of the Scottish Highlands. The coastal shingle succession at Dungeness, England, showed a complete sere; reversion to earlier stages occurred if insufficient humus had developed to support the progression to heath (Scott, 1965). Rock ledge vegetation in southern Illinois, U.S.A. (Winterringer and Vestal, 1956) included thin soil with lichens and mosses liable to cyclic catastrophe. Hale (1959, 1967) examined the dynamics of the growth of *Parmelia conspersa* in cyclic communities and established time-scales. Frey (1959) has quoted an example due to avalanche damage in Switzerland.

Secondary successions may be richer in lichen species than primary successions and are often characterized by a rich *Cladonia* flora. Abandoned fields in North Carolina, U.S.A. (Robinson, 1959) passed through a phase rich in *Cladonia* spp. until the closing canopy of the tree layer made conditions unfavourable. In dune vegetation at St. Cyrus, Scotland, *Cladonia* spp. were prominent in a subsere following burning, but were unimportant in the primary succession (Gimingham, 1951). They also occur in subseres following burning in Scottish heaths (Gimingham, 1964) and in the post-fire recovery sequence of black spruce–lichen woodland in Canada (Maikawa and Kershaw, 1976). The common feature appears to be a requirement for humus and an open environment.

The role of lichens in succession illustrates many of the principles of ecosystem development as discussed for example by Odum (1969), although compared to many organisms they contribute little to energy flow in a community. They are still significant in photosynthesizing in many environments too severe for other forms of life, so that their presence on boulders in heathland, screes or deserts is bound to increase the utilization of incident light energy. Methods have been developed for measuring the energy flow through lichens important in reindeer management (Kärenlampi, 1971b).

Their contribution to community structure is more considerable. The lichen layer may form a sizeable proportion of the productivity of barren areas (see Chapter 6). They increase the intra-biotic fraction of inorganic nutrients: corticolous species trap exudates from the phorophyte and from rainwater, while saxicolous species mobilize and extract mineral nutrients from substrates, rainwater and airborne dust.

As a succession progresses, the lichen species within it in general increase in diversity as do the vascular plants. The occurrence of 20 lichen species has been used to form an Index of Ecological Continuity to apply to oak woodlands in Britain (Rose, 1974). In the early stages of successions dominated by lichens, species diversity increases in general until a closed community is formed. Degelius (1964) found this trend on annual portions of ash twigs; the maximum number of species was found in the 10-year-

TABLE IV.

Distribution of species of lichen and vascular plants in a dune transect at Tentsmuir, Fife, Scotland (P. B. Topham and C. J. B. Hitch, unpublished data).

	\multicolumn{14}{c}{Distance from the strand line (m)}													
	1	21	31	41	51	61	71	81	91	101	111	131
% Cover	6	25	58	30	26·0	26	86	75	33	99	97	96
Mean number of species per quadrat												
Lichens	0	0	1·4	2·0	2·9	3·1	3·3	5·0	1·8	5·5	5·4	3·6
Vascular plants	1·6	1·4	2·3	2·4	3·3	2·5	4·4	4·7	3·3	4·0	4·7	7·0
Cumulative numbers of species												
Lichens	0	0	2	6	8	10	12	16	17	18	20	24
Vascular plants	3	3	8	8	11	11	13	17	19	19	20	26

old zone at about the point that a closed community appeared. On rock in New Zealand (Orwin, 1970) pioneer species were still present on 40-year-old surfaces, so that seral stages could not be defined. The community, however, was still open. The dune succession at Tentsmuir, Scotland, did show a typical staircase pattern of species incidence, perhaps because community development was more rapid. The cumulative number of species and the mean number of species per quadrat (Table IV) were similar for both lichens and vascular plants although the lichens lagged slightly and passed their peak sooner. Species diversity is often considered to reach its peak just before the climax and then decline slightly. This is not necessarily true of lichens considered in relationship to the forest climax of an area; like birds (Johnston and Odum, 1956), they may show a bimodal distribution, diversity being reduced when the canopy closes and increasing again in mature and more open woodland (Robinson, 1959). However, in densely populated areas such as Western Europe, the presence of a rich lichen flora often indicates relict areas of primeval forest, as in the New Forest, England (Rose and James, 1974; Rose, 1976). The equitability component of diversity has been little studied in lichens; Coppins and Shimwell (1971) presented dominance diversity curves for different ages of dry *Calluna* heath showing more equitable distribution and niche specialization at intermediate stages in the cycle.

No lichenologist needs to be persuaded that lichens increase biochemical diversity in an ecosystem. Some aspects of this may be associated with competitive interactions between lichens and other organisms, but perhaps their greatest contribution to diversification is in pattern diversity. The lichen layer or layers supplement whatever other structure the community possesses and are usually capable of subdivision into further layers and niches; the appreciation of this patterning is the hallmark of the skilful field naturalist. A single mature oak may provide accommodation for up to 50 species (Rose, 1974), from *Dermatina quercus* on young twigs to *Coniocybe furfuracea* on the roots, whilst rain tracks may have an elaborate zonation. Barkman (1958) has shown how the epiphytic communities of, for example, exposed wayside trees increase in number as the trees which support them age, and how they become intricately interwoven in space, so that bark fissures support a different flora from those of the intervening plateaux. Niche development and exploitation imply increased pattern diversity.

The presence of lichens also introduces a diversity of accompanying fungal epiphytes and parasymbionts, provides a habitat for a not inconsiderable microfauna (see Chapter 4) and enriches the microflora as succession proceeds (Webley et al., 1973).

The different types of selection to which organisms are subjected at the beginning and end of a succession have been called r and K selection by MacArthur and Wilson (1967) from parameters in a common formulation of the logistic growth equation in which r represents the intrinsic rate of increase and K is the asymptote or limiting value. At the beginning of a succession, the adapted species are those with the greatest rate of increase. At the end, rate of increase is secondary to competitive ability and the production of propagules with a high chance of establishment. Although the majority of lichens are long-lived perennials with perennial reproductive structures, they display what Pianka (1970) has called the r-K-continuum.

The pioneer species in many habitats are "weedy" species with much reduced thalli and abundant production of small ascospores. Examples are *Lecanora dispersa*, *Lecidea granulosa* and *L. uliginosa* on peat; *Buellia*, *Rhizocarpon* and *Acarospora* spp. together with *Bacidia umbrina* on acid rock; and the *Graphis–Ulota* community of smooth bark (Rose and James, 1974). These communities, discussed in detail in Chapter 10, are soon overgrown and replaced by taller-growing life-forms, which also tend to have reproductive bodies of increasing size, with spores, soredia, isidia and thalline lobules and fragments. In general, as the community becomes closed and the ecosystem more mature, diaspore size increases, fewer are produced and the time taken to reproduce increases, although dispersal by thalline fragments is a feature also of many macrolichens of open arctic-alpine communities where climate is the limiting factor.

Many lichen species produce diaspores to a wide range of specifications. *Cladonia* spp. producing spores, soredia and fragmenting thalli can exploit favourable opportunities as they arise. Within this genus the series of related species with scyphi, *C. fimbriata*, *C. chlorophaea* and *C. pyxidata*, show a series from farinose sorediate, granular sorediate to peltate squamules and grow in progressively less open habitats. The *Peltigera* species which can appear quickly and in quantity on disturbed ground such as camping sites and lawns is *P. spuria*, which is relatively small in size, with a larger proportion of its thallus devoted to reproduction than *P. polydactyla* or *P. canina*. A range of parallel reproductive strategies is found in vascular plants (Harper *et al.*, 1970).

The trend for increasing thallus size in climax communities is illustrated both by the reindeer lichens of boreal lichen heaths and woodlands and by the *Lobarion*, a community of large foliose lichens associated with climax oak forest (see Chapter 10). The diaspore size in the latter is not on the whole exceptionally large—perhaps because large propagules are not advantageous to arboreal species dependent on dispersal from tree to tree, with the result that establishment of the community seems to depend on occasional favourable combinations of circumstances.

It has already been mentioned that the pioneer species on twigs are not necessarily crustose; the foliose species involved, such as *Xanthoria lobulata*, *X. polycarpa* and *Physcia* spp., are small for their life-form, and will produce apothecia as quickly as the crustose pioneers. Taking the list given by Degelius (1964) of the annual shoot on which apothecia first appear, and adding to it four crustose pioneer species which are first observed in fruiting condition, comparing this with the age of twig on which soredia and isidia are first observed, the fourteen apothecial species are significantly quicker to reproduce than the seven with soredia or isidia, although they do not differ in the average age of twig on which they first appear. It appears that, just as generalizations have been established for animals relating potential rate of increase, body size and generation time, similar correlations between potential rate of increase, thallus size, diaspore size and age at which reproduction begins could be demonstrated for lichens and would indicate their position in the developing ecosystem.

IV. Competition

The competitive ability of lichens is usually considered to be low—they establish themselves where other organisms cannot. They are nature's pioneers, but also nature's poor relations. However, as Degelius (1940) remarked, competitive ability and succession are closely linked. In any succession each stage competes successfully with its immediate predecessors and succumbs to its immediate successors. Changes in the environment which change the succession will alter the competitive ability of the species present, for instance increasing shade as trees grow (Frey, 1959). Similarly, pollution-resistant species, such as *Lecanora conizaeoides*, *Parmeliopsis ambigua*, *Cetraria chlorophylla* and *Buellia punctata*, may exist in areas of pure air as minor constituents of the flora and yet increase in abundance in zones approaching towns where they are more adapted to the prevailing levels of pollutants than their competitors (Gilbert, 1974).

An interesting aspect of competition lies in the mechanisms by which it becomes effective; the most important, as in successions of vascular plants, seems to be a struggle for light. Generally speaking, crustose species are less competitive than foliose, which in turn are less aggressive than fruticose (Degelius, 1940), at least on rock in northern Europe. Species such as *Platismatia glauca* are intermediate, overgrowing the more adpressed foliose species. A few crustose species, mainly in the genera *Ochrolechia*, *Pertusaria* and *Haematomma*, are exceptions. Competitive ability corresponds in general to the stages in succession, although plants at the edge of their geographical range tend to be less competitive. Mosses seem to behave much as fruticose lichens, but the ability of both to suppress certain crustose species in their vicinity is presumably less due to shading than to their acting as a sponge to absorb moisture and release it gradually, increasing humidity in the substrate and the atmosphere.

The behaviour of two thalli of the same crustose species on rock varies. Degelius states that they fuse, Frey (1959) observed that adjoining thalli of *Lecanora alphoplacum* and of *Haematomma ventosum* fused as their margins met. In other cases, however, closed mosaics with individual thalli distinguishable by prothalline lines may be formed (e.g. *Fuscidea cyathoides*). Beschel (1961) states that the larger thallus of two *Rhizocarpon geographicum* plants will continue to grow forward through the tissues of the other, maintaining its characteristic margin.

Some *Rhizocarpon* spp., notably *R. rittokense*, may be surrounded by "zones of inhibition" 1–5-cm wide in which other species are eroded (Beschel and Weideck, 1973); this might be due to the diffusion of inhibitory substances from the thallus since the effect is more marked downslope, but this requires more critical study. Degelius (1940) and Barkman (1958) suggested that certain lichens exert a chemical influence on other plants in the field. It has been shown that the presence of lichens on granite surfaces caused developing protonemata of a species of *Grimmia* which grows on granite to become brown and unhealthy (Keever, 1957). In the case of vascular plants, *Peltigera canina* extracts have been reported to inhibit the germination and seedling growth in various grasses (Pyatt, 1967, 1968) but this requires confirmation. The relatively rapid growth rates of *Peltigera* spp., e.g. *P. canina*, 3–7 mm year^{-1}, *P. rufescens* 25–27 mm year^{-1} (Frey, 1959), together with their sprawling habit, may enable them to compete successfully with grasses in some habitats. Pure samples of some lichen acids can inhibit the germination of seeds; for example, Dalvi *et al.* (1973) showed that usnic acid inhibited the germination of mung bean seeds by impairing the utilization of food reserves. Some inhibitory action of lichen acids (Ott, 1961) and lichen extracts (Ramaut and Corvisier, 1975) on the seeds of tree species has been described but the presence of *Lecidea granulosa* has been reported to improve the germination of *Picea maritima* in Canada (Gagnon, 1966).

Apart from the fixation of nitrogen by certain pioneer species (see p. 34), the favourable effect of lichens on soil seems to depend on the physical effect of the lichen layer as a protective crust or mulch. A moss–lichen layer of *Pohlia nutans* and *Cladonia pyxidata* may reduce the soil temperature at a depth of 7·6 cm by 10–11°C relative to bare soil on pit heaps in northern England (Richardson, 1958) and so hasten the establishment of pioneer grass species. In the Hudson Bay lowlands lichen-dominated surfaces act as a mulch, so that soil moisture may be 40% higher than in burned areas where this cover has been lost (Rouse and Kershaw, 1971). The effects of such burning are severe, and recovery is slow. In the first example, the amelioration caused by the lichen layers prepares the way for its ultimate destruction in the next stage of the succession. The open lichen-rich woodlands of the subarctic, however, represent a stable climax vegetation extending over vast areas.

The study of competition is also closely linked to the autecology of individual species, since the successful species in a competitive situation is usually the one best adapted to its environment. In the case of *Cladonia stellaris* in north Finland, however, Kärenlampi and Kauhanen (1972) suggested that competition was the reason that the areas of optimum growth and of maximum frequency did not coincide. To characterize the minutiae of the microhabitat is often beyond the capabilities of the instrumentation presently available. Transplant experiments, once the technical difficulties are overcome, permit us to decide whether a species is absent from a situation either because of the accidents of dispersal, or because it is inadequately adapted or because it is less competitive.

Frey (1959), in his observations on lichen growth in the Swiss National Park, used transplant experiments to investigate the competitive ability of various *Cladonia* spp. He concluded that *Cladonia stellaris* had the greatest competitive ability under his conditions. The mat-forming *Cladonia* such as *C. uncialis, C. furcata* and *Cladina* spp. were easier to transplant and more competitive than *Cladonia elongata* or *C. cyanipes*.

The measurement of competition as a characteristic of a species or an environment is not easy. Coppins and Shimwell (1971) used dominance–diversity curves to distinguish which successional stages in *Calluna* heath exhibited competition. A competitive index such as that used by Grime (1973) to ordinate herbaceous plants has not yet been applied to lichens, and one is suggested in Table V. Grime (1974) proposed for herbaceous vegetation a triangular ordination with apices representing stress, disturbance and competition, and for axes a version of his competitive index and a measure of growth rate at the seedling stage. He suggested that the third axis should be some measure of reproductive capacity. Three independent variables, however, cannot be plotted on a two-dimensional triangular diagram. Whilst a full index of reproductive capacity would take into account both propagule size and the proportion of effort devoted to reproduction, propagule size is normally well documented for lichens and a scale is suggested in Table V. In spite of the large amount of information on growth rates discussed earlier in the chapter, it is in general too disjointed to be used to assess the role of growth rates in the competitive ability of lichens.

With these reservations, some communities are assigned places on a

TABLE V.

Scores to assess reproductive capacity and to calculate a competitive index.

Propagule size[a]	Capacity for overgrowth	Thallus cover	Thallus height
1. Spores: under 25 μm	Endolithic or endophloic	Open and sparse (*Bryoria* spp.)	Endolithic etc.
2. Spores: over 25 μm	Growing attached to substrate, no lower cortex	Open but closely spaced (*Cornicularia, Cladonia* spp.)	Up to 1 mm thick
3. Soredia: farinose	Growing adpressed to substrate, with a lower cortex	Discontinuous, areolate, squamulose, or foliose	1–5 mm thick or high
4. Soredia: granular	Overgrowing class (1)	Continuous, coherent (*Peltigera, Ochrolechia* spp.)	5–100 mm thick or high
5. Isidia: growing from soralia or breaking down into soredia	Overgrowing classes (1) and (2)		1–2·5 cm
6. Isidia	Overgrowing foliose lichens and weaker mosses		2·5–5 cm
7. Phyllidia	Overgrowing thick mosses		5–10 cm
8. Schizidia	Overgrowing thick moss and weak vascular plants		Over 10 cm
9. Thallus fragments: differentiated			
10. Thallus fragments: undifferentiated			

[a] cf. Poelt (1974).

3. Colonization, Succession and Competition

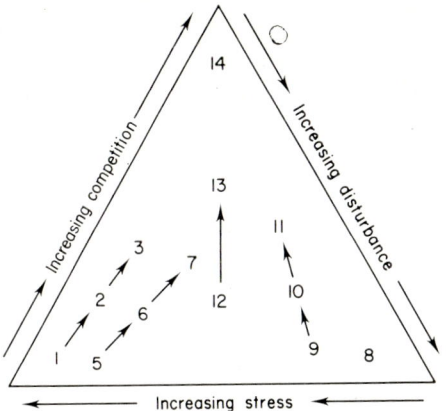

FIG. 3. Triangular ordination of some lichen communities, based on an assessment of the reproductive capacity and competitive ability of their dominant members. *Key:* (1) littoral zone; (2) mesic suppralittoral zone; (3) submesic supralittoral zone; (4) xeric supralittoral zone; (5) pioneer, xeric rock; (6) foliose, submesic rock; (7) fruticose, mesic rock; (8) pioneer, bare peat or humus; (9) pioneer, soil or dunes; (10) fixed dune; (11) dune and other heath; (12) pioneer twigs; (13) mesic forest climax (*Lobarion*); (14) arctic-alpine soil lichens; tundra and fellfield.

triangular ordination (Fig. 3). Compared to most other types of vegetation, all lichen communities are subject to some degree of stress. It is worth noting two successional paths, a lithosere and a psammosere, the first beginning at the stress apex, the second at the disturbance apex, which tend to converge. The competitive pole of this scheme is seen to be one in which K-selection is active and to be on the whole characteristic of mature ecosystems.

References

Acock, A. M. (1940). Vegetation of a calcareous inner fjord region in Spitzbergen. *J. Ecol.* **28**, 81–106.

Ahmadjian, V. and Hale, M. E., eds. (1974) ["1973"]. "The Lichens". Academic Press, New York and London.

Ahti, T. (1961). Taxonomic studies on reindeer lichens (*Cladonia*, subgenus *Cladina*). *Suomal. eläin- ja Kasvit. Seur. van. Tiedon.* **32**, 1–160.

Andersen, J. L. and Sollid, J. L. (1971). Glacial chronology and glacial geomorphology of the glaciers Mitdalsbreen and Nigardsbreen, South Norway. *Norsk Geogr. Tidskr.* **25**, 1–38.

Andreev, V. N. (1954). Prirost kormovykh lishainikov i priemy ego regulirovaniia. [Growth of forage lichens and methods for their regulation.] *Tr. Bot. Inst. AN SSSR*, ser. 3, Geobotanika **9**, 11–74.

Andrews, J. T. and Webber, P. J. (1964). A lichenometrical study of the northwestern margin of the Barnes Ice Cap: a geomorphological technique. *Geogrl Bull.* **22**, 80–104.

Andrews, J. T. and Webber, P. J. (1969). Lichenometry to evaluate changes in glacial mass budgets: as illustrated in North-Central Baffin Island, N.W.T. *Arct. alp. Res.* **1**, 181–194.

Armstrong, R. A. (1973). Seasonal growth and growth rate—colony size relationships in six species of saxicolous lichens. *New Phytol.* **72**, 1023–1031.

Armstrong, R. A. (1974). Growth phases in the life of a lichen thallus. *New Phytol.* **73**, 913–918.

Ascaso, C., Galvan, J. and Ortega, C. (1976). The pedogenic action of *Parmelia conspersa, Rhizocarpon geographicum* and *Umbilicaria pustulata. Lichenologist* **8**, 151–171.

Bailey, R. H. (1976). Ecological aspects of dispersal and establishment in lichens. *In* "Lichenology: Progress and Problems" (D. H. Brown, D. L. Hawksworth and R. H. Bailey, eds), pp. 215–247. Academic Press, London and New York.

Barkman, J. J. (1958). "Phytosociology and Ecology of Cryptogamic Epiphytes". Van Gorcum, Assen.

Benedict, J. B. (1967). Recent glacial history of an alpine area in the Colorado Front Range, U.S.A. I. Establishing a lichen-growth curve. *J. Glaciol.* **6**, 817–832.

Bertalanffy, L. von (1957). Quantitative laws in metabolism and growth. *Q. Rev. Biol.* **32**, 218–231.

Beschel, R. (1954). Eine Flechte als Niederschlagsmesser. *Wett. Leben* **6**, 56–60.

Beschel, R. E. (1958). Flechtenvereine der Städte, Stadtflechten und ihr Wachstum. *Ber. naturw.-med. Ver. Innsbruck* **52**, 1–158.

Beschel, R. E. (1961). Dating rock surfaces by lichen growth and its application to glaciology and physiography (lichenometry). *In* "Geology of the Arctic" (G. O. Raasch, ed.), Vol. 2, pp. 1044–1062. University of Toronto Press, Toronto.

Beschel, R. E. and Weideck, A. (1973). Geobotanical and geomorphological reconnaissance in West Greenland, 1961. *Arct. alp. Res.* **5**, 311–319.

Bioret, G. (1921). Les Graphidées corticoles. *Annls Sci. nat. (Bot.), sér.* 10, **4**, 1–71.

Brightman, F. H. (1959). Some factors influencing lichen growth in towns. *Lichenologist* **1**, 104–108.

Brodo, I. M. (1965). Studies of growth rates of corticolous lichens on Long Island, New York. *Bryologist* **68**, 451–456.

Brodo, I. M. (1974) ["1973"] Substrate ecology. *In* "The Lichens" (V. Ahmadjian and M. E. Hale, eds), pp. 401–439. Academic Press, New York and London.

Bunnell, F. L., Kärenlampi, L. and Russell, D. E. (1973). A simulation model of lichen-Rangifer interactions in Northern Finland. *Rep. Kevo Subarctic Res. Stn* **10**, 1–8.

Burrows, C. J. and Lucas, J. (1967). Variations in two New Zealand glaciers during the past 800 years. *Nature, Lond.* **216**, 467–468.

Burrows, C. J. and Orwin, J. (1971). Studies on some glacial moraines in New Zealand. I. The establishment of lichen growth curves in the Mount Cook area. *N.Z. Jl Sci.* **14**, 327–335.

Cooper, R. and Rudolf, E. D. (1953). Role of lichens in soil formation and plant succession. *Ecology* **34**, 805–807.
Coppins, B. J. and Shimwell, D. W. (1971). Cryptogam complement and biomass in dry *Calluna* heath of different ages. *Oikos* **22**, 204–209.
Crittenden, P. D. (1975). Nitrogen fixation by lichens on glacial drift in Iceland. *New Phytol.* **74**, 41–49.
Dalvi, R. R., Chang, W. C., Salunkhe, D. K. and Campbell, W. F. (1974). Effects of menazon and usnic acid on histochemical and ultrastructural changes during germination of mung beans. *J. Am. Soc. hort. Sci.* **99**, 175–180.
Degelius, G. (1940). Studien über die Konkurrenzverhältnisse der Laubflechten auf naktem Fels. *Acta Horti gothoburg.* **14**, 195–219.
Degelius, G. (1962). Über Verwitterung von Kalk und Dolomitgestein durch Algen und Flechten. Eine Übersicht. *In* "Chemie im Dienst der Archäologie Bautechnik Denkmalpflege" (J. A. Hedvall, ed.), pp. 156–162. Hakam Ohlssons, Lund.
Degelius, G. (1964). Biological studies of the epiphytic vegetation on twigs of *Fraxinus excelsior*. *Acta Horti gotoburg.* **27**, 11–55.
Denton, G. H. and Karlén, W. (1973). Lichenometry: its application to holocene moraine studies in Southern Alaska and Swedish Lapland. *Arct. alp. Res.* **5**, 347–372.
Edwards, R. Y., Soos, J. and Ritcey, R. W. (1960). Quantitative observations on epidendric lichens used as food by caribou. *Ecology* **41**, 425–431.
Faegri, K. (1933). Über die Längenvariationen einiger Gletscher des Jostedalsbre und die dadurch bedingten Pflanzensukzessionen. *Bergens Mus. Årb.* **1933**.
Farrar, J. F. (1974). A method for investigating lichen growth rates and succession. *Lichenologist* **6**, 151–155.
Fletcher, A. (1972). The ecology of marine and maritime lichens of Anglesey. Ph.D. thesis, University of Wales.
Fogg, G. E. and Stewart, W. D. P. (1968). *In situ* determinations of biological fixation in Antarctica. *Bull. Br. Antarct. Surv.* **15**, 39.
Formann, R. T. T. (1975). Canopy lichens with blue-green algae: a nitrogen source in a Columbian rain forest. *Ecology* **56**, 1176–1184.
Francis, G. W. (1842). "The Little English Flora, or a Botanical and Popular Account of all our Common Field Flowers." Simpkin Marshall, London.
Frey, E. (1959). Die Flechtenflora und -Vegetation des Nationalparks im Unterengadin. II. Teil. Die Entwicklung der Flechtenvegetation auf photogrammetrisch kontrollierten Dauerflachen. *Ergebn. wiss. Unters. schweiz. Natn-Parks* **6**, 237–319.
Fry, J. E. (1926). The mechanical action of corticolous lichens. *Ann. Bot.* **40**, 397–417.
Fry, E. J. (1927). The mechanical action of crustaceous lichens on substrata of shale, schist, gneiss, limestone and obsidian. *Ann. Bot.* **41**, 437–460.
Gagnon, J. D. (1966). Le lichen *Lecidea granulosa* constitue un milieu favorable à la germination de l'épinette noire. *Naturaliste can.* **93**, 89–98.
Gilbert, O. L. (1971). Studies along the edge of a lichen desert. *Lichenologist* **5**, 11–17.

Gilbert, O. L. (1974) ["1973"] Lichens and air pollution. *In* "The Lichens" (V. Ahmadjian and M. E. Hale, eds), pp. 443–472. Academic Press, London and New York.

Gilbert, O. L., Jamieson, D., Lister, H. and Pendlington, A. (1969). Regime of an Afghan glacier. *J. Glaciol.* **8**, 51–65.

Gimingham, C. H. (1951). Contributions to the maritime ecology of St. Cyrus, Kincardineshire. Part II. The sand dunes. *Trans. Proc. bot. Soc. Edinb.* **35**, 387–411.

Gimingham, C. H. (1964). Dwarf shrub heaths. *In* "The Vegetation of Scotland" (J. H. Burnett, ed.), pp. 232–287. Oliver and Boyd, Edinburgh.

Grime, J. P. (1973). Competitive exclusion in herbaceous vegetation. *Nature, Lond.* **242**, 344–347.

Grime, J. P. (1974). Vegetation classification by reference to strategies. *Nature, Lond.* **250**, 26–31.

Hakulinen, R. (1966). Über die Wachstumsgeschwindigkeit einiger Laubflechten. *Annls bot. fenn.* **3**, 167–179.

Hale, M. E. (1955). Phytosociology of corticolous cryptogams in the upland forests of southern Wisconsin. *Ecology* **36**, 45–63.

Hale, M. E. (1959). Studies on lichen growth rate and succession. *Bull. Torrey bot. Club* **86**, 126–129.

Hale, M. E. (1967). "The Biology of Lichens". E. Arnold, London.

Hale, M. E. (1970). Single lobe growth rates in the lichen *Parmelia caperata*. *Bryologist* **73**, 72–81.

Hale, M. E. (1974) ["1973"] Growth. *In* "The Lichens" (V. Ahmadjian and M. E. Hale, eds), pp. 473–492. Academic Press, New York and London.

Harper, J. L., Lovell, P. H. and Moore, K. G. (1970). The shapes and sizes of seeds. *A. Rev. Ecol. Syst.* **1**, 327–356.

Harris, G. P. (1972). The ecology of corticolous lichens. III. A simulation model of productivity as a function of light intensity and water availability. *J. Ecol.* **60**, 19–40.

Harris, G. P. and Kershaw, K. A. (1971). Thallus growth and the distribution of stored metabolites in the phycobionts of the lichens *Parmelia sulcata* and *P. physodes*. *Can. J. Bot.* **49**, 1367–1372.

Henssen, A. and Jahns, H. M. (1973)/["1974"]. "Lichenes, eine Einführung in die Flechtenkunde." Thieme, Stuttgart.

Horne, A. H. (1972). The ecology of nitrogen fixation on Signy Island, South Orkney Islands. *Bull. Br. Antarct. Surv.* **27**, 1–18.

Ingram, M. (1958). The ecology of the Cairngorms. IV. The *Juncus* zone: *Juncus trifidus* communities. *J. Ecol.* **46**, 707–737.

Iskandar, I. K. and Syers, J. K. (1971). Solubility of lichen compounds in water: pedogenetic implications. *Lichenologist* **5**, 45–50.

Iskandar, I. K. and Syers, J. K. (1972). Metal complex formation by lichen compounds. *J. Soil Sci.* **23**, 255–265.

Jacks, G. V. (1965). The role of organisms in the early stages of soil formation. *In* "Experimental Pedology" (E. G. Hallsworth and D. V. Crawford, eds), pp. 219–226. Butterworths, London.

Jackson, T. A. (1971). A study of the ecology of pioneer lichens, mosses and algae on recent Hawaiian lava flows. *Pacific Sci.* **25**, 22–32.

Jackson, T. A. and Keller, W. D. (1970). A comparative study of the role of lichens and inorganic processes in the chemical weathering of recent Hawaiian lava flows. *Am. J. Sci.* **269**, 446–466.

Jochimsen, M. (1966). Ist die Grösse des Flechtenthallus wirklich eineb rauchbarer Massstab zur Datierung von glazialmorphologischen Relikten? *Geog. Annaler* **48A**, 157–164.

Johnson, F. L. and Bell, D. T. (1975). Size-class structure of three streamside forests. *Am. J. Bot.* **62**, 81–85.

Johnston, D. W. and Odum, E. P. (1956). Breeding bird populations in relation to plant succession on the Piedmont of Georgia. *Ecology* **37**, 50–62.

Jones, R. J. (1959). Lichen hyphae in limestone. *Lichenologist* **1**, 119.

Kalgutkar, R. M. and Bird, C. D. (1969). Lichens found on *Larix lyalli* and *Pinus albicaulis* in South Western Alberta, Canada. *Can. J. Bot.* **47**, 627–648.

Kappen, L., Lange, O. L., Schulze, E.-D., Evenari, M. and Buschbom, U. (1975). Primary production of lower plants (lichens) in the desert and its physiological basis. *In* "Photosynthesis and Productivity in Different Environments" (J. P. Cooper, ed.), pp. 133–143. Cambridge University Press, Cambridge.

Kärenlampi, L. (1970). Morphological analysis of the growth and productivity of the lichen, *Cladonia alpestris*. *Rep. Kevo Subarctic Res. Stn* **7**, 9–15.

Kärenlampi, L. (1971a). Studies on the relative growth rate of some fruticose lichens. *Rep. Kevo Subarctic Res. Stn* **7**, 33–39.

Kärenlampi, L. (1971b). On methods for measuring and calculating the energy flow through lichens. *Rep. Kevo Subarctic Res. Stn* **7**, 40–46.

Kärenlampi, L. and Kauhanen, H. (1972). A direct gradient analysis of the vegetation of the surroundings of the Kevo Subarctic Station. *Rep. Kevo Subarctic Res. Stn* **9**, 82–98.

Keever, C. (1957). Establishment of *Grimmia laevigata* on bare granite. *Ecology* **38**, 422–429.

Keever, C., Oosting, H. J. and Anderson, L. E. (1951). Plant succession on exposed granite of Rocky Face Mountain, Alexander County, North Carolina. *Bull. Torrey bot. Club* **78**, 401–421.

Kershaw, K. A. (1964). Preliminary observations on the distribution and ecology of epiphytic lichens in Wales. *Lichenologist* **2**, 263–276.

Kershaw, K. A. and Rouse, W. R. (1973). Studies on lichen dominated systems. V. A primary survey of a raised-beach system in north-western Ontario. *Can. J. Bot.* **51**, 1285–1307.

Lambert, J. D. H. and Maycock, P. F. (1968). The ecology of terricolous lichens of the Northern Conifer-Hardwood Forests of central Eastern Canada. *Can. J. Bot.* **46**, 1043–1078.

Lambinon, J. (1968). Anomalies écologiques et accessibilité: l'exemple de quelques lichens de Belgique et du Luxembourg. *Nova Hedwigia* **16**, 403–407.

Lange, O. L. (1953). Hitze- und Trockenresistenz der Flechten in Beziehung zu ihrer Verbreitung. *Flora, Jena* **140**, 39–97.

Lange, O. L. and Evanari, M. (1971). Experimentell-ökologische Untersuchungen an Flechten der Negev-Wüste. IV. Wachstumsmessungen an *Caloplaca aurantia* (Pers.) Hallb. *Flora, Jena* **160**, 100–104.

Laundon, J. R. (1966). Frost damage to *Parmelia caperata*. *Lichenologist* **3**, 273.

Laundon, J. R. (1967). A study of the lichen flora of London. *Lichenologist* **3**, 277–327.

Lechowicz, M. J. and Adams, M. S. (1974). Ecology of *Cladonia* lichens. I. Preliminary assessment of the ecology of terricolous lichen-moss communities in Ontario and Wisconsin. *Can. J. Bot.* **52**, 55–64.

Lindsay, D. C. (1973). Estimates of lichen growth rates in the maritime Antarctic. *Arct. alp. Res.* **5**, 341–346.

Little, S. (1951). Observations on the minor vegetation of the pine barren swamps in southern New Jersey. *Bull. Torrey bot. Club.* **78**, 153–160.

Looman, J. (1964). Ecology of lichen and bryophyte communities in Saskatchewan. *Ecology* **45**, 481–491.

MacArthur, R. H. and Wilson, F. O. (1967). "The Theory of Island Biogeography". University Press, Princeton.

McVaugh, R. (1943). The vegetation of the granitic flat-rocks of the Southeastern United States. *Ecol. Monogr.* **13**, 119–166.

Maikawa, E. and Kershaw, K. A. (1976). Studies on lichen-dominated systems. XIX. The postfire recovery sequence of black spruce-lichen woodland in the Abitau Lake Region, N.W.T. *Can. J. Bot.* **54**, 2679–2687.

Matthews, J. A. (1974). Families of lichenometric dating curves from the Storbreengletschervorfeld, Jotunheimen, Norway. *Norsk Geogr. Tidskr.* **28**, 215–235.

Mellor, E. (1921). "Les Lichens Vitricoles et la Détérioration des Vitraux d'Église." Paris.

Miller, C. D. (1969). Chronology of neoglacial moraines in the Dome Peak area, North Cascade Range, Washington. *Arct. alp. Res.* **1**, 49–66.

Miller, C. D. and Birkeland, P. W. (1974). Probable preneoglacial age of the type Temple Lake Moraine, Wyoming: discussion and additional relative age data. *Arct. alp. Res.* **6**, 301–306.

Miller, G. H. (1973). Variations in lichen growth from direct measurements: preliminary curves for *Alectoria minuscula* from Eastern Baffin Island, N.W.T., Canada. *Arct. alp. Res.* **5**, 333–339.

Miller, G. H. and Andrews, J. T. (1972). Quaternary history of Northern Cumberland Peninsula, East Baffin Island, N.W.T., Canada. Part VI: Preliminary lichen growth curve for *Rhizocarpon geographicum*. *Geol. Soc. Am. Bull.* **83**, 1133–1138.

Morrison, R. G. and Yarranton, G. A. (1974). Vegetational heterogeneity during a primary sand dune succession. *Can. J. Bot.* **52**, 397–410.

Mottershead, D. N. and White, I. D. (1972). The lichenometrical dating of glacier recession, Tunbergsdal, Southern Norway. *Geogr. Annaler* **54A**, 47–52.

Odum, E. P. (1969). The strategy of ecosystem development. *Science, N.Y.* **164**, 262–270.

Olson, J. S. (1958). Rates of succession and soil changes on southern Lake Michigan sand dunes. *Bot. Gaz.* **119**, 125–170.
Oosting, H. J. and Anderson, L. F. (1937). The vegetation of a barefaced cliff in Western North Carolina. *Ecology* **18**, 280–292.
Orwin, J. (1970). Lichen succession on recently deposited rock surfaces. *N.Z. J. Bot.* **8**, 452–477.
Orwin, J. (1972). The effect of environment on assemblages of lichens growing on rock surfaces. *N.Z. J. Bot.* **10**, 37–47.
Ott, E. (1961). Über den Einfluss von Flechtensäuren auf die Keimung verschiedener Baumarten. *Schweiz. Z. Forstw.* **112**, 303–304.
Ovington, J. D. (1965). Nutrient cycling in woodlands. *In* "Experimental Pedology" (E. G. Hallsworth and D. V. Crawford, eds), pp. 208–215. Butterworths, London.
Pentecost, A. and Fletcher, A. (1974). Tufa: an interesting lichen substrate. *Lichenologist* **6**, 100–101.
Phillips, H. C. (1963). Growth rate of *Parmelia isidiosa* (Müll. Arg.) Hale. *J. Tenn. Acad. Sci.* **38**, 95–96.
Pianka, E. R. (1970). On r- and K-selection. *Am. Nat.* **104**, 592–597.
Platt, R. B. and Amsler, F. P. (1955). A basic method for the immediate study of lichen growth rates and succession. *J. Tenn. Acad. Sci.* **30**, 177–183.
Poelt, J. (1974) ["1973"] Systematic evaluation of morphological characters. *In* "The Lichens" (V. Ahmadjian and M. E. Hale, eds), pp. 91–115. Academic Press, New York and London.
Polunin, N. (1936). Plant succession in Norwegian Lapland. *J. Ecol.* **24**, 372–391.
Prince, C. R. (1974). Growth rates and productivity of *Cladonia arbuscula* and *C. impexa* on the Sands of Forvie, Scotland. *Can. J. Bot.* **52**, 431–433.
Pyatt, F. B. (1967). The inhibitory influence of *Peltigera canina* on the germination of graminaceous seeds and the subsequent growth of the seedlings. *Bryologist* **70**, 326–329.
Pyatt, F. B. (1968). The effect of sulphur dioxide on the inhibitory influence of *Peltigera canina* on the germination and growth of grasses. *Bryologist* **71**, 97–101.
Raistrick, A. and Gilbert, O. L. (1963). Malham Tarn House: its building materials, their weathering and colonization by plants. *Fld Stud.* **1**, 89–115.
Ramaut, J.-L. and Corvisier, M. (1975). Effets inhibiteurs des extraits de *Cladonia impexa*, Harm. *C. gracilis* (L.) Willd. et *Cornicularia muricata* (Ach.) Ach. sur la germination des graines de *Pinus sylvestris* L. *Oecol. Pl.* **10**, 259–299.
Reger, R. D. and Péwé, T. L. (1969). Lichenometric dating in the central Alaska range. *In* "The Periglacial Environment" (T. L. Péwé, ed.), pp. 223–247. McGill-Queens University Press, Montreal.
Richards, F. J. (1959). A flexible growth function for empirical use. *J. exp. Bot.* **10**, 290–300.
Richards, F. J. (1969). The quantitative analysis of growth. *In* "Plant Physiology" (F. C. Steward, ed.), Vol. Va, pp. 3–76. Academic Press, New York and London.

Richardson, D. H. S. and Finegan, E. J. (1977). Studies on the lichens of Truelove Lowland. *In* "Truelove Lowland, Devon Island, Canada: A High Arctic Ecosystem". (L. C. Bliss, ed.), University of Alberta Press, Edmonton.
Richardson, J. A. (1958). The effect of temperature on the growth of plants on pit heaps. *J. Ecol.* **46**, 537–546.
Ried, A. (1960). Nachwirkungen der Entquellung auf den Gaswechsel von Krustenflechten. *Biol. Zbl.* **79**, 657–678.
Robinson, H. (1959). Lichen succession in abandoned fields in the Piedmont of N. Carolina. *Bryologist* **62**, 254–259.
Rogers, R. W., Lange, R. T. and Nicholas, D. J. D. (1966). Nitrogen fixation by lichens of arid soil crusts. *Nature, Lond.* **225**, 1253.
Rose, F. (1974). The epiphytes of oak. *In* "The British Oak; its History and Natural History" (M. G. Morris and F. H. Perring, eds), pp. 250–273. Classey, Faringdon.
Rose, F. (1976). Lichenological indicators of age and environmental continuity in woodlands. *In* "Lichenology: Progress and Problems" (D. H. Brown, D. L. Hawksworth and R. H. Bailey, eds), pp. 279–307. Academic Press, London and New York.
Rose, F. and James, P. W. (1974). Regional studies on the British lichen flora. I. The corticolous and lignicolous species of the New Forest, Hampshire. *Lichenologist* **6**, 1–72.
Rouse, W. R. and Kershaw, K. A. (1971). The effects of burning on the heat and water regime of lichen dominated subarctic surfaces. *Arct. alp. Res.* **3**, 291–304.
Rydzak, J. (1961). Investigations on the growth rate of lichens. *Annls Univ. Mariae Curie-Skłodowska*, C, **16**, 1–15.
Rydzak, J. (1967). Badania nad szybkością wzrostu porostów. Czesc II. [Investigations on the growth rate of lichens. Part II.] *Annls Univ. Mariae Curie-Skłodowska*, C, **21**, 167–182.
Sachs, J. (1882). "Textbook of Botany, Morphological and Physical" (S. H. Vines, ed.), 2nd ed., pp. 816–817. Oxford University Press, Oxford.
Schatz, V., Schatz, A., Trelawney, G. S. and Barth, K. (1956). Significance of lichens as pedogenic (soil-forming) agents. *Proc. Pa Acad. Sci.* **30**, 62–69.
Schauer, T. (1969). Die Flechtenvegetation der Kiesfläche auf der Garchinger Haide nördlich von München. *Herzogia* **1**, 181–186.
Schell, D. M. and Alexander, V. (1973). Nitrogen fixation in arctic coastal tundra in relation to vegetation and micro-relief. *Arctic* **26**, 130–131.
Scott, G. A. M. (1965). The shingle succession at Dungeness. *J. Ecol.* **53**, 21–31.
Scotter, G. W. (1963). Growth rates of *Cladonia alpestris*, *C. mitis* and *C. rangiferina* in the Taltson River Regions, N.W.T. *Can. J. Bot.* **41**, 1199–1202.
Seaward, M. R. D. The morphological and ecological performance of *Lecanora muralis* (Schreb.) Rabenh. in an urban environment. *In* "Lichenology: Progress and Problems" (D. H. Brown, D. L. Hawksworth and R. H. Bailey, eds), pp. 323–357. Academic Press, London and New York.

Shields, L. M. (1957). Algal and lichen floras in relation to nitrogen content of certain volcanic and range soils. *Ecology* **38**, 661–663.
Smith, A. L. (1921). "Lichens". Cambridge University Press, Cambridge.
Smith, D. C. (1960). Studies in the physiology of lichens. 2. Absorption and utilization of some simple organic nitrogen compounds by *Peltigera polydactyla*. *Ann. Bot.* **24**, 172–185.
Smith, D. C. (1962). The biology of lichen thalli. *Biol. Rev.* **37**, 537–570.
Stork, A. (1963). Plant immigration in front of retreating glaciers, with examples from the Kebnekajse Area, Northern Sweden. *Geogr. Annaler* **45**, 1–22.
Syers, J. K. (1964). A study of soil formation on Carboniferous limestone with particular reference to lichens as pedogenic agents. Ph.D. thesis, University of Durham.
Syers, J. K. (1969). Chelating ability of fumarprotocetraric acid and *Parmelia conspersa*. *Pl. Soil* **31**, 205–208.
Syers, J. K., Birnie, A. C. and Mitchell, B. D. (1967). The calcium oxalate content of some lichens growing on limestone. *Lichenologist* **3**, 409–414.
Syers, J. K. and Iskandar, I. K. (1973). Pedogenetic significance of lichens. *In* "The Lichens" (V. Ahmadjian and M. E. Hale, eds), pp. 225–248. Academic Press, New York and London.
Tallis, J. H. (1958). Studies in the biology and ecology of *Rhacomitrium lanuginosum* Brid. 1. Distribution and ecology. *J. Ecol.* **46**, 271–288.
Tapper, R. (1976). Dispersal and changes in the local distributions of *Evernia prunastri* and *Ramalina farinacea*. *New Phytol.* **77**, 725–734.
Viereck, L. A. (1966). Plant succession and soil development on gravel outwash of the Muldrow Glacier, Alaska. *Ecol. Monogr.* **36**, 181–199.
Wade, A. E. (1965). The genus *Caloplaca* Th. Fr. in the British Isles. *Lichenologist* **3**, 1–28.
Watt, A. S. (1947). Pattern and process in the plant community. *J. Ecol.* **35**, 1–22.
Webber, P. J. and Andrews, J. T. (1973). Lichenometry: a commentary. *Arct. alp. Res.* **5**, 295–302.
Webley, D. M., Henderson, M. E. K. and Taylor, I. F. (1963). The microbiology of rocks and weathered stones. *J. Soil Sci.* **14**, 102–112.
Wein, R. W. and Bliss, L. C. (1974). Primary production in arctic cotton grass tussock communities. *Arct. alp. Res.* **6**, 261–274.
Whaley, W. G. (1961). Growth as a general process. *In* "Encyclopedia of Plant Physiology" (W. Ruhland, ed.), Vol. 14, pp. 71–112. Springer, Berlin.
Wielgolaski, F. E. (1972). Vegetation types and plant biomass in tundra. *Arct. alp. Res.* **4**, 291–305.
Williams, M. E. and Rudolph, E. D. (1974). The role of lichens and associated fungi in the chemical weathering of rock. *Mycologia* **66**, 648–660.
Williamson, M. (1973). Species diversity in ecological communities. *In* "The Mathematical Theory of the Dynamics of Biological Populations" (M. S. Bartlett and R. W. Hiorns, eds), pp. 325–335. Academic Press, New York and London.
Winterringer, G. S. and Vestal, A. G. (1956). Rockledge vegetation in Southern Illinois. *Ecol. Monogr.* **26**, 105–130.

Woolhouse, H. W. (1968). The measurement of growth rates in lichens. *Lichenologist* **4**, 32–33.

Yarranton, G. A. (1972). Distribution and succession of epiphytic lichens on black spruce near Cochrane, Ontario. *Bryologist* **75**, 462–480.

Yarranton, G. A. (1975). Population growth in *Cladonia stellaris* (Opiz) Pouz. and Vězda. *New Phytol.* **75**, 99–110.

4 | Lichen–Invertebrate Associations

URI GERSON and
MARK R. D. SEAWARD

I.	Introduction	70
II.	The Aquatic Fauna	71
	A. Protozoa	71
	B. Nematoda	72
	C. Rotifera	73
	D. Oligochaeta	73
	E. Tardigrada	73
III.	The Terrestrial Fauna	74
	A. Insecta	75
	B. Mites (Acari)	79
	C. Other Arthropods	82
	D. Mollusca	83
IV.	Distribution	84
	A. Habitats	84
	B. Lichen-locating by Invertebrates	85
	C. Lichen Dispersal by Invertebrates	87
V.	Invertebrate Feeding on Lichens	88
	A. Modes of Feeding	88
	B. Digestion	88
	C. Lichen Resistance to Invertebrate Feeders	90
	D. Energetics of the Lichen Fauna	95
	E. Food-webs and Nutrient Recycling	99
VI.	Lichens as a Protective Environment for Invertebrates	101
	A. Concealment	102
	B. Camouflage	102
	C. Mimicry	104
	D. Industrial Melanism	105

VII.	Lichen–Invertebrate Communities	107
	A. Terrestrial Colonization	107
	B. Competition for Settling Sites	108
	C. Physical and Climatic Effects	109
VIII.	Discussion	110
	A. Evolutionary Aspects	110
	B. Suggested Lines of Approach for Future Enquiry	111
Acknowledgements		112
References		112

I. Introduction

Interest in lichen invertebrates has been flickering for many years, fanned mostly by individual botanists and zoologists often not conversant with the complementary discipline. Soil zoologists (i.e. Kühnelt, 1976; Stebaev, 1963) have contributed their observations, as well as naturalists (Cunningham, 1907), marine biologists (Colman, 1939) and many others. Smith (1921) was the first to sum up and review such associations, devoting a section in her book to lichens as food and shelter for invertebrates. More recent efforts, stimulated by interest in population dynamics (Broadhead, 1958), mite and nematode ecology (Travé, 1963; Gadea, 1974) and beetle taxonomy (Gressitt, 1966), among others, have materially contributed to this still largely unknown field. The effect of industrial air pollution on lichen communities has also focused interest on the resultant changes in distribution of moth (and other insect) morphs (Kettlewell, 1961, 1973). Although interest in the subject is rising, as evinced by the steady publication of lichen–invertebrate papers in the *Lichenologist*, a strictly botanical periodical, such recognition is by no means universal.

The present chapter developed from the senior author's review of arthropods associated with lichens (Gerson, 1973); some subjects treated there will not be repeated here. It must be emphasized that most data were gleaned from a variety of botanical, zoological and ecological publications, and some care must therefore be exercised in applying scientific names.

Travé (1963) made the pertinent observation that there are actually two more or less mutually exclusive faunas on lichens (and mosses): an aquatic fauna and a terrestrial one. Protozoa, Nematoda, Rotifera and Tardigrada belong to the former, whereas Arthropoda and Mollusca belong to the latter. Some overlap between the two faunas exists, and invertebrates living on marine lichens are ignored by this division. Nevertheless, despite certain reservations, it is a very convenient concept and has been adopted here.

II. The Aquatic Fauna

A. Protozoa

Almost all humid habitats harbour protozoans. However, only a few protozoans are actually associated with specific plant taxa; examples are the ciliate *Zoothamnium adamsi* which forms colonies on the alga *Cladophora*, and the flagellate *Phytomonas*, living in the sap of *Euphorbia*. No protozoans are known to be specific to lichens.

Only two of the four major divisions of the Protozoa, namely the rhizopods and ciliates, have been found in association with lichens. Of the former group, the most abundant appear to be the testate or shell-bearing normally (100–150 μm) amoebae, usually known as thecamoebae (which are also the commonest soil protozoans). Several European workers, Decloitre (1962), Gadea (1964a), Heinis (1959) and Laminger (1971), collected lichen thecamoebae from Norway, the Spanish Mediterranean coast and the Alps, respectively. The genera commonly collected were *Archella*, *Centropyxis*, *Euglypha* and *Trinema*. Decloitre (1962) and Laminger (1971) listed the protozoans they collected according to lichen habitat, but no pattern emerges from a perusal of their lists. Species found exclusively in saxicolous lichens by one of these investigators occurs only in terrestrial lichens according to the other. The abundance and diversity of thecamoebae found suggest further research possibilities.

Quantitative data are limited; Laminger (1971) found the terrestrial *Cladonia stellaris* to be richest in thecamoeban numbers, but this may include soil populations. Heinis (1959) stated that pure lichen stands usually have a poor fauna, an opinion shared by Gadea (1964a). Such a poor fauna possibly reflects not only a certain unsuitability of lichens for components of the microfauna, but also, and perhaps more importantly, an early phase in soil and humus formation. As to relative abundance, Gadea (1964b) estimates that thecamoebae account for 27% of the lichen microfauna.

Most free-living protozoans endure dry periods by forming cysts, a state in which they may survive for long periods. Cysts are blown about by wind, and are thus widely transported. Alternatively they may remain on the lichen until it is rewetted; Heinis (1959) found thecamoebans on Mt Macun (2620 m), in the eastern Swiss Alps, in dry thalli of *Cladonia*, *Parmelia* and *Lecanora*.

Among ciliates, the genus most often collected in lichens is the kidney-shaped holotrichid *Colpoda* (Gadea, 1964a, 1974; Thompson, 1960). Members of this genus are common in fresh water, soils and some land slugs, and have a most efficient encystment mechanism (Stout and Heal, 1967).

Members of a third major protozoan division, namely the flagellates, are probably associated with lichens in an indirect way—through being symbiotic in reindeer rumen, and there participating in the digestion of lichens (Scotter, 1965).

B. Nematoda

The nematodes or roundworms are abundant in marine and fresh water, in soil, dung, and as plant and animal parasites. The terrestrial forms are

TABLE I

Relative abundance of lichen-associated nematodes (from Gadea, 1974).

Nematodes	Cladonia foliacea (%)	Xanthoria aureola (%)	Xanthoria parietina (%)
Aphelenchoides parietinus	6	2	4
Ditylenchus intermedius	–	33	6
Ditylenchus sp	18	–	–
Eudorylaimus carteri	–	–	14
Mesodorylaimus bastiani	–	6	–
Panagrolaimus rigidus	–	12	7
Plectus cirratus	20	26	55
Rhabdolaimus terrestris	–	5	–
Tylenchus filiformis	56	16	14

usually small (<1 mm), thin worms which require some water for their activities. Gadea (1964a, b, 1973, 1974) did not find any lichen-specific species; the phytophagous ones are all bryophagous. In his latest report he compared the nematode faunas associated with three lichen species (Table I). It is noteworthy that these faunas, collected from different Mediterranean regions, are actually quite similar, despite variable dominance by different species. Gadea (1964b) concluded that nematodes comprise a substantial part (29%) of the lichen aquafauna.

Few additional reports on lichen nematodes are available. Goodey (1933) noted some species, especially the Araeolaimids *Plectus* and *Wilsonema* (the former, which also figures in Gadea's lists, from house roof lichens). Ali *et al.* (1970 and other papers) have reported on Indian lichen nematodes.

Nematodes survive dry periods by forming resistant resting stages and by encystment. Nielsen (1967), in a study on roof moss-inhabiting nematodes, noted that some species, like *Plectus cirratus*, *Tylenchus filiformis* and *T. davainei*, which also occur in lichens, are quite resistant to desiccation.

C. Rotifera

The Rotifera or Rotatoria (also called "wheel animalcules") are common in freshwater or marine benthic habitats, some species being parasitic on animals, and others living on mosses and lichens. Heinis (1959 and former publications), in his studies on cryptogam "Mikrocoenoses", reported on many rotifers in lichens. Pure stands of *Solorina crocea* harboured *Adineta barbata* and *Ceratrocha cornigera*, and *Macrotrachela ehrenbergii*, *Pleureta alpina* and *Mniobia* spp. were found in dry thalli of *Cladonia pyxidata* and *Parmelia* spp. Gadea (1964b, 1974) collected *Callidina* from *Xanthoria aureola* and some philodincs from other lichens, and estimated that rotifers account for about 8·5% of the lichen aquatic microfauna.

Feeding on lichen ascospores by rotifers was described by Pyatt (1968), who obtained *Philodina* from apothecial surfaces of *Xanthoria parietina* and reported that some ascospore breakdown occurred within the rotifer's body, especially when only a few (3–8) ascospores were ingested.

D. Oligochaeta

The Oligochaeta are represented in the lichen fauna only by members of the family Enchytraeidae. Colman (1939) collected specimens of *Lumbricillus* in *Lichina pygmaea* tufts off the British coast, calculating that there were about 220 individuals per m^2 of rock. Frankland (1974) cited P. M. Latter as rearing enchytraeids on *Cladonia*, as the only source of food.

E. Tardigrada

Tardigrades or "bear animalcules" are small, eight-legged invertebrates which occur in damp soils, on various cryptogams and in both marine and freshwater habitats. They feed on debris and moss cells, and a few are predacious (Cuénot, 1949; Marcus, 1959). Tardigrades are capable of withstanding long dry periods by forming cysts or by anabiosis. Anabiosing animals (tuns) and tardigrade eggs are widely disseminated by winds, often along with small lichen propagules (Cuénot, 1949). No clear reports on lichenophagous tardigrades could be found, but as bryophagous species are frequently collected in pure lichen stands, such feeding habits may be assumed. Argue (1971), Iharos (1968) and Schuster and Grigarick (1966), among others, found tardigrades in various lichens in Canada, Mongolia and the Galápagos Islands, respectively. Argue (1971) reported *Hypsibius oberhaeuseri* and *Pseudechiniscus suillus* only on *Lobaria pulmonaria*, although some mosses were also sampled. The former tardigrade is quite common in lichens, being especially abundant on *Xanthoria* thalli (Cuénot, 1949). Schuster and Grigarick (1966) obtained various tardigrades from *Parmelia*, *Cladonia* and *Ramalina* including the cosmopolitan *Macrobiotus hufelandii* and *Milnesium tardigradum* which occur on many lichens (Heinis, 1959; Marcus, 1959).

Kimmel and Meglitsch (1969) found lichen-inhabiting tardigrades on lichen-covered, and not lichen-free, tree bark in Iowa, suggesting that this substrate is required. Their data indicate some lichen preference among the tardigrades obtained: *M. tardigradum* occurred most often in *Candelaria concolor* and *Physcia millegrana*, and *Hypsibius tuberculatus* preferred *Arthonia caesia*. Furthermore they showed that some tree-lichen combinations were more suitable for tardigrades than others, and that the population density of these animals changes according to the height of the corticolous lichen above ground level.

Barrett and Kimmel (1972) reported that tardigrade density and diversity within colonies of bark-colonizing *Physcia* were much reduced following DDT treatments. They concluded that tardigrade species occupy contiguous, non-overlapping niches within the lichen-bearing tree (*Ulmus*) bark ecosystem. The use of such simple ecosystems was advocated by Barret and Kimmel for evaluating the effects of pesticide stress on forest ecosystems, and for energy flow and mineral recycling studies.

From the limited data available, it would appear that tardigrades are a minor component of the Mediterranean lichen fauna; Gadea (1964b) stated that they accounted for only about 0·5% of the lichen microfauna which he studied.

III. The Terrestrial Fauna

Insects, mites and molluscs constitute important components of the terrestrial fauna and are the main lichen grazers. Although these animals are grouped within the terrestrial fauna (Travé, 1963), many species become active only when the lichen is wet. Marine molluscs, like the periwinkles, are discussed here only for taxonomic convenience. Travé's contention that the two faunas are mutually exclusive would appear to require some modification.

A. Insecta

Insects are the largest class in the animal kingdom, and are proportionally represented in the lichen-associated fauna. Examples of the wingless insects —subclass Apterygota—come from the Thysanura (bristletails), the Collembola (springtails) and the Protura. The latter sometimes occur in lichens (Nosek, 1973). Kühnelt (1976) writes that the thysanuran Machilidae feed mainly on lichens and algae, and Benedetti (1973) reports that the food of *Neomachilis halophila*, a Californian bristletail, contains unicellular algae which were derived from a rock-encrusting lichen under which it lives. A few individuals of *Petrobius* and *Lipura*, other machilids, were found in *Lichina pygmaea* tufts by Colman (1939).

Some springtails feed on lichens (Kühnelt, 1976). Hale (1972) described the feeding of *Hypogastrura packardi* on the corticolous *Parmelia baltimorensis*, noting that the upper cortex and most of the algal layer were completely eaten away. Thousands of collembolans were seen there, causing large areas in the central part of the lichen to disintegrate and fall away. This grazer was estimated to reduce the lichen colony by at least 50%. A xerophilic fauna was found by Hale (1963) in lichens of an eroding blanket bog in England. The lichen colony of *Cladonia* spp. developed on a hag lip, a relatively dry area in such bogs, and harboured large numbers of collembolans. Heavy infestations of collembolans can almost obliterate the crustose lichens beneath: such observations have been made by M. R. D. Seaward (unpublished data) for the collembolan *Hypogastrura tullbergi* on various crustose lichens on saxicolous substrates, and for *Anurophorus laricis* on epiphytic *Pertusaria pertusa* (Seaward, 1975). Byazrov et al. (1976) provide data on collembolan species composition of lichen synusia in Mongolia, and Ridley (1930) mentions the possible dispersal of *Pertusaria amara* by springtails.

The Pterygota—winged or secondarily wingless insects—are divided into those with simple metamorphosis, the Exopterygota (or Hemimetabola), and those with complete metamorphosis, the Endopterygota (or Holometabola). Representatives of eight exopterygote orders have been recorded from lichens. The best known examples are the Psocoptera (psocids or barklice), voracious lichen feeders (Seaward, 1965; Laundon, 1971) whose feeding preferences were extensively studied by Broadhead (1958).

In northern England, the psocids living on larch trees (*Larix decidua*) comprise two distinct feeding groups. Six of the eight commonest psocid species feed primarily on the *Desmococcus*-fungal spore mixture occurring on the bark of the twigs and branches. These species are: *Mesopsocus immunis*, *M. unipunctatus*, *Philotarsus picicornis*, *Amphigerontia bifasciata*,

Elipsocus westwoodi and *E. hyalinus*. None of these species appears to discriminate between *Desmococcus* and fungal spores. When the preferred foods are scarce, these species will take some lichen, but lichen is an unsatisfactory food for them, producing higher mortality rates and lower oviposition rates. The remaining two species, *Elipsocus mclachlani* and *Reuterella helvimacula*, feed primarily on the lichen *Lecanora conizaeoides*, which is the predominant lichen species on larch in northern England. When this lichen is scarce or absent, these lichen-preferring species will nevertheless readily accept *Desmococcus* and fungal spores on which they thrive in the laboratory. The various food components (*Desmococcus*, the spores of honeydew moulds, lichen, bark flakes and structureless detritus) can be identified in the crops of these insects and their relative abundance estimated. Quantitative comparisons of the crop contents of these psocid species have been set out by Broadhead (1958, his Table 4). Photographs of crop contents to show some of these food components are presented in Fig. 1.

Representatives of the Orthoptera often mimic lichens: katydids (Tettigoniidae), grasshoppers and stick insects (Phasmida) include some bizarre lichen-like forms (Chopard, 1938; Cott, 1957). A colour photograph of a Congolese lichen mantid was presented by Farb (1963). Feeding on lichens by grasshoppers was reported by Richardson (1975) from American sources.

The webspinners or Embioptera are little-known insects which construct silken tunnels in which they live. Ross (1966) described a sea-shore habitat in the Galápagos Islands where epiphytic and saxicolous lichens are covered by the silk of *Chelicerca galapagensis*, which apparently feeds on the lichens. A lichen-feeding embiid was also recorded from Australia (Anon., 1970).

Less information is available concerning the relationships of other exopterygote orders with lichens. Some, like the stoneflies (Plecoptera) and the termites (Isoptera), rarely feed on lichens (Coleman and Hynes, 1970; Kalshoven, 1958). Thrips (Thysanoptera) sometimes hide in lichens (Dev, 1964; Kühnelt, 1976). Earwigs (Dermaptera) participated in the lichen-feeding experiments summed up by Smith (1921), and several British bugs (Hemiptera) were listed from lichens by Southwood and Leston (1959). Among these is *Loricula elegantula*, a predacious bug closely associated with *Parmelia*, which feeds on barklice, including *Reuterella helvimacula*, a known lichen feeder (Broadhead, 1958).

Among the holometabolous insects, the butterflies and moths (order Lepidoptera) have evolved the richest associations with lichens. Many feed on these plants (Brightman, 1965; Anon., 1970; Borror and DeLong, 1971), cover their cases with lichen fragments (Weber, 1974), or mimic them (Cott, 1957). Members of the moth family (or subfamily) Lithosiidae

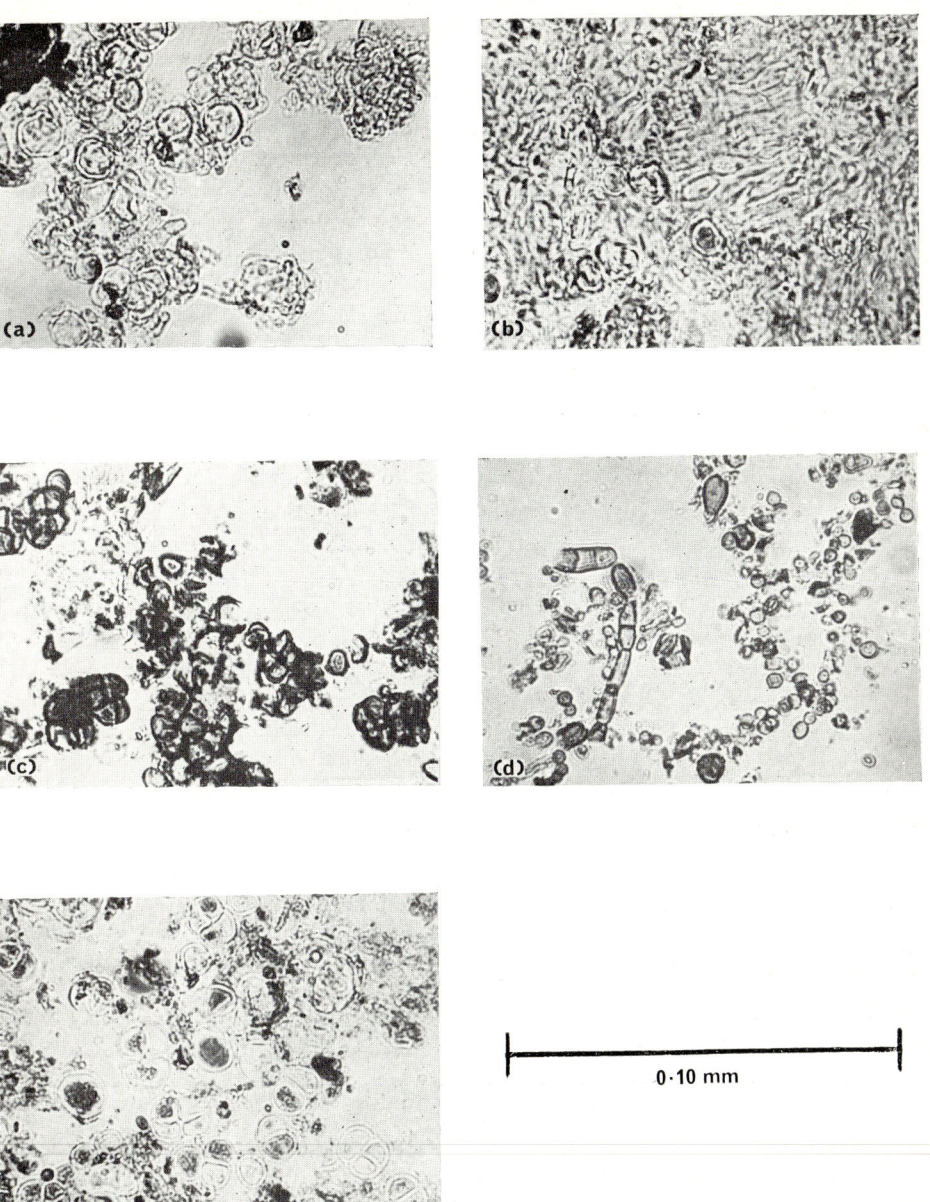

FIG. 1. Portions of the crop contents of various larch-living Psocoptera to show the chief food components: (a) lichen algal cells and hyphae (*Lecanora conizaeoides*) from *Reuterella helvimacula*; (b) lichen hymenium (*Lecanora conizaeoides*) with ascospores from *Elipsocus mclachlani*; (c) the honeydew mould *Sporidesmium* from *Amphigerontia bifasciata*; (d) small fungal spores and hyphae from *Elipsocus westwoodi*; (e) *Desmococcus* from *E. westwoodi* (E. Broadhead, unpublished data).

live in, feed on, and mimic lichens (Richardson, 1975). Their close association has resulted in various appropriate vernacular names (Table II). According to Mani (1962), some Lepidoptera belong to the communities of "yellow, orange and red" saxicolous high-altitude lichens, thus contributing to the developing, colonizing lichen fauna. Caterpillars of the subfamily Meessiinae (Tineidae—clothes moths) are believed to be lichenophagous; they carry cases which imitate their substrate and in which they pupate (Zagulyayev, 1970). Some caterpillars appear to "thrive"

TABLE II

Lichen-associated moths of the family Lithosiidae (Lepidoptera), and their vernacular names (compiled from Holland, 1908).

Moth	Vernacular name
Crambidia pallida	Pale lichen moth
C. casta	Pearly-winged lichen moth
C. allegheniensis	Alleghenian lichen moth
Hypoprepia miniata	Scarlet-winged lichen moth
H. fucosa	Painted lichen moth
Haematomis mexicana	Mexican lichen moth
Comacla simplex	Mouse-coloured lichen moth
Bruceia pulverina	Powdered lichen moth
Clemensia albata	Little white lichen moth
Illice unifascia	Banded lichen moth
I. subjecta	Subject lichen moth
I. nexa	Yellow-blotched lichen moth
Ptychoglene phrada	Druce's lichen moth
P. tenuimargo	Narrow-banded lichen moth
Pygoctenucha funerea	Funereal lichen moth
P. terminalis	Blue-green lichen moth
Lerina incarnata	Crimson-bodied lichen moth

on mosses and lichens (de Boer, 1975). Breakdown of the lichen-mimicking protective crypsis of moths on soot-discoloured walls and tree trunks has often been discussed within the context of industrial melanism (Kettlewell, 1961, 1973).

Beetles (order Coleoptera) also have diverse associations with lichens, but they appear to have been less studied. The fungus weevil *Lichenobius littoralis* lives and excavates tunnels in *Pertusaria graphica* (Holloway, 1970). An elm-leaf beetle was reported by Smith (1921) to resemble apothecia of *Physcia*, and Gressitt (1966) presented photographs of New

Guinea weevils whose dorsum was extensively overgrown with *Parmelia*. These weevils are believed to have evolved special depressions and exudations on their dorsum which facilitate and promote the growth of the cryptogams. Dermestid beetles were obtained from *Parmelia centrifuga* stands in the South Ural Mountains by Stebaev (1963).

Lacewings (order Neuroptera) are predacious insects whose larvae sometimes cover themselves with lichen "packets" (Garrett, 1974; Skorepa and Sharp, 1971). This habit, which probably evolved as camouflage, also serves to disperse soredia when the moulting larvae leave their old, lichen-covered skins behind.

Very little has been reported on the associations of the two remaining orders, Diptera and Hymenoptera, with lichens. Stebaev (1963) obtained up to 135 dipterans (members of the families Empididae and Cecidomyiidae) per dm^2 of *Parmelia*. Séguy (1950) noted gall midges (family Cecidomyiidae) as they were developing in mosses and lichens. Richardson (1975) reported on sciarid larvae which feed on lichens, and Bullock (1966) working in Singapore, noted a dipteran which closely resembles the sporulating bodies of lichens. Concerning Hymenoptera, ants have been found in association with lichens: Béique and Francoeur (1966) collected various Canadian ants from nests which were built underneath *Cladonia* stands, and Bailey (1970) described the dispersal of *Lecanora* soredia by ants. Some unidentified dipterans and hymenopterans were found by Bengtson *et al.* (1974) in Spitsbergen lichen stands.

To conclude, representatives of 15 (i.e. more than half) of the insect orders associate with lichens in one way or another, and further studies will no doubt reveal even more. It is appropriate to conclude this section with the picturesque account given by Cunningham (1907, p. 151):

> The originally smooth surfaces which are left on the stems of many palm-trees by the fall of successive crops of leaves are usually quickly clothed in a coating of crustaceous lichens, painted on in intricate patterns of softly blended colours, and forming a pasture ground on which many different sorts of small insects browse happily.

B. Mites (Acari)

Evans *et al.* (1961) recognized seven orders within the Acari, a classification which, although not universally accepted, is followed here. Four of these orders, namely the Cryptostigmata, the Prostigmata, the Astigmata and the Mesostigmata, have been reported from lichens. A list of Acari collected in the orange lichen intertidal zone in Britain was compiled by Evans *et al.* (1961).

The Cryptostigmata (Oribatei, beetle or moss mites), dark, slow-moving Acari, often graze on or burrow within lichen thalli. Some, like *Pirnodus detectidens*, occur mostly in saxicolous lichens, whereas others, for example *Dometorina plantivaga*, prefer corticolous species; both of these feed exclusively on lichens (Travé, 1963, 1969). Others, like *Oribatula parisi*, are found in mosses as well as lichens. Only immature specimens of *Mycobates parmeliae* live in lichens, the adults being more abundant in mosses and hepatics. Other lichen-frequenting cryptostigmatids, like *Phauloppia lucorum*, appear to be non-specific feeders (Travé, 1963; Seaward, 1974).

Camisia segnis deposits its eggs in lichens. During oviposition some softening substance is excreted by the female which causes the lichen to swell and grow over the egg, thus protecting it (Grandjean, 1950). As the egg hatches, the emerging larva feeds, moults and continues to develop there. Another cryptostigmatid, *Ommatocepheus ocellatus* (Fig. 2) also feeds within the thallus; its periods of activity are correlated with the humidity of the host-lichen. The nymphs (Fig. 2) possess large leaf-like dorsal setae, which probably serve to protect them from desiccation (Travé, 1963). Similar observations are reported for *Dometorina plantivaga* and *Pirnodus detectidens* (Travé, 1969). More than a dozen additional cryptostigmatids associated with lichens were listed by Travé (1963), and numerous other associations are noted by Seyd (1968 and former publications).

Information about lichen-feeding Prostigmata is much rarer. *Tydeus tilbrooki*, possibly the smallest free-living arthropod in Antarctica, feeds on fungal hyphae and lichens (Strong, 1967), and occurs as very dense populations within *Ramalina* and *Caloplaca* thalli (Gressitt and Shoup, 1967). *Tydeus tilbrooki* deposits its eggs on the lichen, where they look like pale-pink spots. Another tydeid, *Paralorryia mali*, includes lichens in its diet (Gerson, 1968). Coineau (1973) found the caeculid *Microcaeculus hispanicus* in rock-encrusting stands of *Parmelia stenophylla* and the moss *Grimmia campestris*. Additional species, representing diverse families, are suspected of being lichen feeders. They include *Cryptognathus lagena*, found on a saxicolous *Parmelia*, whose food may have been extracted from the thallus (Luxton, 1972), *Ledermuelleria* spp., known to feed on mosses (Gerson, 1972) but also collected in lichens, and *Daidalotarsonemus* spp., whose greenish bodies and lichen habitat lend support to the above theory (Gerson, 1971). Representatives of several other prostigmatic families were obtained in sparse numbers from Belgian corticolous lichens by Andre (1975). Prostigmatids found in sea-shore lichens are listed by Evans *et al.* (1961).

"*Tetranychus lapidus*" is a name which has persistently been listed

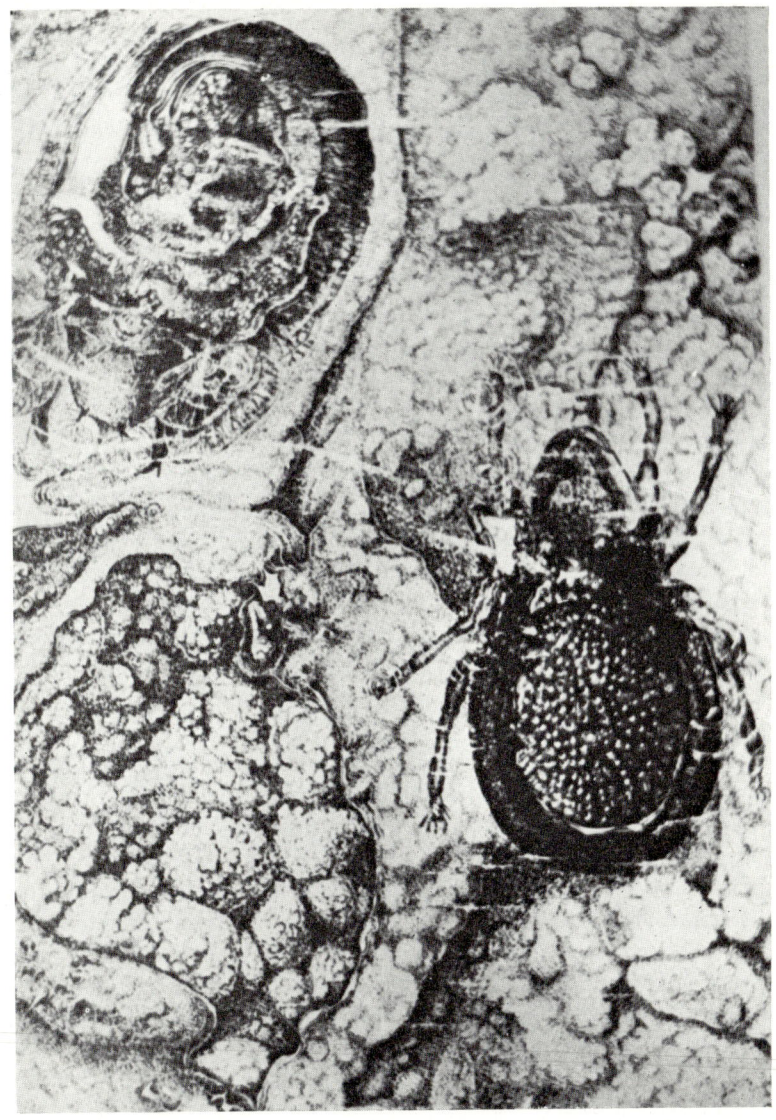

FIG. 2. An adult (right) and a tritonymph (upper left) of *Ommatocepheus ocellatus* (Acari: Cryptostigmata) on a crustaceous lichen covered by a thin water layer ×95. Drawn by Y. Coineau, from Travé (1963).

among the lichen mites (Wheldon, 1914; Smith, 1921; Richardson, 1975). This mite is actually *Petrobia latens*, the brown wheat mite, a member of the prostigmatic family Tetranychidae. It is a pest of many low-growing monocotyledonous plants in the southwestern United States, being known also from Europe, North Africa and Australia (Baker and Pritchard, 1953). *P. latens* deposits its characteristic, whitish, saucer-like diapause egg, with its radially-striated cap, on and under stones and stray bits of wood, their presence on lichens probably being fortuitous. There are no published records of actual lichen-feeding by this mite. Wheldon (1914), who first associated "*Tetranychus lapidus*" with lichens, cautiously wrote that "It is possible that the Acari feed upon these plants", but presented no further proof. Wheldon also noted that "on crushing [the eggs] they exude a thin reddish fluid". This observation was later the basis of the "blood from a stone" hoax entertainingly related by Browning (1950).

Gall mites (Eriophyoidea, tentatively retained in the Prostigmata) form galls on lichens—and lichens are the only Thallophyta on which Acari form galls (Mani, 1964).

Several astigmatids, referable to the families Saproglyphidae and Hyadesidae, were found in lichens colonizing subantarctic islands (Hughes and Tilbrook, 1966; Fain, 1975). At least one of these mites, *Hyadesia halophila*, belongs to a phytophagous genus whose members feed on algae; some lichen feeding may also take place. Unidentified Astigmata were collected by Andre (1975) from corticolous *Lecanora conizaeoides* in Belgium.

The role of mesostigmatic mites in lichen communities is obscure. Some predacious species were listed from British shore lichens by Evans *et al.* (1961). Travé (1963) stated that members of the family Epicriidae are present in dense lichen stands, Tilbrook (1967) recorded several specimens of *Cyrtolaelaps racovitzai* from lichens in maritime Antarctica, and Stebaev (1963) found *Zercon* in South Uralian *Parmelia* material. The association of these mites with lichens is probably quite casual. On the other hand, the occurrence of many predacious Phytoseiidae in Belgian corticolous lichens (Andre, 1975) suggests that they may play some role in that ecosystem.

C. Other Arthropods

Various additional terrestrial and aquatic arthropods are associated with lichens. Among the predacious Arachnida, spiders (Araneae), such as the South American *Azilia*, disguise their snares with bits of lichen, which harmonize with the spider's colours (Cott, 1957), and the giant crab spider from Trinidad (Cooke, *in litt.*) constructs its nest from lichens (mainly

Usnea). Other spiders are to be found in tundra lichens (Bengtson *et al.*, 1974). Pseudoscorpions (Chelonethi) hunt for lichen-grazing barklice in the Chaco region (Morello, 1970). The opilionid (Phalangida, harvestmen) *Phalangium opilio* has an irregular light and dark colour pattern, which makes it rather unlike a phalangid, while blending with a lichen-covered background (Cott, 1957). Unidentified "Myriapoda" were obtained by Stebaev (1963) from *Parmelia centrifuga* in the South Ural Mountains, and the diplopod *Polyxenus* was found in galls of *Ramalina kullensis* collected in Sweden (Zopf, 1907).

Crustaceans settle on, live in, feed upon and compete with lichens in marine environments. Barnacle (Cirripedia) spat may settle on littoral lichens, but very few stay for longer than two weeks, probably due to an imperfect adhesion to some plant surfaces (Fletcher, 1973). Moderate numbers of *Chthamalus stellatus* and *Balanus balanoides* were found amongst *Lichina* tufts (Colman, 1939). Competition for settling sites between lichens and barnacles was described by Fletcher (1973); once established, lichens appear to resist additional colonization by the cirripedes. A scheme of the rocky-shore barnacle-and-lichen zone was presented by Stephenson and Stephenson (1972).

Wieser (1963) studied relationships between *Lichina pygmaea* and the crustacean isopod *Campecopea hirsuta*. This isopod feeds on the lichen by browsing along its branches, eating away the gonidial layer and leaving the mycobiont almost undisturbed. *L. pygmaea* is the isopod's preferred habitat, and Colman (1939) found several thousand specimens in the small tufts of this lichen. *Campecopea* feeds only when its habitat is wet, being, in this respect, like members of the aquafauna. Like them, it is also very resistant to desiccation (Wieser, 1963), another feature amply demonstrated by components of the lichen-associated fauna. Colman (1939) also recorded the amphipod *Hyale nilssoni* in *L. pygmaea* tufts, but its occurrence is probably fortuitous.

D. Mollusca

Representatives of two of the major classes of the Mollusca, namely the Gastropoda and the Bivalvia, associate with lichens. The gastropods—snails, slugs and periwinkles—are terrestrial or intertidal in habit. *Lasaea*, the best-known lichen-associated bivalve, is marine.

Snails feed on lichens (Plitt, 1934; Peake and James, 1967; Coker, 1967), shelter within their stands and disperse their propagules. Shore lichens, like *Arthopyrenia* and *Verrucaria*, settle on living or dead mollusc shells. These associations were reviewed by Peake and James (1967), who also listed British snails and slugs observed with lichens. Holden and Tracey (1950) noted the presence of enzymes which break down the common lichen polysaccharide lichenin in the alimentary tract of snails, and Nielsen (1963) found laminarinase (=lichenase) in the slug *Arion*.

Several species of the periwinkle *Littorina* occur in maritime *Verrucaria* spp. These gastropods are the only conspicuous animals in the middle and lower *Verrucaria* belts in the British Isles (Stephenson and Stephenson, 1972), and are also common on tufts of *Lichina pygmaea* (Colman, 1939). Limpets (*Patella* spp.) graze on littoral lichens in the British Isles, and may even limit the cryptogams' initial colonization, although they avoid firmly established colonies (Fletcher, 1973).

The bivalves *Mytilus edulis* and *Lasaea rubra* both occur on *Lichina*, the former in scant numbers, the latter in large populations (Colman, 1939). *L. rubra* has the remarkable ability to recover from a 12-h exposure to 30°C and dry air. This capability is even more pronounced in *Lasaea*—an obvious adaptation to living in lichens which may dehydrate during midday low tides (Morton *et al.*, 1957). Such a mechanism is reminiscent of the various adaptations to desiccation developed by members of the lichen aquafauna and suggests a degree of eco-physiological convergence. Colman (1939) sampled several species of algae as well as *Lichina* for invertebrates, but *Lasaea* was recovered only from the lichen. Some specific, non-trophic association is thus indicated as *Lasaea*, like other bivalves, is a ciliary feeder and does not use *Lichina* as food. The nature of the association between the bivalve and the lichen remains an open question.

IV. Distribution

A. Habitats

Invertebrates associate with lichens wherever they grow: in primitive soils (Richardson, 1975), on rocks undergoing weathering (Stebaev, 1963), on trees (Broadhead, 1958; Kimmel and Meglitsch, 1969), in the sea and on its shores (Colman, 1939), and even on living beetles (Gressitt *et al.*, 1965) and molluscs (Peake and James, 1967). They associate on mountain tops (Mani, 1962), in the tropics (Kettlewell, 1959) and in Arctic and Antarctic regions (Gressitt and Shoup, 1967). It may thus safely be stated that wherever lichens occur, invertebrates will be there with them: in fact, the two groups of organisms are sometimes dispersed by the same mechanism —wind—and thus settle together in remote habitats (Cuénot, 1949; Kühnelt, 1976).

As already stated, many lichen-associated invertebrates have the physiological means to survive periods of desiccation. This characteristic has great survival value for colonizing animals, and undoubtedly exerts considerable selection pressure on invertebrates deposited on lichens by wind dispersion. This mechanism is also of considerable value to animals living on aquatic lichens (Morton et al., 1957; Wieser, 1963). In fact, the ability to withstand intermittent wet and dry periods is probably the single most important factor determining the ability of invertebrates to survive and flourish on lichens.

B. Lichen-locating by Invertebrates

Winds are a major agency in transporting the lichen fauna to its host plants; this factor accounts for the passive lichen "finding" by most of the unspecific aquafauna. Specific lichen associates, like mites, moths and molluscs, must actively seek out their hosts; the search may be at the level of the habitat, more definitely at that of cryptogams in general, or for one lichen genus or species in particular. Travé (1963) presented evidence for habitat specificity in lichenophagous cryptostigmatids; *Pirnodus detectidens* lives exclusively on rock-encrusting lichens; whereas *Dometorina plantivaga* lives in corticolous ones (Fig. 3). *Ameronothrus maculatus*, a related mite, feeds on various lichens which grow together in the *Aspicilietum calcareae* (syn. *Caloplacetum heppianae*) in London (Laundon, 1967). Observations on mites which prefer lichens to moss, and vice versa, were made by Travé (1963).

A rare case of apparent lichen preference by a grazer due to the former's mechanical properties was described by Plitt (1934). The zebra snail, *Oxystyla undata*, prefers to feed on the crustose *Pertusaria amara*, a lichen producing a bitter taste to man. This preference was thought, but not tested, to stem from the fact that the snail could not hold and rasp foliose and fruticose lichens as effectively as the crustose species. Further evidence for the attraction of invertebrates to specific lichens is lacking, but the frequently voiced observation that lichen-mimicking moths manage to alight on the correct, camouflaging plant is rather suggestive. There are several hundred tree species in eastern Brazil, whose trunks are covered by "white, red and greenish" lichens, and each colour is exploited by insects for camouflage (Kettlewell, 1959). It is inconceivable that moths will indiscriminately come to rest on any of these lichens, because they will quickly be eaten by predators; in order to exploit the multicoloured backgrounds, they must alight on the "correct" lichen(s). This presumes some mechanism by which the moths (as well as other camouflaged arthropods) actively seek out and recognize "their" lichens. Some recognition is probably visual, as moths are known to have colour vision (Wigglesworth,

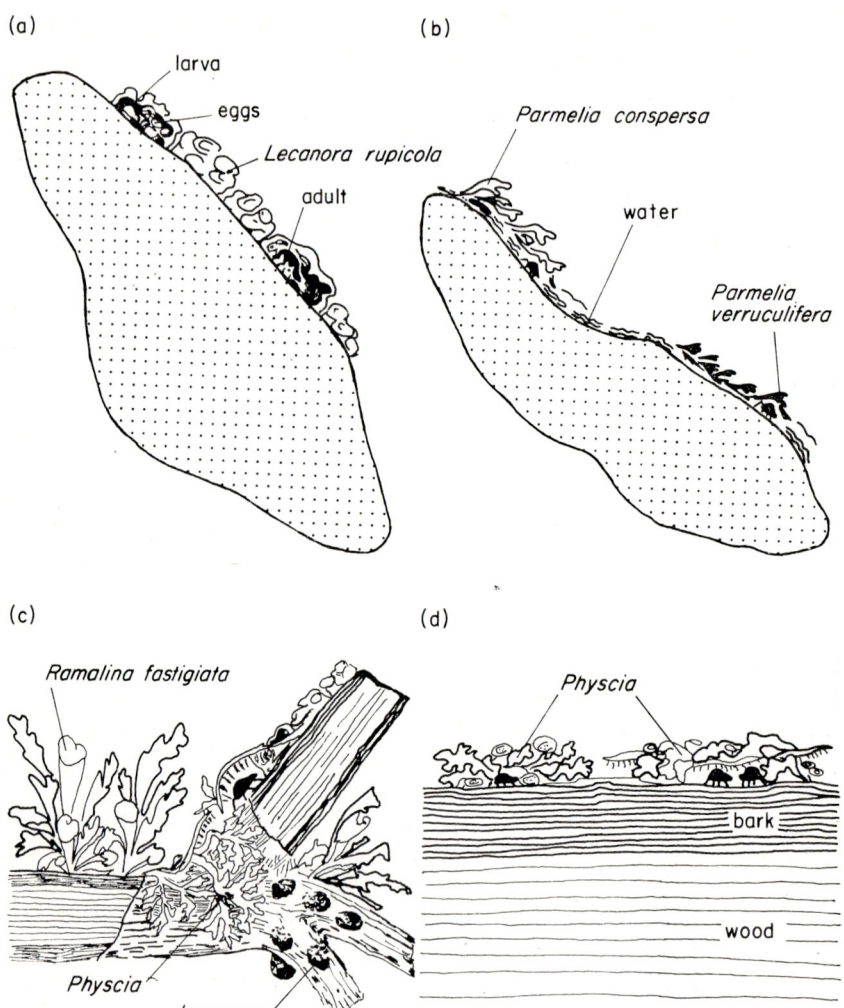

FIG. 3. Simple populations of lichen associated cryptostigmatid mites. (a) *Pirnodus detectidens* on a rock; (b) *Provertex delamarei* on a rock; (c) *Dometorina plantivaga* on an oak branch; (d) *Mycobates parmeliae* on a tree trunk (from Travé, 1963).

1972). However, it is difficult to envisage each moth species finding its "own" lichen during the scanty dawn light. Some supplementary mechanism is a possibility. The "secondary plant substances thesis" of Fraenkel (1959) could be invoked for this purpose and applied to lichens.

Fraenkel (1959) argued that "secondary" plant substances (such as glucosides, saponins, tannins and alkaloids) which only occur inconsistently in plants, have no basic metabolic function in them, and thus are unlikely to have any basic metabolic function in insects also. As many of these substances are inhibitory or toxic towards insects, it may be argued that this was one purpose for which they were evolved or preserved in plants. The fact that many "secondary" substances possess characteristic odours and tastes, as compared to their lack in "primary" (occurring in all plants) substances, lends much support to this theory. Some insects, and possibly other phytophagous invertebrates, have overcome the repellent properties of plants to gain new, and often exclusive, plant hosts. Furthermore, selection has caused the new plant-specific species to become dependent solely on the smell and taste of these secondary plant substances; hence the origins of oligophagy in many insects. We know of no experimental evidence that lichens actually attract certain feeding and mimicking species. Nevertheless, this theory is mentioned because it fits many of the known facts. It is of historical interest that Stahl (1904), who conducted studies on lichen feeding by snails, was actually the first to formulate the theory that plant substances serve to protect them from grazers (cf. Fraenkel, 1959).

C. Lichen Dispersal by Invertebrates

Ants were observed by Bailey (1970) to carry one to three soredia each of *Lecanora conizaeoides*. Beetles with cryptogam gardens, including lichens, were described by Gressitt *et al.* (1965). Mites, *Tricherememaeus serratus*, have recently been found in Scotland with soredia of *Lepraria incana* on the hairs of their backs, and would therefore appear to be aiding the lichen's dispersal (Laundon, *in litt.*). Springtails which overrun colonies of *Pertusaria amara* become powdered with their soredia (Ridley, 1930). Various lacewings carry lichen fragments in their "packets" (Skorepa and Sharp, 1971) or on their entire bodies. In the latter case, as the larvae moult, large groups of soredia are left behind with the cast skins (Garrett, 1974). The cases of moth caterpillars may become covered by lichen fragments (Weber, 1974), as do the slimy bodies of snails and slugs.

Lichen-feeding molluscs may void viable algae, which can grow on the faeces (Kühnelt, 1976). The rotifer, *Philodina*, isolated from *Xanthoria*, was observed to ingest ascospores (Pyatt, 1968). Not all ascospores were digested, and upon splashing, the rotifers could be removed from the lichens, whereupon the animals liberated viable ascospores—another, albeit complicated, method of dispersal.

Further aspects of lichen dispersal by invertebrates are covered by Gerson (1973) and Bailey (1976).

V. Invertebrate Feeding on Lichens

A. Modes of Feeding

Most invertebrate lichen feeders ingest their food by chewing or by scraping. Lichenophagous insects, such as barklice, springtails and moth caterpillars, possess mandibulate mouth parts, with which they bite off their food and chew it.

Acari have chelicerate mouth parts, but the most important lichen feeders, members of the Cryptostigmata, possess a mandibula-like modification, the rutella. These are modified setae which sometimes occur as massive dentate structures capable of scraping solid food off its substrate and particulating it (Evans et al., 1961). Prostigmata (gall mites included) have sharp chelicerae with which they pierce plant cells and suck out their contents.

Gastropods obtain their food by scraping it off by means of a toothed, chitinous radula (Owen, 1966). Snail injury on lichens can usually be recognized by the linear scraping pattern of the radula (Peake and James, 1967). Most feeding consists of tearing or scraping off bits of lichen while the feeder is stationed on the plant. Some species, however, feed within the lichens (Brightman, 1965; Holloway, 1970), where they may excavate intricate burrows (Grandjean, 1950). Feeding (especially by gall mites) may result in abnormal growth forms, usually called galls (Grummann, 1960). Another mechanism of gall formation was described by Schade (1933): some feeders may devour only algal cells and consequently the mycobiont overgrows the phycobiont (or its remnants), causing strange growth shapes and forms to appear. Several new lichen species and varieties, especially in the genus *Rhizocarpon*, were described, which are actually no more than well known species damaged by snails and mites (Schade, 1933). Apothecial regeneration of "*Lecanora subfusca* var. *allophana*" due to the gnawing of hymenia by insects has been recorded by Bouly de Lesdain (1924). Molluscs, especially *Clausilia* spp., show a preference for the green algal morphotype of *Sticta canariensis* rather than the blue-green algal morphotype, "*S. dufourii*" (James and Henssen, 1976).

B. Digestion

Very little has been published about lichen digestion by invertebrates, beyond the observation that some species digest the entire symbiont, whereas others only the phycobiont or the mycobiont. The isopod *Campecopea hirsuta* feeds almost exclusively on the phycobiont of *Lichina pygmaea*, voiding blue-green faecal pellets (Wieser, 1963) which attest to the digestion of the algae. On the other hand, live algal cells may pass unharmed through the alimentary tract of some lichen-feeding snails, dead algae as well as fungi being digested (Schmid, 1929). Peake and James (1967) argue, from a literature survey, that it is very difficult to determine which symbiont of the lichen, if any, is actually preferred by molluscs.

Possession of enzymatic systems which break down lichenin in arthropods, oligochaetes and molluscs was demonstrated by Holden and Tracey (1950) and by Nielsen (1963).

Woodring and Cook (1962) surface-sterilized eggs of the cryptostigmatid *Ceratozetes cisalpinus* and found that the emergent larvae were unable to feed on sterilized food. However, upon being placed in a culture tube with a few live fungal hyphae, the aseptic larvae fed and continued to develop. These results might be interpreted to indicate the necessity of some specific diet attractant for the mite; in view of the fact that surface-sterilized older mite stages did feed on the aseptic lichen, this explanation cannot be accepted. Another possibility is that the hyphae were required to induce the production and secretion of appropriate lichen-digesting enzymes.

Some Cryptostigmata may carry internal symbiotic microorganisms (Woodring, 1963). Their occurrence is more likely in animals which feed exclusively on a single diet, and some strict lichen feeders may indeed possess such organisms. It is only in ruminants, like reindeer, that lichen-digesting symbiotic protozoa are actually known (Scotter, 1965). However, there appear to be no lichen-specific protozoa in the rumen of lichen-feeding ruminants (Hobson, *in litt.*).

The uptake of diverse heavy metals by lichens is well known. It would be of interest to know whether these heavy metals affect lichen edibility and digestibility by invertebrates. Examination of many thalli of *Lecanora muralis* with high metal concentrations failed to reveal any damage due to invertebrates grazing (Seaward, 1972). Only the clover mite, *Bryobia praetiosa* was found to be associated with this lichen in reasonable numbers from a few sites; however, since members of the genus *Bryobia* are known to feed on higher plants, and *B. praetiosa* infests grasses etc., it seems more probable that this species was sheltering in, rather than feeding on, the lichen (see also p. 102).

Most invertebrates feed only on wet lichens. In some cases (e.g. slugs) this is probably connected with the need of these grazers for humidity, but in others it possibly adds to the ease with which the lichen can be scraped or chewed. Wet lichens are perhaps more edible and more easily digested. On the other hand, dry, gelatinous lichens are almost never attacked by invertebrates. The subject of lichen resistance to grazers will be discussed in the next section.

C. Lichen Resistance to Invertebrate Feeders

The controversy regarding the role of lichen products in protecting lichens against various grazers was summarized by Smith (1921, pp. 395–396):

> It has been argued by Zukal (1895) that the great formation of acid substances in lichens is for shielding them against the attacks of animals; Zopf (1896) on the contrary insists that these substances afford the plants no real protection. He made a series of experiments with snails, feeding them with slices of potato smeared with pure lichen acids. Many snails ate the slices with great readiness even when covered with bitter acids such as cetraric, or with those which are poisonous for other animals such as rhizocarpic and pinastrinic. The only acid they refused was vulpinic, which is said to be poisonous for vertebrates. The crystals of the acids passed unchanged through the alimentary canal of the snails, and were found in masses in the excreta. They were undissolved, but, enclosed in slime, their sharp edges did no damage to the digestive tract.
>
> Stahl (1904) however upholds Zukal's theory of the protective function of lichen acids against the attacks of small animals. Some few snails, caterpillars, etc., that are omnivorous feeders consume most lichens with impunity, and the bitter taste seems to attract rather than repel them; but many others he contends are certainly prevented from eating lichens by the presence of the acids. He proved this by soaking portions of the thalli of certain bitter species for about twenty-four hours in a one per cent soda solution, which was sufficiently strong to extract the acids. He found that these treated specimens were in most cases preferred to fresh portions that had been simply moistened with water.
>
> Even the omnivorous snail, *Helix hortensis*, was several times observed to touch the fresh thallus and then creep away, while it ate continuously the soda-washed portion as soon as it came into contact with it. Calcium oxalate, on the other hand, formed no protection; omnivorous feeders ate indifferently calcicolous lichens such as *Aspicilia* [=*Lecanora*] *calcarea* and *Lecanora saxicola* [=*L. muralis*], whether treated with soda or not, but would only accept lichens with acid contents, such as *Parmelia caperata*, *Evernia prunastri*, etc., after they had been duly soaked.
>
> Experiments were also made with wood-lice (*Oniscus murarius*), and with earwigs (*Forficula auricularia*), and the result was the same: they would only eat bitter lichens after the acids had been extracted by the soda method. Stahl therefore concludes that acids must be regarded as eminently adapted to protect lichens which otherwise, owing to their slowness of growth, would scarcely escape extinction.

A few more recent cases which demonstrate some degree of protection against representatives of various invertebrate groups will be briefly discussed below.

The two common lichenophilous psocid species (*Reuterella helvimacula* and *Elipsocus mclachlani*) occurring on larch trees in northern England show an interesting difference in food preference (Broadhead, 1958); both species graze *Lecanora conizaeoides*. The nymphs and adults of *Reuterella helvimacula* graze the entire surface of the lichen, taking both the apothecia and the algal layer of the thallus indiscriminately. In contrast to this, the nymphs and adults of *Elipsocus mclachlani* graze the apothecia only. No indication was obtained either in the field or in the laboratory that this species ever grazes the thallus of the lichen. The hymenium and the algal layer immediately below it are eaten to expose the white medulla. Broadhead considered that a lichen substance rendered the thallus unpalatable to *E. mclachlani*. However, the major substance in this lichen is now known to be fumarprotocetraric acid which is present in both the apothecium and thallus, so the reason for this discrimination remains obscure. The collembolan *Hypogastrura packardi* feeds only on the upper cortex and most of the algal layer of *Parmelia baltimorensis*, leaving the medulla largely untouched (Hale, 1972). Presence of protocetraric acid in the medulla was believed to protect the lichen from its grazer. Gressitt and Shoup (1967), working in the Antarctic, noted that the oribatid *Maudheimia petronia* feeds only in the stem and base of the fruticose *Usnea antarctica*, feeding being apparently restricted to the "white portions" of the lichen.

Browsing on various lichens by the slug *Lehmannia marginata* was reported by Coker (1967). *Lobaria pulmonaria*, *Hypogymnia physodes* and *Pertusaria pertusa* were all grazed, but apothecia were preferred to the cortex in the latter lichen. However, *Pertusaria amara*, growing in association with *P. pertusa*, was not affected.

A list of some invertebrates reported to feed on lichens, together with the latter's secondary products (culled mainly from Culberson, 1969, 1970; Table III), raises more questions. Fumarprotocetraric acid may partially protect *Usnea "barbata"* from *Maudheimia petronia* (Gressitt and Shoup, 1967). On the other hand, the psocid *Reuterella helvimacula* freely grazes on *L. conizaeoides* (Broadhead, 1958; Seaward, 1965), and the mite *Phauloppia lucorum* feeds on *Cladonia fimbriata* (Seaward, 1974), which also contains fumarprotocetraric acid. This substance further occurs in *Cladonia ochrochlora*, which is damaged by an unidentified gall maker (Bachmann, 1929). The lichen acid thus affords, at best, only partial protection against grazers. Another similar example is picrolichenic acid, apparently specific to the genus *Pertusaria* (Culberson, 1970). It probably renders *P. amara* unpalatable to the slug *Lehmannia marginata*, but another molluscan,

TABLE III.

Invertebrates and the lichens they feed upon. Data pertaining to lichen acids, mainly from Culberson (1969, 1970).

Invertebrate	Lichen	Lichen compounds	Source
Cryptostigmata:			
Ameronothrus maculatus	*Candelariella medians*	calycin	Laundon (1967)
A. maculatus	*Physcia caesia*	atranorin, zeorin	Laundon (1967)
A. maculatus	*P. orbicularis*	rhodophyscin ($\pm$)	Laundon (1967)
Dometorina plantivaga	*Caloplaca ferruginea*	emodin, fallacinol, parietin, parietinic acid	Travé (1969)
D. plantivaga	*Lecanora subfusca* agg.	atranorin	Travé (1969)
D. plantivaga	*Pertusaria hymenea*	thiophanic acid, gyrophoric acid ($\pm$)	Travé (1963)
Maudheimia petronia	*Usnea antarctica*[a]	fumarprotocetraric acid, usnic acid	Gressitt and Shoup (1967)
Phauloppia lucorum	*Cladonia fimbriata*	fumarprotocetraric acid	Seaward (1974)
P. lucorum	*Parmelia saxatilis*	atranorin, lobaric acid, salazinic acid	MacNeill (1966)
P. lucorum	*Xanthoria parietina*	atranorin, parietin, parietinic acid	Seaward (1974)
Pirnodus detectidens	*Pertusaria rupicola*	norstictic acid, stictic acid, thiophaninic acid	Travé (1969)

Psocoptera:

Cerobasis guestfalicus	*Xanthoria parietina*	atranorin, parietin, parietinic acid	Seaward (1965)
Elipsocus mclachlani	*Lecanora conizaeoides*[a]	fumarprotocetraric acid	Broadhead (1958)
Mesopsocus immunis	*Lecanora conizaeoides*	fumarprotocetraric acid	Seaward (1965)
Reuterella helvimacula	*Hypogymnia physodes*	atranorin, chloroatranorin, physodalic acid, physodic acid	Laundon (1971)
R. helvimacula	*Evernia prunastri*	atranorin, chloroatranorin, evernic acid, usnic acid	Laundon (1971)

Mollusca:

Lehmannia marginata	*Hypogymnia physodes*	atranorin, chloroatranorin, physodic acid, physodalic acid	Coker (1967)
L. marginata	*Lobaria pulmonaria*	constictic acid, norstictic acid, stictic acid	Coker (1967)
Oxystyla undata	*Pertusaria amara*	picrolichenic acid, fumarprotocetraric acid (±)	Plitt (1934)
O. undata	*Lecanora pallida*	atranorin, norstictic acid, protocetraric acid	Plitt (1934)

[a] Denotes partial feeding.

Oxystyla undata, thrives on this lichen (Plitt, 1934), and *P. albescens*, which also contains picrolichenic acid, is devoured by the mite *Dometorina* (Travé, 1963). It is known that different lichen substances are restricted to different parts of the same lichen (e.g. cortex v. medulla), as reviewed by Hawksworth (1976).

Hale (1972) attributed the protection of the medulla of *Parmelia baltimorensis* from a collembolan grazer to the presence of protocetraric acid in that region. This acid, as well as fumarprotocetraric acid, is a depsidone (Culberson, 1969), and substances belonging to this group could thus be expected to serve as lichen protectants. However, norstictic, physodalic, salazinic and stictic acids are also depsidones, occurring in lichens such as *Pertusaria pertusa*, *Lobaria pulmonaria* and *Hypogymnia physodes*, which are grazed by various invertebrates.

It thus appears that the whole subject of "protection from grazers" is still full of contradictions, and no general theory can as yet be formulated. Lack of experimental data as well as information on the quantitative and qualitative distribution of the acid within the various lichens hinders such a formulation.

Several mechanisms have been proposed to account for the widely reported lichen resistance to invertebrates. These mechanisms include the specific effect of lichen acids, the protective, gelatinous covering of some lichens, and the presence of chelating agents in many lichens (Gerson, 1973). The effect of lichen acids is probably made up of several factors, such as solubility, reduced palatability, outright toxicity and possible indirect toxicity mediated through the antibiotic effect of some acids on the symbiotic microflora of certain invertebrate grazers. A fourth component may well be the acidic environment, in part formed by lichen acids released into the soil substrate. Gadea (1964a, 1973) believed that *Xanthoria* and *Roccella* thus inhibit some soil animals, especially nematodes.

Certain extracts have in fact been found to be active against microorganisms, and have served as the basis of a modest lichen antibiotic industry. The possibility of using such extracts against insect pests was explored by Heal *et al.* (1950), within a broad survey of plants for insecticidal activity. Aqueous extracts of *Cetraria juniperina*, *Letharia vulpina*, *Lecanora rubina*, *Parmelia perforata*, *Peltigera canina* and *Lobaria pulmonaria* were all toxic to American cockroaches (*Periplaneta americana*) when injected into their bodies. But immersing German cockroaches (*Blattella germanica*) and milkweed bugs (*Oncopeltus fasciatus*) into most of these extracts did not affect the insects. Chloroform extracts of *Cetraria juniperina* and *Lecanora rubina* were very toxic to larvae of the black carpet beetle (*Attagenus piceus*), but not to German cockroaches, milkweed bugs and larvae of the webbing clothes moth (*Tineola bisselliella*) and of the mosquito *Anopheles*. On the other hand, the larvae of the mosquito *Aedes* were affected by the chloroform extract of *Lecanora rubina*. One may thus conclude that lichens are not promising source material for insecticides.

Other invertebrates are also known to be affected by lichen products. Richardson (1975), citing Russian sources, states that a sodium salt of usnic acid ("Binan") is used in Russia against the protozoan *Trichomonas* which is parasitic in man and infests the lower reproductive tract. An acidic extract of *Xanthoria parietina* was reported by Grzyb (1964) to inhibit development of the egg of the common pig roundworm (*Ascaris suum*). However, as extracts of other lichens had no effect on these eggs, it was concluded that such extracts have little applied value.

Some lichen chemicals were used extensively in the past, and are to a limited extent used today, as dyestuffs. One wonders whether garments dyed with these lichen extracts were indeed more resistant to clothes moths. This question is of some evolutionary interest, as larvae of certain members of the clothes moth family, the Tineidae, feed on lichens.

D. Energetics of the Lichen Fauna

Studies on the energetics of various plant and animal communities are coming more and more to the forefront of current ecological research. Engelmann (1966) presented a lucid discussion of productivity and the kinds of information needed for the study of animal energetics. Various estimates are required: ingestion, assimilation, egestion, respiration, growth and death rates of the relevant fauna, its numbers and biomass, and the energy units represented by these figures.

Little work on the energetics of lichens has been published to date; most of it has been summed up by Wielgolaski (1975). Thus, maximum lichen biomass in southern alpine Norway comes to about 1200 g m^{-2} in unshaded, not too dry ridges. At the central Hardangervidda plateau the dry weight biomass of lichens, dominated by *Cetraria nivalis*, usually comes to 200–400 g m^{-2}. The annual primary production of lichen heaths totally dominated by lichens is only about 100 g m^{-2} even in dense mats. The production to biomass ratio in lichens in this area is about 0·2. No data are available on specific invertebrate lichen feeders. Several species, or their close relatives, which include lichens in their diet, have been investigated in this context (Slobodkin and Richman, 1961; Engelmann, 1966) and reviewed by Harding and Stuttard (1974). Speculative energy balance sheets could thus perhaps be drawn up, but this appears to be premature. A specific problem is the intermittent, humidity-dependent development of several lichen feeders, as discussed above. Too little is known about the metabolism of these grazers to make any effort at energetics study worthwhile, but it appears useful to compile and summarize one kind of relevant data, namely quantitative information on lichen invertebrates.

Most available data come from ecological studies of Arctic and Antarctic regions, probably because lichens are such an important floral component there. Watson et al. (1966), reporting on Alaskan terrestrial invertebrates, included an unidentified lichen community in their observations. The community was sampled from June to September (Table IV), and Collembola were, on a yearly basis, more than twice as abundant as mites. In another study (Bengtson et al., 1974), soil invertebrates were sampled in several

TABLE IV.

Collembola and Acari (per m^2) obtained from a lichen community, Cape Thompson Region, Alaska (from Watson et al., 1966).

Date	No. of samples	Collembola ($\pm$s.e.)	Acari ($\pm$s.e.)
June 21	4	12,900 $\pm$ 2780	7840 $\pm$ 938
July 10	4	12,800 $\pm$ 3410	10,500 $\pm$ 921
July 24	4	15,200 $\pm$ 4110	5240 $\pm$ 984
August 8	4	37,600 $\pm$ 9440	8340 $\pm$ 1580
August 21	4	4240 $\pm$ 1170	6440 $\pm$ 1390
September 8	4	14,400 $\pm$ 7740	6780 $\pm$ 1430
September 18	3	5330 $\pm$ 2670	5990 $\pm$ 2330

TABLE V.

Collembola and Acari (per m^2) obtained from four different plant communities in Spitsbergen (from Bengtson et al., 1974).

	Collembola ($\pm$s.e.)	Acari ($\pm$s.e.)
Cetraria spp.: vegetation and litter	9510 $\pm$ 784	16,192 $\pm$ 1874
soil (0–3 cm)	11,288 $\pm$ 2604	5117 $\pm$ 1094
Cetraria spp.: vegetation and litter	27,091 $\pm$ 1929	19,195 $\pm$ 1850
soil (0–3 cm)	11,180 $\pm$ 1485	5031 $\pm$ 2521
Moss (*Drepanocladus* spp.)	243,475 $\pm$ 1921	22,090 $\pm$ 1667
Grassland	268,830 $\pm$ 24,727	247,880 $\pm$ 19,606

Spitsbergen plant communities (Table V); Collembola were again more abundant in lichens (*Cetraria* spp.), but were still low in numbers as compared to the moss and grassland fauna. Mites, on the other hand, did not become more abundant in moss; this suggests that the cryptogams have a similar "carrying capacity" for mites, but lack of additional data on the mites and their diets renders such information of limited value from the energetics point of view.

Gressitt (1967) collected invertebrates from rock-encrusting lichens in the maritime Antarctic and obtained somewhat different results (Table VI): Acari (consisting mainly of three cryptostigmatid and one prostigmatid species) outnumbered Collembola.

Working in the South Ural Mountains, Stebaev (1963) sampled invertebrates in sparse and dense rock-encrusting *Parmelia centrifuga* stands with a northern and a southern aspect. His data (Table VII) again show that

TABLE VI.

Average (and maximum) sample counts of arthropods obtained from various microhabitats in the Antarctic Peninsula–South Shetland Islands–South Orkney Islands area, and processed through Berlese funnels (from Gressitt, 1967, his Table 2).

	Usnea	Ramalina	Physcia	Caloplaca/Xanthoria	Umbilicaria
Acari: Mesostigmata					
Cyrtolaelaps racovitzai	–	1	21	–	–
Acari: Prostigmata					
Stereotydeua villosus	–	–	4	–	–
Tydeus tilbrooki	18	150	110	250	–
Acari: Cryptostigmata					
Oppia loxolineata	0·1	–	–	–	–
Alaskozetes antarcticus	0·1	1	115	3	–
Halozetes belgicae	53	2	60	75	–
Magellozetes antarcticus	64	–	138	37	2
Insecta: Collembola					
Cryptopygus antarcticus	0·1	–	–	–	–
Archisotoma brucei	–	–	0·6	–	–
Parisotoma octooculata	–	2	2	–	–
Belgica antarctica	–	–	12·2	–	–
Total	135·3	156	462·8	365	2
Maximum population in single sample	450	209	1158	542	2

TABLE VII.

Arthropods (per dm²,) in sparse (Cycle IIa) and dense (Cycle IIb) colonies of saxicolous Parmelia centrifuga *(from Stebaev, 1963).*

Arthropods	South Cycle IIa	Cycle IIb	North Cycle IIb
Acari:			
Prostigmata	25·0	174·0	28·0
Cryptostigmata	692·5	486·0	900·0
Collembola	237·5	1148·0	410·0
Diptera	50·0	135·0	5·0

Acari (especially Cryptostigmata) and Collembola are the dominant arthropods there. Upon comparing these data with those of Watson *et al.* (1974), it becomes apparent that the South Ural lichen fauna is much richer in numbers. Seaward (1974) estimated about 23 *Phauloppia lucorum* (Cryptostigmata) to be present in 1 cm² of lichen; in other words, about 230,000 m⁻². This is more than ten times the number obtained in Alaskan and Spitsbergen work, and also considerably higher than Stebaev's figures. Although the habitat and climatic conditions are quite dissimilar, one should emphasize the different community sizes studied in the interpretation of these data.

Colman's (1939) paper is still the best source for numerical data on invertebrates living on and in marine lichens (Table VIII—assembled

TABLE VIII.

Invertebrates on Lichina pygmaea *(from Colman, 1939).*

Animal group	Individuals/100 g *Lichina*	Individuals/m² rock surface
Oligochaeta	11	220
Cirripedia	287	5740
Isopoda	2886	57,720
Amphipoda	35	700
Acari	436	8720
Insecta	161	3220
Bivalvia	9447	188,940
Gastropoda	453	9060
Total	13,716	274,320

from sections of Tables XX and XXII in Colman, 1939). He also reported that 1 m² of *Lichina pygmaea* weighs 2 kg and that the most abundant animal on the lichen is the bivalve *Lasaea rubra*. Although Colman's values for *L. rubra* were quite high, Morton (1954) found even denser populations: he reported that 15 cm² of *Lichina pygmaea* tufts yielded close to 900 *Lasaea rubra*. This is equivalent to over half a million individuals per m² of the lichen.

TABLE IX.

Number of species, density, biomass and respiration of Acari and Collembola (excluding carnivores) collected in crustose lichen habitat on Signy Island, Antarctica (from Collins et al., 1975).

		Collembola	Acari	Total
No. of species		1	6	
Density ($\times 10$ m^{-2})	Summer	7	39	
	Winter	<1	40	
	Annual mean	4	40	
Mean annual biomass (mg wet weight m^{-2})		100	2408	2508
Total annual respiration (cm² O$_2$ m^{-2})		105	1021	1126

Weight (or biomass) of many invertebrate groups which occur in lichens is actually available. Healey (1967), Engelmann (1968), and Harding and Stuttard (1974), among others, compiled data on the biomass and respiration of nematodes, mites and springtails. Collins *et al.* (1975) provided similar information on the fauna of the maritime Antarctic; some of their data are given in Table IX. The main components that are still missing are energetics data on the lichens themselves, and physiological studies on their grazers. Barrett and Kimmel (1972) advocated using lichen-bearing tree bark ecosystems for energy-flow and mineral–recycling studies, in view of the simplicity and accessibility of such systems. An early model of the kind of study needed was presented by Golley (1961), who obtained energy values for an entire community, including diverse plants (and their different parts) and invertebrates, on a monthly basis.

E. Food-webs and Nutrient Recycling

Lichens serve as the basis of food webs in which lichenophagous species serve as primary consumers, being in turn devoured by various predators, but very little is known about these relationships. *Loricula elegantula*, a predacious bug (Microphysidae) lives in lichens, especially *Parmelia* spp., where it deposits its eggs; the larvae and adults feed on mites and psocids (Southwood and Leston, 1959) including the lichenophagous barklouse *Reuterella helvimacula*. Pseudoscorpions feed on lichen-grazing psocids which live on shrubs in the Chaco region of Argentina (Morello, 1970). Many general predators which occur in lichen colonies and presumably feed on the indigenous fauna were listed, among others, by Cott (1957), Mani (1962) and Stebaev (1963).

A special case which involves an entomogenous fungus was mentioned by Cunningham (1907) from Bengal. The bark of *Oreodoxa* palms is overgrown by multicoloured lichens, upon which many psocids browse; they, in turn, are heavily attacked by a fungal disease.

Air pollution adversely affects epiphyte populations, and therefore diminishes their grazers. Gilbert (1971), in a study on the fauna of corticolous lichens, bryophytes and algae, found that primary consumers (mostly psocids) were reduced in an air-polluted area. The predators, on the other hand, survived in their usual numbers, probably by feeding on alternative prey. The effect of diminishing lichen "pastures" on specific predators and parasitoids was not studied.

Lichens are well known for their ability to absorb large quantities of heavy metals and radioactive elements (James, 1973; Tuominen and Jaakkola, 1974). Thus it is to be expected that such chemicals would be passed on in food-webs. The lichen–caribou–man food-chain maintains high radionuclide levels in Alaskan residents, but nothing is known concerning the effects of these compounds on invertebrates. Nor has anything been published about the effect of heavy metal concentration in lichens on their grazers.

Pesticides are another group of chemicals which may accumulate in food-webs. A pertinent case was recounted by Shilova *et al.* (1973) from subarctic Siberia: the insecticide Sevin (1-naphthyl *N*-methylcarbamate, also known as carbaryl) was applied there against blood-sucking flies, and was afterwards found at various trophic levels, reaching 0·5 p.p.m. in the lichens assayed.

Barrett and Kimmel (1972) suggested that lichen-bearing tree bark might be a suitable ecosystem for studying energy flow and mineral recycling. This was indeed done by Denison (1973) in Douglas fir forests in Oregon. *Lobaria oregana* is the most abundant lichen in the tops of these trees; it fixes atmospheric nitrogen through one of its phycobionts, the blue-green alga *Nostoc*. Denison (1973) estimated that *L. oregana* contributes 1·8–10·0 lb N acre^{-1} (=2·0–11·1 kg ha^{-1}) per annum to the fir forest ecosystem (cf. Forman, 1975). A major part of this element reaches the soil through the fall and decay of *L. oregana*; another route is by animal feeding, including nematodes, mites and insects.

Frankland (1974), while discussing decomposition of lower plants, stated that nothing is known about lichen decomposition: whether it is by autolysis, weathering or microbial activity. Decomposition, in the broad sense, is considered to have two components: these are mechanical breakdown, often mediated through invertebrates, and chemical decomposition, brought about by microbial action. The significance of breakdown to decomposition stems from the grazing and burrowing activity of invertebrates, which exposes innumerable new, often formerly impenetrable, surfaces to the action of decomposing microorganisms. Invertebrates which graze on lichens thus contribute to the plants' decomposition.

Lichen breakdown is described by Hale (1972): the collembolan *Hypogastrura packardi* completely chews away the upper cortex and most of the algal layer of *Parmelia baltimorensis*. Consequently, large areas in the lichen centres were disintegrating and falling away, resulting in a loss of at least half the lichen colony. Additional cases of lichen breakdown by invertebrates have been described by Schmid (1929), Denison (1973) and Laundon (1967, 1971) among others. Feeding and burrowing by these grazers cause additional, often unprotected lichen surfaces to be exposed to other microorganisms. They also entail some vertical and (when feeding upon terricolous lichens) horizontal redistribution of lichen fragments.

An indirect mechanism of lichen breakdown through insect activity was described by Llano (1956) from Arctic regions. Reindeer which graze in lichen pastures during dry periods cause excessive thallus fragmentation by trampling on them during their short rapid runs windward, intended to shake off blood-sucking insects.

These few examples suffice to show that invertebrates, whether directly or indirectly, play a particular role in lichen breakdown and decomposition. Quantitative data, however, are not available.

VI. Lichens as a Protective Environment for Invertebrates

As one surveys the variety of invertebrate life associated with lichens, and considers the great abundance of these plants on trees, walls, stones and soil, it is not surprising that lichens often serve as shelter for invertebrates. In its simplest form, this merely means that invertebrates hide under lichens; no special adaptations are involved. After prolonged sheltering in this microhabitat, some invertebrates were no doubt favoured by evolving various forms of adaptive coloration. This blending of the animal into its lichen background can be brought about by camouflage and by mimicry. Camouflage may consist of lichen fragments and soredia, or of the entire plant growing on the animal. Evolving along another route, invertebrates mimic lichens by their very appearance. This adaptation, especially well developed in moths, met a severe challenge during the Industrial Revolution, when the ascendancy of melanic morphs provided a fascinating glimpse of evolution in action.

A. Concealment

Lichen colonies on rock and tree bark provide a buffer zone between certain invertebrates and their environment. Economic entomologists have noted the presence of pests underneath lichens and stated that these have to be removed in order to control the pests. The Italian pear scale, *Epidiaspis leperii* (Homoptera, Diaspididae) is a pertinent case. This pest is often overlooked because it is hidden under lichens on fruit trees. Control recommendations include caustic soda, which is added to the spray in order to remove the protecting lichens (Essig, 1958). Similar pest control using a tar oil wash on lilac trees has recently been observed in Ireland (M. R. D. Seaward, unpublished). Colonies of the pear phylloxera, *Aphanostigma pyri*, a serious pest of pears in Israel, were observed to thrive under *Xanthoria parietina* on otherwise clean pear branches in Israel (U. Gerson, unpublished). Some pests, like the hemlock loopers, *Lambdina fiscellaria fiscellaria* and *L.f. lugubrosa* (Lepidoptera), lay their eggs on epiphytic moss and lichen patches (Thomson, 1958; Jobin, 1973; Brodo and Hawksworth, 1977). Pupation of the Assam thrips, *Scirtothrips dorsalis*, takes place in similar habitats (Dev, 1964). Lichens provide winter sanctuary for many phytophagous insects and mites, an example of the latter being *Bryobia* (Tetranychidae) (Travé, 1963). During the later stages of the 1976 ladybird (*Coccinella* spp.) "invasion" in the British Isles, lichens provided an effective shelter, and, for example, counts of more than 50 specimens beneath many thalli of *Pseudevernia furfuracea* in exposed places at higher altitudes of the Pennines were not uncommon during October (M. R. D. Seaward, unpublished).

Marine invertebrates, like the bivalve *Lasaea rubra*, also find shelter in lichens, although they do not feed on them (Morton *et al.*, 1957).

B. Camouflage

Representatives of at least two insect orders employ what Cott (1957) has termed "masks of adventitious material", which consist of lichen particles.

Larvae of the South American brown lacewing *Hemerobius* cover themselves with mosses and lichens. Hingston (cited by Cott, 1957) describes these larvae:

> ... one I met was a perfect circle, it was made out of tiny bits of lichen and ... was an exact image of the round lichenous patches that occur so commonly on forest trees.

This aggressive deception helps *Hemerobius*, a predator, to take its prey by surprise. A North American neuropteron, *Notida pavida*, regularly collects lichen fragments to construct a "packet" which it places on its back; Skorepa and Sharp (1971) present photographs of such lichen-covered larvae. Garrett (1974) found larvae of a Colombian lacewing, *Chrysopa*, covered by soredia of several lichens. The larvae excrete a sticky silk, to which soredia adhere as *Chrysopa* walks across the lichens.

Weber (1974) found two lepidopterous caterpillars in Australia and New Guinea, which were covered with *Parmelia* fragments. In one instance, larval cases of *Cebysa leucoteles*, and in the other, the larvae themselves, of an unnamed Eucleid, were covered with them.

An example of insect camouflage by means of living lichen growth was found by Gressitt (1966) in New Guinea moss forests. This association involves Curculionid beetles (weevils) which support extensive plant growths on their backs. The beetles, mostly of *Gymnopholus* subgen. *Symbiopholus*, are often structurally modified to accommodate the various plants growing on them. These modifications are dorsal depressions surrounded by ridges, specialized setae and scales, and a sticky secretion which apparently promotes plant growth. About 28% of all beetle specimens collected bore conspicuous lichen growths, and another 5–6% primary growths. Ten *Gymnopholus* species supported lichens, the latter being identified as *Parmelia* sp., *P. crinita*, *P. reticulata*, *Anaptychia* sp. and *Physcia* sp. All except the last species were obtained from the aptly named *G. lichenifer*. Lichen colony size on these beetles suggested that the insects were at least 3–5 years old when collected, and this was confirmed by later observations (Gressitt and Sedlacek, 1970). The beetle–cryptogam association is considered to be of mutual advantage, the animals providing a favourable environment for the plants, the latter endowing the insects with protective resemblance, a form of camouflage, to cryptogam-covered parts of vegetation or to inedible plants (Gressitt *et al.*, 1965). The term "epizoic symbiosis" was coined by Gressitt (1966) to describe this association.

Marine lichens, like *Arthopyrenia halodytes* and, more rarely, *Verrucaria* spp., are observed on living molluscs and barnacles. This growth may well be fortuitous, since the lichens are also found on empty shells, and no special adaptations designed to encourage colonization are found. Nevertheless, some survival value of a protective nature may well accrue to the molluscs.

C. Mimicry

As regards appearance, bark-resembling animals lead up to, and intergrade with, other types which in their most specialized form resemble in high degree the lichens growing on wood or on walls. The resemblance is literally superficial, rather than structural—that is to say, it is generally due to the most ingenious and deceptive disruptive patterns, which give the optical impression of irregular processes and deep interstices—even when painted, as they often are, on the flat canvas of a moth's wing or on the ovoid abdomen of a spider—rather than to actual excrescences and irregularities in the animal's form. From the standpoint of camouflage these cases are therefore of the greatest interest.

Lichen-like species are not confined to any particular continent or country: they may be seen on the limestone walls of a Wiltshire garden or on the slopes of an extinct volcano in Lanzarote; in the luxuriant rain-forests of South America, or on the rocky summit of a Scottish mountain. Nor are they restricted to one phylum or to a few families. Wherever lichen forms a characteristic feature of the environment, different animals are found associated with it for purposes of feeding, of shelter, or of predation; and many of these—including such a diverse assemblage as phalangids and spiders; mantids, stick-insects, and grass-hoppers; moths and caterpillars, weevils and longicorn beetles . . . are decorated in colour (and sometimes modified in form) so that they appear to resemble it, and are most difficult, and often almost impossible, to detect in their natural roosting, feeding, or hunting grounds. (Cott, 1957)

Photographic evidence of arthropods blending into their lichen background is shown by Cott (1957), Kettlewell (1959) and Farb (1963). Morphological modifications of the insect body, in addition to appropriate colour and disruptive patterns, intended to resemble the structure of lichens have seldom been documented. Chopard (1938) mentions a tettigoniid (Orthoptera), *Cymatomera*, whose femora are lobe-like, thus closely resembling lichens which occur on tree bark, and Richardson (1975) mentions South American bugs whose flat bodies resemble lichen thalli.

Cott (1957) pointed out that animals which mimic a certain background also live there; they are usually closely associated with their model. Regarding lichens, resemblance may be restricted to one stage in the development, or it may last throughout the animal's cycle. Some invertebrates only use the lichen background as a sanctuary for rest. Various moths may be mentioned in this context, as well as the flies Bullock (1966) saw resting on saxicolous lichens in Singapore, which simulated sporulating bodies until disturbed. Similar observations involving Collembola on corticolous and saxicolous substrates have been made by Seaward (1975 and unpublished data).

How does a moth or fly "recognize" the appropriate background on which to alight so as to become indiscernible to its predators? Cott (1957) wrote that a moth, whose wings resemble the bark of casuarina, "instinctively" settles with its body in a vertical position, thus bringing the pattern of the wings parallel to that of the bark, which they simulate. Cott also notes that moths whose wings resemble bark cracks settle on the bark in alignment with real cracks.

It is reasonable to postulate that any moth which settles otherwise would quickly be eliminated by predators, consequently selection would favour moths which settle in the "correct" position. Lichens, however, occur in a random pattern and are subject to succession and grazing. How then does a moth "recognize" the lichen or lichens which best suit its own wing pattern? This question becomes of special interest if one considers the situation in Brazil, where, according to Kettlewell (1959), multicoloured lichens occur on diverse trees, and all are utilized for defence mechanisms. Visual substrate recognition cannot be the entire answer. Had moths been able to find their resting sites by sight alone, the typical forms of the peppered moths (see below) would have found some other greenish substrates, and industrial melanism might never have evolved. It may well be that moths (and other insects) are attracted to the odour of specific lichen compounds. Specific moths may also settle on specific lichens in response to specific environmental factors required by the lichen. These could include humidity, temperature and aspect, as well as other stimuli resulting from factors which enable the lichen to settle and grow on any particular site.

D. Industrial Melanism

Industrialization, with its concomitant smoke and soot pollution, has brought about a breakdown in the concealment association between lichens and some insects. Lichens are extremely sensitive to such pollution; trees, rocks and soil, formerly covered by lichens in many parts of northern Europe, have become bare and bleak in industrial centres, subsequently assuming a blackened appearance due to incessant soot deposition. As lichen-covered surfaces were the normal resting places for many cryptic moths, the animals became exposed to visually directed predators, especially birds. The complicated, cryptic colour patterns on the moth's wings, which sometimes closely mimic foliose lichens, evolved under selection for concealment, and became a liability when observed against the newly blackened backgrounds. The mode of survival of many of these species, the best known being the peppered moth, *Biston betularia*, through the selection and subsequent success of dark, melanic forms, has become known as "industrial melanism" (Kettlewell, 1961).

Kettlewell (1956) showed that in unpolluted areas of British countryside, supporting a rich lichen flora, *B. betularia* occurs mostly in the form *typica*, which is practically impossible to see against a background of lichen-covered tree trunks. The dark, melanic form, *carbonaria*, also occurs from time to time, but due to its conspicuousness on the same lichen-covered trees, it is rapidly eliminated by bird predation. This form is maintained only by recurrent mutations and migration from polluted areas. When smoke and soot intrude, the trees blacken, the lichens begin to disappear, and the form *typica* becomes the more conspicuous of the two, *carbonaria* the better protected. The latter, or an intermediate melanic form, *insularia*, is consequently the dominant form of *B. betularia* in all industrialized and lichen-depleted regions of Britain (Kettlewell, 1973).

Despite these reports based on the association of high frequencies of melanics and absence of lichens, no detailed evidence was formerly available on the role that lichens actually play in determining the frequency of *carbonaria*. Using a method developed by Lees *et al*. (1973), Bishop *et al*. (1975) sampled lichen taxa on oak trunks along a transect from northwest Manchester to north Wales. They also recorded several physical variables, including the light reflectance off trees, and estimated the frequency of the *carbonaria* morph. All these data were subjected to a multiple regression analysis, which revealed the strong relationship between the frequency of melanic moths and lichen diversity. Furthermore they found that at about 75 km from Manchester the number of lichen taxa began to rise dramatically where the percentage occurrence of *carbonaria* declined significantly.

There is strong evidence here to support the hypothesis that the disappearance of lichens due to the increase in air pollution (mainly sulphur dioxide), rather than the direct blackening of substrate surfaces by the particulate matter of the latter (cf. Kettlewell, 1973), may have been the primary cause for the increase in melanic morphs in this species.

Popescu (personal communication) has studied the distribution of the psocid *Mesopsocus unipunctatus* in Yorkshire. This species occurs in two morphs: a light form which is concealed on lichen-covered bark, and a melanic form which blends well with bark blackened by soot. Results of a multiple regression analysis using 12 environmental variables show a significant relationship between increasing populations of the melanics with decreasing lichen cover and with increasing sulphur dioxide and smoke concentrations (cf. Bishop et al., 1975). Lichen cover thus appears to be critical for the maintenance of light-coloured morphs of *Biston betularia* and *Mesopsocus unipunctatus*.

The implementation of the Clean Air Act (1956) in recent years by the establishment of smokeless zones throughout the British Isles opens up some intriguing ecological contingencies. Should lichens return to at least some of their former diversity and abundance, moth melanics would again be at a serious disadvantage, and may suffer a severe eclipse. Between 1961 and 1974 the frequency of non-melanics of *Biston betularia* had increased from 5·2 to 10·5% in Cheshire (Bishop and Cook, 1975), and similar increases with this (e.g. Clark and Sheppard, 1966) and other insects have been noted in urban areas elsewhere. A reversal of industrial melanism is probably now taking place in the British Isles.

A further complication arises from the establishment of toxi-tolerant algae and the lichen *Lecanora conizaeoides* which create a monovegetational epiphytic cover and hence a further background (light to dark green) in which to blend. Such a condition, or the return of lichens in limited diversity and (or) abundance, may result in the evolution of another morph of *Biston betularia*. Some clue to this line of development may be provided from research on the morph *insularia* which was first noted in the British Isles about the turn of this century and is now considered to be a non-industrial melanic occurring at a low frequency. This morph appears to be admirably suited for resting on boughs covered with algae; background scoring efficiency gives *insularia* a position intermediate between *typica* and *carbonaria*, but *insularia* has never been found higher than 50% in British populations of *Biston betularia* (Kettlewell, 1973). Nevertheless, *insularia* is on the increase in many areas and is found most frequently at the present time on the edge of industrial areas but outside centres with a high frequency of *carbonaria*, in whose presence it is impossible to detect them phenotypically. Other work has shown areas where melanics are on the increase and those where they are decreasing: high *carbonaria* is found in non-industrial East Anglia and high *insularia* in industrial southwest England and South Wales, so other factors, possibly including climate and lichen diversity, are involved.

VII. Lichen–Invertebrate Communities

A. Terrestrial Colonization

Changes in invertebrate number and diversity during various stages of soil development were studied by Stebaev (1963). He divided the process of soil colonization by animals into discrete cycles. His Cycle II consisted of *Parmelia centrifuga* patches on granite rock. As *Parmelia* grew denser (the shift from Cycle IIa to IIb), there was a notable increase in animal numbers (Table VII). Later on, as mosses replaced the lichens, a substantial increase in invertebrate abundance and diversity took place. Similar observations were made by Gadea (1964c). Summing up his studies of high-mountain moss zoocoenoses, he stated that lichens provide a very poor biotope for the microfauna (aquafauna in the present context). It is only when bryophytes begin to colonize such habitats that animal populations increase.

Some invertebrates still play a lichen-mediated role in the early colonization of certain soils. Schmid (1929) noted the snails *Pyramidula rupestris* and *Chondrina avenacea*, which feed on the endolithic *Verrucaria calciseda* and on *Protoblastenia rupestris*. The snails graze the lichens' thalli and apothecia, voiding undigested algal remains in their faeces, which also contain chalk granules and detached radula teeth. These droppings form a suitable substrate for mosses as well as for higher plants. "From this it follows that snails are of great significance in the colonization of rock surfaces and in this way contribute to soil formation" (Kühnelt, 1976, p. 111).

To sum up invertebrates which live in lichen colonies contribute to soil formation by grazing on the plants, enriching the surrounding area by their faeces and cadavers, and by attracting predators to that site. A further contribution is made by animals which only shelter there.

B. Competition for Settling Sites

Within the littoral zone of rocky shores around the British Isles is a wide, black area of encrusting *Verrucaria* spp. The lichens intergrade with a zone of barnacles; as these are also sessile organisms, some competition for settling sites takes place there (Stephenson and Stephenson, 1972; Fletcher, 1973).

Such competition only appears to be of importance in the initial stages of colonization, because once established, lichens successfully resist further barnacle settlement (Fletcher, 1973). Resistance is therefore due mainly to a mechanical mechanism: young, mobile barnacles (spat) settle on lichen tufts, but few stay there for more than 2 weeks. Fletcher believes that this is due to a general imperfect adhesion of the spat to the lichen's surface. A further mechanism involving some repellent or inhibitory substance(s) produced by the lichens is possible. Any repellent effects can be expected to be more pronounced in larger, and thus older, lichen colonies. Indications of such chemical defence mechanisms are: (1) once established, lichens resist barnacle encroachment, and (2) limpets (*Patella*) which graze on colonizing lichens tend to avoid established colonies (Fletcher, 1973).

The term "ectocrine" was coined by Lucas (1947) to describe excretory substances of aquatic organisms which may have biological effects on other marine organisms. A review on algal ectocrines is available (Sieburth, 1968), but nothing seems to be known concerning analogous lichen products.

On the other hand, patches of *Lichina pygmaea* may occur completely surrounded by barnacle colonies (Stephenson and Stephenson, 1972). This suggests that *L. pygmaea* is not inhibitive towards invertebrates, an opinion supported by the large population of arthropods and molluscs obtained from that lichen (Colman, 1939).

C. Physical and Climatic Effects

Aspect of substrate affects the distribution and size of lichen patches, and thus also the populations of their associated invertebrates. The effect may be a simple one: larger lichen patches, growing on preferred aspects, can be expected to support larger and more diversified animal populations. Andre (1975) found the abundance of the cryptostigmatid *Carabodes labyrinthicus* on Belgian trees to be determined by corticolous lichen abundance, whereas the related *Dometorina plantivaga* appeared to prefer the lichens' southern colonies. The specific effect of aspect on certain groups of invertebrates may be more complicated. *Parmelia centrifuga* on granite rocks with a southern aspect maintained larger populations of prostigmatids, springtails and fly larvae than lichens on northern rocks, but the latter had more cryptostigmatids (Table VII; Stebaev, 1963). Tree face relative to compass-bearing did not appear to have any effect on tardigrade populations living in lichen colonies on Iowan trees. Tree height, however, affected the populations of one species, *Hypsibius tuberculatus*, which was most abundant at about 2 m above ground level. The populations of the four other tardigrades obtained were densest at ground level, decreasing with increasing lichen colony height on the trees (Kimmel and Meglitsch, 1969).

Data on the relative abundance of lichen-feeding psocids on larch trees with increasing altitude in Yorkshire were provided by Broadhead (1958). Turner and Broadhead (1974), working on the diversity and distribution of psocids on mango trees in Jamaica, showed that microepiphyte diversity (fungi, lichens and algae) based on percentage cover alone shows no relationship with either altitude or psocid diversity. However, epiphyte diversity calculated from epiphyte volume data increases significantly with increasing altitude and is significantly correlated with psocid diversity. Both the number of psocid species and their relative abundance appear therefore to be responding to changes in the complexity of their microhabitat which in turn are likely to be the product of changes in the climatic conditions prevailing at these various altitudes.

VIII. Discussion

A. Evolutionary Aspects

The prolonged association between lichens and invertebrates has had some evolutionary effects on these two groups of organisms. It is convenient to sum these aspects up under three headings: (1) concomitant evolution, (2) invertebrate adaptations to lichens, and (3) invertebrate effect on lichens.

1. Concomitant Evolution (in Sections II and III)

Many animals have been noted which have developed or were selected for extreme resistance to desiccation. This is also a lichen attribute. Components of the aquafauna are disseminated by winds—another lichen quality. At times a lichen propagule is blown about and settles while its protozoan, tardigradan or nematodan associate adheres to it. Thus it may be said that lichens and many members of its fauna were selected for the same characters, namely wind dispersal and the ability to withstand prolonged, intermittent periods of wetting and drying.

2. Invertebrate Adaptations to Lichens

The theory of lichen finding by invertebrates, whether for food or shelter, presumes that the animals have adapted their "life style" to life in and with lichens. The many mechanisms developed by arthropods to camouflage themselves with lichens and to mimic these cryptogams are all adaptations intended to exploit the protective presence of lichens. The recent rise to dominance of melanic forms of moths and psocids may be said to be an adaptation to the sudden disappearance of lichens.

The evolutionary effect of lichen diet on tineid moths (subfamily Meessiinae) was explored by Zagulyayev (1970). He stated that the transition in larval feeding habits from rotten wood and fungi to lichens must have necessitated a reorganization of the whole digestive system. The caterpillars also had to change their mode of life to carrying (and pupating in) cases which imitate their lichen substrate.

3. Invertebrate Effect on Lichens

Some authors (as summed up by Smith, 1921) believed that lichens produce their specific substances in response to invertebrate grazing. Although many arthropods and molluscs do feed on lichens, and the specific products are probably only by-products of lichen metabolism, this theory should not be discredited. It could probably be modified in the context of the "secondary plant substances" theory of Fraenkel (1959) mentioned above.

Formation of characteristic lichen galls in response to invertebrate feeders (and fungal parasites) suggests that this association is already old enough for lichens to react to their pests by characteristic and consistent growth.

B. Suggested Lines of Approach for Future Enquiry

The foregoing sections provide, at best, a rather incomplete picture of lichen–invertebrate associations. The reason for this fragmentary panorama is a distressing lack of experimental data about most aspects. Some of the more important areas in which information is in demand are:

1. Feeding and sheltering specificity.
2. Effect of lichen products on invertebrate grazers: specificity, mode of action, tolerance and possible detoxification.
3. Enzymatic studies on lichen digestion.
4. Energy budgets of lichen feeders, with associated physiological data.
5. Effect on grazers of radionuclides, heavy metals and pesticides accumulating in lichens.
6. Lichen dispersal through invertebrate agencies.
7. Effect of the pollution-mediated disappearance of lichens on dependent invertebrates.
8. Virgin lands, including deserts, forests and Arctic tundra are increasingly coming under "domestication". There are many indigenous lichens in these habitats harbouring invertebrate faunas. Data on these associations and the changes they undergo are unavailable.
9. Lichens may constitute the only source of energy in extreme-climate regions, such as hot and cold deserts, and yet very little seems to be known regarding food-webs emanating from such lichens. The same may be said about tropical areas, where invertebrates enjoy lush lichen growth.

10. Marine lichens offer opportunities to study some of the above problems under aquatic conditions.

There is a note of urgency in these proposals. Raven *et al.* (1971) point out that pollution, on a world scale, is increasing so rapidly that organisms are already becoming extinct at a high rate. Lichens are among the fastest disappearing organisms, and unless they, with all their unique associations, can be studied in the near future, we run the risk of not studying many of them at all.

Acknowledgements

We wish to express our thanks to Dr P. N. Hobson for placing unpublished data on lichen-feeding ruminants at our disposal, to Mr D. McFarlane for information on lichen-inhabiting mites and on *"Tetranychus lapidus"*, to Dr E. Broadhead for comments on the section dealing with Psocoptera and the provision of Fig. 1, to Dr J. Travé for permission to reproduce Figs 2 and 3, to Miss C. Popescu, Dr L. M. Cook, Dr J. A. L. Cooke and Mr E. L. Seyd for their helpful comments, and to Dr D. L. Hawksworth for valuable criticism of the draft manuscript.

References

Ali, S. M., Farooqui, M. N. and Tejpal, S. (1970). A report of *Laimaphelenchus penardi* (Steine, 1914) Fil. Sch. Stek. 1941 (Nematoda: Aphelenchoididae) from India. *Marathwada Univ. J. Sci.* **9**, 43–44.

Andre, H. (1975). Observations sur les Acariens corticoles de Belgique. Fondation Univ. Luxemburgeoise, Notes Res. No. 4.

Anonymous (1970). "The Insects of Australia". University Press, Melbourne.

Argue, C. W. (1971). Some terrestrial tardigrades from New Brunswick, Canada. *Can. J. Zool.* **49**, 401–415.

Bachmann, E. (1929). Tiergallen auf Flechten. *Arch. Protistenk.* **66**, 61–103.

Bailey, R. H. (1970). Animals and the dispersal of soredia from *Lecanora conizaeoides* Nyl. ex Cromb. *Lichenologist* **4**, 256.

Bailey, R. H. (1976). Ecological aspects of dispersal and establishment in lichens. *In* "Lichenology: Progress and Problems" (D. H. Brown, D. L. Hawksworth and R. H. Bailey, eds), pp. 215–247. Academic Press, London and New York.

Baker, E. W. and Pritchard, A. E. (1953). A guide to the spider mites of cotton. *Hilgardia* **22**, 203–234.

Barrett, G. W. and Kimmel, R. G. (1972). Effects of DDT on the density and diversity of tardigrades. *Proc. Iowa Acad. Sci.* **78**, 41–42.

Béique, R. and Francoeur, A. (1966). Les fourmis d'une pessière à *Cladonia* (Hymenoptera: Formicidae). *Nature Can.* **93**, 99–106.

Benedetti, R. (1973). Notes on the biology of *Neomachilis halophila* on a California sandy beach (Thysanura: Machilidae). *Pan-Pacific Ent.* **49**, 246–249.

Bengtson, S.-A., Fjellberg, A. and Solhöy, T. (1974). Abundance of tundra arthropods in Spitsbergen. *Entomologica scand.* **5**, 137–142.

Bishop, J. A. and Cook, L. M. (1975). Moths, melanism and clean air. *Scient. Am.* January, 90–99.

Bishop, J. A., Cook, L. M., Muggleton, J. and Seaward, M. R. D. (1975). Moths, lichens and air pollution along a transect from Manchester to North Wales. *J. appl. Ecol.* **12**, 83–98.

Borror, D. J. and DeLong, D. M. (1971). "An Introduction to the Study of Insects", 3rd ed. Holt, Rinehart and Winston, New York.

Bouly de Lesdain, M. (1924). Écologie d'une aulnaie dans les Moëres (Nord). *Bull. Soc. bot. Fr.* **71**, 3–25.

Brightman, F. H. (1965). Insect on lichens. *Lichenologist* **3**, 154.

Broadhead, E. (1958). The psocid fauna of larch trees in northern England. An ecological study of mixed species populations exploiting a common resource. *J. anim. Ecol.* **27**, 217–263.

Brodo, I. M. and Hawksworth, D. L. (1977). *Alectoria* and allied genera in North America. *Op. bot. Soc. bot. Lund.* **42**, 1–164.

Browning, E. (1950). "Blood from a stone"—a mite causes a stir. *Ill. Lond. News*, 11/3/1950.

Bullock, J. A. (1966). Observations on the fauna of Pulau Tioman and Pulau Tulai. 9. Introductory report on the terrestrial arthropods. *Bull. natn. Mus. St. Singapore* **34**, 104–128.

Byazrov, L. G., Martynova, E. F. and Medvedev, L. N. (1976). Nogokhvoski (Collembola) v lishainikovykh sinuziyakh Khangaya (MNR). [Collembola in the lichen sinusia of Hangai (Mongolian People's Republic)]. *Byull. Mosk. Obshch. Ispyt. Prir., Otdel Biol.* **81** (3), 66–73 [In Russian].

Chopard, L. (1938). "La Biologie des Orthoptères". Lechevalier, Paris.

Clarke, C. A. and Sheppard, P. M. (1966). A local survey of the distribution of the industrial melanic forms in the moth *Biston betularia* and estimates of the selective values of these in an industrial environment. *Proc. R. Soc., ser. B* **165**, 424–439.

Coineau, Y. (1973). Les Caeculidae (Acariens Prostigmates) quelques aspects de leurs particularités éco-éthologiques. *Bull. Ecol.* **4**, 329–337.

Coker, P. D. (1967). Damage to lichens by gastropods. *Lichenologist* **3**, 428–429.

Coleman, M. J. and Hynes, H. B. N. (1970). The life histories of some Plecoptera and Ephemeroptera in a southern Ontario stream. *Can. J. Zool.* **48**, 1333–1339.

Collins, N. J., Baker, J. H. and Tilbrook, P. J. (1975). Signy Island, Maritime Antarctic. *Bull. Ecol.* **20**, 345–374.

Colman, J. (1939). On the faunas inhabiting intertidal seaweeds. *J. mar. biol. Ass. U.K.* **24**, 129–183.

Cott, H. B. (1957). "Adaptive Coloration in Animals". Methuen, London.

Cuénot, L. (1949). Les Tardigrades. *In* "Traité de Zoologie" (P.-P. Grassé, ed.), Vol. 6, pp. 39–59. Paris.

Culberson, C. F. (1969). "Chemical and Botanical Guide to Lichen Products". University of North Carolina Press, Chapel Hill.
Culberson, C. F. (1970). Supplement to "Chemical and Botanical Guide to Lichen Products". *Bryologist* **73**, 177–377.
Cunningham, D. D. (1907). "Plagues and Pleasures of Life in Bengal". Murray, London.
De Boer, S. (1975). Breeding *Nudaria mundana* L. (Lep., Arctiidae). *Ent. Ber., Amst.* **35** (12), 181–182.
Decloitre, L. (1962). Sur quelques Thécamoebiens récoltés en Norvège. *Hydrobiologia* **19**, 179–182.
Denison, W. C. (1973). Life in tall trees. *Scient. Am.* June, 74–80.
Dev, H. N. (1964). Preliminary studies on the biology of the Assam thrips, *Scirtothrips dorsalis* Hood on tea. *Indian J. Ent.* **26**, 184–194.
Engelmann, M. D. (1966). Energetics, terrestrial field studies, and animal productivity. *Adv. ecol. Res.* **3**, 73–115.
Engelmann, M. D. (1968). The role of soil arthropods in community energetics. *Am. Zool.* **8**, 61–69.
Essig, E. O. (1958). "Insects and Mites of Western North America". Macmillan, New York.
Evans, G. O., Sheals, J. G. and Macfarlane, D. (1961). "The Terrestrial Acari of the British Isles", Vol. I. British Museum (Nat. Hist.), London.
Fain, A. (1975). Acariens récoltés par le Dr. J. Travé aux Iles Subantarctiques. I. Familles Saproglyphidae et Hyadesidae. *Acarologia* **16**, 684–708.
Farb, P. (1963). "Ecology". Time-Life International, New York and Amsterdam.
Fletcher, A. (1973). The ecology of marine (littoral) lichens on some rocky shores of Anglesey. *Lichenologist* **5**, 368–400.
Forman, R. T. T. (1975). Canopy lichens with blue-green algae: a nitrogen source in a Colombian rain forest. *Ecology* **56**, 1176–1184.
Fraenkel, G. S. (1959). The *raison d'être* of secondary plant substances. *Science, N.Y.* **129**, 1466–1470.
Frankland, J. C. (1974). Decomposition of lower plants. *In* "Biology of Plant Litter Decomposition" (C. H. Dickinson and G. J. F. Pugh, eds), pp. 3–36. Academic Press, London and New York.
Gadea, E. (1964a). Sobre la nematofauna muscicola y liquenicola de las islas Pitiusas. *Publ. Inst. Biol. Apl.* **37**, 73–93.
Gadea, E. (1964b). El poblamiento animal liquenicola en pequeños islotes del Mediterraneo español. *Bol. Real Soc. Españ. Hist. Natur., Secc. Biol.* **62**, 333–336.
Gadea, E. (1964c). La zoocenosis muscicola en los biotopos altimontanos. *Publ. Inst. Biol. Apl.* **36**, 113–120.
Gadea, E. (1973). Sobre la nematofauna liquenicola de Lanzarote (Islas Canarias). *Miscelenea zool.* **2** (6), 3–6.
Gadea, E. (1974). Nematodes liquenicolas de Columbretes. *Miscelenea zool.* **3** (4), 3–8.
Garrett, R. M. (1974). A species of *Chrysopa* of special interest to lichenologists. *Revue bryol. lichén.* **40**, 283–286.

Gerson, U. (1968). Five tydeid mites from Israel (Acarina: Prostigmata). *Israel J. Zool.* **17**, 191–198.
Gerson, U. (1971). The mites associated with citrus in Israel. *Israel J. Ent.* **6**, 5–22.
Gerson, U. (1972). Mites of the genus *Ledermuelleria* (Prostigmata: Stigmaeidae) associated with mosses in Canada. *Acarologia* **13**, 319–343.
Gerson, U. (1973). Lichen-arthropod associations. *Lichenologist* **5**, 434–443.
Gilbert, O. L. (1971). Some indirect effects of air pollution on bark-living invertebrates. *J. appl. Ecol.* **8**, 77–84.
Golley, F. B. (1961). Energy values of ecological materials. *Ecology* **42**, 581–584.
Goodey, T. (1963). "Soil and Freshwater Nematodes". Methuen, London.
Grandjean, F. (1950). Observations éthologiques sur *Camisia segnis* (Herm.) et *Platynothrus peltifer* (Koch) (Acariens). *Bull. Mus. Hist. nat., Paris*, **22**, 224–231.
Gressitt, J. L. (1966). Epizoic symbiosis: the Papuan weevil genus *Gymnopholus* (Leptopiinae) symbiotic with cryptogamic plants, oribatid mites, rotifers and nematodes. *Pacif. Insects* **8**, 221–280.
Gressitt, J. L. (1967). Notes on arthropod populations in the Antarctic Peninsula–South Shetland Islands–South Orkney Islands area. *In* "Entomology in Antarctica" (J. L. Gressitt, ed.), pp. 373–391. Washington, D.C.
Gressitt, J. L. and Sedlacek, J. (1970). Papuan weevil genus *Gymnopholus*: second supplement with studies in epizoic symbiosis. *Pacif. Insects* **12**, 753–762.
Gressitt, J. L. and Shoup, J. (1967). Ecological notes on free-living mites in North Victoria Land. *In* "Entomology in Antarctica" (J. L. Gressitt, ed.), pp. 307–320. Washington, D.C.
Gressitt, J. L., Sedlacek, J. and Szent-Ivany, J. J. H. (1965). Flora and fauna on backs of large Papuan moss-forest weevils. *Science, N.Y.* **150**, 1833–1835.
Grummann, V. J. (1960). Die Cecidien auf Lichenen. *Bot. Jb.* **80**, 101–144.
Grzyb, Z. S. (1964). Wpływ wyciagow porostowych na żywotność i tempo rozwoju jaj *Ascaris*. *Wiad. parazyt.* **10**, 69–77.
Hale, M. E. (1972). Natural history of Plummers Island, Maryland. XXI. Infestation of the lichen *Parmelia baltimorensis* Gyel. & For. by *Hypogastrura packardi* Folsom (Collembola). *Proc. biol. Soc. Wash.* **85**, 287–296.
Hale, W. G. (1963). The Collembola of eroding blanket bog. *In* "Soil Organisms" (J. Doeksen and J. van der Drift, eds), pp. 406–413. North-Holland, Amsterdam.
Harding, D. J. L. and Stuttard, R. A. (1974). Microarthropods. *In* "Biology of Plant Litter Decomposition" (C. H. Dickinson and G. J. F. Pugh, eds), Vol. II, pp. 489–532. Academic Press, London and New York.
Hawksworth, D. L. (1976). Lichen chemotaxonomy. *In* "Lichenology: Progress and Problems" (D. H. Brown, D. L. Hawksworth and R. H. Bailey, eds), pp. 139–184. Academic Press, London and New York.
Heal, R. E., Rogers, E. F., Wallace, R. T. and Starnes, O. (1950). A survey of plants for insecticidal activity. *Lloydia* **13**, 89–162.

Healey, I. N. (1967). The population metabolism of *Onychiurus procampatus* Gisin (Collembola). *In* "Progress in Soil Zoology" (O. Graff and J. E. Satchell, eds), pp. 127–134. North-Holland, Amsterdam.

Heinis, F. (1959). Beitrag zur Mikrobiocoenose der Schneetälchen auf Macun (Unterengadin). *Ber. geobot. Forsch. Inst. Rübel* **1958**, 110–123.

Holden, M. and Tracey, M. V. (1950). A survey of enzymes that can break down tobacco-leaf components. 2. Digestive juice of *Helix* on defined substances. *Biochem. J.* **47**, 407–414.

Holland, W. J. (1908). "The Moth Book". Doubleday, New York.

Holloway, B. A. (1970). A new genus of New Zealand Anthribidae associated with lichens (Insecta: Coleoptera). *N.Z. Jl Sci. Technol.* **13**, 435–446.

Hughes, A. M. and Tilbrook, P. J. (1966). A new species of *Calvolia* (Acaridae, Acarina) from the South Sandwich Islands. *Bull. Br. Antarct. Surv.* **10**, 45–53.

Iharos, G. (1968). Ergebnisse der Zoologischen Forschungen von Dr. Z. Kaszab in der Mongolei 162. Tardigrada, II. *Opusc. zool. Bpest* **8**, 31–35.

James, P. W. (1973). The effect of air pollutants other than hydrogen fluoride and sulphur dioxide on lichens. *In* "Air Pollution and Lichens" (B. W. Ferry, M. S. Baddeley and D. L. Hawksworth, eds), pp. 143–175. Athlone Press, London.

James, P. W. and Henssen, A. (1976). The morphological and taxonomic significance of cephalodia. *In* "Lichenology: Progress and Problems" (D. H. Brown, D. L. Hawksworth and R. H. Bailey, eds), pp. 27–77. Academic Press, London and New York.

Jobin, L. (1973). "L'Arpenteuse de la Pruche". Ministry of the Environment, Quebec.

Kalshoven, L. G. E. (1958). Observations on the black termites, *Hospitalitermes* sp., of Java and Sumatra. *Insectes soc.* **5**, 9–30.

Kettlewell, H. B. D. (1956). Further selection experiments on industrial melanism in the Lepidoptera. *Heredity* **10**, 287–301.

Kettlewell, H. B. D. (1959). Brazilian insect adaptations. *Endeavour* **18**, 200–210.

Kettlewell, H. B. D. (1961). The phenomenon of industrial melanism in Lepidoptera. *A. Rev. Ent.* **6**, 245–262.

Kettlewell, H. B. D. (1973). "The Evolution of Melanism". Oxford University Press, Oxford.

Kimmel, R. G. and Meglitsch, P. A. (1969). Notes on Iowa tardigrades. *Proc. Iowa Acad. Sci.* **76**, 454–462.

Kühnelt, W. (1976). "Soil Biology, with special reference to the Animal Kingdom", 2nd ed. Faber, London.

Laminger, H. (1971). Über das Vorkommen von Schalenamoeben (Rhizopoda, Testacea) in alpinen Flechten. *Zool. Anz.* **186**, 335–337.

Laundon, J. R. (1967). A study of the lichen flora of London. *Lichenologist* **3**, 277–327.

Laundon, J. R. (1971). Lichen communities destroyed by psocids. *Lichenologist* **5**, 177.

Lees, D. R., Creed, E. R. and Duckett, J. G. (1973). Atmospheric pollution and industrial melanism. *Heredity* **30**, 227–232.

Llano, G. A. (1956). Utilization of lichens in the Arctic and Subarctic. *Econ. Bot.* **10**, 367–392.
Lucas, C. E. (1947). The ecological effects of external metabolites. *Biol. Rev.* **22**, 270–295.
Luxton, M. (1972). A re-description of *Cryptognathus lagena* Kramer 1879 (Acari: Prostigmata: Cryptognathidae). *Acarologia* **14**, 591–594.
MacNeill, N. (1966). Mites (Acari) on lichen. *Ir. Nat. J.* **15**, 242–243.
Mani, M. S. (1962). "Introduction to High Altitude Entomology". Methuen, London.
Mani, M. S. (1964). "Ecology of Plant Galls". W. Junk, The Hague.
Marcus, E. (1959). Tardigrada. *In* "Fresh-water Biology" (W. T. Edmonson, ed.), pp. 508–521. Wiley, New York.
Morello, J. H. (1970). Ecologia del Chaco. *Boln Soc. argent. Bot.* **II** (Supp.), 161–174.
Morton, J. E. (1954). The crevice faunas of the upper intertidal zone at Wembury. *J. mar. biol. Ass. U.K.* **33**, 187–224.
Morton, J. E., Boney, A. D. and Corner, E. D. S. (1957). The adaptations of *Lasaea rubra* (Montagu), a small intertidal lamellibranch. *J. mar. biol. Ass. U.K.* **36**, 383–405.
Nielsen, C. O. (1963). Laminarinases in soil and litter invertebrates. *Nature, Lond.* **199**, 1001.
Nielsen, C. O. (1967). Nematoda. *In* "Soil Biology" (A. Burges and F. Raw, eds), pp. 197–211. Academic Press, London and New York.
Nosek, J. (1973). "The European Protura, their Taxonomy, Ecology and Distribution, with Keys for Determination". Muséum d'Histoire Naturelle, Genève.
Owen, G. (1966). Feeding. *In* "Physiology of Mollusca" (K. M. Wilbur and C. M. Yonge, eds), Vol. II, pp. 1–51. Academic Press, New York and London.
Peake, J. F. and James, P. W. (1967). Lichens and Mollusca. *Lichenologist* **3**, 425–428.
Plitt, C. C. (1934). A lichen-eating snail. *Bryologist* **37**, 102–104.
Pyatt, F. B. (1968). The occurrence of a rotifer on the surfaces of apothecia of *Xanthoria parietina*. *Lichenologist* **4**, 74–75.
Raven, P. H., Berlin, B. and Breadlove, D. E. (1971). The origins of taxonomy. *Science, N.Y.* **174**, 1210–1213.
Richardson, D. H. S. (1975). "The Vanishing Lichens. Their History, Biology and Importance". David and Charles, Newton Abbot.
Ridley, H. N. (1930). "The Dispersal of Plants Throughout the World". Reeve and Co., Ashford.
Ross, E. S. (1966). A new species of Embioptera from the Galápagos Islands. *Proc. Calif. Acad. Sci.* **34**, 499–504.
Schade, A. (1933). Flechtensystematik und Tierfrass. *Ber. dt. bot. Ges.* **51**: 168–192.
Schmid, G. (1929). Endolithische Kalkflechten und Schneckenfrass. *Biol. Zbl.* **49**, 28–35.

Schuster, R. O. and Grigarick, A. A. (1966). Tardigrada from the Galápagos and Cocos Islands. *Proc. Calif. Acad. Sci.* **34**, 315–328.

Scotter, G. W. (1965). Chemical composition of forage lichens from northern Saskatchewan as related to use by barren-ground caribou. *Can. J. Pl. Sci.* **45**, 246–250.

Seaward, M. R. D. (1965). Lincolnshire psocids. *Trans. Lincs. Nat. Un.* **16**, 99–100.

Seaward, M. R. D. (1972). Aspects of urban lichen ecology. Ph.D. thesis, University of Bradford.

Seaward, M. R. D. (1974). A note on *Phauloppia lucorum* C. L. Koch (Acari: Oribatei) and lichens. *Lichenologist* **6**, 126–127.

Seaward, M. R. D. (1975). Contributions to the lichen flora of southeast Ireland —I. *Proc. R. Ir. Acad.* **75B**, 185–205.

Séguy, E. (1950). "La Biologie des Diptères". Lechevalier, Paris.

Seyd, E. L. (1968). Studies on the moss mites of Snowdonia (Acari: Oribatei). I. Moel Hebog. *Entomologist* **101**, 37–41.

Shilova, S. A., Denisova, A. V., Sedykh, E. L. and Efron, K. M. (1973). After-effects of insecticide treatment in the Subarctic. *Zool. Zh.* **52**, 1008–1012 [In Russian].

Sieburth, J. McN. (1968). The influence of algal antibiosis on the ecology of marine microorganisms. *Adv. Sea Microbiol.* **1**, 63–94.

Skorepa, A. C. and Sharp, A. J. (1971). Lichens in "packets" of lacewing larvae (Chrysopidae). *Bryologist* **74**, 363–364.

Slobodkin, L. B. and Richman, S. (1961). Calories/gm. in species of animals. *Nature, Lond.* **191**, 299.

Smith, A. L. (1921). "Lichens". Cambridge University Press, Cambridge.

Southwood, T. R. E. and Leston, D. (1959). "Land and Water Bugs of the British Isles". Warne, London.

Stahl, G. E. (1904). Die Schutzmittel der Flechten gegen Tierfrass. In "Festschrift z. 70. Geburtstage von Ernst Haeckel", pp. 357–375. G. Fischer, Jena.

Stebaev, I. V. (1963). Die Veränderung der Tierbevölkerung der Boden im Laufe der Bodenentwicklung auf Felsen und auf Verwitterungsprodukten im Wald-Wiesenlandschaften des Süd-Urals. *Pedobiologia* **2**, 265–309 [In Russian].

Stephenson, T. A. and Stephenson, A. (1972). "Life Between Tidemarks on Rocky Shores". Freeman, San Francisco.

Stout, J. D. and Heal, D. W. (1967). Protozoa. In "Soil Biology" (A. Burges, and F. Raw, eds), pp. 149–195. Academic Press, New York and London.

Strong, J. (1967). Ecology of terrestrial arthropods at Palmer Station, Antarctic Peninsula. In "Entomology in Antarctica" (J. L. Gressitt, ed.), pp. 357–371. Washington, D.C.

Thompson, J. C. (1960). Ciliated Protozoa from tree borne mosses and lichens. *J. Protozool.* **7**(Supp.), 17.

Thomson, M. G. (1958). Egg sampling for the western hemlock looper. *For. Chron.* **34**, 248–256.

Tilbrook, P. J. (1967). The terrestrial invertebrate fauna of the Maritime Antarctic. *Phil. Trans. R. Soc. B*, **252**, 261–278.

Travé, J. (1963). Écologie et biologie des Oribates (Acariens) saxicoles et arboricoles. *Vie Milieu* Supp. 14, 1–267.

Travé, J. (1969). Sur le peuplement des lichens crustacés des Iles Salvages par les Oribates (Acariens). *Rev. Ecol. Biol. Sol.* **6**, 239–248.

Tuominen, Y. and Jaakkola, T. (1974) "[1973"]. Absorption and accumulation of mineral elements and radioactive nuclides. *In* "The Lichens" (V. Ahmadjian and M. E. Hale, eds), pp. 185–223. Academic Press, New York and London.

Turner, B. D. and Broadhead, E. (1974). The diversity and distribution of psocid populations on *Mangifera indica* L. in Jamaica and their relationship to altitude and micro-epiphyte diversity. *J. anim. Ecol.* **43**, 173–190.

Watson, G., Davis, J. J. and Hanson, W. C. (1966). Terrestrial Invertebrates. *In* "Environment of the Cape Thompson Region, Alaska" (N. J. Wilimovsky and J. N. Wolf, eds), pp. 565–584. U.S. Atomic Energy Commission.

Weber, W. A. (1974). Two lichen-arthropod associations in Australia and New Guinea. *Lichenologist* **6**, 168.

Wheldon, J. A. (1914). Stone mites in West Lancashire. *Lancs. Chesh. Nat.* 7, 31–32.

Wielgolaski, F. E. (1975). Functioning of Fennoscandian tundra ecosystems. *In* "Fennoscandian Tundra Ecosystems. Part 2: Animals and Systems Analysis" (F. E. Wielgolaski, ed.), pp. 300–326. Springer Verlag, Berlin.

Wieser, W. (1963). Adaptations of two intertidal isopods. II. Comparison between *Campecopea hirsuta* and *Naesa bidentata* (Sphaeromatidae) *J. mar. biol. Ass. U.K.* **43**, 97–112.

Wigglesworth, V. B. (1972). "The Principles of Insect Physiology", 7th ed. Chapman and Hall, London.

Woodring, J. P. (1963). The nutrition and biology of saprophytic Sarcoptiformes. *Adv. Acarol.* **1**, 89–111.

Woodring, J. P. and Cook, E. F. (1962). The biology of *Ceratozetes cisalpinus* Berlese, *Scheloribates laevigatus* Koch, and *Oppia neerlandica* Oudemans (Oribatei), with a description of all stages. *Acarologia* **4**, 101–137.

Zagulyayev, A. K. (1970). Two new primitive species of lichenophagous moths (Lepidoptera, Tineidae) from the wet forests of Azerbaidzhan. *Ent. Rev., Wash.* **49**, 408–411.

Zopf, W. (1896). Uebersicht der auf Flechten schmarotzenden Pilze. *Hedwigia* **35**, 312–366.

Zopf, W. (1907). Biologische und morphologische Beobachtungen an Flechten. III. Durch tierische Eingriffe hervorgerufene Gallenbildungen an Vertreten der Gattung *Ramalina*. *Ber. dt. bot. Ges.* **25**, 233–237.

Zukal, H. (1895). Morphologische und biologische Untersuchungen über die Flechten. *Sber. K. Böhm. Ges. Wiss. Math.-nat. Kl.* **104**, 1303–1395.

5 | *Lichens and Vertebrates*

DAVID H. S. RICHARDSON and
COLIN M. YOUNG

I. Introduction	121
II. Amphibians	122
III. Reptiles	124
IV. Birds	125
V. Mammals	129
A. Reindeer and Caribou	129
B. Other Large Mammals	133
C. Rodents and Bats	134
D. Man	136
Acknowledgements	138
References	139

I. Introduction

The important relationships that exist between lichens and many vertebrates have been largely ignored since the time of the natural historians in the nineteenth century who first made observations on this subject. However, the recent stress on the ecosystem approach has resulted in further scattered reports. The exception to this general statement concerns studies on the interrelation between lichens and reindeer. The literature here is so extensive that only a summary of the present position will be given together with references to direct the reader further. The purpose of this chapter is to review the observations that have been made in the hope that it will stimulate research into this interesting aspect of lichenology.

II. Amphibians

There are no recorded cases of these animals eating lichens (with one possible exception), but several have evolved an effective camouflage which resembles lichen-covered trees or rocks.

The common or gray tree-frog of North America, *Hyla versicolor*, is reported to be so well camouflaged that it might easily be mistaken for a stone or piece of bark covered with lichen (Wright and Wright, 1949); others consider it patterned to resemble the lichened bark of old trees on which it rests during the heat of the day (Cochran, 1961). This frog (Fig. 1), about 5 cm long, is found from southern Canada to the northern parts of Florida, and its ability to remain absolutely still, except for the vibration of the throat during breathing, renders its camouflage highly successful. A related species, the bird-voiced tree-frog *H. avivoca*, is very similar in appearance but smaller and is found in the Mississippi Valley.

Another example of camouflage is that of the green salamander *Aneides aeneus* which looks like a piece of lichen as it lurks among lichen-covered rocks (Fig. 2). It is the only green salamander from the U.S.A., being found in the Appalachians and from Ohio and Pennsylvania to Oklahoma. Another species, the arboreal salamander *A. lugubris*, is not so strikingly camouflaged, but is unusual in breathing through its toes and climbing 18 m up trees in search of fungus (and perhaps lichens) growing in hollows in the tree where it may also lay its eggs. The fungus supplements a diet of invertebrates and the arboreal salamander has evolved large sabre teeth in the lower jaw, useful in scraping off the fungi (Cochran, 1961).

Some of the flying frogs (*Rachophorus* spp.) of Asia live in the topmost branches of the forest, leaping and gliding among the trees in search of insects. Many have considerable facilities for changing their hues, being green as a leaf on one occasion and grey and spotted on another (Lydekker, 1896).

The bony-headed tree-frogs, from Central and South America, also live in tall trees, concealing themselves during the day so that they are seldom seen except by scientific collectors. *Trachycephalus lichenatus*, from Jamaica, lives in the wild pines of that region and has a long, bony casque at the top of its head which makes it look as if overgrown with lichen (Wood, 1863). Finally, some of the Asiatic horned frogs (*Megophrys* spp.) include several odd-looking members that mimic the shapes and colours of leaves (and perhaps foliicolous lichens and algae).

FIG. 1. *Hyla versicolor*, the common or gray tree-frog of North America, reproduced from Dickerson (1920).

Fig. 2. *Aneides aeneus*, the green salamander from North America, reproduced from Cochran (1961).

III. Reptiles

The coloration of many small lizards makes them difficult to spot against rocks and walls. The most exquisite example of a camouflaged reptile is *Agama atricolis* which lives in the rain-forest belt around Catuane in Mozambique. This lizard suns itself on tree stumps or runs on the rough trunks of the trees where its coloration harmonizes wonderfully with the lichen-encrusted bark (Cott, 1957). The ground colour of the animal is brown or olive to silvery grey, broken firstly by irregular patches of silvery green and sepia, and secondly by scattered blue-green and yellow scales plus a conspicuous black area on the shoulder. The head ranges from grey to blue or emerald, while the tail is similar but broken with silvery bands. A similar colour camouflage is found among the iguanas, e.g. *Anolis ortonii* from the Amazon Valley.

Another group of reptiles, the geckos, includes one from Queensland, Australia, which most perfectly resembles lichen and moss-covered bark. This animal, a leaf-tailed gecko, is up to 0·3 m long and lives in the rain forest (Fig. 3; Beste and Beste, 1974). A relative, the Malagassy bark gecko, *Uroplates fimbriatus*, which has lateral flap-like extensions to the tail, is similarly camouflaged (Cott, 1957), and also shows close adpression of the body to the bark.

FIG. 3. The leaf-tailed gecko from Australia, reproduced from Beste and Beste (1974).

The giant land tortoise *Geochelone elephantopus* was discovered in 1964 to serve as host for the lichen *Dirinaria picta* (Hendrickson and Weber, 1964). This observation is the only recorded instance of lichens growing on a living land animal. The lichen was only found on the upper rear carapace of the males. This is because *D. picta* cannot develop on the lower part of the shell since the animal has a habit of lying, for long periods, in shallow freshwater pools during the rainy season; also the lichen is abraded from the front parts of the carapace as the tortoises move through the vegetation. Finally the lichen is removed from the rear part of the females during attempts at copulation. These tortoises, so colonized, live on the island of Santa Cruz of the Galapagos group.

IV. Birds

The interrelation between lichens and birds is manifest in terms of (1) material for nest building; (2) camouflage; and (3) feeding behaviour. In every continent (except perhaps Antarctica), the number of birds which

TABLE I.

American birds which make their nests partly or mainly of lichens.

Common merganser	*Mergus merganser*
Red-shouldered hawk	*Buteo lineatus*
Red-bellied hawk	*Buteo lineatus elegans*
Swainson's hawk	*Buteo swainsoni*
Ruby-throated hummingbird	*Archilochus colubris*
Black-chinned hummingbird	*Archilochus alexandri*
Anna's hummingbird	*Calytpe anna*
Broad-tailed hummingbird	*Selasphorus platycercus*
Rufous hummingbird	*Selasphorus rufus*
Allen's hummingbird	*Selasphorus sasin*
Rivoli's hummingbird	*Eugenes fulgens*
Buff-bellied hummingbird	*Amazilia yucatanensis*
White-eared hummingbird	*Hylocharis leucotis*
Broad-billed hummingbird	*Cynanthus latirostris*
Coue's flycatcher	*Contopus pertinax*
Eastern wood pewee	*Contopus virens*
Olive-sided flycatcher	*Nuttallornis borealis*
Vermilion flycatcher	*Pyrocephalus rubinus*
Boreal chickadee	*Parus hudsonicus*
Common bushtit	*Psaltriparus minimus*
Black-eared bushtit	*Psaltriparus melanotis*
Wrentit	*Chamaea fasciata*
Varied thrush	*Ixoreus naevius*
Swainson's thrush	*Hylocichla ustulata*
Gray-cheeked thrush	*Hylocichla minima*
Blue-gray gnatcatcher	*Polioptila caerulea*
Black-tailed gnatcatcher	*Polioptila melanura*
Golden-crowned kinglet	*Regulus satrapa*
Silky flycatcher	*Phainopepla nitens*
White-eyed vireo	*Vireo griseus*
Yellow-throated vireo	*Vireo flavifrons*
Solitary vireo	*Vireo solitarius*
Black-whiskered vireo	*Vireo altiloquus*
Philadelphia vireo	*Vireo philadelphicus*
Parula warbler	*Parula americana*
Olive warbler	*Peucedramus taeniatus*
Townsend's warbler	*Dendroica townsendi*
Hermit warbler	*Dendroica occidentalis*
Cerulean warbler	*Dendroica cerulea*
Blackburnian warbler	*Dendroica fusca*
Yellow-throated warbler	*Dendroica dominica*

5. Lichens and Vertebrates

TABLE I—continued

Blackpoll warbler	*Dendroica striata*
American redstart	*Setophaga ruticilla*
Rusty blackbird	*Euphagus carolinus*
White-winged crossbill	*Lotia leucoptera*

From Davie (1898) and Headstrom (1970); names according to the A.O.U. Checklist of North American Birds (fifth edition, 1957).

use lichen for nest-building is surprisingly large. Table I lists some North American examples while those from Europe and Australia may be found by consulting Harrison (1975) and Campbell (1900) respectively. Lichens are not incorporated in nests (Fig. 4) merely because they are abundant, easily plucked and carried by birds. Hawksworth (personal communication) has observed that long-tailed tits (*Aegithalos caudatus*) in south Devon selectively collect *Evernia prunastri*. Furthermore, in central Ireland finches which build nests where lichen epiphytes are scarce were found to visit widely distributed trees to obtain sufficient amounts of *Ramalina farinacea* and *Parmelia perlata* for their nests (Seaward, *in litt.*).

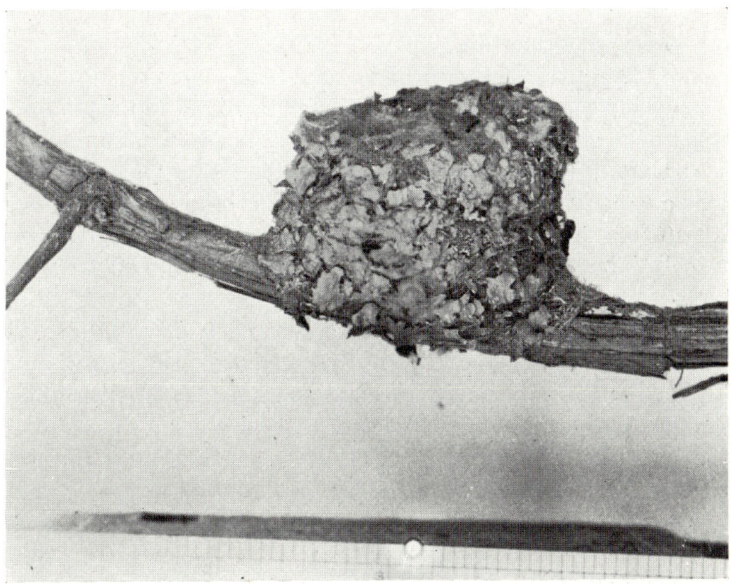

FIG. 4. Lichen-covered hummingbird nest, *c*. 5 cm diameter, from Central America.

In more tropical areas the silky flycatchers make nests using considerable quantities of lichen (Skutch, 1965), whereas hummingbirds generally make theirs from other plant materials and then attach fragments of lichen on the outside using cobweb (Fig. 4). Thus hummingbirds around Pueblo, Mexico, construct a nest of tree-fern scales and attach fragments of *Physcia setosa* to the outside (Riba and Herrera, 1973). It appears that the use of lichen provides camouflage which makes the nests difficult to find so that predators are less likely to eat the eggs. The swiftlets of Southeast Asia incorporate various types of fibrous vegetation including lichens in their nests (Medway, 1969). Perhaps the latter contribute to the flavour of the famous soup made from the nests.

The golden plovers of St. Lawrence Island in the Bering Sea not only make their nests from the lichen *Thamnolia vermicularis* s.l. and select spots where conspicuous bright and varied lichens surround the nest, but they also possess a coloration which perfectly camouflages them when sitting on the nest (Sauer, 1962). Sanderlings also use the same lichen for nest building (Parmelle, 1970). The Newfoundland willow ptarmigans frequently use the ground lichen *Cladonia* as nesting cover (Bergerud and Huxter, 1969). The forest goatsuckers (Caprimulgidae) of North America have a coloration and pattern which faithfully represent the intricate bark and lichen patterns of the surface of fallen logs. The nighthawks which perch in the open on lichen-covered rocks and tree boughs show similar features of cryptic coloration (Thayer, 1909).

It is not likely that lichens form a useful food source for birds although in very adverse conditions these plants may be found in the stomach. For example a stonechat, during a harsh winter in Wales, was found to contain *Parmelia saxatilis* (Bolam, 1913). However, the feeding behaviour of birds provides a selective pressure resulting in the evolution of many beautiful examples of cryptic coloration and camouflage among invertebrates (see Chapter 4). The efficiency of predators often depends on how well the food source matches the background. A change in the latter, caused, for example, by the death of lichens due to increasing air pollution, can lead to a marked drop in the population of particular food species. The reduction in numbers and diversity of food sources may in turn affect bird populations. Only where an insect can respond quickly to the absence of lichens by evolving a melanic form can the population of the insect be maintained. An excellent account of the evolution of industrial melanism following the loss of lichens due to air pollution is provided by Kettlewell (1973) and Bishop and Cook (1975). In other instances the removal of lichens from trees leads to the loss of food sources or egg-laying sites for invertebrates, and this may affect the food-web on which bird populations depend.

V. Mammals

A. Reindeer and Caribou

1. General

Reindeer and caribou belong to the same species, *Rangifer tarandus*, which has been subdivided into several subspecies (Banfield, 1961). The term "reindeer" refers to the Eurasian subspecies, while "caribou" is retained for those that are native to North America. Lichens form the primary winter food for these animals (Klein, 1971; Scotter, 1965a) and are consumed in quantities of 3–5 kg day^{-1} (Skuncke, 1958, 1969; Bunnell et al., 1975). The proportion of lichens in the winter diet usually exceeds 50% but varies with the geographic location of the herd (Kelsall, 1968; Bergerud, 1974). Further details on the productivity of the lichen stands of reindeer and caribou winter ranges are provided in Chapter 6.

Animals dig through the snow with their forefeet to expose the lichens, forming conspicuous feeding craters in the process. A wide variety of terricolous lichens are eaten, including *Cladonia arbuscula*, *C. stellaris*, *C. amaurocrea*, *C. mitis*, *C. rangiferina*, *C. uncialis*, *Cetraria islandica*, *C. nivalis*, *C. cucullata* and *Stereocaulon paschale* (e.g. Aleksandrova et al., 1964; Ahti and Hepburn, 1967; Kelsall, 1968). Arboreal lichens are extremely important for woodland caribou (Cringan, 1957) and reindeer in Scandinavia (Skuncke, 1958). The species consumed by the North American woodland caribou include *Alectoria* and *Bryoria* spp. (see Brodo and Hawksworth, 1977, for details), *Evernia mesomorpha*, *Hypogymnia physodes* and *Usnea* spp. In areas with rock outcrops where species of *Umbilicaria* are abundant, these too are often eaten to some extent. The caribou expose the rock surface, knocking the lichens loose with sweeping motions of the forefeet. They then pick up the lichen fragments from the snow surface (Kelsall, 1968). Data on the number of wild reindeer and caribou in different countries are summarized in Table II.

The important forage lichens on which caribou and reindeer survive the winter grow about 3·5–6·0 (2·2–7·3) mm year^{-1} (Ouzilleau and Payette, 1975; Andreev, 1954; Pegau, 1968a,b). Kärenlampi (1970, 1971a,b) developed methods for growing these lichens in plastic boxes in the natural habitat and measuring the dry weight and size increase periodically. He found that precipitation among other factors can critically affect the production rate of the lichens. Another factor is the destruction of the pastures by fires, which is especially important in the wintering grounds of the barren-ground caribou in the taiga of northern Canada. Scotter (1970) has studied the change in standing crop of usable forage lichens.

In the absence of wild fires, there is a 20-fold increase in the total amount of lichen and a predominance of the lichens preferred by the caribou. These differences are reflected in the distribution of winter-feeding craters which, when surveyed from the air, were found to be largely confined to mature forests. The importance of fire depends on the location. In some areas of Alaska and Newfoundland, fire is much less harmful to the lichen ranges and can occasionally improve them by making conditions more suitable for the growth of lichens rather than mosses which the caribou do not utilize. The exploitation of northern Canada and changes in the summer weather leading to more forest fires pose a threat to the lichen ranges and hence the caribou population.

TABLE II.

The estimated number of wild caribou or reindeer in various countries.

Country	Estimated number	Reference
Alaska[a] and Yukon	2,000,000	Cahalane (1958)
Idaho, U.S.A.	100	Banfield (1961)
Canada	357,000	Banfield (1974)
Finland	30	Scotter (1965a)
Norway	40,000	Scotter (1965a)
Spitzbergen	1500	Banfield (1961)
U.S.S.R.	over 350,000	Heptner and Naumov (1966)

[a]1976 survey estimates 220,000 for Alaska (Klein, *in litt.*).

The nutritive value of lichens is rather low, being especially deficient in protein, e.g. Scotter (1965b). It is believed that domestic animals and black-tailed deer require a diet with close to 5% protein in order to survive and the majority of lichens contain less than this. However, according to Steen (1968) the Scandinavian reindeer only require 170 g of digestible protein day^{-1} in winter (for an average body weight of 80 kg). The deficiency in protein usually results in a decrease in body weight of the reindeer and caribou over the winter as they break down muscle to compensate for the low nitrogen in the diet. It is presumed that microorganisms in the rumen break down lichenin and isolichenin which are the complex polysaccharides of lichens. They may also convert the sugar alcohols (which make up the greatest part of the nitrogen-free extract) into sugars which can be metabolized by the animals. The ciliate fauna of the rumen is also able to ingest lichens (Luick, 1971; Westerling, 1970).

Lichens are also usually rather poor in several mineral elements including potassium, magnesium, copper and zinc. The introduced reindeer in Canada periodically develop symptoms of malnutrition characterized by bones breaking when the animals walk. Scotter and Miltimore (1973) postulate that this may be caused by copper deficiency exacerbated by the high ratio of molybdenum to copper in the higher plant *Eriophorum vaginatum*. The reindeer from the Taymyr Peninsula in U.S.S.R. have a different problem in that the forage contains excessive amounts of lead and nearly excessive amounts of nickel (Podkorytov, 1969).

TABLE III.

The estimated number of domesticated reindeer in various countries.

Country	Estimated number	Area of range (km^2)	Reference
Alaska	36,000	54,000	Pegau (1968a)
Canada	2800	47,000	Scotter (1972)
Finland	220,000	130,000	Scotter (1965a)
Norway	200,000	110,000	Scotter (1965a)
Sweden	250,000	160,000	Scotter (1965a)
South Georgia	1300	1450	Lindsay (1973)
U.K.	90	40	Gilbert (1974)
U.S.S.R.	2,400,000	3,148,000	Andreev (1968)

In addition, small numbers survive in Mongolia, Kerguelen Islands and probably Greenland and Iceland.

2. Introduced and Domesticated Reindeer

Reindeer have been introduced from Scandinavia into several countries with greater (Alaska and South Georgia—Hadwen and Palmer, 1972) or lesser success (Canada, U.K., Iceland, Greenland, and Kerguelen Islands —see Table III). The proper utilization of the lichen ranges is of prime importance and the lack of competent herding has been responsible for many failures. According to Scotter (1965a), the Russians recommend a 3-year rotation of pastures, but he suggests that an 8 to 10-year rotation might be more appropriate. It is critical that the lichens be only lightly grazed and that a significant proportion of the range remains undamaged by herding. Pegau (1970a) found that herding a group of 500 animals over an ungrazed area of lichen-rich meadow resulted in 68% of the lichens being dislodged and left lying on the surface after only one season. Pegau (1970b) further observed that some portions of the reindeer range, from which the animals had been excluded by fences, had not regenerated even after 30 years. Elsewhere recovery had occurred by regeneration of the old plants rather than by recolonization, but even here the lichens had not attained their former abundance. Recently, computer simulation has been used to study *Rangifer*-range relations in Canada and Finland (Bunnell *et al.*, 1974, 1975). These studies show the value of balancing reindeer populations, available forage and supplemental feeding. They also have the added merit of pinpointing areas where knowledge of the physiology and ecology of many lichens is lacking.

In South Georgia where reindeer were introduced 60 years ago, half of the macrolichen species found in ungrazed parts of the island are no longer found where reindeer abound. Lindsay (1973) is of the opinion that *Cladonia rangiferina* and *Cetraria islandica* will only survive over the long term in isolated inaccessible places. In Scotland, Gilbert (1974) discovered that the reindeer had not affected the diversity of the lichen flora but that the biomass of lichens in the 1000-acre enclosure, where part of the herd was kept, was about five times less than outside. The reindeer grazed terricolous, corticolous and saxicolous macrolichens. Thus, an important factor as regards the effect on the lichen flora, is the density of the reindeer population. In South Georgia, there is one animal per $0.7-1.1$ km^2, and Lindsay (1973) states there is little hope for the recovery of the lichens, or indeed certain flowering plants, unless the numbers of reindeer are drastically reduced.

3. Reindeer, Man and Radioactive Fallout

Industrial man has had an impact on the lichen ranges and on the reindeer and caribou in numerous ways. These have been reviewed recently by Klein (1971), James (1973) and Richardson (1975). Of particular interest has been the impact of nuclear fallout and this topic will be briefly summarized.

Lichens possess ion exchange mechanisms whereby ions are rapidly and efficiently taken up from rain, melted snow and windblown dust (Puckett *et al.*, 1973; Tuominen and Jaakkola, 1974; Tomassini *et al.*, 1976). Lichens therefore accumulate substantial levels of radioactive nuclides, if these are present in the environment (Hoffman, 1972). This accumulation can be over a wide area following an atmospheric nuclear test or localized as was found when a U.S. bomber crashed near Thule in Greenland, scattering, but not exploding its nuclear weapons (Hanson,

1972). Many radionuclides are absorbed by lichens from the above sources, including ^{90}Sr, ^{137}Cs, ^{55}Fe, ^{22}Na and ^{237}Pu.

As lichens form a major food source for reindeer and caribou, the hazard of radioactive contamination of reindeer meat, and subsequently of man is obviously high. Liden (1961), who first discovered this danger, showed that reindeer meat contained 280 times the ^{137}Cs level of beef produced in the same general area. Kelsall (1968) and Richardson (1975) provide a general discussion of this topic, but the most detailed source of information up to the year 1967 is the compilation of Aberg and Huneal (1967).

The effects of the treaty banning nuclear tests are seen in studies by Persson (1971) and Mattsson (1972). The former showed that ^{90}Sr levels in the Swedish lichen *Cladonia stellaris* reached a maximum of 13 nCi (kg dry wt.)$^{-1}$ during the period 1963–1965. This had decreased to 5 nCi kg^{-1} during 1967–1969. It was estimated that the half-life of ^{90}Sr in the lichen carpet was $2 \cdot 5 \pm 0 \cdot 8$ years. The concentration of this isotope reached a peak of 700 pCi g^{-1} in reindeer meat and 1000 pCi g^{-1} in bone during 1965; since then it has fallen to about 400 pCi g^{-1} in both tissues. It was estimated that the total dose to which Lapps were exposed over the period 1950–1969 was 30 mrad to the endosteum (cells lining the bone) and 20 mrad for the bone marrow.

Although the level of radioactivity in lichens has fallen, a recent study (Book *et al.*, 1972) showed that *Usnea* and *Ramalina* species in California contained 6–14 pCi ^{137}Cs (g dry wt.)$^{-1}$ in 1969, which was 10–140 times the concentration found in other vegetation and largely accounted for the levels of radioactivity found in black-tailed deer from that region.

The potential hazard of this increased radiation to people such as Lapps who eat large quantities of reindeer meat has been assessed and thought small. This is firstly because atmospheric tests have largely ceased, and secondly because the Lapp families who eat meat from this source are not a genetically closed group. Intermarriages between these families and those not directly involved in herding are common, so that it is unlikely that mutations caused by this increased radiation will be detected.

B. Other Large Mammals

The Columbian black-tailed deer (*Odocoileus hemionus columbianus*) which live in the chaparral and oak woodlands of California use lichens as an important dietary item during the winter. These deer eat *Ramalina menziesii* and species of *Usnea* which blow off the trees in considerable quantities (Book *et al.*, 1972). The black-tailed deer *Odocoileus hemionus* of western Canada and the roosevelt elk *Cervus elaphus rooseveltii* also consume considerable quantities of lichen in winter time (F. L. Bunnell, personal communication).

In the East Sayan Mountains of Central Asia, the musk deer, *Moschus moschiferus*, is found in the dark coniferous forest where the largest reserves of its main food, *Usnea* spp., occur (Ustinov, 1969; Lydekker, 1896). These animals, which yield the musk of commerce, lack antlers in either sex but have short tusks which perhaps help in gathering the lichens. A number of other ungulates are reported to consume lichens as part of their diet; examples include *Capreolus capreolus* (roe deer), *Rupricapra rupricapra* (chamois), *Alces alces* (elk), *Cervus elaphus* (red deer), *Capra ibex* (ibex) and *Oreamnos americanus* (mountain goat) (T. Ahti, personal communication; Lindahl, 1971). Musk-oxen *Ovibos moschatus* have been observed to nibble *Umbilicaria* spp. from exposed rocks during winter. However, lichens such as *Cetraria nivalis* and *Alectoria ochroleuca* appear to be eaten only incidentally along with other food (Tener, 1965). The same probably applies to *Cetraria islandica* in polar-bear droppings. The lichen was not digested and passed through the animal in a determinable form (data of R. Russell via Brodo, *in litt.*).

Domesticated animals have long been (and are locally still) fed with lichens in the mountain districts of Norway, Sweden and Iceland. The lichens are fed to the cattle, sheep, goats and pigs, together with hay and fishmeal. The latter is generally added to make up for the low protein and fat content of the lichens (Porsild, 1954; Lindahl, 1971). Camels (*Camelus* spp.), llamas (*Lama* spp.) and guanacos (*L. huanacos*) may also consume lichens (Follmann, 1964).

C. Rodents and Bats

The lemmings which live in the Russian tundra eat away the bases of thick mats of *Cetraria nivalis* and *C. cucullata* (V. N. Andreev, personal communication). When lemming numbers are high, the biomass of lichens over several square kilometres in extent can be considerably reduced by the actions of these animals. However, in feeding experiments both in summer and winter, Kalela (1961) found that captive lemmings almost completely rejected *Cetraria islandica*, *Stereocaulon paschale* and various species of *Cladonia* while readily eating mosses.

Holisova (1966) reports finding lichens in the stomachs of voles during February, but considers that these plants do not play an important part in their diet over the whole year. Denison (1973) reports that the red tree mouse (*Phenacomys longicaudus*) supplements its diet with *Lobaria oregana* which grows on 60-m Douglas fir trees in western Oregon, and another species, *P. intermedius* (the heather vole), is also reported to eat lichens

(Banfield, 1974). Riewe (1973) found toothmarks of the vole *Microtus pennsylvanicus terraenovae* on the thalli of *Peltigera* spp. in Newfoundland but he was uncertain if the animals consumed the lichens. In food preference tests he found these animals rejected *Cladonia stella*.

Cladonia spp. form part of the diet of the hoary marmot (*Marmota caligata*) on the Kenai Peninsula, Alaska (Hansen, 1975). An interesting early record of *Ramalina fraxinea* forming part of a squirrel's diet is provided by Drummond (1844).

FIG. 5. *Lasiurus cinereus*, the hoary bat from California which roosts in festoons of the lichen *Ramalina menziesii* L. (Photo by Laidlaw-Williams.)

The hoary bat (*Lasiurus cinereus*) roosts in festoons of *Ramalina menziesii* at Point Lobos, California. Here, this interesting migratory bat spends the winter. Almost all the roosting sites were partly concealed with *Ramalina* streamers similar to those in Fig. 5 (L. Laidlaw-Williams, unpublished study). Brock Fenton (*in litt.*) also reports that bats are very reluctant to take *Hypoprepia fucosa* or *H. miniata*, two moths which feed on lichens. It is possible that absorbed lichen acids make these moths distasteful to the bats.

D. Man

Man's impact on the lichen flora varies from region to region. From the time of Egyptian civilization, lichens have been used by man as food, and for the production of dyes, perfumes, medical preparations and alcohol. They are also widely used in model-building and in Scandinavia for making wreaths. These economic uses of lichens are discussed by Smith (1921), Kursanov and D'yachkov (1945), Llano (1944, 1948, 1950, 1956), Aleksandrova *et al.* (1964), Lindahl (1971, 1972, 1973, 1975), Henssen and Jahns (1973) and Richardson (1975).

Today the indirect action of air pollution from industrial activity and of habitat change due to modern agricultural and forestry practices have had a most profound effect on the lichen flora of many areas. These aspects form the bases for recent books (Ferry *et al.*, 1973; Richardson, 1975; Hawksworth and Rose, 1976) and are briefly summarized below and discussed further in Chapters 9 and 10.

1. Food

In times of need lichens have been used as food in many parts of the world but particularly in Arabia, the Arctic and boreal regions and Japan. In the semi-arid lands in South-west Asia, south-east Europe and North Africa, the crustose lichens *Aspicilia esculenta*, *A. desertorum* and *A. jussufii* have been used for food by man (see also Chapter 8). These lichens form thick wrinkled yellowish crusts on rocks. As the thalli become older they tend to become detached and carried around by the wind, becoming piled up behind bushes and rocks. The occasional rains then wash these thalli into depressions in such quantity that a man may collect 4–6 kg in a day (Kerner and Oliver, 1897). The lichen is ground into a meal from which a kind of bread, "Shirsad", is made. No recent estimates of the quantity of lichens used for this traditional food are available. In boreal and Arctic regions, lichens such as *Umbilicaria muehlenbergii*, *Cladonia stellaris*, *Cetraria islandica* and *Bryoria fremontii* have been used in times of need

(Turner, 1976; Richardson, 1975). However, in Japan *Umbilicaria esculenta* is regarded as a delicacy in areas around Hiroshima (Sato, 1968). This lichen is collected from the central mountainous areas from very steep granite rock surfaces using climbing ropes. About 800 kg are eaten annually; the lichen is sold under the name "Iwatake" and is also called the food of Sennin. Sennin was a mythical superhuman with a white beard and stick who ate clouds. Because of the recognized long life of this plant, it is traditionally served during marriage ceremonies. It is eaten as a vegetable, in soups, salads or deep-fried served with rice. The practice of collecting the large amount of lichen required probably has a marked effect on the lichen communities in these mountain areas.

2. Dyes, Perfumes and Medicines

The collection of lichens for the preparation of lichen-dyed tweed has now largely ceased, but was carried out on a large scale prior to 1860 (Kok, 1966). Even the ancient Egyptians used a dye made from lichens as a cosmetic rouge. Today about 25 tons of *Roccella tinctoria* and *R. montagnei* are utilized per annum for the manufacture of orcinol and litmus (see Richardson, 1975). The sources of supply are the Cape Verde Islands and Madagascar but the effects of this collection on the standing crop of these lichens are not fully known. It would appear that the commercial companies in Madagascar, which also produce erythritol from these lichens, are already concerned about the reduction in supply and are in need of new sources.

Much larger quantities of the lichens *Evernia prunastri* and *Pseudevernia furfuracea* are used to make "Oakmoss extracts" which are important fixatives in high quality perfumes and soaps. It appears difficult to obtain estimates of the absolute quantities of lichen which are collected annually from central France, Italy and eastern Europe.

Lichens have been the bases of many traditional medicines both in ancient China and Europe since the Middle Ages, when the concept of the doctrine of signatures held sway. The effectiveness of lichen medicines was substantiated by the discovery of the antibiotic properties of usnic acid, a common constituent of lichens (Vartia, 1974). For a time, commercial preparations of this substance were available in several European countries, especially Finland, where about 1000 kg of *Cladonia stellaris* were processed annually for the manufacture of the salve "Usno". However, production has now ceased due to rising labour costs and decreased abundance of the raw material and heavy competition from other antibiotics produced by industrial fermentations; in the Soviet Union a similar preparation (see Lazarev and Savicz, 1957) is probably still in use.

3. Air Pollution and Habitat Modification

Lichens have very efficient uptake mechanisms and readily accumulate radionuclides and other metal ions from industrial activity (Nieboer et al., 1972; Seaward, 1973). Generally, accumulation of these ions appears not to harm lichens, probably because most are taken up by an ion exchange process on sites which are on the cell wall and hence external to the cell membranes (Nieboer et al., 1976a). Sulphur dioxide, on the other hand, is highly toxic to lichens, so that around most urban and industrial centres there is a lichen desert. A little further out from these centres a few pollution tolerant species may be found, and the lichen flora then improves until at a distance a wide variety of species occurs. The documentation and importance of these changes are discussed in Ferry et al. (1973), and a current review of literature on this aspect of lichenology is provided by Hawksworth (1974, 1975a,b, 1976). Research is currently aimed at assessing the level of sensitivity of particular species (e.g. Türk et al., 1974) and their response to ameliorating conditions (e.g. Seaward, 1976), trying to understand the mechanisms of damage and recovery (e.g. Nieboer et al., 1976b). The value of lichens as indicators of both heavy metal (Tomassini et al., 1976) and gaseous air pollution is now recognized, and these plants are being used, for example, to monitor emissions around newly established industrial plants in many countries.

Habitat modification is one of the more insidious aspects of man's activity and is proceeding at a largely unrecognized pace throughout the world. The various aspects of habitat modification, particularly with reference to the British Isles, are covered in more detail in Chapter 11. The mechanization of agriculture and forestry together with more efficient drainage techniques, plus the use of fertilizers and sprays, all lead to the loss of diversity in the lichen flora. The felling of tropical forests, underplanting deciduous forest with conifers, or conversion of forests to agricultural land have effects both on the relatively small number of vertebrates depending on lichens, and also on numerous invertebrates (see Chapter 4). These changes may in turn have far-reaching consequences to the ecosystem and thus involve animals with no obvious interrelationships with lichens.

Acknowledgements

We would especially like to acknowledge the help of Dr T. Ahti who kindly read and made many valuable suggestions on the section dealing with mammals. We would also like to thank Dr D. L. Hawksworth and Dr M. R. D. Seaward for their helpful comments and for sending additional references for incorporation into this chapter.

References

Aberg, B. and Huneal, F. P., eds. (1967). "Radioecological Concentration Processes". Pergamon Press, Oxford.
Ahti, T. and Hepburn, R. L. (1967). Preliminary studies on woodland caribou range, especially on lichen stands in Ontario. *Ont. Dept. Lands Forests Res. Rep.* **74**, 1–80.
Aleksandrova, V. N., Andreev, V. N., Vakhtina, T. V., Dydina, R. A., Karev, G. I., Petrovsky, V. V. and Shamurin, V. F. (1964). Kormovaya kharakteristika rasteniy Kraynego Severa. *Rastitel'nost' Kraynego Severa SSSR i ee osvoenie* **5**, 1–484.
Andreev, V. N. (1954). Prirost kormovykh lishaynikov i priemy ego regulirovaniya. *Trudy Bot. Inst. AN SSSR*, ser. III, **9**, 11–74.
Andreev, V. N. (1968). Problemy ratzional'nogo ispol'zovaniya i uluchsheniya olen'iks pastbishch. *Problemy Severa* **13**, 76–87.
Banfield, A. W. F. (1961). A revision of the reindeer and caribou, genus *Rangifer*. *Natn Mus. Can. Bull.* **177**, 1–137.
Banfield, A. W. F. (1974). "The Mammals of Canada", pp. 193, 388. National Museum of Canada, Ottawa.
Bergerud, A. T. (1974). Decline of caribou in North America following settlement. *J. Wildl. Mgmt* **38**, 757–774.
Bergerud, A. T. and Huxter, D. S. (1969). Breeding season habitat utilization and movement of Newfoundland willow ptarmigan. *J. Wildl. Mgmt* **33**, 967–974.
Beste, J. and Beste, J. (1974). The green world of Queensland's rainforest. *Animals* **16**, 64–78.
Bishop, J. A. and Cook, L. M. (1975). Moths, melanism and clean air. *Scient. Am.* **232**, 90–99.
Bolam, G. (1913). "Wildlife in Wales". Frank Palmer, London.
Book, S. A., Connolly, G. E. and Lonehurst, W. M. (1972). Cs^{137} accumulation in two adjacent populations of northern California deer. *Hlth Phys.* **22**, 379–385.
Brodo, I. M. and Hawksworth, D. L. (1977). *Alectoria* and allied genera in North America. *Op. bot. Soc. bot. Lund.* **42**, 164.
Bunnell, F. L., Kärenlampi, L. and Russell, D. E. (1974). A simulation model of lichen-*Rangifer* interactions in northern Finland. *Rep. Kevo Subarctic Res. Stn* **10**, 1–8.
Bunnell, F. L., Daulphine, D. C., Hilborn, R., Miller, D. R., Miller, F. L., McEwan, E. H., Parker, G. R., Petersen, R., Scotter, G. W. and Walters, C. T. (1975). Preliminary report on computer simulation of barren-ground caribou management. *Proc. 1st Int. Reindeer/Caribou Mgmt Symp.* Fairbanks, Alaska.
Cahalane, V. H. (1958). "Mammals of North America". Macmillan, New York.
Campbell, A. J. (1900). "Nests and Eggs of Australian Birds". Pawson and Brailsford, Sheffield.

Cochran, D. M. (1961). "Living Amphibians of the World". Doubleday, New York.
Cott, H. B. (1957). "Adaptive Coloration in Animals". Methuen, London.
Cringan, A. T. (1957). History, food habits and range requirements of the woodland caribou of continental North America. *Trans. N. Am. Wildl. Nat. Resour. Conf.* **22**, 485–501.
Davie, O. (1898). "Nests and Eggs of North American Birds". Musson, Toronto.
Denison, W. C. (1973). Life in tall trees. *Scient. Am.* **228**, 75–80.
Dickerson, M. C. (1920). "The Frog Book", p. 119. Doubleday, New York.
Drummond, J. (1844). Lichens eaten by squirrels. *Gdnrs' Chron.* Jan. **1844**, 28.
Ferry, B. W., Baddeley, S. M. and Hawksworth, D. L. (1973). "Air Pollution and Lichens". Athlone Press, London.
Follmann, G. (1964). Nebelflechten als Futterpflanzen des Küstenguanacos. *Naturwissenschaften* **51**, 19–20.
Gilbert, O. L. (1974). Reindeer grazing in Britain. *Lichenologist* **6**, 165–167.
Hadwen, S. and Palmer, L. J. (1972). Reindeer in Alaska. *U.S. Dept. Agric. Bull.* **1089**, 1–75. Washington, D.C.
Hansen, R. M. (1975). Foods of the Hoary marmot on Kenai Peninsula. *Am. Midl. Nat.* **79**, 348–353.
Hanson, W. C. (1972). Plutonium in lichen communities of the Thule, Greenland region during the summer of 1968. *Hlth Phys.* **22**, 39–42.
Harrison, C. (1975). "A Field Guide to the Nests, Eggs and Nestlings of British and European Birds". Collins, London.
Hawksworth, D. L. (1974–1976). Literature on air pollution and lichens I–V. *Lichenologist* **6**, 122–125; **7**, 62–66, 173–177; **8**, 87–91, 179–182.
Hawksworth, D. L. and Rose, F. (1976). "Lichens as Pollution Monitors". Studies in biology, no. 66. Edward Arnold, London.
Headstrom, R. (1970). "A Complete Field Guide to Nests in the United States". Ives Washburn, New York.
Hendrickson, J. R. and Weber, W. M. (1964). Lichens on Galapagos giant tortoises. *Science, N.Y.* **144**, 1463.
Henssen, A. and Jahns, H. M. (1973) ["1974"]. "Lichenes". Thieme, Stuttgart.
Heptner, V. G. and Naumov, N. P. (1966). "Die Säugetiere der Sowjetunion". I. Fischer Verlag, Jena.
Hoffman, G. R. (1972). The accumulation of Cesium-137 by cryptogams in a *Liriodendron tulipifera* forest. *Bot. Gaz.* **133**, 107–111.
Holisova, V. (1966). Food of an overcrowded population of the bank vole, *Clethrionomys glareolus* Schreb, in a lowland forest. *Zoo. Listy* **15**, 207–224.
James, P. W. (1973). The effects of air pollutants, other than hydrogen fluoride and sulphur dioxide, on lichens. *In* "Air Pollution and Lichens" (B. W. Ferry, M. S. Baddeley and D. L. Hawksworth, eds), pp. 143–175. Athlone Press, London.
Kalela, O. (1961). Seasonal changes of habitat in the Norwegian lemming, *Lemmus lemmus*. *Ann. Acad. Scient. Fenn. Ser. A. iv Biol.* **55**, 1–72.
Kärenlampi, L. (1970). Morphological analysis of the growth and production of the lichen genus *Cladonia*. *Rep. Kevo Subarctic Res. Stn* **7**, 9–15.

Kärenlampi, L. (1971a). Studies on the relative growth rates of some fruticose lichens. *Rep. Kevo Subarctic Res. Stn* **7**, 33–39.
Kärenlampi, L. (1971b). On methods for measuring and calculating the energy flow through lichens. *Rep. Kevo Subarctic Res. Stn* **7**, 40–46.
Kelsall, J. P. (1968). "The Migratory Barren-ground Caribou of Canada". Canadian Wildlife Service, Ottawa.
Kerner, A. and Oliver, F. W. (1897). "The Natural History of Plants", Vol. 6, pp. 810–812. Blackie, London.
Kettlewell, H. B. D. (1973). "The Evolution of Melanism". Clarendon Press, Oxford.
Klein, D. R. (1971). Reaction of reindeer to obstructions and disturbances. *Science, N.Y.* **173**, 393–398.
Kok, A. (1966). A short history of Orchil dyes. *Lichenologist* **3**, 248–271.
Kursanov, A. L. and D'yachkov, N. N. (1945). "Lishayniki i ikh prakticheskoe ispol'zovanie". Akad. Nauk SSSR, Leningrad.
Lazarev, N. V. and Savicz, V. P. (1957). "Novyy antibiotik binan ili natrievaya sol' usninovoy kisloty (botanicheskie i meditsinskie issledovaniya)". Akad. Nauk, SSSR, Leningrad.
Liden, K. (1961). Caesium^{137} burdens in Swedish Laplanders and reindeer. *Acta Radiol.* **56**, 237–240.
Lindahl, P.-O. (1971). Lavar och djur. *Fauna och Flora* **66**, 159–167.
Lindahl, P.-O. (1972). Lavar till mat och dryck. *Fauna och Flora* **67**, 123–129.
Lindahl, P.-O. (1973). Lavar som medicinalväxter. *Fauna och Flora* **68**, 49–55.
Lindahl, P.-O. (1975). Lavar som Fargväxter. *Fauna och Flora* **70**, 233–238.
Lindsay, K. (1973). Effects of reindeer on plant communities in the Royal Bay area of South Georgia. *Bull. Br. Antarct. Surv.* **35**, 101–109.
Llano, G. A. (1944). Lichens—their biological and economic significance. *Bot. Rev.* **10**, 1–65.
Llano, G. A. (1948). Economic uses of lichens. *Econ. Bot.* **2**, 15–45.
Llano, G. A. (1950). Economic uses of lichens. *A. Rep. Smithsonian Inst.* **1950**, 385–422.
Llano, G. A. (1956). Utilization of lichens in the Arctic and Subarctic. *Econ. Bot.* **10**, 367–392.
Luick, J. R. (1971). Studies on the nutrition and metabolism of reindeer–caribou in Alaska with special interest in nutritional and environmental adaptations. Technical Progress Report, 1970–71, U.S. Atomic Energy Commission.
Lydekker, R. (1896). "The Royal Natural History", Vol. 5, p. 269. Warne, London.
Mattsson, L. J. S. (1972). Sodium^{22} in the food chain: Lichen–reindeer–man. *Hlth Phys.* **2**, 223–230.
Medway, L. (1969). Studies on the edible-nest swiftlets of South-east Asia. *Malay Nat. J.* **22**, 57–63.
Nieboer, E., Ahmed, H. M., Puckett, K. J. and Richardson, D. H. S. (1972). The heavy metal content of lichens in relation to distance from a nickel smelter in Sudbury, Ontario. *Lichenologist* **5**, 292–304.

Nieboer, E., Puckett, K. J. and Grace, B. (1976a). The uptake of nickel by *Umbilicaria muehlenbergii*: a physicochemical process. *Can. J. Bot.* **54**, 724–733.

Nieboer, E., Richardson, D. H. S., Puckett, K. J. and Tomassini, F. D. (1976b). Lichen response studies. *In* "Effects of Air Pollutants on Plants" (T. A. Mansfield, ed.), pp. 61–85. Cambridge University Press, Cambridge.

Ouzilleau, J. and Payette, S. (1975). Croissance de quelques lichens à caribou de genre *Cladonia* (*sensu* genre *Cladonia*) en milieu subarctique, Nouveau-Québec. *Naturaliste Can.* **102**, 597–602.

Parmelle, D. C. (1970). Breeding behaviour of sanderlings in the Canadian high arctic. *Birds* **97**, 146.

Pegau, R. E. (1968a). Reindeer range appraisal in Alaska. Unpubl. M.Sc. thesis, Univ. Alaska, College, Alaska.

Pegau, R. E. (1968b). Growth rates of important reindeer forage lichens on the Seward Peninsula, Alaska. *Arctic* **21**, 255–259.

Pegau, R. E. (1970a). Effect of reindeer trampling and grazing on lichens. *J. Range Mgmt* **23**, 95–97.

Pegau, R. E. (1970b). Succession in two enclosures near Unalaklett, Alaska. *Can. Fld Nat.* **84**, 175–177.

Persson, B. R. (1971). ^{90}Sr in northern Sweden: relationship and annual variations from 1961 to 1969 in lichens, reindeer and man. *Hlth Phys.* **20**, 393–402.

Podkorytov, F. M. (1969). Trace element content in fodder plants on reindeer pastures in the Taimyr forest-tundra. *Proc. Rec. Inst. Agr. Far North* **17**, 27–50.

Porsild, A. E. (1954). Land use in the Arctic. Part II. *Can. Geog. J.* **49**, 20–35.

Puckett, K. J., Nieboer, E., Gorzynski, M. J. and Richardson, D. H. S. (1973). The uptake of metal ions by lichens: a modified ion exchange process. *New Phytol.* **72**, 329–342.

Riba, R. and Herrera, T. (1973). Ferns, lichens and hummingbirds' nests. *Am. Fern J.* **63**, 128.

Richardson, D. H. S. (1975). "The Vanishing Lichens". David and Charles, Newton Abbot.

Riewe, R. R. (1973). Food habits of insular meadow voles, *Microtus pennsylvanicus terraenovae* (Rodentia: Cricetidae), in Notre Dame Bay, Newfoundland. *Can. Fld Nat.* **87**, 5–13.

Sato, M. (1968). An edible lichen of Japan, *Gyrophora esculenta* Miyoshi. *Nova Hedwigia* **16**, 505–509.

Sauer, E. G. F. (1962). Ethology and ecology of golden plovers on St. Lawrence Island, Bering Sea. *Psychologische Forsch.* **26**, 399–470.

Scotter, G. W. (1965a). Reindeer ranching in Fennoscandia. *J. Range Mgmt* **18**, 301–305.

Scotter, G. W. (1965b). Chemical composition of forage lichens from northern Saskatchewan as related to use by barren-ground caribou in the taiga of northern Canada. *Can. J. Pl. Sci.* **45**, 246–250.

Scotter, G. W. (1970). Wildfires in relation to the habitat of barren-ground caribou in the taiga of northern Canada. *Proc. Ann. Tall Timbers Fire Ecol. Congr.* **1970**, 85–105.
Scotter, G. W. (1972). Reindeer ranching in Canada. *J. Range Mgmt* **25**, 167–174.
Scotter, G. W. and Miltimore, J. E. (1973). Mineral content of forage plants from the Reindeer Preserve, Northwest Territories. *Can. J. Pl. Sci.* **53**, 263–268.
Seaward, M. R. D. (1973). Lichen ecology of the Scunthorpe heathlands. I. Metal accumulation. *Lichenologist* **5**, 423–433.
Seaward, M. R. D. (1976). Performance of *Lecanora muralis* in an urban environment. *In* "Lichenology: Progress and Problems" (D. H. Brown, D. L. Hawksworth and R. H. Bailey, eds), pp. 323–358. Academic Press, London and New York.
Skuncke, F. (1958). "Reindeer Grazing and Grading in Sweden". Wiksell, Uppsala.
Skuncke, F. (1969). Reindeer ecology and management in Sweden. *Univ. Alaska Biol. Pap.* **8**, 1–82.
Skutch, A. F. (1965). Life history of the long-tailed silky-flycatcher, with notes on related species. *Auk* **82**, 375–426.
Smith, A. L. (1921). "Lichens". Cambridge University Press, Cambridge.
Steen, E. (1968). Some aspects of the nutrition of semi-domestic reindeer. *In* "Comparative Nutrition of Wild Animals" (M. A. Crawford, ed.), pp. 117–128. Academic Press, London and New York.
Tener, J. S. (1965). "Muskoxen in Canada". Canadian Wildlife Service, Ottawa.
Thayer, G. H. (1909). "Concealing Coloration in the Animal Kingdom", p. 35. Macmillan, New York.
Tomassini, F. D., Puckett, K. J., Nieboer, E. and Richardson, D. H. S. (1976). Determination of copper, iron, nickel, and sulphur by X-ray fluorescence in lichens from the Mackenzie Valley, Northwest Territories, and the Sudbury District, Ontario. *Can. J. Bot.* **54**, 1591–1603.
Tuominen, Y. and Jaakkola, T. (1974) ["1973"]. Absorption and accumulation of mineral elements and radioactive nuclides. *In* "The Lichens" (F. Ahmadjian and M. E. Hale, eds), pp. 185–219. Academic Press, New York and London.
Türk, R., Wirth, V. and Lange, O. L. (1974). Carbon exchange measurements for determination of sulphur dioxide resistance of lichens. *Oecologia (Berl.)* **15**, 33–64 [N.R.C.C. Technical Translation].
Turner, N. J. (1976). "Food Plants of the British Columbia Indians: Interior Peoples". British Columbia Provincial Museum Handbook, Victoria, B.C.
Ustinov, S. K. (1969). On the feeding of *Moschus moschiferus* L. and its adaptations to conditions of food searches. *Zool. Zhurn. Moscow* **48**, 1558–1563.
Vartia, K. O. (1974) ["1973"]. Antibiotics in lichens. *In* "The Lichens" (V. Ahmadjian and M. E. Hale, eds), pp. 547–561. Academic Press, New York and London.
Westerling, B. (1970). Rumen ciliate fauna of semi-domestic reindeer (*Rangifer tarandus* L.) in Finland: composition, volume and some seasonal variations. *Acta zool. fenn.* **127**, 1–76.

Wood, J. G. (1863). "The Illustrated Natural History: Reptiles, Fishes and Insects", p. 171. Routledge, Warne, London.
Wright, A. H. and Wright, A. A. (1949). "Frogs and Toads of the United States and Canada." Cornell University Press, Ithaca.

6 | Lichens of the Boreal Coniferous Zone

TEUVO AHTI

I.	Lichenological Exploration of Boreal Zone	147
II.	Origin of Boreal Lichen Flora	148
III.	Distributional Types of Boreal Lichens	149
	A. Circumpolar (or Almost Circumpolar)	149
	B. Oceanic, Incompletely Circumpolar (Amphi-Atlantic and Amphi-Pacific)	150
	C. Amphi-Atlantic	151
	D. Amphi-Pacific	151
	E. Western Eurasian	152
	F. Eastern Eurasian	152
	G. Interior Eurasian	152
	H. Western North American	153
	I. Eastern North American	153
	J. Interior North American	153
	K. Western Eurasian–Western North American	154
	L. Eastern North American–Eastern Eurasian	155
	M. Interior Eurasian–Interior North American	155
IV.	Ecoclimatic Differentiation	156
	A. Subzonal Differentiation	156
	B. Sectorial Differentiation	158
V.	Edaphic–Topographic Differentiation	159
VI.	Differentiation due to Man	160
VII.	Effect of Air Pollution	161
VIII.	Speciation of Boreal Lichens	162
IX.	Terrestrial Lichen Communities	164
X.	Epiphytic Lichen Communities	167

XI. Lignicolous Lichen Communities 169
XII. Saxicolous Lichen Communities 170
XIII. Conclusions 171
References 172

The boreal zone is a major bioclimatic vegetation zone (biotic zone, biome or life zone), and stretches in a continuous circumpolar band, about 1000 km broad, across the northern parts of Eurasia and North America (Hare, 1954; Sjörs, 1963a; see Fig. 1). Because most of the boreal regions

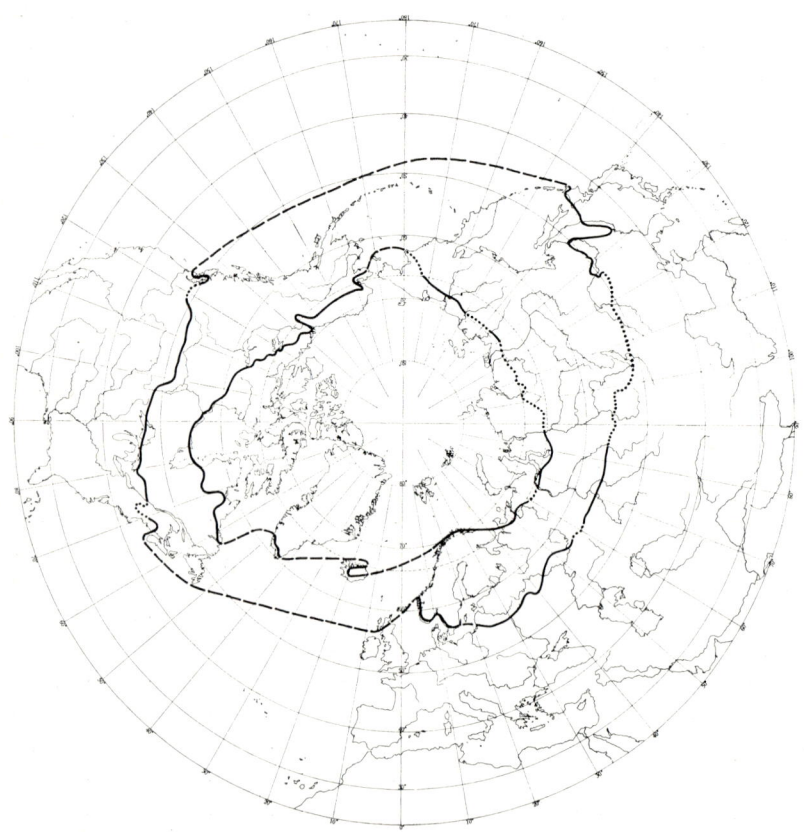

FIG. 1. The northern and southern boundary of the boreal zone proper as delimited in this chapter. The dotted stretches of the boundaries indicate the major mountain areas where the lines have been drawn very schematically. Thus the isolated oroboreal outliers and most of the southerly oroboreal extensions have been omitted.

are dominated by coniferous (*Picea, Larix* or *Pinus*) forests, the zone is also called the boreal coniferous zone or *taiga*. However, the boreal zone also includes extensive tracts where deciduous hardwood species prevail, such as the mountain birch (*Betula pubescens* ssp. *tortuosa*) woods in Scandinavia, the *Betula ermanii* woods in Kamchatka and the *Populus tremuloides* –*P. balsamifera* woods (i.e. parts of the "Aspen Parkland") in the interior of western Canada. Various deciduous second-growth woods (*Betula, Populus, Alnus*) are likewise common. Broad-leaved deciduous trees, such as *Quercus* and *Acer*, are found only along the southern fringe of the zone or as planted ornamental trees.

In addition to the circumpolar latitudinal or horizontal main boreal zone, the boreal zone in the broad sense also includes various altitudinal counterparts ("montane" or "subalpine zones") in numerous more or less isolated mountain outliers (e.g. the European Alps) or southern extensions (e.g. the Rocky Mountains) in areas where the sea-level (basal) zone belongs to the temperate or meridional (warm temperate) major zone (Hämet-Ahti et al., 1974). However, in this context these separate mountain regions, which may be called *oroboreal* (Ahti et al., 1968), are not generally taken into account.

I. Lichenological Exploration of Boreal Zone

The intensity of floristic exploration of the boreal zone varies greatly. The only fairly well studied boreal region is Scandinavia and Finland, but even in this area the knowledge of the crustose lichen flora is very far from complete. Indeed macrolichens, which are the best collected fungi in many herbaria, are fairly well known even elsewhere, as in the Soviet Karelia and Kola Peninsula, in parts of the Ural Mountains, around Lake Baykal, on Kamchatka, in Alaska, Ontario, Quebec and in Newfoundland. But the rest, most of European north Russia, Siberia and the Far East, the Yukon, northern British Columbia, northern Alberta, Saskatchewan, Manitoba and the Northwest Territories are still very sporadically and superficially explored and almost no truly comprehensive local lichen floras have been published, even on macrolichens. There are some larger, important unpublished collections from Siberia and Canada, however, but in general it almost seems that the true Arctic areas are relatively better explored.

Yet many lichens of the boreal zone belong to the best-known species in the world, because they often have very extensive ranges and have been observed and studied by numerous ecologists, plant geographers and phytosociologists in addition to lichen taxonomists. As will be shown below, the boreal lichen flora is so homogeneous that a very detailed knowledge of all regions is not necessary for reliable generalizations concerning the distribution and history of the flora.

II. Origin of Boreal Lichen Flora

Based on fossil record of phanerogams it is generally assumed that in the Early Tertiary a continuous circumpolar zone of deciduous broad-leaved (nemoral) forests ("mixed mesophytic forest") was dominant throughout the northerly parts of the Northern Hemisphere. Only in the Late Pliocene and especially in the Pleistocene extensive conifer-dominated boreal forests were developed in connection with climatic deterioration (e.g. Hultén, 1937, 1962; Heusser, 1960; Frenzel, 1968a,b). The geographical position of the boreal zone has shifted in accordance with the vicissitudes of the Pleistocene glaciations, extending down as far as the present temperate or meridional regions during periods of maximal extensions of the ice-sheets and the associated cold periods. The vegetation was also periodically greatly diminished and fragmented due to the presence of excessively dry climates in the more southern latitudes. In most of Eurasia the flora was evidently impoverished much more severely than in North America.

Plant geographers also commonly accept the idea that the cradle of the boreal flora is somewhere in the North Pacific region, in the mountain and coastal areas of western North America and eastern Asia. The decisive role of the Bering Straits region as a major Pleistocene glacial refugium for numerous plants is often emphasized, particularly by Hultén (1937). The problem of the so-called amphi-Atlantic element, i.e. whether some plants have been able to migrate across the North Atlantic and when and how this might have been possible, has been much discussed (Löve and Löve, 1963).

There are many papers on the assumed boreal (often referred to as Arctic) lichen glacial relics in central Europe or other more southern areas (Suza, 1936, 1937; Minyaev, 1940; Hakulinen, 1965; Wirth, 1972), but the distributions of various boreal lichens as related to their glacial survival have also been discussed (e.g. Degelius, 1935, 1952, 1957; Dahl, 1946, 1950; Ahlner, 1948; Ahti, 1961b; Černohorský, 1963; Makarevich, 1963; Krog, 1968; Brodo, 1968; Thomson, 1972). As expected, the broad features of boreal lichen distribution are the same as in phanerogams. However, some clear differences are obvious, like the higher number of circumpolar species in lichens and rarity of local or even regional endemics (see also Ahti, 1964). From the point of view of phytogeography the boreal lichens may be related in many respects to other cryptogams, notably macrofungi, bryophytes and pteridophytes, rather than phanerogams.

III. Distributional Types of Boreal Lichens

Although the distributions of the common boreal lichens are relatively well known, there are very few adequate maps of their total ranges available. In fact, because of their wide ranges the mapping procedure is very tedious, requiring the study of numerous herbaria and publications written in several languages. It should also be admitted that insufficient lichenological exploration of the Soviet Union and Canada has been undertaken to allow the drawing of distribution maps of the highly useful type compiled by Hultén (1958, 1962, 1971) of boreal phanerogams. However, many of the circumpolar phanerogam maps are also very incomplete, while some lichens have been mapped in detail. Good total range maps of boreal bryophytes and fungi are also very rare. The lack of adequate distribution maps has resulted in contradictory statements about the ranges of many boreal lichens (compare the lists by Makarevich, 1963; Brodo, 1968; and Krog, 1968, for instance) and hence many statistics of lichen elements are almost worthless.

In spite of this a classification of the main distribution types of the boreal lichens is made here. The grouping is not primarily based on the total ranges of the species but on their ranges within the boreal zone. The scheme is mainly based on the macrolichen species, since their ranges are much better known than those of the microlichens.

A. Circumpolar (or Almost Circumpolar)

This group constitutes the majority of the boreal lichens, about 60–70% of the macrolichen flora. In northern Ontario about 85% of the macrolichen flora (c. 205 spp.) are the same as in the climatically corresponding part of Fennoscandia of similar size, while only about 30% of the indigenous vascular plants are the same species (and almost all of these are represented by different intraspecific taxa in the two areas; Ahti, 1964).

Good examples of completely circumpolar boreal lichens are many terricolous *Cladonia* species, such as *C. rangiferina*, *C. mitis* (both mapped by Ahti, 1961b), *C. cornuta*, *C. crispata*, *C. phyllophora*, *C. uncialis*, *Stereocaulon tomentosum*, *S. paschale* (Lamb, 1951), *Cetraria nivalis* (Rassadina, 1950; Thomson, 1972) and *Lecidea granulosa*. Many saxicolous lichens may also be brought into this group, although their ranges are restricted by a complete absence of rock outcrops and stones in parts of north Russia, Siberia, central Canada and the Hudson Bay Lowlands. Many species on granitic and other oligotrophic rocks are not able to grow on highly calcareous rocks and vice versa. However, species like *Parmelia saxatilis*, *Umbilicaria vellea*, *Xanthoria elegans* and *Rhizocarpon obscuratum* have wide ranges, which are at least almost circumpolar. Among the corticolous and lignicolous species there are numerous widespread species, like *Hypogymnia physodes*, *Parmelia sulcata*, *P. septentrionalis* (Ahti, 1966), *Lecanora fuscescens* and *Buellia disciformis*.

Very many of the circumpolar species also extend both into the Arctic and the temperate zones, like *Hypogymnia physodes*, but their major range may in fact be in one of those zones, like *Cetraria nivalis* in the Arctic and *Cladonia coniocraea* in the temperate zone. They are also commonly present in the oroboreal outliers of southern mountain ranges, but in those areas there are many peculiar anomalies, too (like the absence of *Cladonia rangiferina* from Colorado).

Most of the widespread and common boreal lichens have such a wide ecological amplitude that it may be fairly safely assumed that they survived in many different refugia around the glaciers, though very little in the extensive, arid, arctic to boreal steppes of the glacial periods. During the glaciations the species also generally extended far south with the retreating bioclimatic zonation, and there are still scattered remnants to be seen in the higher mountains down to about 35–40°N and even further. However, the disturbing and confusing influences of the glaciations are also evidenced in the lichen flora, although more weakly than seems the case in higher plants. Perhaps the lichens suffered most from arid periods of climate, although it should be remembered that in present-day steppe conditions there is generally a rich lichen flora of crustose and foliose species on rock outcrops. An example of apparent influence of postglacial history on the present-day distribution is the range of *Parmelia olivacea* as mapped by Ahti (1966). In Canada it is much less frequent and vigorous than would be expected from its ecology and occurrence in northern Europe and Siberia.

B. Oceanic, Incompletely Circumpolar (Amphi-Atlantic and Amphi-Pacific)

This group includes those species that are present in all the four major areas with pronounced oceanic climates in the boreal zone, namely northwestern Europe (especially Scandinavia and Scotland), eastern Canada, western Canada, Alaska and the Soviet Far East (and Hokkaido). The last area is clearly the poorest known, but the fairly well known Japanese mountain flora may give some hint as to what can be expected in the northern Sikhote-Alin Range and elsewhere. In any case the North Pacific coast is considered here as only one area.

A good example of a truly boreal, oceanic species is provided by *Platismatia norvegica*, mapped by Ahlner (1948) and Culberson and Culberson (1968), which is common on *Picea abies* especially in Tröndelag, central Norway and adjacent Sweden, though it also grows on rocks down to the northern temperate zone. In the same way it grows in the strongly oceanic parts of the island of Newfoundland and in southern Alaska (Krog, 1968). Another lichen with similar disjunctions is *Cavernularia hultenii* (see Degelius, 1952; Ahti and Henssen, 1965), which, however, ranges down to the temperate zone along the west coast of North America, although it keeps to the boreal zone elsewhere. Other lichens with three to four major subareas in the boreal zone include *Cladonia tenuis* (Ahti, 1961b; essentially temperate), *C. bellidiflora*, *Alectoria sarmentosa* (Hawksworth, 1972), *Platismatia glauca* (Culberson and Culberson, 1968), *Hypogymnia tubulosa*, *Leptogium cyanescens* (Degelius, 1935), *Usnea longissima* (Ahlner, 1948), *Peltigera horizontalis* (Thomson, 1950) and *P. collina* (Thomson, 1950; also in eastern North America, Ahti, 1964). The extensions of their ranges vary but they do not presumably possess true transcontinental ranges.

Although this group is relatively small, containing something like 15 macrolichens, it should be noted that several microlichens apparently belong here and in any case this group is larger in lichens than in vascular plants.

C. Amphi-Atlantic

This is a small group of species, including for example *Erioderma pedicellatum*, *Pycnothelia papillaria*, *Cladonia floerkeana*, *C. strepsilis*, *Stereocaulon dactylophyllum* and *Lasallia pustulata*. Additional European species are expected to occur in the relatively poorly known coastal areas of eastern Canada and the northern Appalachian Mountains (cf. Degelius, 1940b; Brodo, 1968).

D. Amphi-Pacific

There are apparently a fairly large number of lichen taxa belonging to this group, but most of them are essentially temperate and only a small fragment of their range extends to what is included in the boreal zone here (cf. the maps of amphi-Pacific bryophytes compiled by Schofield, 1965: his maps 1–8). Such species include very few macrolichens, however, but *Cladonia pseudomacilenta*, *Hypogymnia pseudophysodes* and *Stereocaulon sasakii* may belong here. Species such as *Stereocaulon intermedium* (Lamb, 1951), *Nephroma helveticum* var. *sipeanum* (Wetmore, 1960) and *Cladonia pseudostellata* are slightly more widespread in the true boreal zone. Other species, like *Cladonia kanewskii* (Ahti, 1973), *Cetraria kamczatica* and *Asahinea scholanderi* (Culberson and Culberson, 1965; Thomson, 1972) are again essentially arctic but may be found in the boreal zone near the timberline. All the listed species are more or less coastal but some other species protrude far inland even though they are still clearly Beringian radiants, such as *Cetraria richardsonii* (Rassadina, 1950; Thomson, 1972), *C. laevigata* (Rassadina, 1950) and *Cladonia alaskana*.

E. Western Eurasian

The number of truly boreal species in this group must be very low, even if we include the species extending far eastwards to Siberia. However, a good number of essentially or largely temperate species of this group reach the southern subzones of the boreal zone. Among them, *Pseudevernia furfuracea* is quite widespread, while *Parmelia acetabulum*, *P. tiliacea*, *Cladonia foliacea*, *C. rangiformis*, *Ramalina fraxinea* and *R. fastigiata* (the latter two species are probably erroneously reported from eastern North America) are only marginal species in the boreal zone.

F. Eastern Eurasian

The number of truly boreal lichens seems small here too, but, in the same way as in Europe, a number of essentially more southern species protrude into the boreal zone especially in the coastal regions, e.g. *Hypogymnia mundata*, *H. fragillima*, *Parmelia pseudosaxatilis*, *Cetraria pseudocomplicata*, *Lobaria sachalinensis*, and *L. japonica* (this mapped by Yoshimura, 1971).

G. Interior Eurasian

This group includes the continental species that do not reach the coastal areas. They are particularly likely to occur in Yakutia and adjacent regions, a continuation of the oroboreal mountains of Mongolia and adjoining regions. *Umbilicaria pertusa* is in this group with the boreal zone, though it is also present in East Africa (Krog, 1973). Other possible examples, most of them little known species, include *Cetraria annae*, *C. komarovii*, *Lecidea (Psora) pulcherrima*, *Parmelia ryssolea* (Karavaev and Skryabin, 1971) and

P. sibirica. Among vascular plants in north-east Siberia this xerophytic element is well developed (Yurtsev, 1972). Similar outliers of xerothermic species are found along the southern fringe of the boreal zone in Europe, e.g. western Estonia (Trass, 1965a, 1968), Gotland and Oeland in Sweden (Albertson, 1950a,b) and Gudbrandsdalen, South Norway (Ahlner, 1949).

H. Western North American

There are a good number of species that are restricted to western North America and reach the boreal zone, especially the southern parts right along the coast, where they usually also extend as far down south as Oregon and California. Typical representatives are *Cavernularia lophyrea*, *Platismatia herrei* (Culberson and Culberson, 1968), *Hypogymnia duplicata*, *Lobaria oregana* (Jordan, 1973), *Pseudocyphellaria anomala*, *P. anthraspis* and *Parmelia multispora* (Ahti, 1966), and in more inland areas (reaching the Yukon or northern interior of British Columbia), *Cetraria canadensis* and *C. merrillii*.

I. Eastern North American

There are some widespread boreal (to temperate) species here, which may range from the coast over to Alberta or even Alaska in the west, like *Cladonia multiformis*, *C. cristatella*, *Parmelia trabeculata* (Ahti, 1966) and *Lasallia papulosa* (Llano, 1950; very rare in the west). Much more numerous are those essentially temperate species that extend along the Atlantic coast northward to Newfoundland or to the north shore of Lake Superior in Ontario, reaching the southern boreal areas at most. They include *Peltigera evansiana* (Thomson, 1950), *Physcia millegrana* (Thomson, 1963), *Parmeliopsis placorodia* (Culberson, 1956; Ahti, 1964), *Parmelia rudecta*, *Pseudevernia consocians* (Hale, 1968), *Cetraria aurescens* and *Platismatia tuckermanii*. The more oceanic species *Cladonia boryi* (Ahti, 1973) ranges throughout the boreal zone along the seaboard.

J. Interior North American

Like the respective element in Asia, this has a poorly developed lichen flora, connected with essentially montane elements in the south. A typical range for members of this element is that mapped for *Parmelia albertana* (Ahti, 1969; but see Section M, below); *Parmelia wyomingiaca* and *P. subolivacea* (Ahti, 1966) also seem to belong here. The range of this group may also resemble that of *Buellia epigaea* in North America (Thomson, 1972), which has a characteristic outlier in the arid boreal parts of the Yukon, namely Kluane Lake (Hoefs and Thomson, 1972).

K. Western Eurasian–Western North American

There are some conspicuous species in this group, including *Bryoria fremontii*, *Cetraria chlorophylla*, *Letharia vulpina* and *Tholurna dissimilis* (Østhagen, 1974). *Evernia prunastri* was formerly included here, but is now known as rare in eastern North America, too.

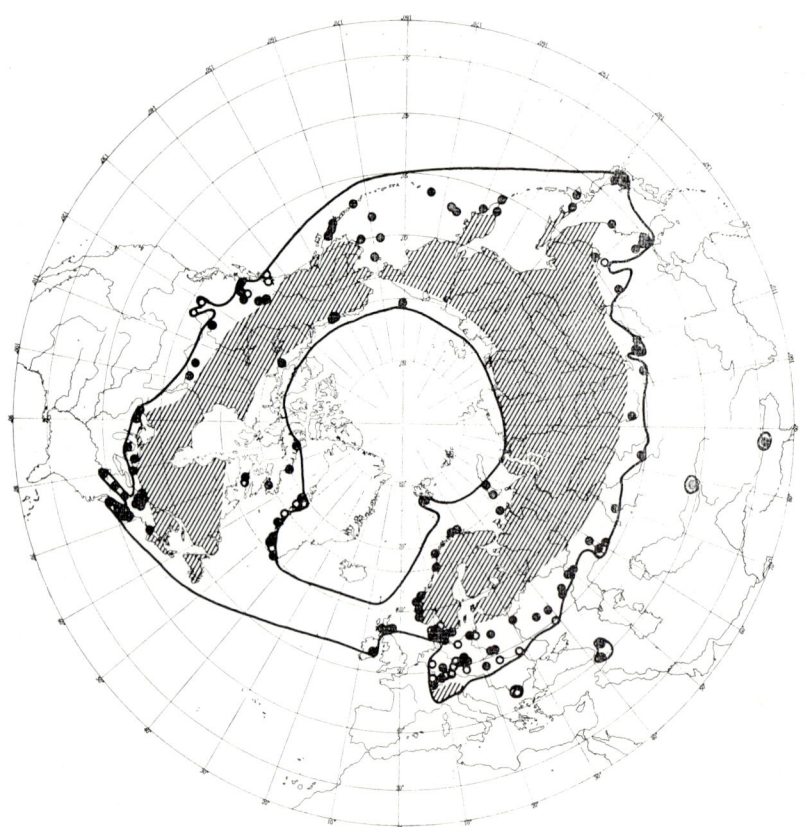

FIG. 2. Total distribution of a typical boreal lichen species, *Cladonia* (*Cladina*) *stellaris* (Opiz) Pouzar and Vězda, including the psoromic acid race called *C. aberrans* (Abb.) Stuck. Hatching indicates common occurrence, black circles marginal localities verified by herbarium specimens, and open circles literature records considered to be reliable. Chiefly after Ahti (1961, Figs 6 and 7 united), with several additions and corrections.

L. Eastern North American–Eastern Eurasian

This element much discussed in other plants is also represented by some lichens (Yoshimura, 1968; Kurokawa, 1972; Culberson, 1972). However, the species common to these regions are temperate rather than boreal, although some species, like *Lobaria quercizans* (Yoshimura, 1970; Jordan, 1973) do reach the marginal parts of the boreal zone. Indeed, there are a few lichens, such as *Collema pulcellum*, *Cladonia pseudorangiformis* (Ahti, 1962), *C. stellaris* var. *aberrans* (Ahti, 1961b; see also Fig. 2), *Umbilicaria muehlenbergii* and *Lasallia pensylvanica* (Llano, 1950), which occur in eastern North America and Asia and are widespread in the boreal zone, but they seem to be transcontinental in North America. *Parmelia squarrosa* is similar, but is almost certainly absent between the Great Lakes and the Rocky Mountains.

M. Interior Eurasian–Interior North American

This continental group includes *Parmelia ulophyllodes* (Rassadina, 1959), *P. albertana* (Ahti, 1969; recently detected in Mongolia), *Buellia epigaea* (Mattick, 1940; Thomson, 1972) and perhaps several grassland lichens, some not boreal, discussed by Looman (1964a). They are essentially confined to the arid interior regions of the continents. There is a group of other continental species which are incompletely circumpolar, being absent from the Bering Strait region and around the North Atlantic (in eastern Canada they may reach the coast, perhaps because of the thermally continental climate that prevails there); these are *Evernia mesomorpha*, *Cetraria ciliaris* agg. (Culberson and Culberson, 1967; Thomson *et al.*, 1969), *Bryoria furcellata*, *Hypogymnia bitteri*, *Ramalina minuscula*, *R. sinensis*, *Cornicularia odontella* and *Stereocaulon saxatile*. Some of the species listed belong to the dominant lichens in some "subcontinental" boreal regions (see Ahti and Hepburn, 1967).

According to the author's preliminary calculation there are about 500 species of macrolichens in the boreal zone (a number of very rare and marginal species omitted). Among them, *c.* 300 or 65% represent the circumpolar or almost circumpolar distribution group, while all the other groups are quite small. The most deviant area is the coastal region of the North Pacific, to which about 5–10% of the lichen species present in the boreal zone are confined. It is also expected that the highly oceanic coastal areas, like north-western British Columbia, southern Alaska, south-eastern Newfoundland, southern Sakhalin and middle Norway, have the greatest species diversity in the boreal zone. The arid regions of Yakutia, the Yukon

District of Mackenzie, Alberta and Saskatchewan have a very small number of species, especially in areas where no bare rock outcrops are found. The total number of lichen species occurring in the boreal zone may be estimated to be about 2500–3000.

IV. Ecoclimatic Differentiation

The treatment above is mainly based on the geographic distributional types of the species, which are essentially dependent on historical (florogenetic) factors. The influence of present ecological conditions on the ranges and composition of the lichen flora will be briefly illustrated below.

There are three principal climatic gradients in the boreal and temperate zones, namely:

1. The horizontal or *zonal* gradient, which is essentially based upon the amount of heat during the growing season; it is generally a south–north gradient;

2. The vertical or *altitudinal* gradient, which causes the zonation on mountains; it is also essentially based on the amount of heat;

3. The *sectorial* (sectional) gradient, which shows the gradation between the oceanic and continental regions or sectors; it is generally an east–west gradient, but may also appear as a north–south or even vertical gradient, depending on topography.

A. Subzonal Differentiation

Along the zonal gradient we may distinguish units, which may be conveniently called subzones. Ahti *et al.* (1968) distinguished four circumpolar subzones in the boreal zone: hemiboreal (=boreonemoral of Sjörs, 1963a), southern boreal, middle boreal and northern boreal (the hemiarctic forest–tundra was referred to the Arctic main zone). Quite obviously there are numerous "plurisubzonal" lichen species, i.e. they are so wide-ranging ecologically that in broad features they show no differentiation in distribution in the boreal subzones. A good example is *Hypogymnia physodes*, which is one of the most frequent and abundant epiphytic species throughout almost the whole of the boreal zone. However, weak sectorial differentiation is evident even in this species; e.g. it is rare in parts of the northern boreal oceanic birchwoods (as in Finnmark, North Norway). *Cladonia rangiferina* is a similar species with wide subzonal amplitude. Examples of weak subzonal differentiation are shown by *Cladonia stellaris* (see Fig. 2) and *Stereocaulon paschale*, which range throughout the boreal subzones but are most abundant in the middle and northern boreal subzones.

6. Lichens of the Boreal Coniferous Zone

Solorina crocea is a species of the northern boreal subzone (ranging further to the Arctic zone), having few localities in the middle boreal subzone. *Ramalina fraxinea* is common in the European hemiboreal subzone, scattered in the southern boreal and practically absent from the middle boreal. *Cladonia cristatella* in Canada is common up to the middle boreal but is rare in the northern boreal.

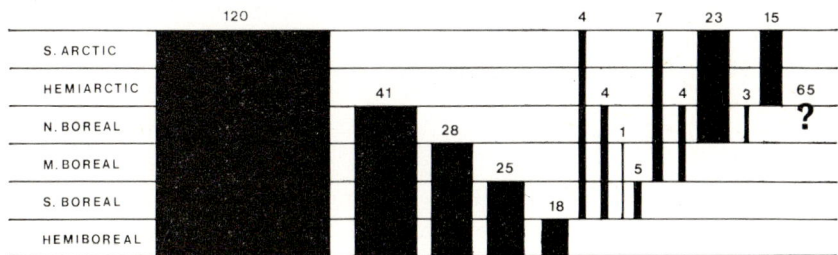

FIG. 3. Schematic subzonal spectrum of the ecoclimatic distribution of the macrolichen species of Finland (363 species). For instance, 120 species have a range covering all the subzones, while 41 species only reach from the hemiboreal to the northern boreal subzone. N = northern, M = middle, S = southern. The hemiarctic and southern Arctic subzones are only found as altitudinal in the country. The question mark represents 65 species, omitted because of rarity, taxonomic confusion, insufficient data on distribution or specific habitats not well represented in all the subzones (e.g. the strictly littoral species are excluded).

Figure 3 shows the subzonal differentiation of the macrolichen flora in Finland, which is almost totally boreal. About 30% of the species range throughout the country, without real boundaries, from the hemiboreal to the orohemiarctic subzone (the other oroarctic subzones are omitted here). About 35% have a northern boundary in the boreal zone, especially if very disjunct occurrences are excluded. Most of these restricted species reach the middle and northern boreal subzones, while 40% of them are confined to the hemiboreal (transitional to temperate) and the southern boreal subzones, representing extensions of the temperate lichen flora. On the other hand, 15% of the boreal macrolichens have a southern boundary in the boreal zone, usually along the line between the middle and northern boreal zones. In addition, there are a number of poorly known taxa, very rare species or some exceptional substrate specificities (marine and maritime species, extreme limestone dwellers etc.), which are difficult to place in this scheme. It is expected that approximately the same relationships prevail in other boreal areas even when the microlichens are included.

True boreal endemics among lichens are rare; these may include some crustose species, but very rarely macrolichens. This is probably due to the fact that the boreal flora is generally very young *in situ*. However, several species are concentrated in the boreal subzones, though they usually also reach either the northern temperate subzone or the southern Arctic subzone. These essentially boreal species, which have very few occurrences outside the truly boreal zone (oroboreal outliers in the southern mountains are not really outside) are generally most frequent in the somewhat continental parts of the northern boreal subzone, e.g. *Nephroma bellum, Hypogymnia bitteri, Cladonia bacilliformis, C. alaskana, Lecidea elabens, Lecanora coilocarpa* and *L. fuscescens*. Many of the lichens discussed by Ahlner (1948) and Hasselrot (1953) belong to this group.

B. *Sectorial Differentiation*

Sectorial differentiation is actually an item much investigated by European lichenologists and other plant geographers, who have particularly studied the influence of oceanicity on the lichen flora. However, the Scandinavian oceanic element (Degelius, 1935) for instance, esentially falls outside the boreal zone. That element is not only oceanic, but at the same time mainly confined to the northern outliers of the northern temperate subzone, which extends farthest north along the coast of Norway. Many come up to the hemiboreal or southern boreal subzones though. The distribution types (referred to above as western Eurasian, amphi-Atlantic etc.) are mainly due to the fact that many species require more or less oceanic conditions, being dependent on high relative moisture content of the air or else they are unable to stand the extreme low temperatures of the continental sectors. The climatic dependence of species like *Platismatia norvegica, Cavernularia hultenii* and *Pseudocyphellaria crocata*—all characteristic of damp, oceanic spruce forests in the boreal zone—is obvious to anyone who has observed their habitats in the field, both in Europe and North America.

As a rule the oceanic lichens in Europe are oceanic also in Asia or America. If this is not so, there is usually something wrong with the taxonomic delimitation of the species. A rare exception to this rule is probably *Cladonia scabriuscula*, which is a suboceanic, temperate to hemiboreal lichen in Europe, but in Canada it is not uncommon throughout the boreal zone, for instance in Ontario, and it is also reported in Baffin Island and Greenland.

Slightly oceanic species may reach the central parts of the continents, which are floristically least known both in Eurasia and Canada. About 30% of the boreal macrolichen species seem to have climatically dependent oceanic tendencies. Many examples were listed in connection with the geographic elements but in addition there are species which are only less frequent in the interior areas of the continents, but still probably transcontinental, like *Parmelia olivacea* and *Cladonia arbuscula*. Perhaps most circumpolar species show some differences in abundance along the circumpolar sectorial gradients, but they will become obvious only by means of detailed mapping.

On the other hand, continental species have been much less studied. Some outstanding examples of continental lichens of conifers were extensively discussed by Ahlner (1948). For instance, *Evernia mesomorpha* and *Cetraria ciliaris* (including *C. halei*) are rare species in Fennoscandia but eastwards in parts of Siberia and interior Canada they compete with *Hypogymnia physodes* in abundance. Incidentally, *Usnea longissima*, *Ramalina roesleri* and *R. thrausta*, which Ahlner also included in "continental, eastern species" in Fennoscandia, seem to be rather oceanic from the circumpolar point of view, being commoner near coasts and are not present in the highly continental areas at all. It should also be noted that in Newfoundland (oceanic in the hygcri sense but mostly continental in the thermal sense), some species behave differently from those in Europe and western North America. The clearly continental *Evernia mesomorpha* is extremely rare, while the likewise clearly continental *Bryoria furcellata* and *Cetraria ciliaris* are extremely common there right down to the coast, growing even with *Cavernularia hultenii*. In Fennoscandia and Alaska, *Cetraria ciliaris* and *Cavernularia* would never meet. Some species are apparently more dependent on temperature factors, others on moisture factors.

V. Edaphic–Topographic Differentiation

Of course, the climate cannot explain the whole ecological distribution of a lichen species. The qualities and distribution of various substrates are decisive for many species.

In boreal Europe there is a great Baltic or Fennoscandian shield of Precambrian rocks and in eastern Canada a still wider Canadian shield, most of the remainder being formed by sedimentary rocks. A most important fact is that while the Precambrian areas are mainly oligotrophic, the areas of sedimentary rocks are eutrophic. This is also seen in the lichen flora of base soil, rock outcrops etc., since the flora of calcareous habitats is considerably different from that of acidic and oligotrophic habitats. The Precambrian areas are generally also richer in bare, hard rocks.

There seem to be many lichens which have a wide climatic amplitude but which are strictly substrate-specific, like *Placynthium nigrum*, *Solorina saccata* and many Collemataceae—all outstanding calcicole lichens. There are scattered calcareous or otherwise eutrophic areas even in the Precambrian regions, and some calcicole lichens (e.g. *Cladonia pocillum*) have been able to spread also to small calcareous spots, but generally the maps of such species clearly reflect the edaphic conditions of each region. The acidophilous saxicolous and terricolous lichens behave in the same way in predominantly calcareous regions. Epiphytic and epixylic species often show more continuous distributions, because they are not so much dependent on substrate. However, the frequency of some substrate trees or phorophytes, like *Populus*, may cause anomalies in the distribution maps of certain lichens (e.g. *Lecanora populicola*), which are almost host-specific to this tree genus.

VI. Differentiation due to Man

The activities of man play a very important role in the distribution of vascular plants, a factor often considerably underrated. The influence of man's activities on lichen distribution is not so great, although in certain heavily populated areas, like the Netherlands and Great Britain (Hawksworth *et al.*, 1974) the observed changes have been drastic, mostly due to air pollution and destruction of forest communities. But apparently the "weedy" flora is not as well developed in lichens as it is in the phanerogams. In any case the native forest communities of the boreal zone have been preserved much better than in the temperate zone. This is because agriculture is only practised to a small extent except for some hemiboreal areas like southern Sweden. On the other hand, the forests have been used extensively in the wood industry so that considerable areas are not in a virgin state. This has affected the terricolous and saxicolous flora relatively little, but it may be that a number of epiphytic and epixylic species have been badly reduced in the boreal zone, except in the most inaccessible areas of Siberia and Canada.

However, a great number of lichen species in the boreal zone are exclusively or almost exclusively found on man-made or man-influenced habitats. Several of them may not be really native in that zone. Linkola (1938, 1940) made a study on some Finnish bark lichens in relation to their possible dispersal by human agencies (anthropochorous). For instance, he regarded *Parmelia acetabulum* to be only anthropochorous in that country, since it never occurs outside parks and yards and almost always on planted southern broad-leaved trees by old manors or in town parks. *Ramalina fraxinea* and *R. fastigiata* are also mainly confined to man-made habitats, being presumably native only right along the south-west (hemiboreal) coast of Finland. *Xanthoria fallax* and *Anaptychia ciliaris* (excluding var. *melanosticta* = *A. mamillata*) are also considered to be mainly of anthropochorous origin in the country. In the hemiboreal subzone of Sweden saxicolous northern lichens such as *Umbilicaria cylindrica*, *U. proboscidea* and *U. hyperborea* have numerous southern outposts, but these are almost exclusively on man-made habitats, e.g. stone walls, stones in fields etc. (Hasselrot, 1941, 1953). These lichens cannot be relics in the area, and the only explanation is that these essentially arctic lichens have spread southwards into these relatively new and very open habitats.

Xanthoria parietina became common in wide areas in Finland only after the period 1850–1880 (Linkola, 1938, 1940). For instance, the very reliable lichenologists J. P. Norrlin and E. A. Vainio published several local floras on lichens in central, eastern and northern Finland around 1870, without reporting this lichen in many areas where it is now common (though mainly in villages). In Lapland, *X. parietina* has apparently also spread in recent times, after Linkola's study.

Räsänen (1926), in his study-area in west-central Finland (mainly middle boreal), reported 22% of the lichen taxa (589 in all, including several obvious ecads) to be clearly hemerophilous, i.e. growing mainly in man-made habitats or otherwise drawing benefit from man's activities in nature. However, most of the species were considered to be native to the region (apophytes), and only few, such as *Physcia orbicularis*, *P. nigricans* var. *tremulicola*, *Caloplaca saxicola* and *Lecanora dispersa* (*L. hagenii*) were called synanthrope (aliens). In fact, the number of "aliens" may be even higher, since when completely uninhabited boreal forest regions are compared to inhabited (agricultural) areas, the lichen flora of the latter seems to include a considerable number of species absent from the former. Some idea of the distribution of various lichens along the "anthropogenic gradient" may be obtained from the informative table published by Koskinen (1955, Table 1); see also Räsänen (1927, his p. 182).

VII. Effect of Air Pollution

Air pollution caused by sulphur dioxide or other pollutants is not at all as widespread in the boreal zone as it is in many temperate areas. This is simply because the industrial districts are relatively small in extent. However, severe local pollution commonly occurs in cities and around various industry. In the hemiboreal subzone, studies on the effect of air pollution upon lichens have been made in several places, e.g. Oslo, Stockholm, Uppsala, Turku, Helsinki and Tartu in Europe, and Sudbury, Wawa and Arvida in Canada (see LeBlanc and Rao, 1973; Trass, 1973). As to the southern boreal subzone, papers on Köpmanholmen (Moberg, 1968) and Örnsköldsvik (Westman, 1975) in Sweden; and on Tampere (Sahrakorpi, 1973; Ranta, 1974) and Harjavalta (Laaksovirta, 1973) in Finland have been published, with several more towns under study in this subzone. No extensive studies in the northern boreal subzone appear to have been conducted. Also the studies in the U.S.S.R. are remarkably few, but in the Siberian cities Krasnoyarsk and Irkutsk, for instance, the effect of pollution is very clear (in Irkutsk no lichens were detected by Ahti in the city in 1970 and according to a local botanist they only appear 7 km from the centre).

It is clear from the published studies and otherwise that the biological scale for mapping atmospheric pollution on the basis of lichen sensitivity proposed by Hawksworth and Rose (1970) for Britain does not work well in the boreal zone, not even in Europe, except perhaps in the suboceanic parts of the hemiboreal subzone. In fact, such general scales must be constructed separately for each major sector of the biotic zones (sometimes even subzones). This is because the lichen flora and the indicator value of each species is often somewhat changed from one climatic region to another.

For instance, *Lecanora conizaeoides*, *Buellia canescens* and epiphytic *Parmelia saxatilis* (see Hawksworth and Rose, 1970) are absent or rare in the boreal zone, and *Lecanora expallens* and *Lepraria incana* less abundant. On the other hand, *Bacidia chlorococca* is very abundant and toxitolerant (Ahti and Vitikainen, 1974), and *Hypogymnia physodes*, *Parmelia sulcata*, *P. olivacea*, *Parmeliopsis ambigua*, *Lecidea scalaris*, *Platismatia glauca*, *Pseudevernia furfuracea*, *Cetraria chlorophylla* and *Ochrolechia turneri* are other important species of non-eutrophicated bark in the "struggle zones" of city environments. In Ontario (Rao and LeBlanc, 1967), for example, characteristic species include some additional taxa, such as *Cetraria ciliaris* and several *Bryoria* spp. In Siberia, and generally in towns of more or less arid areas, the trees are commonly devoid of any lichens, perhaps not only because of air pollution, but because of excessive drought and severe winter conditions.

In the boreal zone the studies on the relations of lichens to air pollution must often be based on the epiphytes of pine, spruce and birch trees (birch is often the dominant planted and native tree in the northern and middle boreal towns) which are very poor in lichen species, and therefore a detailed mapping of pollution zones on a floristic basis alone may be impossible. The pollutant contents and the degree of injury of *Hypogymnia physodes* (Lundström, 1968) often become decisive criteria in delimitation of the zones.

VIII. Speciation of Boreal Lichens

It is characteristic of the boreal lichen flora, particularly of the foliose macrolichens, that many apomictic species are common. Such lichen species only or almost only reproduce in a vegetative manner, i.e. by soredia, isidia or thallus fragmentation (Thomson, 1972). Poelt (1963, 1970, 1972) called such species secondary and they seem to be especially widespread in the areas of Pleistocene glaciations, which are now largely boreal. Their ancestors, the regularly fertile primary species, rarely reach the boreal zone, but are usually confined to regions which are now temperate or still warmer and as a rule also more oceanic. Since the primary species are considered to be Tertiary relics, it is understandable that they are unable to thrive in the boreal conditions, which presumably were not widespread in the Late Tertiary. However, in largely unglaciated coastal areas, many primary species like *Hypogymnia enteromorpha* (Alaska), *Pannaria rubiginosa* (Newfoundland, Norway), *Evernia esorediosa* (East Asia, Alaska?) and *Cavernularia lophyrea* (Alaska) are found within the boreal zone. It is true that a few primary species among those listed by Poelt (1970) are widespread in this zone (e.g. *Physcia ciliata*, *Xanthoria parietina*, *X. elegans* and *Lecanora intricata*), but they are nitrophilous species, which seem to be able to spread more easily than the non-nitrophilous lichens.

Apomixis explains the remarkably complete morphological and ecological similarity of many boreal lichen populations (e.g. *Hypogymnia physodes* and *H. tubulosa*) in widely different areas (like Newfoundland and Finland). These species are not expected to have much morphological or ecotypic differentiation. Such differentiation is often clearly discernible in most sexual vascular plants in the same regions, even when the populations represent the same taxonomic subspecies or variety. Also lichens which regularly produce spores are apt to show some variability in appearance as well as chemistry, when disjunct populations are carefully compared. *Cladonia rangiferina* displays such a regional variation in colour, branching pattern etc.—perhaps also in chemistry, since Krog (1968) reported an atranorin-deficient strain in Alaska—that it is apparently not a typical apomictic secondary species (apothecia are actually common though inconspicuous). Similarly the regularly fertile *Parmelia olivacea* is not fully homogeneous throughout its range (Ahti, 1966).

On the other hand, *Ramalina farinacea* agg. is a typical widespread secondary species (richly sorediate, almost never fertile), but it has several chemical races in the boreal zone. If all the chemical apomictic entities really derive from one fertile ancestral species, as should be perhaps assumed, then *R. farinacea* agg. must be polyphyletic and the races cannot be very young. Yet one is not inclined to give specific status to these races (excluding saxicolous *R. subfarinacea*), unlike Culberson (1966), because they do not seem to show significant ecological and geographical differentiation in the field. Their total ranges are still unknown, however.

Even the regularly spore-producing boreal species are usually very homogeneous throughout the zone. However, although the race formation of most species is poorly studied, it is known that some species have distinct intraspecific regional taxa, which could be called subspecies or varieties. They may be recognized morphologically (*Cladonia uncialis* ssp. *dicraea*, *C. arbuscula* ssp. *beringiana*, *Collema occultatum* var. *populinum*, *Parmelia olivacea* var. *albopunctata*), chemically (*Cladonia stellaris* var. *aberrans*; the races often unnamed) or by both characters (*Cladonia gracilis* var. *nigripes*, *Parmelia omphalodes* var. *discordans*). In some cases there are closely related species in different areas (vicariants), e.g. *Cetraria juniperina–C. canadensis* (*–C. viridis*); *Cladonia impexa–C. pacifica–C. terraenovae*; *Cladonia uncialis–C. pseudostellata*, which might rather be called subspecies in some cases.

IX. Terrestrial Lichen Communities

In the boreal zone the role of the terrestrial (epigeic) lichens and bryophytes in the dominant regional (zonal) plant communities is generally much more conspicuous than in the warmer zones (Rübel, 1927). In addition to various quite temporary successional stages, lichens may be found as predominant plants in the understorey vegetation even in what could be called climax communities. In part, the abundance of lichens in the boreal zone is due to the widespread occurrence of glacial sandy deposits and hard rock outcrops covered by a thin veneer of soil. Such heavily podzolized, usually very oligotrophic and fairly dry soils are unable to support any vegetation other than open tree stands and their successions after fire and other disturbance are much slower than on mesic and wetter sites, thus preventing effective competition by vascular plants and bryophytes with lichens.

The lichen woodlands on sand are most abundant in the northern boreal subzone (Hare, 1950, 1954; Hustich, 1951; Lavrenko and Soczava, 1956; Fraser, 1956; Ritchie, 1959; Ahti, 1961a; Hämet-Ahti, 1963; Ahti and Hepburn, 1967; Ahti *et al.*, 1968), but in certain regions they are common also in the middle boreal subzone, while in the more southern subzones they are confined to the driest sandy areas only, and are always pine woods. The dominant tree in the northern boreal lichen woodlands is either spruce (*Picea mariana* or *P. glauca* in Canada and Alaska; *P. abies* ssp. *obovata* in north Russia and west Siberia), larch (*Larix laricina* in Canada; *L. gmelinii* or *L. sibirica* in Siberia) or pine (*Pinus banksiana* or *P. contorta* ssp. *latifolia* in Canada; *P. sylvestris* in Russia and Fennoscandia). The field layer in these woods displays considerable regional, subzonal and sectorial differentiation, but many ericaceous dwarf shrubs and *Empetrum* are widespread.

It is highly noteworthy that the ground layer, including both lichens and bryophytes, is very uniform throughout the boreal zone (Kornaś, 1972). It includes about 25–40 species of lichens (phytosociologists generally overlook and misidentify several species in this community type) and 5–15 bryophytes. However, the only really abundant species are the so-called reindeer lichens ("caribou mosses"—*Cladonia* subgen. *Cladina*): *Cladonia rangiferina*, *C. mitis*, *C. arbuscula* and *C. stellaris*. As a rule very few other lichens, notably *C. uncialis*, *C. amaurocraea* (chiefly on rocks), *Stereocaulon paschale* and *Cetraria islandica* agg., are able to compete successfully with *Cladina* spp. in the boreal lichen woodlands. The rest of the flora is mainly composed of cup lichens (*Cladonia* subgen. *Cladonia*), and several species of *Cetraria* and *Peltigera* are also frequent, but in mature stands their degree of cover is usually below 1%. In Canada, *Cetraria laevigata*, *Cladonia cristatella*, *C. pseudorangiformis* and *C. alaskana* are almost the only lichens in this community type that are not present in Europe. In Siberia, *Cetraria laevigata* (coupled with the absence of *C. ericetorum*) may be the only common non-European species.

Fire is a very important ecological factor all over the boreal forest, especially in the continental areas. In lichen forests, fires used to be, and still widely are, particularly destructive to dry vegetation. The succession of boreal lichen woodlands after fire has been studied in Europe (Kujala, 1926a; Sarvas, 1937; Jalas, 1953; Korchagin, 1954; Uggla, 1958), Asia (Rabotnov, 1936) and North America (Hustich, 1951; Lutz, 1956; Ahti, 1957, 1959; Scotter, 1964, 1970; Rouse and Kershaw, 1971; Bergerud, 1971; Rowe and Scotter, 1973). The following rough pattern of succession has been established over almost all the boreal forest:

1. Bare soil stage, 1–3 years after fire.
2. Crustose lichen stage, 3–10 years after fire; *Lecidea oligotropha*, *L. uliginosa* and *L. granulosa* dominant.
3. Cup lichen stage, 10–30(–50) years after fire; *Cladonia* subgen. *Cladonia* dominant (e.g. *C. cornuta* var. *cornuta*, *C. gracilis* var. *dilatata*, *C. crispata*, *C. gonecha*).
4. First reindeer lichen stage, 30(–50)–80(–120) years after fire; *Cladonia mitis*, *C. arbuscula*, *C. rangiferina* and *C. uncialis* dominant.
5. Second reindeer lichen stage, 80(120) or more years after fire; *Cladonia stellaris* dominant.

It should be noted that the time-table of this succession is greatly dependent upon the moisture regime and the climatic position of the stand, and different rates of succession may be encountered side by side (Jalas and Valpas, 1962). In somewhat mesic lichen forests (recognized by a thicker humus layer) there may be a stage of very dense, young forest, when lichens are temporarily in decline and even absent, although they appear again when the climax is approached. It also seems that in oceanic areas and in the hemiboreal subzone in general, *Cladonia stellaris* is not able to attain dominance at all, but the first reindeer lichen stage is the long-persistent "final" phase. In addition, there is a *Stereocaulon paschale* stage in some continental areas, such as western Lapland (Ahti, 1961a) and northern Manitoba (Ritchie, 1959), but its ecological background is not well understood. In early phases of succession on thin soils of rock outcrops, some deviations in the above pattern may be discerned (Häyrén, 1914).

Since the lichen woodlands are important winter ranges to both domesticated reindeer and wild reindeer or caribou (all races of *Rangifer tarandus*), the biomasses of these communities are also fairly well known (e.g. Andreev, 1954; Skuncke, 1958, 1959, 1963, 1969; Bevis and Krueger, 1964; Scotter, 1964; Trass, 1968; Bergerud, 1971; Kärenlampi, 1973; Kallio and Kärenlampi, 1975). In the *Cladonia stellaris* stage the standing crop of palatable reindeer lichen may exceed 3000 kg ha^{-1} (height 12 cm; Kärenlampi, 1973) and form as much as 50% of the total above-ground vegetation biomass, but the annual growth of such an "overmature" lichen land may be negligible. The most productive lichen-rich forest for the reindeer industry only has a 4–6-cm tall lichen stand, which may support about 500–1500 kg ha^{-1} of forage, with an annual growth of 60–160 kg ha^{-1}. Such an optimal range is in the first reindeer lichen stage, having little of *Cladonia stellaris* present (Ahti, 1957, 1959, 1961a). For sufficient regeneration it only needs about 3–5 years (Andreev, 1954, 1960; Kärenlampi, 1973), which means that pasture rotation is necessary in areas with high grazing pressure. If the range management is not well organized, the reindeer pastures may become badly overgrazed, leading to slower regeneration. This is the situation in much of northern Finland, where the reindeer lichen stands are often only about 2 cm high, thus being far from the climax phase.

In recent years detailed experimental analyses on the role of *Cladonia* mats in the northern ecosystems have been conducted in Finnish Lapland (e.g. Kärenlampi, 1970a,b; Bunnell *et al.*, 1973), northern Ontario (e.g. Kershaw and Rouse, 1971a,b; Lechowicz and Adams, 1974) and Labrador (Kershaw and Field, 1975).

The *Cladonia stellaris* phase has been classified as "the association *Cladonietum alpestris*" (Klement, 1955, 1959; Looman, 1964a,b) or the "*Cladonia alpestris* sociation" (Trass, 1968) by workers who emphasize the relative independence of the terrestrial cryptogamic communities on the accompanying vascular vegetation. Although such a one-layer community is almost circumpolar and occurs seemingly independently on the floors of various forests, the present writer would rather classify it in connection with the other plants, the whole phytocoenosis, and in any case the associated bryophytes should always be included.

Very considerable terrestrial lichen stands, also generally dominated by reindeer lichens but with a slightly different composition are developed in certain mire communities. Raised bogs, blanket bogs and palsa bogs in particular (Ahti, 1959; Ruuhijärvi, 1960; Ritchie, 1960; Eurola, 1962; Sjörs, 1963b; Ahti and Hepburn, 1967) may have good stands of lichens on hummocks, or sometimes even in hollows (*Cetraria delisei*, *C. islandica*, *Cladonia subfurcata*, *C. squamosa* agg. and *Ochrolechia frigida* are common in wet places). The lichen flora of boreal bogs is discussed in some detail by Räsänen (1919) and Paasio (1931).

In less dry and mesic woods, mosses take over the dominance in the boreal forest (Kujala, 1926b; Lambert and Maycock, 1968), but even then some lichens are fairly regularly found immixed. Besides a few "mesic" *Cladonia* spp. (*Cladonia furcata*, *C. multiformis*, *C. scabriuscula*, *C. gracilis* agg., *C. rangiferina* and *C. arbuscula*), some *Peltigera* spp. (*P. aphthosa*, *P. polydactyla* agg., *P. scabrosa*, *P. malacea* and *P. canina*) and *Nephroma arcticum* are widespread boreal terrestrial lichens. The presence of nitrogen-fixing blue-green algae as phycobionts or cephalodia in *Stereocaulon*, *Solorina*, *Peltigera* and *Nephroma* is expected to be of importance to the general ecology of boreal forest vegetation (Kallio et al., 1972; Kallio and Kallio, 1975).

X. Epiphytic Lichen Communities

With the exception of some continental regions, lichen epiphytes are plentiful on boreal trees (note that bryophyte epiphytes are usually scarce) and commonly form distinctive communities. These have been studied to some extent, though much less than the terrestrial communities, and with many different approaches. Therefore the overall picture is quite fragmentary and the general phytogeographical features are not precisely known at all. Of course, this situation only reflects the highly controversial state of phytosociology of epiphytic communities (Barkman, 1973).

Du Rietz (1945) and Krusenstjerna (1945) proposed a very useful basic scheme for the epiphytic communities of the common boreal trees in (primarily hemiboreal) Sweden. They distinguished the federation *Physodion*, which is a community type found on acid, oligotrophic barks. It could be divided into four unions as follows:

1. The *Euphysodetum*, characterized by *Hypogymnia physodes*, *Platismatia glauca* and *Pseudevernia furfuracea*, and occurring mainly on trunks and twigs of pine and spruce trees;

2. The *Physodeto-Sulcatetum*, characterized by *Parmelia sulcata*, *Evernia prunastri* and others, and occurring on trunks of *Betula*, *Alnus*, *Quercus*, *Salix* and *Sorbus*;

3. The *Parmeliopsidetum*, characterized by *Parmeliopsis ambigua*, *P. hyperopta* and *Cetraria pinastri*, and confined to the bases of conifers and the above listed hardwoods, where it also marks the average thickness of winterly snow cover in the woods (snow-tolerant community);

4. The *Ptilidietum*, characterized by some bryophytes (e.g. *Ptilidium pulcherrimum*, *Hypnum cupressiforme*) and *Cladonia coniocraea*, and found at shady bases of trees, primarily birch.

This scheme, even if not complete and not showing the geographic variation, is still applicable in broad features throughout the boreal forest, from Europe to Canada, irrespective of which names the species' assemblages are given. A more complete synusial system was outlined by Trass (1968) for Estonia (see also Almborn, 1955, though his area is northern temperate). Du Rietz (1945) also recognized the federation *Xanthorion* (on *Populus*, *Acer* etc.), but its boreal lichen unions have not been well classified. It may be noted that perhaps all boreal *Abies* trees, as well as *Picea glauca*, seem to have the union *Physodeto-Sulcatetum* (with "mesotrophic" bark) rather than the *Euphysodetum*, which is clearly the normal union of *Picea mariana* (cf. Raup, 1930a; Ahti and Hepburn, 1967).

Barkman (1958) cited only a few examples of epiphytic communities in the boreal zone (see also Barkman, 1954), for example, the associations of *Alectorietum fremontii* (on pines in dry northern boreal woods), *Parmeliopsidetum ambiguae* (see above) and *Parmelietum olivaceae* (especially on mountain birch, above the snow-cover line; Hämet-Ahti, 1963). All these entities are easily recognizable and thus acceptable, while some of the common communities (many included in the *Pseudevernietum furfuraceae*) may be placed into Barkman's units only with difficulty. However, all the boreal epiphytic unions have very wide ranges (often circumpolar), something very rare in the case of vascular plant communities.

Sõmermaa (1972) made a thorough study on the composition and ecology of the arboreal lichen flora in relation to various forest types in (hemiboreal) Estonia. Koskinen (1955) worked with a somewhat similar method in central Finland. They showed that the flora of each tree species generally differs from the others and that the surrounding forest is also an important factor. Koskinen's study revealed how great the number of little-known crustose lichens is in the boreal epiphytic communities; only after considerable taxonomic and floristic work will it be possible to classify the epiphytic communities properly. Another problem in the classification is that all the common boreal trees (spruces, pines, larches and birches) have a rather unstable, flaking bark surface, except in the case of very old trees, and therefore the communities are often rather fragmentary and temporary assemblages.

In North America, the species relations and ecology of corticolous lichen communities have been studied on *Populus tremuloides* (Jonescu, 1970) in the southern boreal (to hemiboreal) subzone of Saskatchewan (Jesberger and Sheard, 1973) and on *Picea mariana* (Yarranton, 1972) in Ontario by means of ordination and other quantitative methods. Records of epiphytic communities in Ontario and Quebec have also been published by Ahti and Hepburn (1967, mainly on conifers), Lambert and Maycock (1968, mainly temperate) and in several papers on air pollution (e.g. LeBlanc et al., 1972a,b; Rao and LeBlanc, 1967). Some of the temperate unions of Quebec (LeBlanc, 1963) are also found in the boreal areas. Raup (1930a,b) made some observations on the boreal epiphytic communities in the Great Slave Lake area and northern Alberta. The community group called by LeBlanc (1963) the *Evernia mesomorpha* union seems to be widespread, but these data are too scanty for phytogeographic generalizations.

The productivity of the epiphytic lichens has been studied mainly in connection with reindeer and caribou range surveys. In northern boreal Saskatchewan, Scotter (1962) reported 359–572 kg acre^{-1} (145–231 kg ha^{-1}) of lichens (air-dry) on mature *Picea mariana* and 217–1167 kg acre^{-1} (88–472 kg ha^{-1}) on *Pinus banksiana* (see also Scotter, 1964). On *Picea* more than 50% of the lichen loads occurred within 3 m above ground (within reach of caribou), while on *Pinus* only 20% occurred within this area. Most of the loads consisted of "*Alectoria jubata* agg.", *Evernia mesomorpha* and *Usnea hirta*. Similar loads (114–1332 kg acre^{-1}; 46–539 kg ha^{-1}) mainly of *Alectoriae* were collected on trees in the more oceanic, oroboreal woods in British Columbia by Edwards et al. (1960). In Kamchatka, Trass (1965b) recorded 220 kg ha^{-1} of lichens on *Betula ermanii* and in Estonia 402–480 kg ha^{-1} on *Pinus sylvestris*. The regeneration of these stands is probably slower than in terrestrial stands.

XI. Lignicolous Lichen Communities

The lignicolous lichen flora includes many characteristic boreal lichens, notably members of the Caliciales, and the genera *Lecidea*, *Catillaria*, *Micarea* and *Xylographa*. If the flora of soft, rotten logs and stumps is included, there are also several *Cladonia* spp., e.g. *C. cenotea*, *C. bacilliformis*, *C. carneola*, *C. botrytes* and *C. bacillaris*. Such habitats are very common in the boreal forest, and in the northern boreal zone they are especially persistent because of the low decomposition rate of dead trees. Very characteristic for northern boreal Europe are desiccated standing pines, the distribution of which in Finland has been mapped by Kalliola (1966). These lichenologically interesting trees are now much used in the construction of summer cottages etc., and therefore in danger of gradual extinction. Their flora frequently includes species like *Embolidium italicum*, *Lecidea elabens* and *Cyphelium tigillare*.

It is surprising how few phytosociological and ecological studies have been undertaken on epixylic lichens and the associated epixylic fungi (e.g. *Agyrium rufum*). However, in central Sweden, Sernander (1936) distinguished six, and Arnborg (1943) eight, stages of decomposition ("necrotization") of logs, stumps etc. in woods; each stage may have a characteristic lichen or bryophyte community (see also Krusenstjerna, 1945). Other useful data are found in papers by Krohn (1924; in part unreliable), Räsänen (1926, 1927), Kujala (1926b), Koskinen (1955) and Lundström (1970). The conditions in Asia and North America are almost unknown, although lists of lignicolous species have been published for certain boreal regions (Thomson, 1951, 1953; Thomson et al., 1969).

XII. Saxicolous Lichen Communities

The numerous rocky outcrops and loose stones encountered in the extensive Precambrian areas in the boreal zone are normally covered by a rich lichen flora, forming relatively stable communities, since the rocks are mainly composed of hard, slowly weathering granites and gneisses.

The lichen flora of different boreal rock types is surprisingly little studied. Among the very characteristic lichens of bare siliceous rocks may be mentioned the yellow species of *Rhizocarpon* (e.g. *R. geographicum* agg.), the brown and grey species of the same genus (e.g. *R. obscuratum* and *R. grande*), numerous black-fruited species of *Lecidea* (e.g. *L. lapicida* and *L. lithophila*), *Lepraria neglecta* agg., *Umbilicaria hyperborea*, *U. deusta* and *Parmelia saxatilis*. The flora of calcareous rocks is quite different (cf. Degelius, 1955). In northern Finland, Räsänen (1953) recognized six groups of rock types according to their lichen flora, each group distinguished by a characteristic colour: (1) anorthosite and ossipite, (2) serpentine-peridotite, (3) quartzite, (4) granite, (5) gabbro, greenstone, basic schists and slates and (6) dolomite.

In addition to the epilithic (on naked rock) communities there are well developed, often extensive, chomophytic (on thin soil) communities on rocks. The chomophytic communities are often treated in connection with lichen woodlands on deep soils, but in careful analysis the communities on rocks are slightly different in composition. Their mature stages are usually dominated by *Cladonia* subgen. *Cladina*; see for instance the relevés of

Cladinetum described from hemiboreal Sweden by Krusenstjerna (1945) and from western Ontario by Ahti and Hepburn (1967). Degelius (1940a) gave an example of the sequence of successional stages on boreal rocks in Uppland, Sweden:
1. *Trentepohlia iolithus* stage
2. First crustose lichen stage: *Bacidia umbrina, Rhizocarpon lecanorinum, Candelariella vitellina, Lecanora intricata* etc.
3. Second crustose lichen stage: *Aspicilia cinerea, Lecidea fuscoatra, Parmelia conspersa* etc.
4. Foliose lichen stage: *Parmelia saxatilis, P. taractica, Hypogymnia physodes* etc.
5. Moss and fruticose lichen stage: *Hedwigia ciliata, Cladonia* spp. etc.

There must be many variations of this pattern which are not yet known. As a rule, there are considerable geographic and substrate-dependent differences in the saxicolous vegetation, but they have been little studied. There are several studies on the zonation along boreal sea-shore rocks (e.g. Häyrén, 1914; Du Rietz, 1921, 1932; Kärenlampi, 1964) and also on the lacustrine zonation (e.g. Sernander, 1912; Häyrén, 1914; Santesson, 1939a,b) and other special ecological habitats (Hakulinen, 1962a,b), but the ordinary rocks and their communities have received little attention. However, some data have been published (e.g. Sernander, 1912; Du Rietz, 1921, 1932; Räsänen, 1927; Krusenstjerna, 1945, 1965; Klement, 1959; Trass, 1968) and many of the communities described by Wirth (1972) in central Europe seem to be present in the boreal zone too. The literature on boreal epilithic communities in Asia and North America is negligible (e.g. Lambert and Maycock, 1968, included some epilithic and chomophytic lichens in their "terricolous" lichens).

XIII. Conclusions

It should be emphasized how exceptional the boreal lichen flora is in the world: how a very homogeneous flora is able to cover tremendous areas, from Norway through Kamchatka and Alaska to Newfoundland, forming an essentially continuous, circumpolar band. This is because the ecological conditions in this area vary relatively slightly, the influence of florogenetic isolations is small, and the evolutionary rate of the component flora is slow. It will be highly interesting to see the situation in the future, when the distributions of the boreal lichen species (as well as bryophyte, fungus and alga species) and their communities have been adequately described and mapped. Then these plants can be used to better effect as ecological and geobotanical indicators than is possible now, and they may perhaps prove more valuable in some respects than vascular plants, which often display confused and variable patterns.

References

Ahlner, S. (1948). Utbredningstyper bland nordiska barrträdslavar. (Verbreitungstypen unter fennoskandischen Nadelbaumflechten.) *Acta Phytogeog. Suec.* **22**, 1–257.

Ahlner, S. (1949). Contributions to the lichen flora of Norway. I. *Solorinella asteriscus* Anzi new to Scandinavia. *Svensk bot. Tidskr.* **43**, 157–162.

Ahti, T. (1957). Poronjäkäliköistä peurojen asuma-alueina. *Luonnon Tutk.* **61**, 76–79.

Ahti, T. (1959). Studies on the caribou lichen stands of Newfoundland. *Annls bot. Soc. zool.-bot. fenn. Vanamo* **30**(4), 1–44.

Ahti, T. (1961a). Open boreal woodland subzone and its relation to reindeer husbandry. *Arch. Soc. zool.-bot. fenn. Vanamo* **16**(suppl.), 91–93.

Ahti, T. (1961b). Taxonomic studies on reindeer lichens (*Cladonia*, subgenus *Cladina*). *Annls. bot. Soc. zool.-bot. fenn. Vanamo* **32**(1), 1–160.

Ahti, T. (1962). Notes on the lichen *Cladonia pseudorangiformis* Asah. *Arch. Soc. zool.-bot. fenn. Vanamo* **17**, 36–41.

Ahti, T. (1964). Macrolichens and their zonal distribution in boreal and arctic Ontario, Canada. *Annls. bot. fenn.* **1**, 1–35.

Ahti, T. (1966). *Parmelia olivacea* and the allied non-isidiate and non-sorediate corticolous lichens in the Northern Hemisphere. *Acta bot. fenn.* **70**, 1–68.

Ahti, T. (1969). Notes on brown species of *Parmelia* in North America. *Bryologist* **72**, 233–239.

Ahti, T. (1973). Taxonomic notes on some species of *Cladonia*, subsect. *Unciales*. *Ann. bot. fenn.* **10**, 163–184.

Ahti, T. and Henssen, A. (1965). New localities for *Cavernularia hultenii* in eastern and western North America. *Bryologist* **68**, 85–89.

Ahti, T. and Hepburn, R. L. (1967). Preliminary studies on woodland caribou range, especially on lichen stands, in Ontario. *Ont. Dept. Lands Forests Res. Rep. (Wildlife)* **74**, 1–134.

Ahti, T. and Vitikainen, O. (1974). *Bacidia chlorococca*, a common toxitolerant lichen in Finland. *Mem. Soc. fauna Flora fenn.* **49**, 95–100.

Ahti, T., Hämet-Ahti, L. and Jalas, J. (1968). Vegetation zones and their sections in northwestern Europe. *Ann. bot. fenn.* **5**, 169–211.

Albertson, N. (1950a). *Heppia lutosa* (Ach.) Nyl. i öländsk alvarvegetation. (*Heppia lutosa* (Ach.) Nyl. in lavar vegetation on the island of Öland (SE-Sweden).) *Svensk. bot. Tidskr.* **44**, 113–124.

Albertson, N. (1950b). Das grosse südliche Alvar der Insel Öland. *Svensk. bot. Tidskr.* **44**, 269–331.

Almborn, O. (1955). Lavvegetation och lavflora på Hallands Väderö. (Lichen vegetation and lichen flora in the island Hallands Väderö, S. Sweden) *K. svenska VetenskAkad. Avh. Naturskydd.* **11**, 1–92.

Andreev, V. N. (1954). Prirost kormovykh lishaynikov i priemy ego regulirovaniya. *Trudy bot. Inst. AN SSSR, Ser. III* **9**, 11–74.

Andreev, V. N. (1960). Reindeer pastures and meadows of the Far North of the U.S.S.R. and some problems of their improvement and utilization. *Proc. 8th int. Grassland Congr.* **1960**, 166–168.

Arnborg, T. (1943). Granberget. En växtbiologisk undersökning av ett sydlappländskt granskogsområde. (Granberget. Eine pflanzenbiologische Untersuchung eines südlappländischen Fichtenwaldgebietes). *Norrländs. Handbibl.* **14**, 1–282.

Barkman, J. J. (1954). Zur Kenntniss einiger Usneion-Assoziationen in Europa. *Vegetatio* **4**, 309–333.

Barkman, J. J. (1958). "Phytosociology and Ecology of Cryptogamic Epiphytes". Van Gorcum, Assen, Netherlands.

Barkman, J. J. (1973). Synusial approaches to classification. *In* "Handbook of Vegetation Science" (R. H. Whittaker, ed.), Vol. 5, pp. 435–491. W. Junk, The Hague.

Bergerud, A. T. (1971). Abundance of forage on the winter range of Newfoundland caribou. *Can. Fld Nat.* **85**, 39–52.

Bevis, F. B. and Krueger, R. J. (1964). Phytosociology of lichen vegetation on a sandy, glacial outwash plain in Northern Michigan. *Mich. Tech. Univ. Ford Forestry Center Bull.* **10**, 1–9.

Brodo, I. M. (1968). The lichens of Long Island, New York: A vegetational and floristic analysis. *N.Y. St. Mus. Sci. Serv. Bull.* **410**, 1–330.

Bunnell, F. L., Kärenlampi, L. and Russell, D. E. (1973). A simulation model of lichen-*Rangifer* interactions in Northern Finland. *Rep. Kevo Subarctic Res. Stn* **10**, 1–8.

Černohorský, Z. (1963). Survival of lichens during the Glacial Age in the North Atlantic basin. *In* "North Atlantic Biota and their History" (Á. Löve and D. Löve, eds), pp. 233–240. Pergamon Press, Oxford.

Culberson, W. L. (1956). Note sur la nomenclature, répartition et phytosociologie du *Parmeliopsis placorodia* (Ach.) Nyl. *Rev. Bryol. Lichénol.* **24**, 334–337.

Culberson, W. L. (1966). Chimisme et taxonomie des lichens du groupe *Ramalina farinacea* en Europe. *Rev. Bryol. Lichénol.* **34**, 841–851.

Culberson, W. L. (1972). Disjunctive distributions in the lichen-forming fungi. *Ann. Mo. bot. Gdn* **59**, 165–173.

Culberson, W. L. and Culberson, C. F. (1965). *Asahinea*, a new genus in the Parmeliaceae. *Brittonia* **17**, 182–190.

Culberson, W. L. and Culberson, C. F. (1967). A new taxonomy for the *Cetraria ciliaris* group. *Bryologist* **70**, 158–166.

Culberson, W. L. and Culberson, C. F. (1968). The lichen genera *Cetrelia* and *Platismatia* (Parmeliaceae). *Contr. U.S. natn. Herb.* **34**, 449–585.

Dahl, E. (1946). On different types of unglaciated areas during the Ice Ages and their significance to phytogeography. *New Phytol.* **45**, 225–242.

Dahl, E. (1950). Studies in the macrolichen flora of South West Greenland. *Meddr Grønland* **150**(2), 1–176.

Degelius, G. (1935). Das ozeanische Element der Strauch- und Laubflechtenflora von Skandinavien. *Acta Phytogeog. Suec.* **7**, 1–411.

Degelius, G. (1940a). Studien über die Konkurrenzverhältnisse der Laubflechten auf nacktem Fels. *Acta Horti gothoburg.* **14**, 195–219.

Degelius, G. (1940b). Contributions to the lichen flora of North America. I. Lichens from Maine. *Ark. Bot.* **30A**(1), 1–62.

Degelius, G. (1952). On the lichen *Cavernularia hultenii* Degel. and the problem of the glacial survival of spruce in Scandinavia. *Svensk bot. Tidskr.* **46**, 53–61.

Degelius, G. (1955). The lichen flora on calcareous substrata in southern and central Nordland (Norway). *Acta Horti gothoburg.* **20**, 35–56.

Degelius, G. (1957). The epiphytic lichen flora of the birch stands in Iceland. *Acta Horti gothoburg.* **22**, 1–51.

Du Rietz, G. E. (1921). "Zur Methodologischen Grundlage der Modernen Pflanzensoziologie." Adolf Holzhausen, Wien.

Du Rietz, G. E. (1932). Zur Vegetationsökologie der ostschwedischen Küstenfelsen. *Beih. bot. Zbl.* **49**(Ergänzungsband), 61–112.

Du Rietz, G. E. (1945). Om fattigbark- och rikbarksamhällen. *Svensk bot. Tidskr.* **39**, 147–150.

Edwards, R. Y., Soos, J. and Ritcey, R. W. (1960). Quantitative observations on epidendric lichens used as food by caribou. *Ecology* **41**, 425–431.

Eurola, S. (1962). Über die regionale Einteilung der südfinnischen Moore. *Ann. bot. Soc. zool.-bot. fenn. Vanamo* **33**, 1–243.

Fraser, E. M. (1956). The lichen woodlands of the Knob Lake area of the Quebec-Labrador. *McGill Subarct. Res. Pap.* **1**, 1–28.

Frenzel, B. (1968a). "Grundzüge der Pleistozänen Vegetationsgeschichte Nord-Eurasiens." Franz Steiner, Wiesbaden.

Frenzel, B. (1968b). The Pleistocene vegetation of northern Eurasia. *Science, N.Y.* **161**, 637–639.

Hakulinen, R. (1962a). Flechtenökologische Beobachtungen an Klippensteinen in Süd- und Mittelfinnland. *Arch. Soc. zool.-bot. fenn. Vanamo* **17**, 4–12.

Hakulinen, R. (1962b). Ökologische Beobachtungen über die Flechtenflora der Vogelsteine in Süd- und Mittelfinnland. *Arch. Soc. zool.-bot. fenn. Vanamo* **17**, 12–15.

Hakulinen, R. (1965). Über einige nördliche Flechtenarten im südöstlichen Fennoskandien. *Ann. bot. fenn.* **3**, 180–198.

Hale, M. E. (1968). A synopsis of the lichen genus *Pseudevernia*. *Bryologist* **71**, 1–11.

Hämet-Ahti, L. (1963). Zonation of the mountain birch forests in northernmost Fennoscandia. *Ann. bot. Soc. zool.-bot. fenn. Vanamo* **34**(4), 1–127.

Hämet-Ahti, L., Ahti, T. and Koponen, T. (1974). A scheme of vegetation zones for Japan and adjacent regions. *Ann. bot. fenn.* **11**, 59–88.

Hare, F. K. (1950). Climate and zonal divisions of the boreal forest formation in eastern Canada. *Geogrl Rev.* **40**, 615–635.

Hare, F. K. (1954). The Boreal conifer zone. *Geogrl Stud.* **1**, 4–18.
Hasselrot, T. E. (1941). Till kännedomen om några nordiska umbilicariacéers utbredning. (Zur Kenntnis der Verbreitung einiger Umbilicariaceen in Fennoskandia.) *Acta Phytogeog. Suec.* **15**, 1–75.
Hasselrot, T. E. (1953). Nordliga lavar i Syd- och Mellansverige. (Nördliche Flechten in Süd- und Mittelschweden.) *Acta Phytogeog. Suec.* **33**, 1–200.
Hawksworth, D. L. (1972). Regional studies in *Alectoria* (Lichenes). II. The British species. *Lichenologist* **5**, 181–261.
Hawksworth, D. L. and Rose, F. (1970). Qualitative scale for estimating sulphur dioxide air pollution in England and Wales using epiphytic lichens. *Nature, Lond.* **227**, 145–148.
Hawksworth, D. L., Coppins, B. J. and Rose, F. (1974). Changes in the British lichen flora. *In* "The Changing Flora and Fauna of Britain" (D. L. Hawksworth, ed.), pp. 47–78. Academic Press, London and New York.
Häyrén, E. (1914). Über die Landvegetation und Flora der Meeresfelsen von Tvärminne. *Acta Soc. Fauna Flora fenn.* **39**(1), 1–193.
Heusser, C. J. (1960). "Late-Pleistocene Environments of North Pacific North America." American Geographical Society, New York.
Hoefs, M. and Thomson, J. W. (1972). Lichens from the Kluane Game Sanctuary, S.W. Yukon Territory. *Can. Fld Nat.* **86**, 249–252.
Hultén, E. (1937). "Outline of the History of Arctic and Boreal Biota during the Quarternary Period". Thule, Stockholm.
Hultén, E. (1958). The amphi-atlantic plants and their phytogeographic connections. *K. svenska VetenskAkad. Handl., Ser. 4* **7**(1), 1–340.
Hultén, E. (1964). The circumpolar plants. I. Monocotyledons. *K. svenska VetenskAkad. Handl., Ser. 4* **8**(5), 1–280.
Hultén, E. (1971). The circumpolar plants. II. Dicotyledons. *K. svenska VetenskAkad. Handl., Ser. 4* **13**(1), 1–463.
Hustich, I. (1951). The lichen woodlands in Labrador and their importance as winter pastures for domesticated reindeer. *Acta Geog. Helsinki* **13**(2), 1–50.
Jalas, J. (1953). Rokua, suunnitellun kansallispuiston kasvillisuus ja kasvisto. (Vegetation und Flora des geplanten Nationalparks von Rokua in Mittelfinnland.) *Silva fenn.* **81**, 1–98.
Jalas, J. and Valpas, A. (1962). Flechtenheide oder Heidewald. Analyse eines Grenzfalls. *Arch. Soc. zool.-bot. fenn. Vanamo* **16**, 67–74.
Jesberger, J. A. and Sheard, J. W. (1973). A quantitative study and multivariate analysis of corticolous lichen communities in the southern boreal forest of Saskatchewan. *Can. J. Bot.* **51**, 185–201.
Jonescu, M. E. (1970). Lichens on *Populus tremuloides* in west-central Canada. *Bryologist* **73**, 557–578.
Jordan, W. P. (1973). The genus *Lobaria* in North America north of Mexico. *Bryologist* **76**, 225–251.
Kallio, S. and Kallio, P. (1975). Nitrogen fixation in lichens at Kevo, North-Finland. *In* "Fennoscandian Tundra Ecosystems" (F. E. Wielgolaski, ed.), pp. 291–304. Springer-Verlag, Berlin.

Kallio, P. and Kärenlampi, L. (1975). Photosynthesis in mosses and lichens. In "Photosynthesis and Productivity in Different Environments" (J. P. Cooper, ed.), pp. 393–423. Cambridge University Press, Cambridge.

Kallio, P., Suhonen, S. and Kallio, H. (1972). The ecology of nitrogen fixation in *Nephroma arcticum* and *Solorina crocea*. *Rep. Kevo Subarctic Res. Stn* **9**, 7–14.

Kalliola, R. (1966). The reduction of the area of forests in natural condition in Finland in the light of some maps based upon national forest inventories. *Annls Bot. fenn.* **3**, 442–448.

Karavaev, M. N. and Skryabin, S. Z. (1971). Ovsetzovye stepi s *Helictotrichon krylovii* (Pavl.) Henrard na kraynem severo-vostoke Sibiri. (Steppes with *Helictotrichon krylovii* (Pavl.) Henrard in the extreme north-eastern Siberia.) *Bot. Zh. SSSR.* **56**, 1436–1443.

Kärenlampi, L. (1964). The succession of the lichen vegetation on the rocky shore geolittoral and adjacent parts of the epilittoral in the southwestern archipelago of Finland. *Annls bot. fenn.* **3**, 79–85.

Kärenlampi, L. (1970a). Morphological analysis of the growth and productivity of the lichen *Cladonia alpestris*. *Rep. Kevo Subarctic Res. Stn* **7**, 9–15.

Kärenlampi, L. (1970b). "Kohti poronjäkälien ekologisen alasysteemin mallia". (On the ecological subsystem model of lichens in the tundra ecosystems.) Turku University, Turku.

Kershaw, K. A. and Field, G. F. (1975). Studies on lichen-dominated systems. XV. The temperature and humidity profiles in a *Cladina alpestris* mat. *Can. J. Bot.* **53**, 2614–2620.

Kershaw, K. A. and Rouse, W. R. (1971a). Studies on lichen-dominated systems. I. The water relations of *Cladonia alpestris* in spruce-lichen woodland in northern Ontario. *Can. J. Bot.* **49**, 1389–1399.

Kershaw, K. A. and Rouse, W. R. (1971b). Studies on lichen-dominated systems. II. The growth pattern of *Cladonia alpestris* and *Cladonia rangiferina*. *Can. J. Bot.* **49**, 1401–1410.

Klement, O. (1955). Prodromus der mitteleuropäischen Flechtengesellschaften. *Feddes Reprium Beih.* **135**, 5–194.

Klement, O. (1959). Zur Soziologie subarktischer Flechtengesellschaften. *Nova Hedwigia* **1**, 131–156.

Korchagin, A. A. (1954). Vliyanie pozharov na lesnuyu rastitel'nost' i vosstanav-lenie ee posle pozhara na evropeyskom severe. *Trudy bot. Inst. AN SSSR, Ser. III* **9**, 75–149.

Kornaś, J. (1972). Corresponding taxa and their ecological background in the forests of temperate Eurasia and North America. In "Taxonomy, Phytogeography and Evolution" (D. H. Valentine, ed.), pp. 37–59. Academic Press, London and New York.

Koskinen, A. (1955). "Über die Kryptogamen der Bäume, besonders die Flechten. I." Privately printed, Helsinki.

Krog, H. (1968). The macrolichens of Alaska. *Norsk Polarinst. Skrift.* **144**, 1–180.

Krog, H. (1973). On *Umbilicaria pertusa* Rass. and some related lichen species. *Bryologist* **76**, 550–554.
Krohn, V. (1924). Über die Vegetation der Baumstümpfe in Südfinnland. I. Die Stirnvegetation. *Ann. Acad. Sci. fenn.*, Ser. A **23**(2), 1–68.
Krusenstjerna, E. von (1945). Bladmossflora och bladmoss-vegetation i Uppsalatrakten. (Moss flora and moss vegetation in the neighbourhood of Uppsala.) *Acta Phytogeog. Suec.* **19**, 1–250.
Krusenstjerna, E. von (1965). The growth on rock. *Acta Phytogeog. Suec.* **50**, 144–148.
Kujala, V. (1926a). Untersuchungen über den Einfluss von Waldbränden auf die Waldvegetation in Nord-Finnland. *Communs Inst. Quaest. Forest Finland* **10**(5), 1–41.
Kujala, V. (1926b). Untersuchungen über die Waldvegetation in Süd- und Mittelfinnland I. C. Flechten. *Communs Inst. Quaest. Forest Finland* **10**(3), 1–61.
Kurokawa, S. (1972). Probable mode of differentiation of lichens in Japan and eastern North America. In "Floristics and Paleofloristics of Asia and Eastern North America" (A. Graham, ed.), pp. 139–146. Elsevier, Amsterdam.
Laaksovirta, K. (1973). Ilman saastuminen Harjavallassa. *Ympäristö Terveys* **4**, 539–552.
Lamb, I. M. (1951). On the morphology, phylogeny, and taxonomy of the lichen genus *Stereocaulon*. *Can. J. Bot.* **29**, 522–584.
Lambert, J. D. H. and Maycock, P. F. (1968). The ecology of terricolous lichens of the Northern Conifer-Hardwood forests of central Eastern Canada. *Can. J. Bot.* **46**, 1043–1078.
Lavrenko, E. M. and Soczava, V. B., eds. (1956). "Rastitel'nyy pokrov SSSR." (Descriptio vegetationis URSS.) I. AN SSSR, Moskva and Leningrad.
LeBlanc, F. (1963). Quelques sociétés ou unions d'épiphytes du sud du Québec. *Can. J. Bot.* **41**, 591–638.
Le Blanc, F. and Rao, D. N. (1973). Evaluation of the pollution and drought hypotheses in relation to lichens and bryophytes in urban environments. *Bryologist* **76**, 1–19.
LeBlanc, F., Rao, D. N. and Comeau, G. (1972a). The epiphytic vegetation of *Populus balsamifera* and its significance as an air pollution indicator in Sudbury, Ontario. *Can. J. Bot.* **50**, 519–528.
LeBlanc, F., Rao, D. N. and Comeau, G. (1972b). Indices of atmospheric purity and fluoride pollution pattern in Arvida, Quebec. *Can. J. Bot.* **50**, 991–998.
Lechowicz, M. J. and Adams, M. S. (1974). Ecology of *Cladonia* lichens. I. Preliminary assessment of the ecology of terricolous lichen-moss communities in Ontario and Wisconsin. *Can. J. Bot.* **52**, 55–64.
Linkola, K. (1938). Eräiden asutusta suosivien kaarnajäkälien levinneisyydestä maassamme. *Luonnon Ystävä* **42**, 92–104; 132–144.
Linkola, K. (1940). Über den Einfluss der menschlichen Haushaltung auf die Verbreitung gewisser Epiphytenflechten in Finnland. *Sber. finn. Akad. Wiss.* **1937**, 39–40.

Llano, G. A. (1950). "A Monograph of the Lichen Family Umbilicariaceae in the Western Hemisphere". Office Naval Res., Washington, D.C.

Looman, J. (1964a). Ecology of lichen and bryophyte communities in Saskatchewan. *Ecology* **45**, 481–491.

Looman, J. (1964b). The distribution of some lichen communities in the Prairie Provinces and adjacent parts of the Great Plains. *Bryologist* **67**, 209–224.

Löve, Á and Löve, D., eds. (1963). "North Atlantic Biota and their History." Pergamon Press, Oxford.

Lundström, H. (1968). Luftföroreningars inverkan på epifytfloran hos barrträd i Stockholmsområdet. *Stud. Forest Suec.* **56**, 1–55.

Lundström, H. (1970). Epixyler på impregnerade trästolpar i Bogesund. (Epixyls on treated wooden posts at Bogesund.) *Swed. Wood Preserv. Comm. Rep.* **101**, 1–9.

Lutz, H. J. (1956). Ecological effects of forest fires in the interior of Alaska. *U.S. Forest Serv. Techn. Bull.* **1133**, 1–121.

Makarevich, M. F. (1963). "Analiz likhenoflory Ukrainskykh Karpat." AN URSR, Kyiv.

Mattick, F. (1940). *Buellia epigaea* (Pers.) Tuck., eine mitteleuropäisch-kontinentale Erdflechte. *Ber. dt. bot. Ges.* **58**, 328–345.

Minyaev, N. A. (1940). Reliktovye elementy v sovremennoy flore lishaynikov vostochnoy Pribaltiki. (Relict elements in the recent lichen flora of the regions adjoining the eastern part of the Baltic Sea.) *Bot. Zh. SSSR* **25**, 415–437.

Moberg, R. (1968). Luftföroreningars inverkan på epifytiska lavar i Köpmanholmen. *Svensk bot. Tidskr.* **62**, 169–196.

Østhagen, H. (1974). The macrolichens *Cladonia luteoalba* and *Tholurna dissimilis* new to Central Europe. *Norw. J. Bot.* **21**, 161–164.

Paasio, I. (1931). Pohjois-Satakunnan soiden jäkälistä. (Über die Flechten der Moore in Nord-Satakunta.) *Ann. Soc. zool.-bot. fenn. Vanamo* **15**, 133–151.

Poelt, J. (1963). Flechtenflora und Eiszeit in Europa. *Phyton, Graz* **10**, 206–215.

Poelt, J. (1970). Das Konzept der Artenpaare bei den Flechten. *Vortr. GesGeb. Bot. N.F.* **4**, 187–198.

Poelt, J. (1972). Die taxonomische Behandlung von Artenpaaren bei den Flechten. *Bot. Notiser* **125**, 77–81.

Rabotnov, T. A. (1936). Ekologicheskie nablyudeniya nad lishaynikami v yuzhnoy Yakutii. *Sov. Bot.* **1936**, 149–153.

Ranta, P. (1974). Tampereen jäkälävyöhykkeet ja ilman saastuminen. (Comparisons between air pollution and bark lichen zones of Tampere City.) *Terra* **86**, 7–13.

Rao, D. N. and LeBlanc, F. (1967). Influence of an iron-sintering plant on corticolous epiphytes in Wawa, Ontario. *Bryologist* **70**, 141–157.

Räsänen, L. K. (1953). Eri kivilajien jäkäläkasvistosta Kivaloiden Ala-, Keski- ja Ylä-Penikalla Lapinläänin eteläosassa. (On the lichen flora of different rocks in the Ala-, Keski- and Ylä-Penikka districts of the Kivalo Hills in southern Lapland.) *Kuopion Luon. Ystäv. Yhdist. Julk., Ser. B*, **3**(1), 1–64.

Räsänen, V. (1919). Pohjanmaan nevojen jäkäläkasvisto. *Luonnon Ystävä* **23**, 123–126.
Räsänen, V. (1926). Die Flechtenflora des Gebiets Ostrobottnia borealis. *Ann. Soc. zool.-bot. fenn. Vanamo* **3**, 268–349.
Räsänen, V. (1927). Über Flechtenstandorte und Flechtenvegetation im westlichen Nordfinnland. *Ann. Soc. zool.-bot. fenn. Vanamo* **7**, 1–202.
Rassadina, K. A. (1950). Tzetrariya (*Cetraria*) SSSR. *Trudy bot. Inst. AN SSSR, Ser. II* **5**, 171–304.
Rassadina, K. A. (1959). O gruppe *Parmelia caperata* v SSSR. (De stirpe *Parmeliae caperatae* in URSS notula.) *Trudy bot. Inst. AN SSSR, Ser. II* **12**, 5–17.
Raup, L. C. (1930a). An investigation of the lichen flora of *Picea canadensis*. *Bryologist* **33**, 1–11.
Raup, L. C. (1930b). The lichen flora of the Shelter Point region, Athabasca Lake. *Bryologist* **33**, 57–66.
Ritchie, J. C. (1959). The vegetation of northern Manitoba. III. Studies in the Subarctic. *Tech. Pap. Arct. Inst. N. Am.* **3**, 1–56.
Ritchie, J. C. (1960). The vegetation of northern Manitoba. VI. The lower Hayes River region. *Can. J. Bot.* **38**, 769–788.
Rouse, W. R. and Kershaw, K. A. (1971). The effects of burning on the heat and water regimes of lichen-dominated subarctic surfaces. *Arct. alp. Res.* **3**, 291–304.
Rowe, J. S. and Scotter, G. W. (1973). Fire in the boreal forest. *Quaternary Res., N.Y.* **3**, 444–464.
Rübel, E. (1927). Einige skandinavische Vegetationsprobleme. *Veröff. geobot. Inst. Zürich* **4**, 19–41.
Ruuhijärvi, R. (1960). Über die regionale Einteilung der nordfinnischen Moore. *Ann. bot. Soc. zool.-bot. fenn. Vanamo* **31**(1), 1–360.
Sahrakorpi, S. (1973). Tampereen kaarnajäkälävyöhykkeet. (Bark lichen zones in the town of Tampere.) *Luonnon Tutk.* **77**, 25–31.
Santesson, R. (1939a). Über die Zonationsverhältnisse der lakustrinen Flechten einiger Seen im Anebodagebiet. *Medd. Lunds Limnol. Inst., Lund* **1**, 1–70.
Santesson, R. (1939b). Die Flechtenvegetation der Strandblöcke. *Acta Phytogeog. Suec.* **12**, 51–54.
Sarvas, R. (1937). Havaintoja kasvillisuuden kehityksestä Pohjois-Suomen kuloaloilla. (Beobachtungen über die Entwicklung der Vegetation auf den Waldbrandflächen Nord-Finnlands.) *Silva fenn.* **44**, 1–64.
Schofield, W. B. (1965). Correlations between the moss floras of Japan and British Columbia, Canada. *J. Hattori bot. Lab.* **28**, 17–42.
Scotter, G. W. (1962). Productivity of arboreal lichens and their possible importance to barren-ground caribou (*Rangifer arcticus*). *Arch. Soc. zool.-bot. fenn. Vanamo* **16**, 155–161.
Scotter, G. W. (1964). Effects of forest fires on the winter range of barren-ground caribou in northern Saskatchewan. *Wildl. Mgmt Bull., Ottawa Ser. 1* **18**, 1–111.

Scotter, G. W. (1970). Wildfires in relation to the habitat of barren-ground caribou in the taiga of northern Canada. *Proc. Ann. Tall Timbers Fire Ecol. Congr.* **1970**, 85–105.
Sernander, R. (1912). Studier öfver lafvarnes biologi. I. Nitrofila lafvar. *Svensk bot. Tidskr.* **6**, 803–883.
Serander, R. (1936). Granskär och Fiby urskog. (The primitive forests of Granskär and Fiby.) *Acta Phytogeog. Suec.* **8**, 1–232.
Sjörs, H. (1963a). Amphi-Atlantic zonation, Nemoral to Arctic. *In* "North Atlantic Biota and their History" (Á. Löve and D. Löve, eds), pp. 109–125. Pergamon Press, Oxford.
Sjörs, H. (1963b). Bogs and fens on Attawapiskat River, northern Ontario. *Bull. natn Mus. Can.* **186**, 45–133.
Skuncke, F. (1958). Renbeten och deras gradering. *Lappväs. Renforskn. Medd.* **4**, 1–204.
Skuncke, F. (1959). Gradering av lavhedar och lavrika skogar. *Lappväs. Renforskn. Medd.* **5**, 1–8.
Skuncke, F. (1963). Renbetet, marklavarna och skogsbruket *Lappväs. Renforskn. Medd.* **8**, 1–264.
Skuncke, F. (1969). Reindeer ecology and management in Sweden. *Univ. Alaska Biol. Pap.* **8**, 1–81.
Sõmermaa, A. (1972). Ecology of epiphytic lichens in main Estonian forest types. *Scr. Mycol.* **4**, 1–117.
Suza, J. (1936). Das arktische Element als Glazialrelikt in der Flechtenflora der alpinen Vegetationsstufe der Westkarpathen (CSR), bzw. Mitteleuropas. *Sber. K. Böhm. Ges. Wiss., Ser. II* **1935**, 1–30.
Suza, J. (1937). Einige wichtige Flechtenarten der Hochmoore im Böhmischen Massiv und in den Westkarpathen. *Sber. K. Böhm. Ges. Wiss., Math.-nat. Kl. Ser. II* **1937**, 1–33.
Thomson, J. W. (1950). The species of *Peltigera* of North America north of Mexico. *Am. Midl. Nat.* **44**, 1–68.
Thomson, J. W. (1951). Some lichens from Keweenaw Peninsula, Michigan. *Bryologist* **54**, 17–53.
Thomson, J. W. (1953). Lichens of Arctic America. I. Lichens from west of Hudson's Bay. *Bryologist* **56**, 8–36.
Thomson, J. W. (1963). The lichen genus *Physcia* in North America. *Beih. Nova Hedwigia* **7**, 1–172.
Thomson, J. W. (1972). Distribution patterns of American Arctic lichens. *Can. J. Bot.* **50**, 1135–1156.
Thomson, J. W., Scotter, G. W. and Ahti, T. (1969). Lichens of the Great Slave Lake region, Northwest Territories, Canada. *Bryologist* **72**, 137–177.
Trass, H. (1965a). Lishayniki al'varov Estonii. (Lichens on the alvars of Estonia.) *IV symp. pribaltiysk. mikol. likhenol.* 199–202. AN ESSR, Tartu.
Trass, H. (1965b). Lishaynikovye sinuzii kak komponent biogeotzenozov (ekosistem). (Lichen synusiae as components of biogeocoenosis (ecosystems).) *IV symp. pribaltiysk. mikol. likhenol.* 207–211. AN ESSR, Tartu.
Trass, H. (1968). "Analiz likhenoflory Estonii". Bot. Inst. AN SSSR, Tartu.

Trass, H. (1973). Lichen sensitivity to the air pollution and index of poleotolerance (I.P.). *Folia cryptogam. eston.* **3**, 19–22.
Uggla, E. (1958). Skogsbrandfält i Muddus Nationalpark. (Forest fire areas in Muddus National Park, Northern Sweden.) *Acta Phytogeog. Suec.* **41**, 1–116.
Westman, L. (1975). Air pollution and vegetation around a sulphite mill at Örnsköldsvik, North Sweden. *Wahlenbergia* **2**, 1–146.
Wetmore, C. M. (1960). The lichen genus *Nephroma* in North and Middle America. *Publ. Mich. St. Univ. Mus., ser. biol.* **1**(11), 369–452.
Wirth, V. (1972). Die Silikatflechten-Gemeinschaften im ausseralpinen Zentraleuropa. *Dissnes Bot.* **17**, 1–306.
Yarranton, G. A. (1972). Distribution and succession of epiphytic lichens on black spruce near Cochrane, Ontario. *Bryologist* **75**, 462–480.
Yoshimura, I. (1968). The phytogeographical relationships between the Japanese and North American species of *Cladonia*. *J. Hattori bot. Lab.* **31**, 227–246.
Yoshimura, I. (1971). The genus *Lobaria* of eastern Asia. *J. Hattori bot. Lab.* **34**, 231–364.
Yurtsev, B. A. (1972). Phytogeography of northeastern Asia and the problem of Transberingian floristic interrelations. *In* "Floristics and Paleofloristics of Asia and Eastern North America" (A. Graham, ed.), pp 19–54. Elsevier, Amsterdam.

7 | *Lichens of Cold Deserts*

DENNIS C. LINDSAY

I. Introduction 183
II. Some Ecological Aspects of Polar Cold Desert Lichens . . . 185
III. Distribution Patterns of Lichens in Polar Cold Deserts . . . 188
   A. Present Distribution Patterns of Arctic Lichens . . . 189
   B. Present Distribution Patterns of Antarctic Lichens . . . 193
   C. The Evolution of Distribution Patterns 198
   D. Dispersal Mechanisms 204
References 206

I. Introduction

Lichens are one of the most widely distributed groups of organisms in the world, ranging from Pole to Pole and from below low-tide level on rocky shores to near the summits of the highest mountains. They thus exhibit one of the widest latitudinal and altitudinal world distributions of any organism, and although *individual* species may show a very restricted habitat range, lichens generally occupy the broadest range of habitats. However, their significance as biogeographical indicators, especially in cold desert regions, has been largely unappreciated until recently. This is a result of two main factors. Firstly, most lichen collections from cold desert regions, particularly the Antarctic, were made in the past by non-lichenologists and so were often unrepresentative and poor taxonomic material. A sound taxonomic basis is an absolute necessity before any meaningful discussion can proceed upon biogeographical problems. Secondly, poor taxonomy, especially of the Antarctic species, has hampered discussion of the biogeography of cold desert lichens.

However, recent visits by lichenologists to many of the cold desert regions of the world have resulted in a more soundly based taxonomy, which, coupled with recent advances in Quaternary research, has enabled more purposeful discussions on the present and presumed past geographical distributions of lichens in relation, for example, to the Late Cenozoic glaciations of the polar regions.

The term "cold desert" appears to have been applied rather loosely to a number of regions around the world which occasionally have totally different characteristics. Smiley and Zumberge (1971) defined a "polar desert" as a "glacier-free terrestrial area wherein mean annual precipitation is less than 250 mm and the mean temperature for the warmest month is less than 10°C." This definition is, however, rather restrictive since it excludes much of the maritime Antarctic which has a mean annual precipitation of 300–400 mm water equivalent. It is perhaps better to consider other features in delimiting cold deserts, such as composition of the flora and variation of climatic parameters other than temperature and precipitation (so that alpine areas of temperate and tropical regions may be included). Employing this broad definition, a cold desert has low annual precipitation, with great diurnal and seasonal variation in one or all of the factors: temperature, precipitation and insolation. Its flora is characterized by the lack of timber-producing plants, by the high ratio of cryptogams to phanerogams and by the influence of glacial activity during Recent times. Glacial advances may have isolated populations, thus preventing gene exchange. Finally periglacial processes and wind erosion may play a significant role.

Although in some regions, such as the maritime Antarctic, precipitation may be fairly high, as much as 400 mm annual water equivalent, the prevailing low temperatures mean that much of this falls as snow. Thus many plants of such regions live in a state of more or less continual physiological drought except when melt processes are at their peak.

Cold deserts occur near mountain summits in tropical and temperate regions, but this chapter will be concerned mainly with the lichens of the cold deserts surrounding the Poles, especially the Antarctic which has been mentioned by numerous authors (e.g. Hooker, 1847; Du Rietz, 1940; Lamb, 1951) as a probable centre of evolution of many plant groups before the onset of the Late Cenozoic glaciations. However, many of the environmental, physiological and dispersal problems encountered by the lichens of the polar cold deserts are also faced by those in the cold deserts of lower latitudes.

Although it might be assumed that there are many features in common, the Arctic and Antarctic are quite dissimilar. The physiographic differences are certainly the most striking: the Arctic is essentially an ocean basin surrounded by land masses of continental proportions whereas the Antarctic is an elevated, dome-shaped continent isolated by the vast Southern Ocean, and, except for the northernmost tip of the Antarctic Peninsula, is over 2000 km from the nearest continent. These basic differences mean that there are considerable climatic and vegetational variations between what might be considered comparable areas in the two polar regions. The Late Cenozoic glaciations have also imposed other differences. In the Arctic, the advancing ice-sheets isolated several areas in Alaska and Siberia but extensive plant migration was possible southwards across North America, Europe and Asia. In the Antarctic, however, the ice-sheets covered the whole continent, apart from small isolated ranges of nunataks, and spread up to 180 km out to sea (Adie, 1964). The only opportunity for plant survival was on the nunataks. The semi-tropical floras of the Antarctic Peninsula region (Adie, 1964; Barton, 1963) could not withstand the increasingly harsh conditions and could not readily migrate northwards because of the massive barrier of the Southern Ocean, and so became extinct. Perhaps a number were successful in transoceanic dispersal to other southern continents. The nunataks that occurred in Antarctica at the height of the Late Cenozoic glaciations, to judge from the areas of unglaciated ground today, were very small and in no way comparable to the apparently well vegetated unglaciated regions that existed in the Arctic at the height of glacial advances there. Certainly the nunataks in the Antarctic possessed, and still possess, a restricted cryptogamic flora whereas those in the Arctic were probably refugia not only for a wide range of cryptogams but also for vascular plants. Owing to its great geographical isolation, plant recolonization of the Antarctic has been slow, whereas the more varied floras of the Arctic nunataks, coupled with proximity to northern land masses, resulted in more rapid recolonization of deglaciated regions.

The distribution patterns of lichens in the polar regions thus possess different degrees of significance in relation to certain factors, since those in the Antarctic were undoubtedly isolated on nunataks under far harsher conditions for gene exchange than those in the Arctic and this has led to differences between the two floras. The present patterns will be outlined and compared for the lichens of both polar regions, and some attempt will be made to reconstruct changes in the lichen flora of Antarctica in relation to the Late Cenozoic glaciations and to show how present-day distribution patterns were attained. However, prior to this, it is necessary to outline briefly some aspects of the ecology of cold desert lichens, since these have had a strong influence on the development of distribution patterns.

II. Some Ecological Aspects of Polar Cold Desert Lichens

The polar cold desert environment, especially that of continental Antarctica, imposes stresses that are rarely encountered elsewhere, such as great diurnal temperature range, physiological drought and exposure to intense ultra-violet radiation. For instance, in continental Antarctica, Siple (1938) noted that during sunny periods with an air temperature of $-40°C$, black bulb thermometers reached $+30°C$. Black pigmented lichen thalli may therefore experience great extremes of temperature. In the maritime Antarctic, D. C. Lindsay (unpublished data) recorded daytime temperatures of $+30°C$ in lichen colonies in moss banks while air temperature was $+2°C$, both the lichen and air temperatures falling to $-3°C$ at night. Whether the fact that many continental Antarctic lichens have dark-coloured thalli signifies an adaptation to the cold desert environment is not clear. Certainly the dark pigment, which is deposited in the cortex of the thallus, would absorb radiation and raise the thallus temperature. A second function may possibly be the absorption of ultra-violet radiation, preventing damage to the algal cells underlying the cortex. In species which do not have a pigmented cortex, a necrotic layer of varying thickness overlies the algal cells (Dodge, 1965).

In the maritime Antarctic, with far less harsh environmental characteristics than continental Antarctica, dark-coloured species are less prominent and are less heavily pigmented. For example, *Usnea sulphurea* has a heavily to almost entirely blackened thallus in continental Antarctica but in the maritime Antarctic it is lightly variegated with only branch tips being blackened.

While the thickened cortical or necrotic layer may act as a barrier to water absorption, it may also act as a barrier to water loss. Furthermore, the dark colouring of many species may assist in melting snow through absorption of radiation. However, the cold desert environment imposes a state of severe physiological drought. The prevailing low ambient temperature combined with the low precipitation means that much of the water is present in non-available form, as snow and ice. In continental Antarctica this severely restricts the choice of habitats occupied by lichens (Lange and Kappen, 1972; Schofield, 1972). In the McMurdo Sound region the lack of available water means that many species are confined to habitats where snow collects, such as sheltered situations in scree. *Usnea antarctica* appears to be confined there to the undersides of boulders on stable scree, where snow accumulates and melts regularly during the summer season (Schofield, 1972). Similarly Llano (1959) noted that *Buellia frigida* was confined to a narrow belt on mountain sides near McMurdo Sound, where snow flurries collected daily and provided a meagre water supply.

Water supply in continental Antarctica is thus certainly one of the major factors controlling local lichen distribution. In the maritime Antarctic, the slightly warmer climate means that water is more available than in continental Antarctica, and many species consequently exhibit an increased range of habitats; *Usnea antarctica*, for example, occupies a far wider range of habitats, from dry bird-perching stones to slightly nitrogenous melt-water runnels and boulders in sheltered to exposed situations.

The altitudinal range of lichens may be profoundly influenced by water supply. In continental Antarctica and on the east coast of the Antarctic Peninsula, lichens may be absent or poorly developed near sea-level but grow quite luxuriantly at 500 m (Lamb, 1970; Schofield, 1972). This is probably due to low cloud providing moisture towards the summits of mountains while their bases remain dry. Temperature inversions where sea-level temperature may be considerably lower than that at several hundred metres altitude may also be important.

Wind is another important factor in influencing the habitats occupied by lichens in cold deserts, acting in three ways: by desiccation, wind-chill and ice-blast action. The first two are to some extent inseparable, in that wind dries and cools at the same time in the polar cold desert. The ice-blast action is especially pronounced in continental Antarctica and parts of the Arctic. With decrease in temperature, the relative hardness of snow and ice particles increases. When blown by strong winds these particles become extremely forceful agents of erosion (Fristrup, 1951; Juckes, 1969). The stunting and deformation of lichens by exposure in the Arctic has been known for a long time (Lindsay, 1871). However, many species avoid the rigours of the Arctic winter and exposure to wind by occupying sites which accumulate enough snow cover for protection (Lynge, 1937).

The fierce winds experienced in some parts of continental Antarctica result in some localities remaining both snow- and lichen-free. In the Antarctic some summer days may be as harsh as Arctic winter nights; stunting and deformation of thalli are thus more frequent there and the distribution of the various life-forms tends to reflect the effects of wind.

In the maritime Antarctic, fruticose and foliose lichens are the dominant growth form, but they decrease proportionally in species numbers and abundance with the increase in continentality of climate. In such situations they become dwarfed and increasingly adpressed to the substratum to avoid the effects of wind. Dodge (1965) noted that many thalli of lichens from high latitudes in Antarctica were affected to some degree by wind erosion and recorded that many crustose species adopted an endolithic habit to avoid exposure. The differences in latitudinal limits between macrolichens and microlichens in Antarctica may be due to wind or a combination of factors. Macrolichens reach their southernmost limits at a latitude of about 80°S (Siple, 1938), but microlichens penetrate as far as 86°20′S (Claridge *et al.*, 1971). This difference in latitudinal limits is probably due to a boundary layer effect, the layer of air close to the substratum providing a warmer, less turbulent micro-environment, decreasing with increase in latitude and altitude. In the maritime Antarctic this layer, judging from the growth of macrolichens, may be in the order of several centimetres in thickness, whereas in continental Antarctica it is reduced to only a few millimetres. Certainly this boundary layer would seem to decrease in thickness with increase in latitude and thus explain the different latitudinal limits between macro- and microlichens (cf. Larson and Kershaw, 1976). The lichen thallus would appear to grow only within the limits of this more favourable layer, which influences the morphology of the lichen thallus and its reproductive behaviour (Larson and Kershaw, 1976).

In fruticose lichens the apothecia tend to be borne towards the upper parts of the thalli, and in many species towards the tips of the branches. In the Antarctic many fruticose species have been found to fail to produce ascospores when approaching their limits of distribution, even though apothecia may be produced and appear normal (Lindsay, 1975). For example, *Usnea fasciata* may produce apothecia abundantly throughout its range in the Antarctic Peninsula (see Fig. 2) but mature ascospores are produced only infrequently, and then by plants growing in exceptionally favourable habitats. This appears to be a case of arrested development and is known in a number of species in the maritime Antarctic, such as *Ochrolechia antarctica*, *O. frigida* (Lindsay, 1971), *Placopsis contortuplicata* and *Umbilicaria* spp. *Ochrolechia frigida* exhibits arrested development of apothecia in the Antarctic. In the South Orkney and South Shetland Islands thalli may be totally sterile or produce only a few small, cupuliform apothecia, which do not develop asci. However, further south on the west coast of the Antarctic Peninsula in localities which experience a slightly milder climate (Smith and Corner, 1973), thalli which have abundant apothecia are found with fully developed ascospores. The species that suffer from arrested development seem to be those that occur mainly in the Antarctic Peninsula region and may be presumed to be relatively recent immigrants. Other species, notably in the genus *Buellia*, which apparently evolved in the Antarctic Peninsula region and may thus be better adapted to the environment, do not appear to suffer from this problem (Lamb, 1968).

The significance of success or failure of reproduction will be discussed later.

III. Distribution Patterns of Lichens in Polar Cold Deserts

These have long been the subject of comment and speculation, especially regarding the survival of species on nunataks during the Late Cenozoic glaciations. So far, most of the published data refer to distribution patterns in the Arctic, and discussions of the significance of present lichen distribution patterns and nunatak survival have been presented for example by Lynge (1933, 1939a), Dahl (1946, 1955) and Thomson (1972). However, since the Antarctic is becoming both more accessible and of greater interest, data are accumulating which provide more useful comparisons and contrasts with the Arctic.

A. Present Distribution Patterns of Arctic Lichens

Modern distribution patterns of Arctic lichens are much better known than those of the Antarctic, but even so, the lichen floras of some Arctic regions are still poorly known. Reviews of Arctic lichen distributions have been given by Lynge (1932, 1933, 1934) and Thomson (1972), for example. The interpretation of distribution maps presented by these authors may be open to doubt on a number of points. Under-recording and changes in taxonomic concepts appear to be the major factors to consider in the interpretation of such distribution patterns. As Thomson (1967) has shown, a sound taxonomic basis is an absolute necessity before any phytogeographical significance can be given to lichen distribution patterns. For example, Lynge (1932) presented a taxonomic review of the genus *Rhizocarpon* in Greenland, defining a number of narrowly delimited species, each with a restricted geographical range. Lynge's work was revised by Runemark (1956), who showed that Lynge had used a species concept which did not recognize the wide range of variation exhibited by many polar species. As a result, many of the Arctic *Rhizocarpon* species were proved to be circumpolar in distribution, and a number of Lynge's species were placed in synonymy.

Thomson (1972) has mapped and reviewed a number of distribution patterns shown by lichens in Arctic North America, defining eight categories:

1. Circumpolar broad ranging,
2. Boreal forest outliers,
3. Appalachian extensions,
4. Arcto–Pacific (Beringian),
5. Amphi-Atlantic disjuncts,
6. Arctic endemics,
7. Great Plains–Arctic,
8. Western States–eastern Arctic disjuncts.

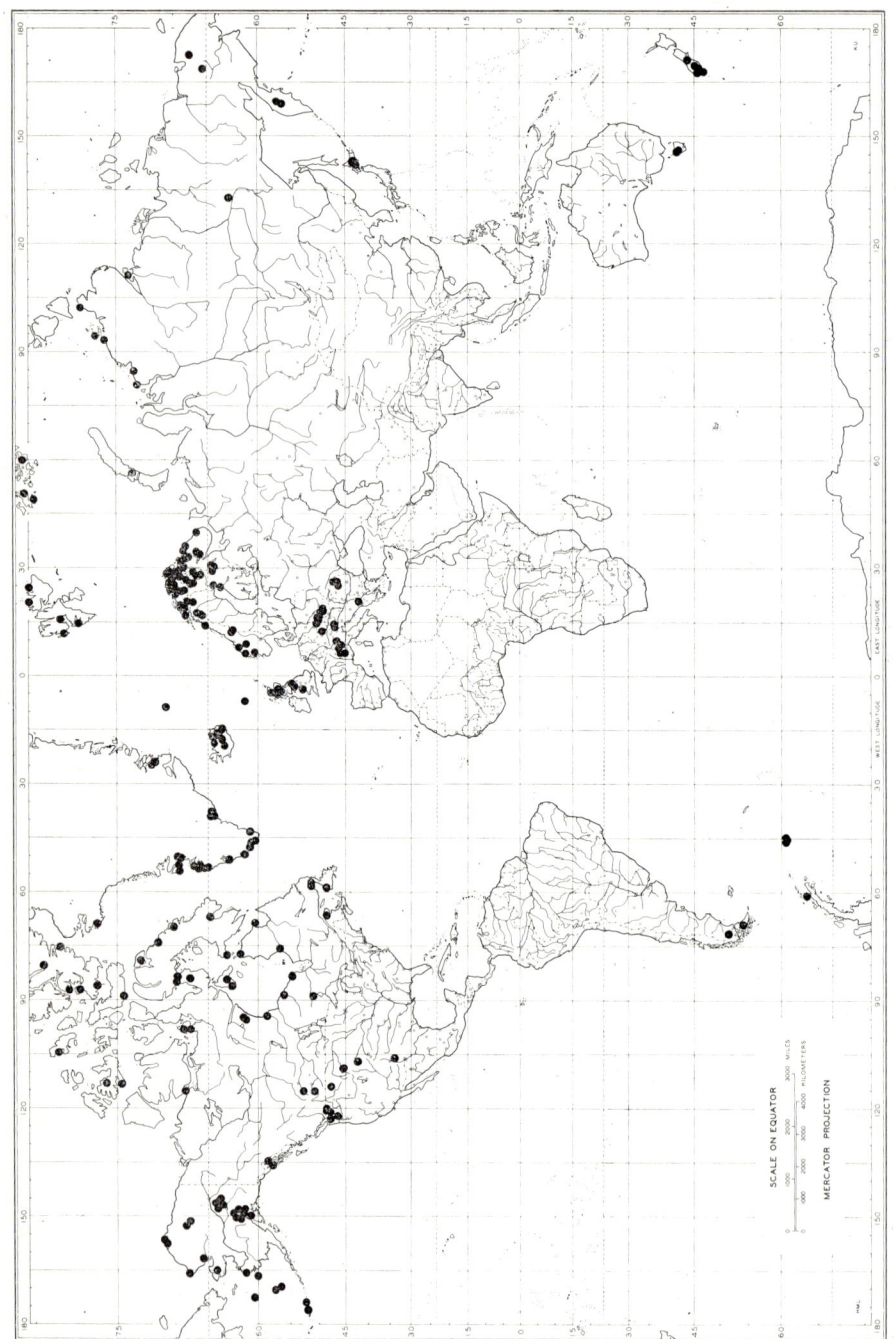

FIG. 1. World distribution of *Alectoria nigricans*, a bipolar species. After Hawksworth (1972).

1. Circumpolar Broad Ranging

As expected, the group containing the largest number of species is the circumpolar broad-ranging category, in which the species exhibit relatively wide ecological amplitudes. A number may be termed omnicolous, since they grow on a very wide range of substrata. Many species are also bipolar in their world distribution, occurring along the Rocky Mountain and Andean chain, although often discontinuous, into southern South America and occasionally Antarctica. Species showing this broad-ranging circumpolar distribution pattern which also occur in the Southern Hemisphere are *Alectoria nigricans* (Fig. 1), *Cornicularia aculeata*, *Platismatia glauca*, *Sphaerophorus globosus* and *Thamnolia vermicularis*.

2. Arcto–Pacific

A second distinct, fairly large group of species shows an Arcto–Pacific or Beringian distribution pattern. Here the species involved occur on both sides of the Bering Straits. It is believed such a pattern arose from plants surviving the Late Cenozoic glaciations in unglaciated areas of Siberia, then spreading across the Beringian land bridge to Arctic North America after the retreat of the ice. Thomson (1972) has produced several distribution maps which he considers illustrate various stages in dispersal from the Bering Straits area across North America.

3. The Remaining Distribution Patterns

The other patterns distinguished by Thomson (1972) appear to be represented by only a few species each. Some appear to be relict patterns resulting from the influences of fairly recent historical events. Thomson (1972) cites the occurrence of *Icmadophila ericetorum*, a boreal forest outlier, beyond the tree line in Arctic tundra, in localities where forest existed centuries ago. The remaining groups contain species which also seem to exhibit more exacting ecological requirements than the omnicolous circumpolar lichens. The Appalachian extension pattern, shown by *Baeomyces roseus* and *Umbilicaria caroliniana*, may be relict or a result of rapid postglacial migration. Further distribution records are required before the significance of this pattern can be evaluated. The amphi-Atlantic group, containing species such as *Placopsis gelida*, occurs on the Atlantic coasts of Europe and North America. The components appear to be oceanic in their requirements and to be of some antiquity. However, *Usnea sulphurea*, placed in this group by Thomson (1972), is continental in its distribution in Antarctica (Fig. 2). The Great Plains–Arctic pattern

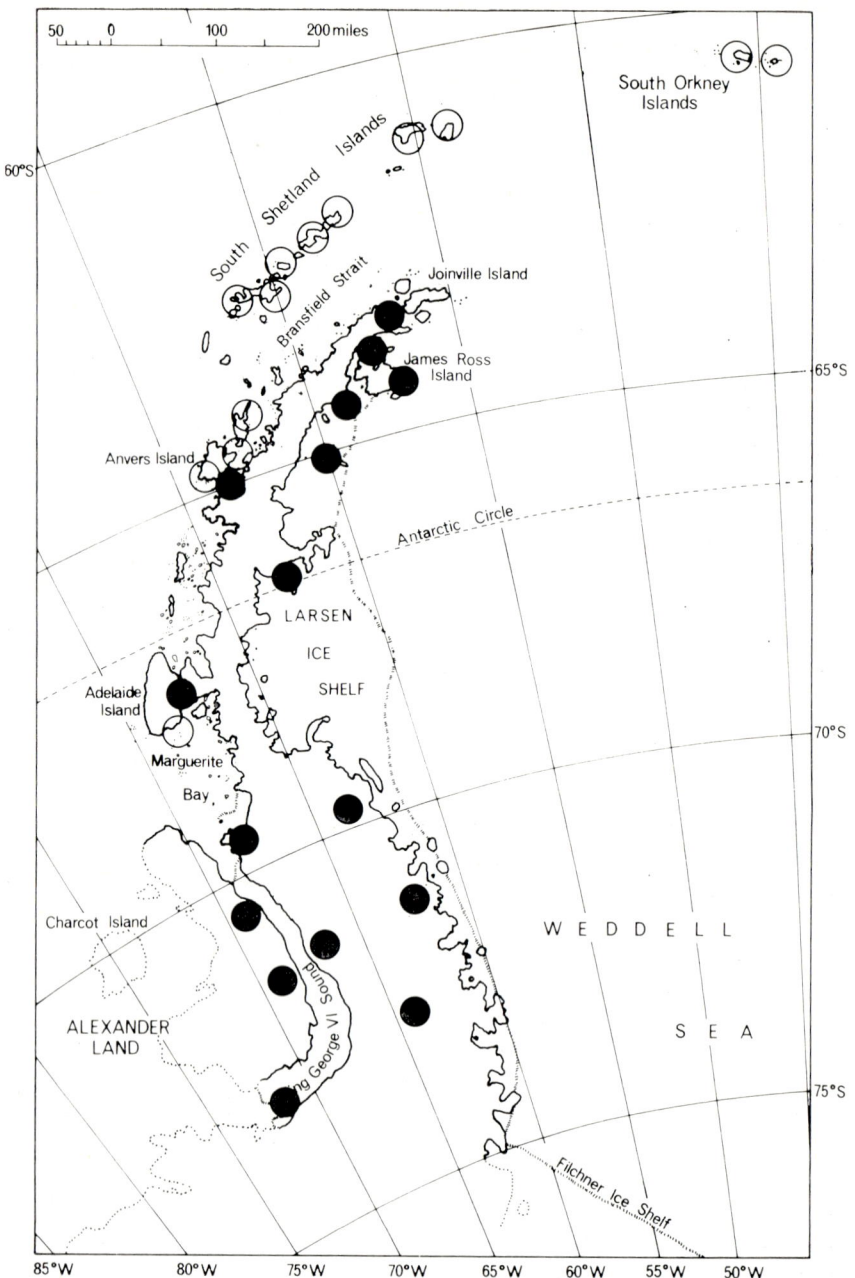

FIG. 2. Antarctic Peninsula distribution of *Usnea fasciata*, Antarctic Peninsula (west coast) element (open circles) and *Usnea sulphurea*, Antarctic Peninsula (east coast)–continental Antarctic element (closed circles).

contains a number of species found in the mid-west States and scattered throughout the Arctic, and which are believed to have critical temperature and water supply relationships (Thomson, 1972). The western States–eastern Arctic pattern may only indicate under-collecting and have no significance. Endemics appear to be few in the American Arctic, and a knowledge of their restricted geographical ranges may again be due to under-collecting, although some species are apparently restricted to areas that were never glaciated.

Although a number of articles have appeared which discuss the distribution patterns of lichens in Arctic Europe and Asia, e.g. Dahl (1955) and Lynge (1933), there has been no recent review of such patterns, which, however, appear to be similar to those outlined for lichens in the American Arctic.

B. Present Distribution Patterns of Antarctic Lichens

In contrast to the Arctic, very little has been published on lichen distribution patterns in the Antarctic. In this latter region, however, fewer distribution types have been recognized, possibly because of the very restricted nature of its lichen flora in comparison with the Arctic. Distribution maps of species of *Usnea* and *Buellia* in the Antarctic Peninsula region have been published by Lamb (1964, 1968) and Rudolph (1967) has outlined a number of distribution patterns for the whole of Antarctica. The latter author illustrates three main patterns, i.e. ubiquitous, Antarctic Peninsula and continental Antarctica, the last two having several subdivisions. However, some of the categories, especially for continental Antarctica, are very poorly known, having been based on only a few collecting localities. Recently there have been considerable extensions of the known geographical ranges of several species in continental Antarctica. For instance, *Acarospora gwynnii* was placed in the continental Antarctic group, but it is now known from a number of localities along the Antarctic Peninsula as far north as the subantarctic island of South Georgia. I have thus rejected several of Rudolph's (1967) distribution patterns since they may be liable to radical alteration through future collecting. It has been shown (Rudolph, 1967, his Plate 2) that lichen collecting in Antarctica has been centred on only a few regions such as the Antarctic Peninsula and Victoria Land; the floras of many inland mountain ranges are scarcely known. For this reason discussion of lichen distribution patterns in Antarctica will be centred on the better known and floristically richer Antarctic Peninsula.

The distribution patterns in the Antarctic outlined below follow in some respects those defined by Rudolph (1967), but only four categories are recognized here:

1. Antarctic Peninsula (west coast),
2. Antarctic Peninsula (east coast)–continental Antarctica,
3. Antarctic Peninsula endemic,
4. Circumpolar Antarctic (ubiquitous).

1. The Antarctic Peninsula (West Coast)

This category is represented by a large number of species which have their distribution patterns centred on the Scotia Arc, i.e. Antarctic Peninsula, South Shetland and South Orkney Islands and South Georgia, occasionally with outlying populations in the Falkland Islands, Tierra del Fuego and Andean Patagonia. Typical of this group is *Ramalina terebrata* (Fig. 3), which occurs along the west coast of the Antarctic Peninsula to lat. 64°S, with its northernmost station in Tierra del Fuego. *Usnea fasciata* shows a similar pattern (Fig. 2) but extends further south along the Antarctic Peninsula. The origins of this type of distribution pattern are of some interest, since a number of specialized endemics have evolved in this region. These are stipitate members of otherwise crustose genera, e.g. *Bacidia stipata*, *Catillaria corymbosa*, *Caloplaca regalis*, *Lecania brialmontii* and *Rinodina petermannii*, which are restricted to the west coast of the Antarctic Peninsula and the South Shetland and South Orkney Islands.

Among the species that may be placed in this Antarctic Peninsula (west coast) pattern are *Cladonia rangiferina*, *Cornicularia aculeata*, *Cystocoleus niger*, *Himantormia lugubris*, *Massalongia carnosa*, *Pannaria hookeri*, *Parmelia gerlachei*, *P. ushuaiensis*, *Pseudephebe pubescens*, *Sphaerophorus globosus* and *Stereocaulon glabrum*. The whole group is one of the most distinctive elements in the Antarctic lichen flora and its origins will be discussed later.

Although most of the species in this group show a strong maritime tendency in their distribution, some, such as *Himantormia lugubris*, show a slightly continental preference in their choice of habitat, in this case nunataks and exposed ground, being more abundant at higher altitudes than at sea-level. However, the distribution patterns of these species occasionally overlap with those placed in the second category, described below.

2. The Antarctic Peninsula (East Coast)–Continental Antarctica

This pattern is represented by species which have their distribution centred on an area which experiences a much more continental climate

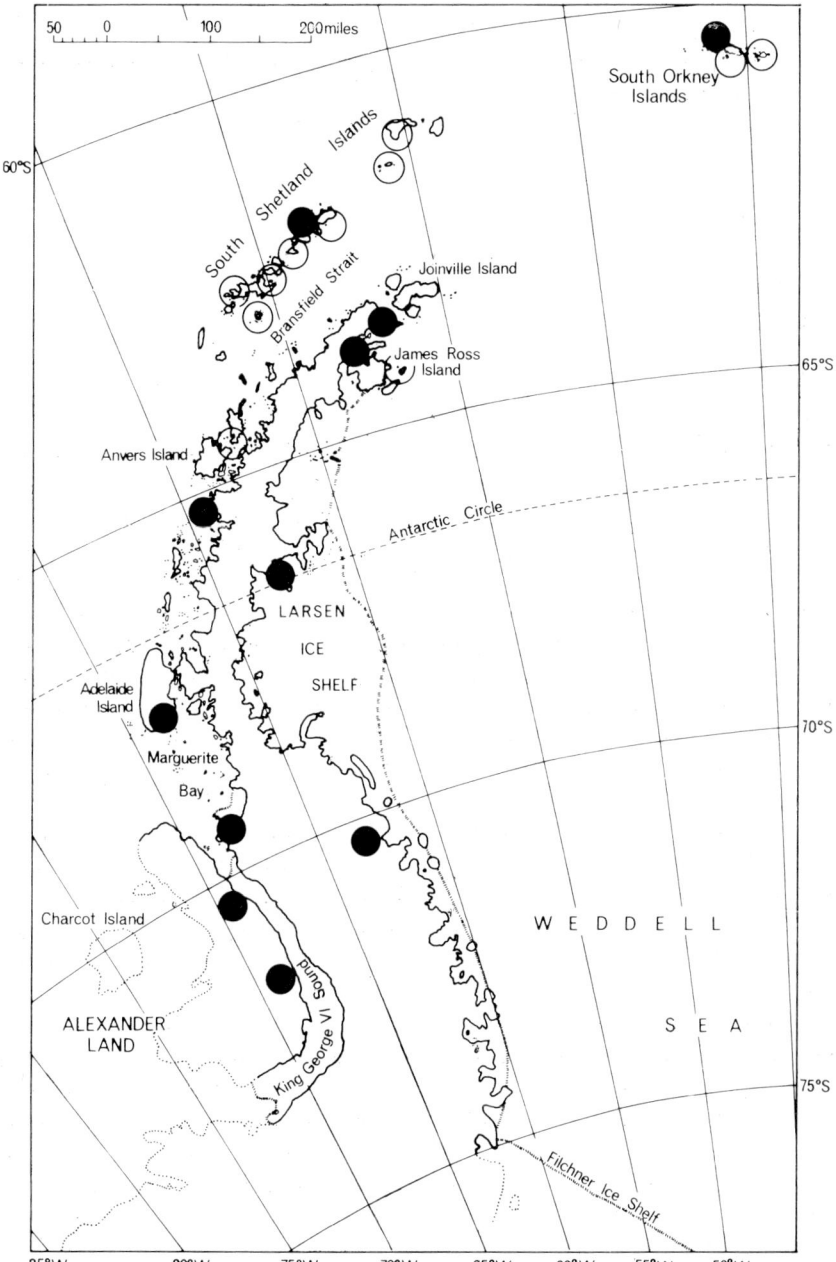

Fig. 3. Antarctic Peninsula distribution of *Ramalina terebrata*, Antarctic Peninsula (west coast) element (open circles) and *Pseudephebe minuscula*, Antarctic Peninsula (east coast)–continental Antarctic element (closed circles).

than the west coast. For most of its length the Antarctic Peninsula possesses a continuous mountainous backbone, varying between 1500 m and 2500 m in height; this range forms a climatic divide between the two coasts, which in places may be only a few kilometres apart. The west coast has a strongly maritime climate with relatively warm, wet winds, but the east coast is abutted by an extensive ice shelf and has a continental climate with cold, relatively dry winds, approaching that of continental Antarctica. A species typical of this type of distribution pattern is *Usnea sulphurea* (Fig. 2), which is absent from the northern half of the Antarctic Peninsula and the South Shetland and South Orkney Islands. It overlaps with *Usnea fasciata* in the central part of the west coast of the Peninsula and becomes the dominant species further south. Another species showing this pattern is *Pseudephebe minuscula* (Fig. 3), which has a slightly greater range,

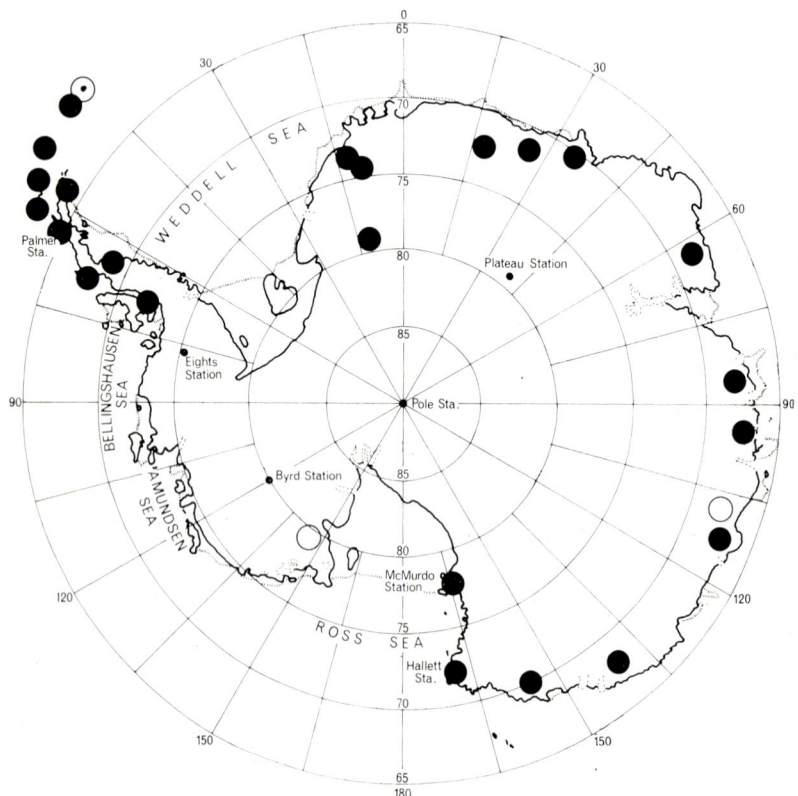

FIG. 4. Outline of Antarctic distribution of *Xanthoria elegans*, a widely recorded circumpolar species (closed circles) and the entire distribution of *Umbilicaria cristata*, an under-recorded, probably circumpolar species (open circles).

having been found, although sparsely, in the South Shetland and South Orkney Islands. In continental Antarctica, both *P. minuscula* and *Usnea sulphurea* show a distribution pattern similar to that of *Xanthoria elegans* (Fig. 4), which is regarded as circumpolar Antarctic (see below).

3. The Antarctic Peninsula Endemic

This category consists mainly of a number of stipitate members of otherwise crustose genera. These species are restricted to the west coast of the Antarctic Peninsula and the South Shetland and South Orkney Islands and have apparently originated in this region, no trace of them having been found further north in the Scotia Arc (Lindsay, 1975).

4. The Circumpolar Antarctic

Within this group the species show a scattered distribution around the continent and may also occur in the Antarctic Peninsula region. The Antarctic endemic *Buellia frigida* is typical of this pattern, being found all round continental Antarctica with an outlying population near the base of the Antarctic Peninsula (Fig. 5). *Xanthoria elegans* (Fig. 4) may be placed in this group or may better be referred to a more world-wide element, but only a few species show such a wide range throughout Antarctica.

Rudolph (1967) has recognized a number of distribution patterns within the continental Antarctic type, such as interior mountains–Ross Sea, Antarctic Peninsula–Eights Coast–Victoria Land and Ross Sea–Byrd Land–Dronning Maud Land groups. However, recent collecting has shown that some of these patterns are based on inadequate survey, as in the case of *Acarospora gwynnii*. Because of this, subdivisions of the major distribution categories of lichens in Antarctica have not been discussed in this chapter. For example, *Umbilicaria cristata* is probably circumpolar Antarctic in its distribution, but at present is known only from a few scattered localities in continental Antarctica and the South Orkney Islands (Fig. 4); it almost certainly occurs in intermediate localities but so far has remained unrecorded.

Dodge (1965, 1973) considers that the lichen flora of the Antarctic is essentially ancient in nature, having survived the Late Cenozoic glaciation on nunataks, then spreading throughout the continent with simultaneous evolution into different endemic species as the climate became milder and more ground was exposed. He contends that most of the geographic regions of Antarctica support their own peculiar endemic floras, admitting the existence of only a very small circumpolar element. It seems more probable, however, that apparent differences in populations of nunataks in

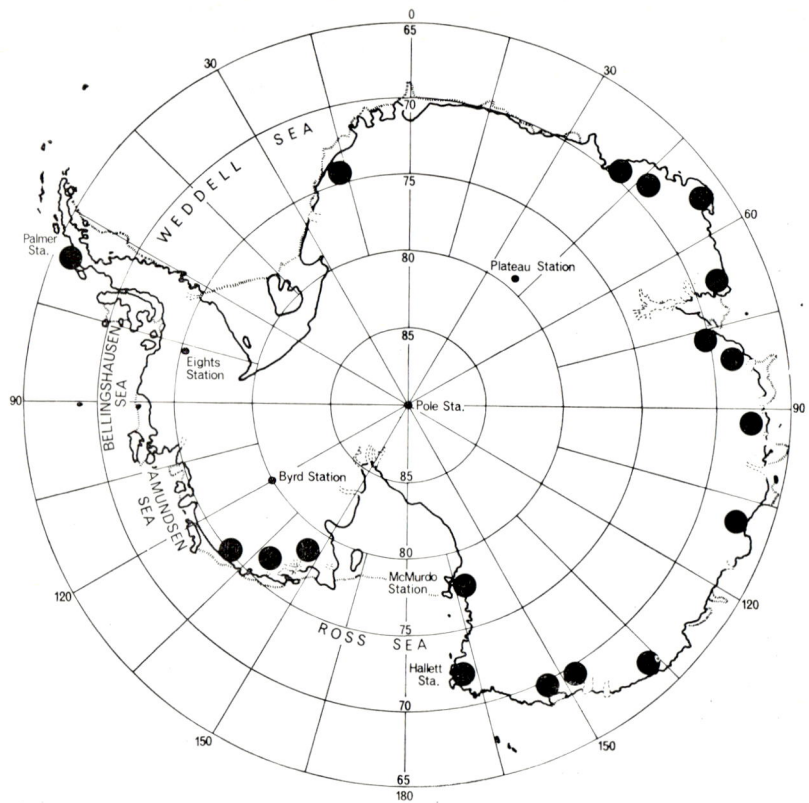

Fig. 5. Distribution of *Buellia frigida*, continental Antarctic element.

continental Antarctica are due to under-collecting, although environmental modification could also play a significant part. At best, Dodge's distribution patterns appear to reflect the degree of environmental influence on the development of lichen thalli and to have no relationship to their real geographical distributions.

C. The Evolution of Distribution Patterns

In the absence of any fossil record for lichens from the polar regions, any remarks on past distribution patterns and their transformation into modern types may be considered speculative. However, with recent advances in knowledge of continental rafting and other events in geological time, some discussion may be useful in proposing origins for modern distribution

TABLE I.
Numbers of genera and species of lichens listed for various polar and subpolar regions.

Region	Genera	Species	Source
ANTARCTIC CONTINENT			
(S of lat. 60°S)	86	424	Dodge (1973)
Continental Antarctica:			
MacRobertson Land			
(67–70°S, 60–72°E)	16	26	Filson (1966)
Dronning Maud Land			
(68–73°S, 20°W–45°E)	15	25	Lindsay (1972)
Shackleton Range			
(80–81°S, 19–31°W)	6	7	Lindsay (1974)
Maritime Antarctic:			
South Orkney Islands			
(60–61°S, 44–46°W)	40	98	Lindsay (unpublished)
Subantarctic:			
Tierra del Fuego			
(52–56°S, 63–68°W)	59	267	Grassi (1950)
Falkland Islands			
(52–53°S, 58–62°W)	32	235	Imshaug (1968)
South Georgia			
(54–55°S, 37–38°W)	50	152	Lindsay (1975)
Prince Edward Islands			
(45°S, 39°E)	27	47	Lindsay (in press)
Archipel de Kerguelen			
(48–49°S, 68–70°E)	51	116	Dodge (1948)
ARCTIC			
Canada (Reindeer Preserve)			
(68–70°N, 128–135°W)	72	356	Ahti *et al.* (1973)
Canada (Baffin Island)			
(62–73°N, 61–90°W)	50	257	Hale (1954)
West Greenland			
(61–80°N, 50–70°W)	63	359	Lynge (1937)
East Greenland			
(61–80°N, 20–45°W)	67	406	Lynge (1940)
Jan Mayen			
(71°N, 8°W)	40	144	Lynge (1939b)
Bear Island			
(74°N, 19°E)	45	185	Lynge (1926)

TABLE 1—continued

Region	Genera	Species	Source
ARCTIC			
Franz Josef Land (80–82°N, 42–65°E)	26	94	Lynge (1931)
Novaya Zemlya (72–77°N, 52–67°E)	?	456	Dahl (1954)
Subarctic:			
Canada (Great Slave Lake) (60–64°N, 107–115°W)	73	343	Thomson et al. (1969)

patterns. Both Lamb (1970) and Rudolph (1967) consider that polar lichens evolve extremely slowly, genetic recombination being a rare event, so that little evolution may have been possible in the Antarctic lichen flora as a whole.

The origins of the polar lichen floras are of interest since they exhibit many similarities. To emphasize these similarities, the distribution patterns of various genera as well as species will be discussed, since the generic unit is considered to be older and will have more relevance to distribution patterns in relation to geological time.

The more restricted nature of the Antarctic lichen flora compared with that of the Arctic is illustrated in Table I; the reasons for this are discussed later. In both areas, regions of continental size generally have a more diverse flora than island groups, reflecting the more recent nature of many of these islands as well as the problems of long distance dispersal over water. However, there are many similarities between the two polar floras. Lynge (1938), for instance, lists 205 species from Spitsbergen, of which only 29 are found in the Antarctic. Of 45 genera known from this area, 20 are also known from the Antarctic. For West Greenland, Lynge (1937) records 328 species, of which only 47 are found in the Antarctic, whereas of 65 genera listed from the former region, 46 are also found in the latter. The greater number of genera in common between Arctic and Antarctic supports the idea that at one time the polar lichen floras were basically similar, but the geographical isolation, especially that of Antarctica from other southern hemisphere continents, has led to differences through speciation. For example, Lamb (1964) notes that the bipolar *Usnea sulphurea* is diverging into incipient species in the Antarctic, and Antarctic populations of *Bryoria chalybeiformis* and *Psoroma hypnorum* also appear to be developing many differences from Arctic populations (Hawksworth, *in litt.*; Lamb, *in litt.*). However, a strong bipolar element exists which still links the polar lichen floras, e.g. *Alectoria nigricans, Buellia coniops, Cladonia rangiferina, Cornicularia aculeata, Cystocoleus niger, Mastodia tesselata, Ochrolechia frigida, Pannaria hookeri, Pseudephebe minuscula, P. pubescens, Rinodina archaeoides, R. nimbosa, Sphaerophorus globosus, Stereocaulon alpinum, Umbilicaria aprina, U. decussata* and *Verrucaria microspora*. Although similarities between floras can be shown by the use of Jaccard and Sørensen coefficients, such data are at present misleading for polar lichen floras because of different taxonomic concepts employed and inadequate sampling, especially in Antarctica.

Many theories have sought to explain the basic similarities of the polar floras on the basis of migration from one region to another (cf. Du Rietz, 1940). While this may apply to a very few species, it apparently does not apply to genera. The basic similarities between the polar floras are too great to be explained simply by migration from the Arctic to the Antarctic or vice versa. At present the distribution patterns of nearly all bipolar lichens are discontinuous between the temperate zones of both hemispheres and the land bridges between the continents were probably not in existence when the migration was supposed to have occurred (cf. Brodo, 1973). A more plausible explanation is perhaps that the ancestors of modern lichens, i.e. the basic generic unit, were already in existence in Pangaean (Permo-Triassic) times. Possibly this ancient flora was fairly uniform before continental rafting split Pangaea into the forerunners of the modern continents (Smith *et al.*, 1973). Since that event, the continents have rafted slowly to their present positions, but their ancient geographical affinities are stamped upon their lichen floras. It is possible that by Pangaean times many lichen genera had differentiated enough to be recognized as modern genera. Certainly a consideration of the distribution patterns of a number of lichen genera (especially from the southern hemisphere) shows the geographical relationships between the modern continents and the ancient land masses. *Pseudocyphellaria*, widespread and represented by numerous species in the southern hemisphere, only occurs as a few species in the northern hemisphere. Similarly, *Usnea* subgen. *Neuropogon* (Lamb, 1939, 1948) and *Placopsis* (Lamb, 1947), each represented by a wide range of species in the southern hemisphere, have only two species in the northern hemisphere (Fig. 6). It would appear that these taxa evolved after the split of Pangaea and arose from elements isolated in Euramerica; the few species occurring in the northern hemisphere probably being examples of migration along the Andean mountain chain or the Malaysian–Papuan link. However, there are large gaps in distribution, e.g. *Usnea sulphurea* was not known from the North American mainland until recently when it was collected on Mount Orizaba in Mexico (Lamb, *in*

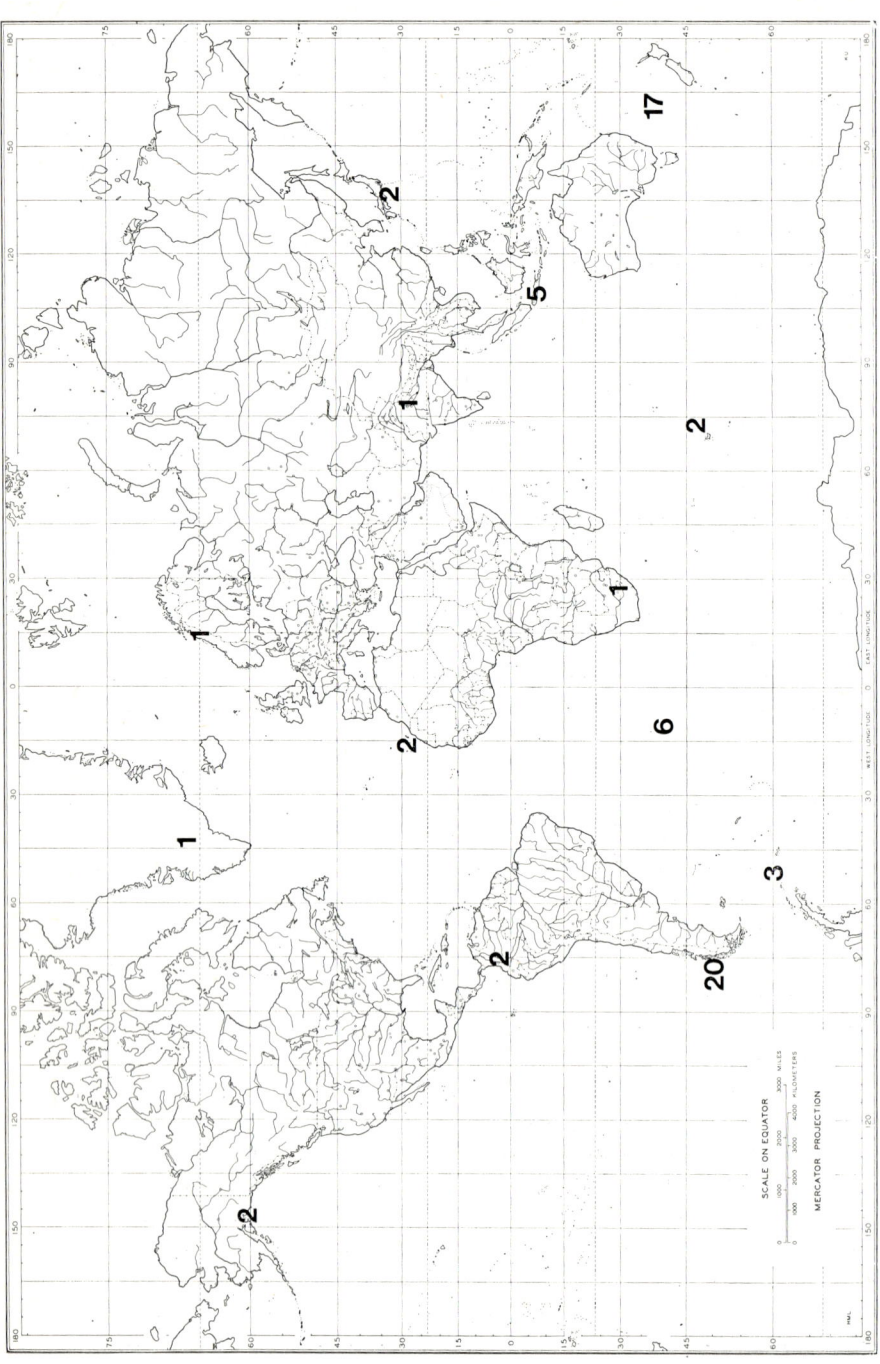

FIG. 6. Distribution of species of *Placopsis* throughout the world. Numbers indicate total of species recorded from that particular area. Based on Hertel (1970, his Fig. 6).

litt.). Such large gaps in the distribution map are probably due to under-recording and may have no phytogeographical significance. A number of other genera, e.g. *Cladonia, Lecanora* and *Lecidea,* appear to have been in existence before the split of Pangaea and have numerous representatives on the resulting continents. Evolution occurred within these genera so that there are now a large number of species each with a somewhat restricted range, but also with a substantial number of cosmopolitan species in each genus. Singer (1954) also postulated a basic uniform flora in past geological time to explain the similarities of the Arctic and Antarctic fungal floras.

The factors influencing the polar floras are much more recent in geological time than continental rafting and thus have had much more influence on the distribution of species rather than genera. The major factor that has affected the composition of both the Arctic and Antarctic lichen floras has been the Late Cenozoic glaciations. The Arctic flora was able to retreat southwards in front of the advancing ice or survive on large ice-free areas that existed in Siberia and Alaska. In the Antarctic, however, continental rafting had already moved that continent to its isolated position near the Pole and the rich subtropical and temperate floras that existed in many localities declined and perished (Barton, 1963). It may be assumed that the lichen flora was also drastically affected (in some localities becoming extinct) and changed from temperate to continental Antarctic at the height of the glaciations. As a result, the plant life of the Antarctic regions has been profoundly modified and still shows, through its depauperate nature, the effect of these massive glaciations to a much greater extent than that of the Arctic. In the Arctic, conditions were much less severe during the Late Cenozoic glaciation, and the geographical proximity of continents in the northern hemisphere in Recent times has meant that sources for the present Arctic flora are richer and nearer, with fewer barriers to dispersal. Thomson (1972) places emphasis on these points by stating that the distribution patterns of many American Arctic lichens today exhibit ecological, rather than historical, factors, i.e. species are not absent from areas because of dispersal problems. In the Antarctic regions, with vast oceans as effective barriers to dispersal, it is believed the present distribution patterns show the converse, historical rather than ecological factors, i.e. species are absent because of dispersal problems.

Lynge (1933, 1938) called attention to the disjunct distributions of a number of Arctic lichens, e.g. species of *Dactylina,* and expressed astonishment at what he termed the small dispersal power of these lichens. This may be true of only a few Arctic lichens but seems to be typical of many Antarctic species. Attention has been drawn previously (p. 194) to the peculiar nature of the endemic lichen flora of the Antarctic Peninsula region, particularly the development of the fruticose habit from crustose thalli and the production of well developed thalline stipes. This is surprising in a flora that is otherwise characterized by strong bipolar and cosmopolitan elements. Although fruticose development in crustose species has been found in other regions of the world under conditions of extreme ornithocoprophily (Lamb, 1968) or unstable dry soils (Weber, 1968; see also Chapter 2), such states arise mainly as a result of ecological conditions and are not primarily genetically controlled as in the Antarctic species. These Antarctic stipitate species exhibit the same basic distribution pattern as the maritime Antarctic species *Usnea fasciata* (Fig. 2) and *Ramalina terebrata* (Fig. 3). Although related species have been reported from continental Antarctica by Dodge (1973), their main centre of distribution is along the west coast of the Antarctic Peninsula, the South Shetland Islands and the South Orkney Islands. Since the evolution of the fruticose growth form in a number of unrelated crustose genera must have taken some considerable time, it is significant that they have not managed to spread northwards to Fuegia, the Falkland Islands or South Georgia (Lindsay, 1975), where suitable habitats exist. Other species, such as *Himantormia lugubris*, which are presumed to have originated in the Antarctic Peninsula region, have managed to spread northwards across ocean barriers. Why some species have achieved this, whereas others have failed, is puzzling, since all species mentioned so far in this context rely on the same methods for dispersal, i.e. ascospores or thallus fragmentation.

D. Dispersal Mechanisms

Some data on the efficiency and efficacy of long distance dispersal in various Antarctic lichens can be gained by a comparison of the floras of geologically new islands, such as the South Sandwich Islands (lat. 58°S, long. 28°W) with those of geologically old islands, such as the South Orkney Islands. Such a comparison shows that the new islands have an impoverished flora and that amongst the macrolichens, sorediate species are much more common than other species which rely on ascospores or thallus fragmentation for dispersal. However, the reverse is true for the microlichens. It is especially noticeable that the stipitate "crustose" species mentioned above and other lichens such as *Parmelia saxatilis* and *Umbilicaria antarctica*, all widespread and abundant in the Antarctic Peninsula region, have failed to appear on these new islands, even though more than sufficient time for colonization has been available. Substratum preferences, which may be thought to play a role here, appear to be negligible, since all are found on various volcanic rocks in the maritime Antarctic.

In the microlichens, soredial production is almost unknown in the

Antarctic species; only the more ecologically aggressive species, particularly *Buellia russa, Caloplaca cirrochrooides, Lecanora polytropa* and *Ochrolechia frigida*, are found on the geologically new islands. All these species appear to rely on ascospores for long distance dispersal. Certainly, while both soredia and ascospores appear to be successful in the long-distance dispersal of many lichens, only the ecologically aggressive species seem to benefit. The presence of less competitive species can be taken as indicating that their particular distribution patterns show areas that have been available for colonization for a long time. Dispersal and colonization by these species appears to be an extremely slow process where trans-oceanic dispersal is involved.

Thomson (1972) indicates that problems of this nature have not occurred in the Arctic, and considers that dispersal and colonization after the retreat of the ice-sheets was a comparatively rapid process since there are no major geographical barriers. Lynge (1933) considered, however, that the reproductive powers of many lichens were seriously reduced as a result of the climatic rigours imposed by the Late Cenozoic glaciations, and despite the long periods of milder climate since, these lichens have not been able to regain their supposedly former greater reproductive and dispersal powers. Lynge (1933, 1938) used this argument to explain why species of *Dactylina*, for example, which were obviously widespread prior to the Late Cenozoic glaciations, now have a fragmented relict pattern. However, it is possible that such species never possessed great dispersal powers anyway, and that their former wide geographical distribution was attained after many millions of years of slow, inefficient dispersal. When they first evolved, their dispersal mechanisms were probably effective under the prevailing conditions, but with the evolution of more aggressive readily dispersed species they have become increasingly less competitive.

A number of dispersal methods are available to cold desert lichens, all of which seem to be effective for the more ecologically aggressive species. Lynge (1939b) discussed a number of these, namely sea birds, wind, drift-wood and drift-ice, in relation to lichen distribution patterns in Spitsbergen, but found none of them could be related to the patterns recognized. This was probably because the patterns are ancient and are relicts from the glaciations, since a number of species involved are not particularly aggressive and appear to be slow evolvers. On the other hand, Westman (1973) has shown that the more aggressive *Lecanora symmicta* var. *sorediosa* may be dispersed throughout the Arctic by birds or drift-wood. Drift-ice may also play a minor role since Polunin (1955) noted clumps of moss on ice off the coast of Alaska which presumably had their origins in the Greenland region. Though this method may be used infrequently in the Arctic, the geographical distribution of land masses and ocean currents in the Antarctic would render successful dispersal by drift-wood or drift-ice a very rare event. A more common method is probably by wind. In the Antarctic, entire thalli of *Umbilicaria antarctica* and *Usnea antarctica* may be carried several kilometres over snow and ice surfaces by strong gusts. Although these two species probably rely on other methods, *Pseudephebe pubescens* and *P. minuscula* must rely on wind dispersal of thallus fragments in the Antarctic, since they produce no other diaspores there.

Despite such a range of available methods, it seems that only the more ecologically aggressive species are capable of using them for long-distance dispersal. The less competitive species may be successful in spreading short distances, but are apparently unable to surmount any major geographical barriers. Lynge (1933) interpreted this as a reduced power of reproduction induced by severe glaciations, but this remains to be proved.

Much more ecological work is required on cold desert lichens to establish the requirements for each species before the reasons behind the distribution patterns outlined above are known. Furthermore, the achievement of a stable taxonomy, especially in the Antarctic lichen flora, is necessary before patterns can be more clearly defined.

References

Adie, R. J. (1964). Geological history. *In* "Antarctic Research" (R. E. Priestly, G. de Q. Robin and R. J. Adie, eds), pp. 118–162. Butterworths, London.
Ahti, T., Scotter, G. W. and Vänskä, H. (1973). Lichens of the Reindeer Preserve, Northwest Territories, Canada. *Bryologist* **76**, 48–76.
Barton, C. M. (1963). The significance of two separate Tertiary plant assemblages from King George Island, South Shetland Islands. *Polar Rec.* **11**, 784–785.
Brodo, I. M. (1973). The lichen genus *Coccotrema* in North America. *Bryologist* **76**, 260–270.
Claridge, G. G. C., Campbell, I. B., Stout, J. D. and Dutch, M. E. (1971). The occurrence of soil organisms in the Scott Glacier region, Queen Maud Range, Antarctica. *N.Z. Jl Sci. Technol.* **14**, 306–312.
Dahl, E. (1946). On different types of unglaciated areas during the Ice Ages and their significance to phytogeography. *New Phytol.* **45**, 225–242.
Dahl, E. (1954). The cryptogamic flora of the Arctic. Lichens. *Bot. Rev.* **20**, 463–476.
Dahl, E. (1955). Biogeographic and geologic indicators of unglaciated areas in Scandinavia during the glacial ages. *Bull. geol. Soc. Am.* **66**, 1499–1519.
Dodge, C. W. (1948). Lichens and lichen parasites. *B.A.N.Z. Ant. Res. Exped. Rep., Ser. B* **7**, 1–276.

Dodge, C. W. (1965). Lichens. *In* "Biogeography and Ecology in Antarctica" (J. Van Meigham and P. Van Oye, eds), Monographiae Biologicae, Vol. 15, pp. 194–200. W. Junk, The Hague.

Dodge, C. W. (1973). "Lichen Flora of the Antarctic Continent and Adjacent Islands". Phoenix Publishing, Canaan, New Hampshire.

Du Rietz, G. E. (1940). Problems of bipolar plant distribution. *Acta phytogeogr. suec.* **13**, 215–282.

Filson, R. B. (1966). The lichens and mosses of MacRobertson Land. *A.N.A.R.E. Sci. Rep.*, Ser. B (2) **82**, 1–169.

Fristrup, B. (1951). Wind erosion in the Arctic deserts. *Geogr. Tidsskr.* **52**, 51–65.

Hale, M. E. (1954). Lichens from Baffin Island. *Am. Midl. Nat.* **51**, 232–264.

Hawksworth, D. L. (1972). Regional studies in *Alectoria* (Lichenes) II. The British species. *Lichenologist* **5**, 181–261.

Hertel, H. (1970). Trapeliaceae—eine neue Flechtenfamilie. *Vortr. Ges. Geb. Bot. N.F.* **4**, 171–185.

Hooker, J. D. (1847). *In* "A Voyage of Discovery and Research in the Southern and Antarctic Oceans during the years 1839–43" (J. C. Ross, ed.), Vol. 2, pp. 288–302. John Murray, London.

Imshaug, H. A. (1968). Expedition to Falkland Islands, 1968. *Antarctic Jl U.S.* **4**, 247–248.

Juckes, L. M. (1969). Weathering hollows in charnockite at Mannefaulknausane, Dronning Maud Land. *Bull. Brit. Antarct. Surv.* **22**, 97–98.

Lamb, I. M. (1939). A review of the genus *Neuropogon* (Nees et Flot.) Nyl., with special reference to the Antarctic species. *J. Linn. Soc. (Bot.)* **52**, 199–237.

Lamb, I. M. (1947). A monograph of the lichen genus *Placopsis* Nyl. *Lilloa* **13**, 151–288.

Lamb, I. M. (1951). On the morphology, phylogeny and taxonomy of the lichen genus *Stereocaulon*. *Can. J. Bot.* **29**, 522–584.

Lamb, I. M. (1964). Antarctic lichens: I. The genera *Usnea, Ramalina, Himantormia, Alectoria, Cornicularia*. *Brit. Antarct. Surv. Sci. Rep.* **38**, 1–34.

Lamb, I. M. (1968). Antarctic lichens: II. The genera *Buellia* and *Rinodina*. *Brit. Antarct. Surv. Sci. Rep.* **61**, 1–129.

Lamb, I. M. (1970). Antarctic terrestrial plants and their ecology. *In* "Antarctic Ecology" (M. W. Holdgate, ed.), Vol. 2, pp. 733–751. Academic Press, London and New York.

Lange, O. L. and Kappen, L. (1972). Photosynthesis of lichens from Antarctica. *Antarctic Res. Ser., Washington* **20**, 83–95.

Larson, D. W. and Kershaw, K. A. (1976). Studies on lichen-dominated systems. XVIII. Morphological control of evaporation in lichens. *Can. J. Bot.* **54**, 2061–2073.

Lindsay, D. C. (1971). Notes on Antarctic lichens: V. The genus *Ochrolechia* Massal. *Bull. Brit. Antarct. Surv.* **26**, 77–80.

Lindsay, D. C. (1972). Lichens from Vestfjella, Dronning Maud Land. *Meddr. norsk Polarinst.* **101**, 1–21.

Lindsay, D. C. (1974). Notes on Antarctic lichens: VIII. Lichens from the Shackleton Range. *Bull. Brit. Antarct. Surv.* **39,** 115–118.

Lindsay, D. C. (1975). The macrolichens of South Georgia. *Brit. Antarct. Surv. Sci. Rep.* **89,** 1–91.

Lindsay, D. C. (in press). The lichens of the Prince Edward Islands, southern Indian Ocean. *Nova Hedwigia*.

Lindsay, W. L. (1871). Observations on the lichens collected by Dr. Robert Brown, M.A., F.R.G.S., in West Greenland in 1867. *Trans. Linn. Soc. Lond.* **27,** 305–368.

Llano, G. A. (1959). Antarctic plant life. *Trans. Am. geophys. Un.* **40,** 200–203.

Lynge, B. (1926). Lichens from Bear Island (Bjørnøya) collected by Norwegian and Swedish Expeditions. *Res. Norsk. Stats. Spitsbergeneksp.* **1** (9), 1–78.

Lynge, B. (1931). Lichens collected on the Norwegian Scientific Expedition to Franz Josef Land, 1930. *Skr. Svalbard Ishavet* **38,** 1–31.

Lynge, B. (1932). A revision of the genus *Rhizocarpon* (Ram.) Th. Fr. in Greenland. *Skr. Svalbard Ishavet* **47,** 1–30.

Lynge, B. (1933). On *Dufourea* and *Dactylina*, three Arctic lichens. *Skr. Svalbard Ishavet* **59,** 1–62.

Lynge, B. (1934). Some general results of recent Norwegian research work on Arctic lichens. *Rhodora* **38,** 133–171.

Lynge, B. (1937). Lichens from West Greenland, collected chiefly by Th. M. Fries. *Meddr. Grønland* **118,** 1–225.

Lynge, B. (1938). Lichens from the west and north coasts of Spitsbergen and the North-East Land, collected by numerous expeditions. I. The macrolichens. *Skr. Norske Vidensk.-Akad., Mat.-Naturv. Kl.* **1938,** No. 6, 1–136.

Lynge, B. (1939a). On the survival of plants in the Arctic. *Norsk Geogr. Tidsskr.* **7,** 233–241.

Lynge, B. (1939b). Lichens from Jan Mayen collected on Norwegian expeditions in 1929 and 1930. *Skr. Svalbard Ishavet* **76,** 1–55.

Lynge, B. (1940). Lichens from Northeast Greenland, collected on the Norwegian Scientific Expeditions in 1929 and 1930. II. Microlichens. *Skr. Svalbard Ishavet* **81,** 1–143.

Polunin, N. (1955). Long distance plant dispersal in the north polar regions. *Nature, Lond.* **176,** 22–24.

Rudolph, E. D. (1967). Lichen distribution. In Bushnell, V., ed., Terrestrial Life in Antarctica. *Antarct. Map Fol. Ser.* **5,** 9–11.

Runemark, H. (1956). Studies in Rhizocarpon. I. *Op. bot. Soc. bot. Lund.* **2,** 1–152.

Schofield, E. (1972). Preserving the scientific value of cold desert ecosystems: past and present practices and a rationale for the future. *In* "Proceedings of the Colloquium on Conservation Problems in Antarctica" (B. C. Parker, ed.), pp. 193–227. Allen Press, Lawrence, Kansas.

Schofield, E., and Ahmadjian, V. (1972). Field observations and laboratory studies on cold desert cryptogams. *Antarctic Res. Ser., Washington* **20,** 97–142.

Singer, R. (1954). The cryptogamic flora of the Arctic. Fungi. *Bot. Rev.* **20,** 451–462.

Siple, P. A. (1938). The Second Byrd Antarctic Expedition—I. Ecology and geographic distribution. *Ann. Mo. bot. Gdn* **25,** 384–514.

Smiley, T. L. and Zumberge, J. H. (1971). Polar deserts. *Science, N.Y.* **174,** 79–80.

Smith, A. G., Briden, J. C. and Drewry, G. E. (1973). Phanerozoic world maps. *In* "Organisms and Continents through Time" (N. F. Hughes, ed.), Special Papers in Palaeontology, Vol. 12, pp. 1–42. Palaeontological Association, London.

Smith, R. I. L. and Corner, R. W. M. (1973). Vegetation of the Arthur Harbour–Argentine Islands region of the Antarctic Peninsula. *Bull. Brit. Antarct. Surv.* **33–34,** 89–122.

Thomson, J. W. (1967). Notes on *Rhizocarpon* in the Arctic. *Nova Hedwigia* **14,** 421–481.

Thomson, J. W. (1970). Lichens from the vicinity of Coppermine, Northwest Territories, Canada. *Can. Fld Nat.* **84,** 155–164.

Thomson, J. W. (1972). Distribution patterns of American Arctic lichens. *Can. J. Bot.* **50,** 1135–1156.

Thomson, J. W., Scotter, G. W. and Ahti, T. (1969). Lichens of the Great Slave Lake region, Northwest Territories, Canada. *Bryologist* **72,** 137–177.

Weber, W. A. (1968). Environmental modifications in crustose lichens. II. Fruticose growth forms in *Aspicilia*. *Aquilo, ser. Bot.* **6,** 43–51.

Westman, L. (1973). Notes on the taxonomy and ecology of an Arctic lichen: *Lecanora symmicta* var. *sorediosa* Westm. *Lichenologist* **5,** 457–460.

8 | Lichens of Hot Arid and Semi-arid Lands

RODERICK W. ROGERS

I. Introduction	211
II. Distribution	214
A. World Patterns	214
B. Regional Patterns	223
III. Adaptive Response to Aridity	226
A. Morphological	226
B. Physiological	232
IV. Factors Controlling Distribution	239
V. Lichens in Arid Ecosystems	242
VI. The Impact of Man	243
Acknowledgements	245
References	245

I. Introduction

The lichens of arid regions have long been the subject of investigation. As early as 1771, Pallas (cited in Elenkin, 1901b) collected and commented upon desert lichens, but it is not until the present century that very much effort has been directed to their study, and even then the effort has been rather fragmented. It is certainly not possible to produce the type of integrated account of the ecology and phytosociology of desert lichens today that Barkman (1958) produced for western European epiphytes. There is, however, sufficient information available to make an overall account worthwhile.

The term "desert" is a vague and emotive one: to define the study area for this account, therefore, the maps prepared by Meigs (1953) have been followed. Meigs mapped areas using the index of aridity proposed by

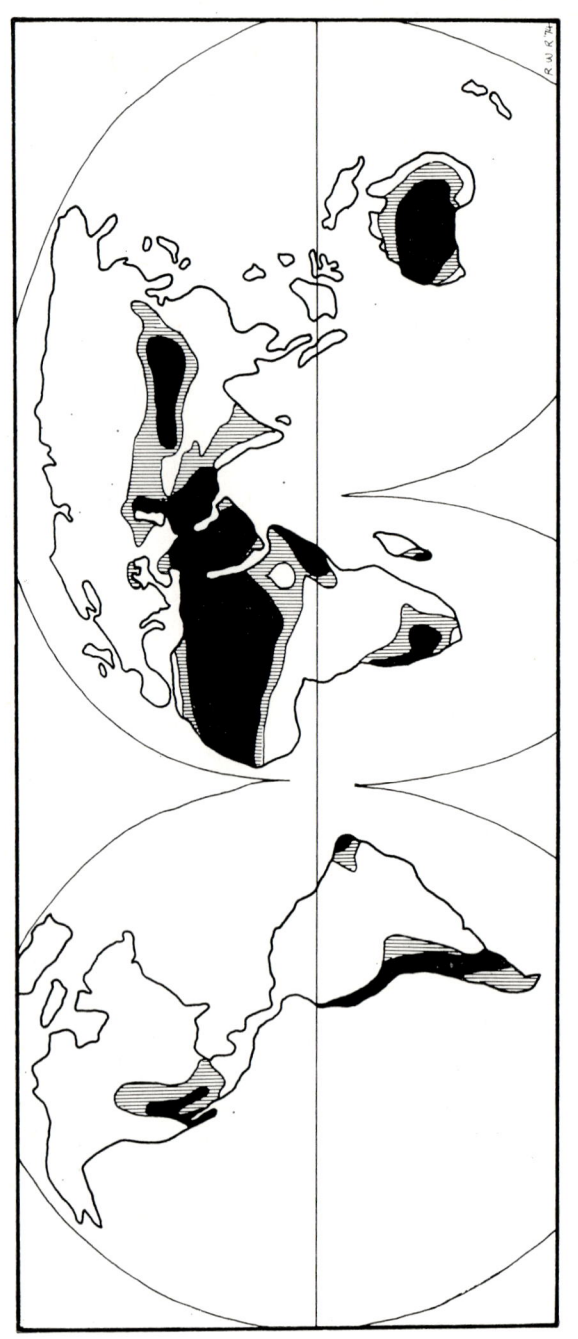

FIG. 1. The distribution of hot arid (black) and semi-arid (shaded) lands in the world.

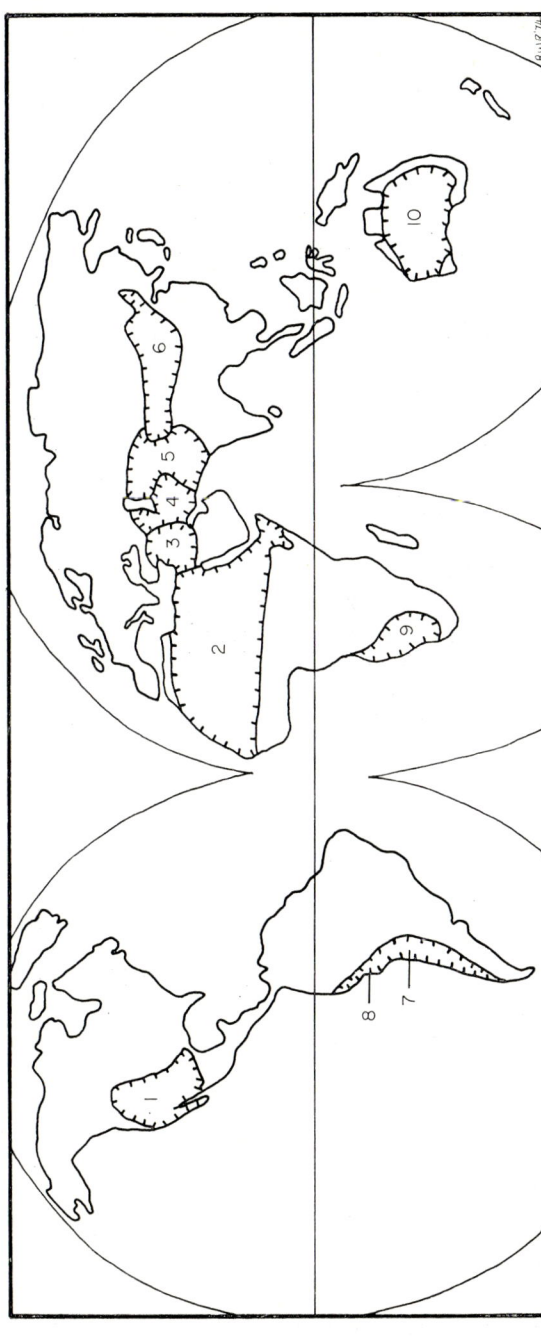

FIG. 2. The ten regions defined for the study of lichen distributions. (1) North America, (2) North Africa, (3) Western Asia, (4) Southwestern Asia, (5) Central Asia, (6) Eastern Asia, (7) South America (Atacama), (8) South America (fog oases), (9) Southern Africa, and (10) Australia.

Thornthwaite (1948), which is a complex index based on both rainfall and temperature regime, and used the descriptive terms "extreme arid", "arid" and "semi-arid". These terms are equivalent of extreme desert, desert and steppe or semi-desert respectively, and are used synonymously throughout this study. Hot arid or semi-arid lands have been interpreted as any region in which the mean temperature of the warmest month exceeds 20°C. This means that areas such as the Mongolian Altai with very low winter temperatures and which might also be considered a cold desert, are included in the study.

Around the world in the band between latitude 15° and 40° is found a band of arid and semi-arid lands in both hemispheres. Notable areas in the northern hemisphere are the Sahara in North Africa; the Negev, Arabian and other western Asian deserts; the Thar in India; the Gobi in central Asia; and the Mohave of North America. In the southern hemisphere there are extensive arid areas in western and central Australia, South America and South Africa (Fig. 1). Characteristically these areas have few low trees with sparse ground cover and, as a result, large areas of bare earth or rock pavement. The environment is harsh. Rainfall is low and erratic, usually less than 250 mm per annum, and in places it is less than 100 mm per annum. Because cloud cover is low, insolation is high; atmospheric temperatures in excess of 40°C and soil surface temperatures in excess of 65°C are commonplace (Rogers, 1971). Because of low cloud cover, night temperatures fall rapidly and temperatures below 0°C are common. Such conditions lead to considerable atmospheric turbulence resulting in strong winds (Geiger, 1950), and since so much of the soil is bare, sand-blasting of the vegetation results. Because many arid regions are far from the moderating influence of the sea, seasonal climatic variation can also be extreme, compounding the diurnal variation. No area, with the possible exception of the tundra, offers such a variable and inhospitable environment.

II. Distribution

A. World Patterns

1. The Regions Examined

There have been a great number of studies which have listed the lichens found in arid lands. After a preliminary survey of this literature, the arid lands of the world have been divided into the ten geographical regions indicated in Fig. 2. Since the vast bulk of literature refers to the Asian land mass, Asia has been treated as four regions, namely: western Asia (Israel, Syria, Iraq), including the Negev and Syrian deserts; southwestern Asia (Iran, southern U.S.S.R. and the Azerbaijan S.S.R.), including the Dasht-i-Kavir and Dasht-i-lut; central Asia (Afghanistan, the Kazahk S.S.R., Turkmen S.S.R. and Uzbek S.S.R.), including the Kara-kum and Kyzyl-kum deserts; and eastern Asia (Mongolia, Sinkiang and Inner Mongolia), including the Gobi and part of the Taklamakan deserts. The deserts of North Africa, North America and Australia are each considered as single, separate entities. The deserts of South Africa (the Kalahari and Namib) are treated as a unit, but very little information is available. From South America a little information is available about the

TABLE I.

Number of species and genera recorded in the literature surveyed for each of the regions considered.

Region	No. of Species	No. of Genera
North America	155	39
North Africa	191	40
Western Asia	262	40
South western Asia	147	33
Eastern Asia	292	35
Central Asia	122	31
South America (Atacama)	27	9
South America (fog oases)	139	55
South Africa	46	23
Australia	66	29

Atacama desert, but none from the desert regions of Brazil and Argentina. However, a great deal of information is available about fog oases; as a result two geographic units have been defined for South America, the fog oases in the Atacama desert and the remainder of the Atacama desert itself.

The number of species and genera recorded for each region is shown in Table I. A discussion of the sources of information (see also Appendix A, pp. 437) for each region follows.

(*a*) *North America.* The lichen flora of North America is fairly well known, many desert species being recorded in Fink (1935). Information in such a compilation is not easily accessible, nor can it be decided with certainty which species do in fact occur in deserts. Similarly, while many of the species listed by Bouly de Lesdain (1932), Imshaug (1956), Shushan and Anderson (1969) and Weber (1963) are probably from arid regions, insufficient information is given to be certain which are arid, which are from unusually sheltered locations and which are alpine. For biogeographic studies, consideration is limited to those reports which deal with lichens which are known to be from deserts. A number of primarily ecological studies involving soil-surface species (Looman, 1964a, b; Shields, 1957; Shields et al., 1957) list species from desert and semi-desert areas, and have been included in the sources used to compile a list of species from North American desert and semi-desert regions. Some primarily taxonomic papers have also been of value (Darrow, 1950; Fink, 1909a, b; Herre, 1911; Rudolph, 1953; Rundel et al., 1972; Wetmore, 1970) and have therefore been abstracted.

(b) *North Africa*. Lichens from North Africa have been of interest to European taxonomists for a long time. The work done on lichens from the Algerian Sahara has been summarized by Faurel et al. (1953). Additional sources consulted include species lists by Flagey (1901), Müller (1880b, 1884b), Romano (1914), Steiner (1895) and Werner (1966).

(c) *Western Asia*. The western Asian region is probably the most intensively studied of all. Numerous studies covering the Negev have been published (Alon and Galun, 1971; Galun, 1963, 1970; Galun and Reichert, 1960). Werner (1954, 1955, 1956, 1957, 1958, 1959, 1963, 1966) has produced a series of studies on lichens from Syria and Lebanon, some of them from desert and semi-desert lands. Other works examined to prepare a species list for the area include Müller (1884a), Santesson (1942) and Steiner (1921).

(d) *Southwestern Asia*. The basic reference for lichens from southwestern Asia is Szatala (1957). This work apparently follows the tradition established by Magnusson (1940, 1944) in that it has a very narrow species concept. Other references consulted included Lamb (1936) and Steiner (1896).

(e) *Central Asia*. A single work dominates the information from this area: Poelt and Wirth (1968) compiled an extensive account of the lichens from northern Afghanistan. Other sources include an ecological discussion of some Afghanistan lichens by Jacquemin-Roussard and Kilbertus (1971) and accounts of species from the U.S.S.R. by Keller (1930), Elenkin (1901a) and Steiner (1919).

(f) *Eastern Asia.* Although Schubert and Klement (1971) have produced an account of the lichens of Mongolia, reports by Magnusson (1940, 1944) dominate work on the lichens of eastern Asia. Magnusson's exhaustive descriptions of very narrowly circumscribed species have, however, probably greatly inflated the lichen flora of the area. The above works record many species collected from desert areas, as does the report by Klement (1966). Climatically the eastern Asian desert is rather unlike any other desert, with the possible exception of central Asia. Mongolia is a land of extreme heat and cold, and could possibly be considered as a cold desert.

(g) *Australia.* Information concerning lichens from the Australian deserts is based primarily on a compilation by Rogers (1970). Additional sources include ecological and floristic studies by Rogers and Lange (1972) and Rogers (1974), species lists presented by Johnson and Baird (1970) Willis (1951), and an account of central Australian lichens by Müller (1893).

(h) *South America.* Apart from the fog oases on the Chilean coast, very little information is available concerning the lichen species of South American deserts. Zahlbruckner (1925) listed a few species from the Atacama desert, and Follmann (1965) some more. Dodge (1966), Follmann (1967b) and Follmann and Redón (1972) have produced accounts of the lichens from fog oases (see Section IV). Thomson and Iltis (1968) produced similar information concerning a fog oasis in Peru. South American data is therefore segregated into two portions, the fog oases and the Atacama desert.

(i) *South Africa.* Even less information is available about lichens from the South African deserts than from South America. A paper by Zahlbruckner (1926) lists some species from the Namib, and Mattick (1970) lists a few more. Further information has been extracted from the list of South African lichens compiled by Doidge (1950).

2. *Composition of the Lichen Flora*

The literature searched revealed a desert lichen flora of 947 species in 111 genera; representing about 5% of the world's lichen flora of some 20,000 species [estimates from Zahlbruckner (1922–1940) and Lamb (1963)].

Table II shows the ten genera with the largest number of species occurring in arid lands, the percentage of species in each of those genera recorded in arid lands, the percentage contribution to the total number of

Table II.
The ten largest genera of lichens in arid lands.

Genus	% of Genus in Arid Lands	% Arid Flora	% World Flora
Lecanora	10	13	7
Caloplaca	15	10	3
Acarospora	16	8	3
Lecidea	3	6	8
Buellia	10	5	2
Verrucaria	7	4	3
Heppia	36	4	<1
Dermatocarpon	30	4	<1
Rinodina	21	3	1
Parmelia	3	3	5
		60%	33%

The ten genera of lichens with the largest number of species occurring in arid lands, listed in order of species numbers in arid lands. The first column shows the percentage of the total number of species in the genus that occur in arid lands, the second column shows the percentage of species in each genus known to occur in arid lands and the third column indicates the percentage of the total world lichen flora in each genus.

Table III.
The ten largest genera of lichens on a world basis.

Genus	% of Genus in Arid Lands	% Arid Flora	% World Flora
Lecidea	3	6	8
Lecanora	10	13	7
Parmelia	3	3	5
Bacidia	<1	<1	4
Pertusaria	1	1	3
Caloplaca	15	10	3
Verrucaria	7	4	3
Acarospora	16	8	2
Usnea	<1	<1	2
Buellia	10	5	2
		50%	39%

Columns as in Table II.

species in the arid lands each genus makes, and the percentage of the world's lichen species in each genus. Table III shows the same information for the ten genera with the largest number of species on a world basis. It is notable that seven of the ten biggest genera in the world are amongst the ten biggest in arid lands.

Some genera show a much greater development in arid lands than in the rest of the world. Whereas arid lands have only 5% of the world's lichen species, if the genera with more than ten species in the arid lands are considered, two (*Heppia* and *Dermatocarpon*) have 30% or more of their species in arid lands, two (*Xanthoria* and *Rinodina*) have 20% or more in arid lands, three (*Endocarpon*, *Acarospora* and *Caloplaca*) have 15% or more of their species in arid lands, and four genera (*Toninia*, *Diploschistes*, *Buellia* and *Lecanora*) have 10–14% of their species in arid lands. These named genera thus make a disproportionate contribution to the floristic richness of arid lichen floras and so may be regarded in that sense as characteristic of arid lands. The large genera *Bacidia*, *Pertusaria* and *Usnea* are markedly under-represented in arid lands.

3. Relationship between Regions

Because of the great disparity in the intensity of study in the various zones, comparisons at the species level are difficult. It appears from Table I that knowledge of the genera in the regions is more uniform than knowledge of species, most regions having 30–40 genera recorded. Genus concepts are perhaps more stable than are species concepts in relatively poorly known regions, hence problems with synonymy rarely arise.

The similarity between floras of the various regions at a genus level was computed using the Jaccard coefficient (Jaccard, 1912) and the results presented as a dendrogram in Fig. 3. This shows that there are striking similarities in the genera present in North Africa and throughout Asia. The lichen genera in the deserts of North America are also similar to those of Asia and Africa. The floras of Australia and South Africa are more like each other than other regions, but still show a considerable likeness to those of the northern hemisphere desert regions. The South American floras, however, are apparently distinctive.

The differences may be accounted for in part by the relative intensities of study in the various regions. The information available from the South American deserts (as distinct from the fog oases) is so slight as to preclude a high similarity; the floras of Australian and South African deserts differ from the northern hemisphere area mainly in the absence of records for genera, rather than the presence of distinctive genera.

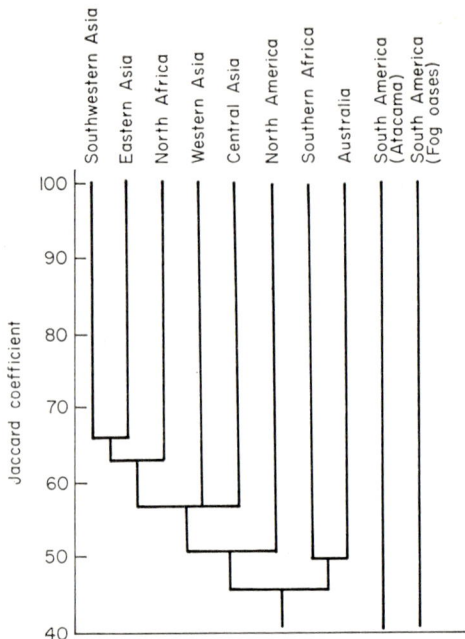

Fig. 3. Dendrogram formed by linking regions in descending order of Jaccard coefficients. The lower the level of linkage, the less similar the composition of the flora.

While comparisons at the species level are less reliable than comparisons at the genus level, some items of interest emerge from such a study. Table IV lists 18 species which occur in five or more regions. One species, *Lecidea decipiens*, is recorded in all desert regions except those in Chile where Follmann (*in litt.*) asserts it does not occur. The information in Table IV reinforces the concept of a single desert lichen flora for most of the world at the genus level. The similarities between Eurasia and America have been commented on by Weber (1963) and Looman (1964b). Rogers and Lange (1972) commented on the similarities in desert soil-surface lichens around the world and Rogers (1974) noted the similarity between an Australian arid lichen flora and other arid lichen floras, particularly those of North America. The South American flora alone appears distinctive.

The erratic *Parmelia* species are of special interest biogeographically, as they present a number of well documented disjunctions. Follmann (1967a) pointed out that *P. vagans* is known from Chile and Asia, but not North America, where *P. chlorochroa* is widespread (Looman, 1964a). Mattick

TABLE IV.

Some species widespread in arid lands.

Region	1	2	3	4	5	6	7	8	9	10
Acarospora bella	+	+		+			+	+		
A. cervina	+	+	+	+	+					+
A. strigata	+	+	+		+	+				
Aspicilia calcarea	+	+	+	+	+	+				+
A. esculenta		+	+	+	+	+				
Buellia subalbula		+	+	+					+	+
Caloplaca cerina	+	+	+	+	+	+		+	+	
C. saxicola	+	+	+	+	+	+				+
Dermatocarpon hepaticum	+	+	+	+						+
D. miniatum	+	+	+	+		+				
Diploschistes scruposus	+	+	+	+	+					+
Fulgensia bracteata			+	+	+	+				+
F. fulgens	+	+	+		+			+		
Lecanora muralis	+	+	+	+	+	+	+			
Lecidea decipiens	+	+	+	+	+	+			+	+
Parmelia conspersa	+		+			+			+	+
Squamarina lentigera	+	+	+	+	+	+				
Toninia coeruleonigricans	+	+	+	+	+					+

Species of lichens, which according to the literature surveyed, occur in at least five of the ten regions studied. The regions are: (1) North America, (2) North Africa, (3) Western Asia, (4) Southwestern Asia, (5) Central Asia, (6) Eastern Asia, (7) South America (Atacama), (8) South America (fog oases), (9) South Africa, and (10) Australia.

(1970) reported *P. convoluta* from Namibia and Rogers and Lange (1972) reported its wide distribution in Australia.

The disjunction of *P. convoluta* and *P. amphixantha* from South Africa to Australia, and of *Chondropsis semiviridis* and *Parmelia reptans* from Australia to New Zealand (Rogers, 1971; Baker *et al.*, 1973) suggests the possibility of dispersal of desert lichens with the westerly flow of the earth's air masses in the southern hemisphere.

Rogers and Lange (1972) have discussed the possibility of long-range dispersal of desert lichens, and conclude that such dispersal is possible. The reports of fertile apothecia on desert lichens not previously known to produce spores (Filson, 1967; Kappen and Schulze, 1972) coupled with reports of long-range fungal spore dispersal (Hirst and Hurst, 1967) support such an argument. Looman (1964b) inferred continental drift as

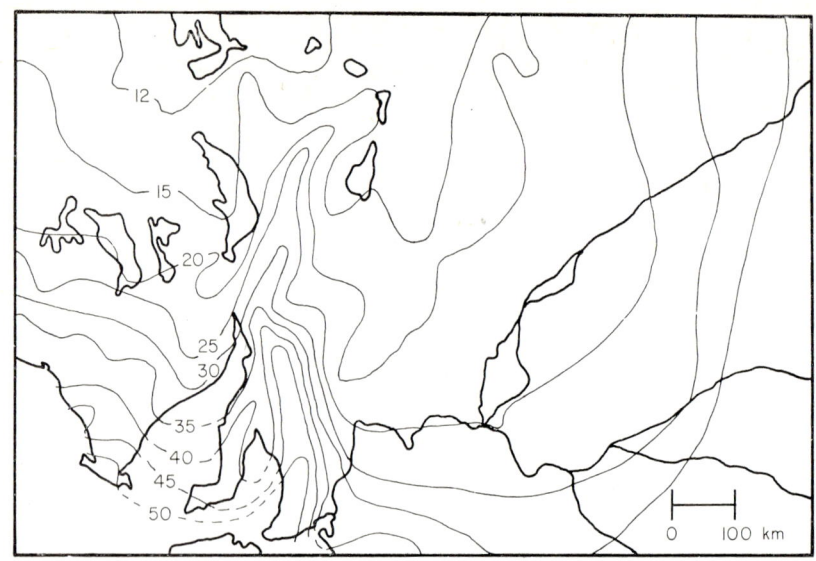

Fig. 4. Mean annual rainfall isohyets (cm) for the part of South Australia studied by Rogers (1972a).

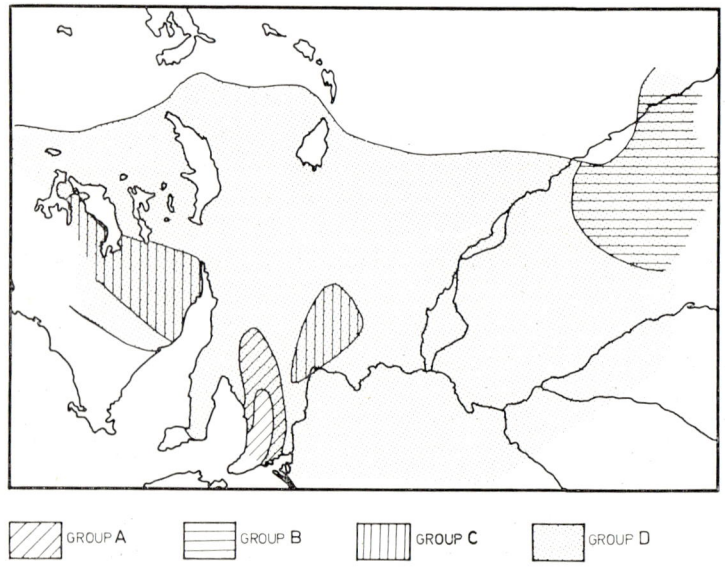

Fig. 5. Distribution of four species groups of soil-surface lichens in arid and semi-arid southeastern Australia. (Reproduced from Rogers, 1972a, by permission.)

an explanation for the close similarities of the lichen floras of steppe and semi-desert of North America and Asia. Such an argument is satisfactory for species showing special disjunctions between two continents, but cannot explain the distribution of species which occur widely in deserts of both hemispheres (Table IV).

It must be remembered, however, that critical taxonomic studies in the lichen floras of arid lands are badly needed. Until consistent studies have been completed embracing taxa across the world, biogeographic findings must be treated as tentative.

B. Regional Patterns

There have been very few studies on the distribution of lichens within desert and semi-desert regions. The most detailed are those by Rogers (1972a,b) which deal only with soil-surface lichens. Since these studies essentially support opinions expressed by other workers, they are useful as a basis for discussion.

An area of about one million square kilometres in southeastern Australia was studied. Most of the area was either arid or semi-arid (Fig. 4), and within this region 42 lichen taxa were reported from soil surfaces. These taxa were arranged into five species groups by consideration of their mutual occurrence and distribution patterns: one group contained only three rare species, so could not be discussed in a geographic context; the other four groups, however, showed an interesting pattern (Fig. 5). Species in group A were confined to humid areas, rarely penetrating to semi-arid regions. Species of group D were widespread through the desert and semi-desert regions, not occurring commonly, however, in the most extreme desert regions, nor in humid regions. In semi-arid regions, species groups B and C formed regional elaborations to the background flora provided by group D species. As the rainfall decreased below 200 mm per annum, so too did the number of species found on the soil surface. Similarly, if the rainfall rose above 300 mm per annum, the soil-surface lichen flora decreased in species richness (Fig. 6).

This demonstrates that there are, at least for soil-surface lichens, distinct arid and semi-arid floras. It is also apparent that arid lands are distinguished from semi-arid lands by the absence of species, rather than by the presence of species. This is perhaps typified by comparing the distribution of *Diploschistes ocellatus* and *D. scruposus* (semi-arid lichens) with *Lecidea decipiens* (a lichen of arid and semi-arid lands) in southern Australia (Fig. 7). It is apparent from Fig. 7 that *D. ocellatus* is confined to the arid–semi-arid margin, mostly in the 200–250-mm annual rainfall band, whereas *D. scruposus* has a wider tolerance (150–200-mm annual rainfall),

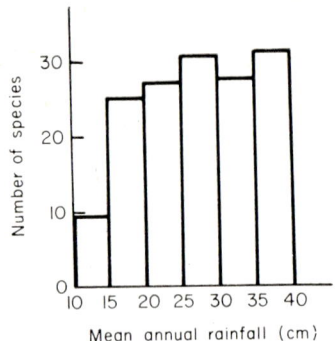

FIG. 6. Histogram showing the decreasing number of soil-surface lichen species with decrease in annual rainfall in southeastern Australia (data from Rogers, 1970).

but is rare below 200-mm rainfall. *Lecidea decipiens*, however, occurs throughout the arid and semi-arid region.

Similarly, Faurel *et al.* (1953) also found that the delimitation of the desert in Algeria was marked by the disappearance of lichens. Reichert (1937a,b, 1940) believed that steppe could be differentiated from desert by the presence of *Diploschistes* in the former. Keller (1930) and Reichert (1937a,b) also suggested that the presence of erratic lichens (see p. 230) indicated an area to be steppe (semi-arid) rather than desert (arid). Galun (1963), however, did find that some saxicolous species in the Negev were characteristic of desert regions, while some were present in both steppe and desert but none were confined to the steppe. Galun also found that there were no soil-surface species characteristic of the desert, whereas there were some which occurred only in steppe regions.

Little published information is available concerning corticolous and lignicolous species. My own studies in Australian arid lands show that the corticolous flora in desert and semi-desert regions is a simplification of that found in adjoining humid or subhumid areas. Of the eleven corticolous species recorded for the Koonamore Vegetation Reserve in South Australia (Rogers, 1974), all but three are known from humid regions of the state. Darrow (1950) made similar observations on the corticolous and lignicolous flora of southeastern Arizona, observing that the species encountered in the desert shrubland and grassland were mostly cosmopolitan.

There appears, therefore, to be a distinctive soil-surface lichen flora in arid and semi-arid regions, the arid zone having a simplified form of the semi-arid flora. The saxicolous lichen flora is apparently distinctive in arid

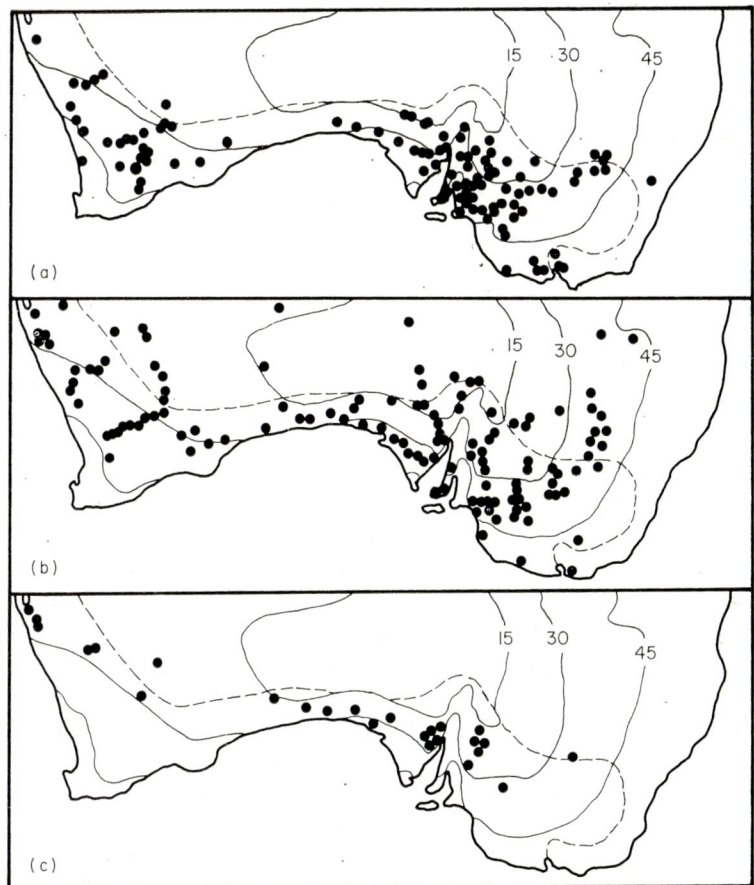

FIG. 7. The distribution of (a) *Diploschistes scruposus*, (b) *Lecidea decipiens* and (c) *Diploschistes ocellatus* in Australia, south of latitude 25°S. The broken line represents the northern and eastern limits of the area with a seasonal maximum of rainfall in winter; the solid lines are rainfall isohyets in cm.

regions but not the semi-arid regions, and the corticolous lichen flora of arid lands is essentially cosmopolitan.

Phytosociological studies of desert lichens have been few. Some workers have, however, formally proposed lichen associations. The earliest work has been summarized by Klement (1958) in his review of lichen associations proposed to that date. The studies of Riemers (1951a,b) and Klement (1955) are pertinent to Europe and Asia, the study by Looman (1964b) refers in part to arid and semi-arid North America, and a study by Mattick

(1970) describes associations from southern Africa. Follmann (1961, 1962, 1965a, 1967a,c) has described and analysed desert lichen associations from South America.

Schubert and Klement (1971) placed a number of desert lichens from Mongolia not in lichen associations, but in associations with phanerogams. Other workers, e.g. Galun (1963) and Rogers (1972a), have discussed lichen communities, but have not introduced syntaxal names. Until taxonomic problems are sorted out and further phytosociological studies undertaken, no comparison of the sociology of arid zone lichens is feasible.

III. Adaptive Response to Aridity

A. Morphological

1. Some General Adaptations

(a) *Life-form*. When considering morphological adaptation to arid conditions, it is appropriate to consider first the life-form spectra of lichen populations from various parts of the world. The significance of the ratio of crustose : foliose : fruticose lichens has been noted by a number of writers. Renaut *et al.* (1968) studied the value of this ratio in assessing microclimatic conditions. Others (Fink, 1909b; Herre, 1911; Magnusson, 1940, 1944; Galun, 1963; Rogers, 1974) have commented on the dominance of crustose species in desert regions. It is generally accepted that the reduced surface area of crustose species is an adaptation which permits survival in arid conditions. It cannot be argued that the crustose growth form is a response to arid conditions: the crustose growth form appears also in rainforests where loss of water by evaporation is not a real problem for the lichens. The desert climate, however, does not permit other growth forms to develop. Table V contains life-form spectra compiled from a number of sources for arid and semi-arid areas, with comparative values shown for South Australia and the United States.

(b) *Coloration*. Some writers have ascribed adaptive value to variations in coloration. Fink (1909b) noticed an abundance of black spots and lines over the thallus; these he believed to protect the algae from intense insolation. However, Herre (1911) found such spots to be no more common than in other areas, and also that algae were not particularly aggregated there, as they might have been if the pigmentation was protective. Contrasting with Fink's observation, Galun (1963) noted that desert species tended to be light in colour, and hence highly reflective, darker-

TABLE V.

Life-form spectra.

Location	Source	Crustose	Foliose	Fruticose	Climate
Skhrirate	Renaut et al. (1968)	60	20	20	S.H.
Jbel Mouch	Renaut et al. (1968)	63	26	10	S.H.
EnJemra	Renaut et al. (1968)	62	28	9	S.A.
Souk Jemaa	Renaut et al. (1968)	78	17	4	S.A.
Skour des Rehamma	Renaut et al. (1968)	73	19	7	A.
Mongolia	Magnusson (1940)	98	1	1	A.
Koonamore	Rogers (1974)	58	42	0	A.
Reno	Herre (1911)	91	9	0	A.
Negev	Galun and Reichert (1960)	91	6	3	A.
Sahara	Faurel et al. (1953)	97	3	0	A.
Arid South Australia	Rogers (1970)	75	23	2	A.
Semi-arid South Australia	Rogers (1970)	57	35	8	S.A.
Temperate South Australia	Rogers (1970)	37	41	22	S.H.
All of South Australia	Rogers (1970)	45	36	19	—
All of the United States	Fink (1935)	73	16	12	—

The percentage crustose, foliose, and fruticose species at each location are shown along with the climate indicated as S.H.—subhumid, S.A.—semi-arid, A.—arid. For comparative purposes, spectra for South Australia and the United States are included.

coloured species usually having a coating of light-coloured pruina. Follmann and Follmann-Schrag (1964) also noted the bright, reflective nature of "Fensterflechten" in Chile.

(c) *Cortical Anatomy.* Anatomical adaptations suggested include a generally parenchymatous thallus, which tends to reduce evaporation (Fink, 1909b; Herre, 1911) and a thickened upper cortex (Blum, 1974; Zukal, 1895). Observations on *Chondropsis semiviridis* show, however, that the thickness of the upper cortex (about 87 μm) is essentially the same in arid and semi-arid sites in South Australia as it is in a subalpine location in New Zealand (R. W. Rogers, unpublished). Another variation in the nature of the upper surface noted by Galun (1963) was the presence of a thick amorphous layer over the surface of *Buellia canescens* from Israel, but not over specimens from Scandinavia. Discussing "Fensterpflanzen" from the southern African deserts, Vogel (1955) considered the thickness of the upper cortex, and produced figures indicating that temperate-zone lichens from sunny habitats have an upper cortex varying from 20–45 μm thick, whereas the desert lichens he studied had an upper cortex from 175–200 μm thick, except *Heppia* (30 μm) and *Eremastrella tobleri* (600–1500 μm). It appears that, as a rule, lichens which occur in deserts have a thicker protective upper cortex than other lichens, although the thickness may be more or less constant for any one species throughout its range.

A thick, cellular upper cortex acts in three ways to protect the thallus. It reduces light intensity, frequently becoming opaque as it dries (Vogel, 1955; Ertl, 1951), and, because of the high water potential that may be developed (Vogel, 1955; Barkman, 1958; Follmann and Follmann-Schrag, 1964), reduces evaporative loss and also permits direct absorption of water vapour from drier air.

A study of *Chondropsis semiviridis* (R. W. Rogers, unpublished) shows that although the thickness of the cortex cannot be related to environment for that species, the breadth of the lobes may (Fig. 8). It is apparent that in desert areas this species has a thallus so much more robust than in semi-desert areas, that the two extremes have been described as varieties (Bibby, 1955). The more robust form may be better suited to the battering such a thallus must take during wind-storms.

2. Specialized Growth Forms

(*a*) *Succulence.* Galun (1963) considered that "succulence" as shown by the homiomerous lichens (e.g. *Collema* and *Psorotichia*) was an adaptation to desert conditions, the thalli absorbing much water, and then slowly releasing it. That gelatinous species occur commonly on the driest of desert soils (Rogers, 1972b) suggests they are indeed well adapted to such places, although Fink (1909b) believed they occurred only in the most protected parts of the desert. The gelatinous species do absorb a great deal of water (Smith, 1962), but have virtually no cortical development and will therefore lose water quite as rapidly as any gel. The adaptation of gelatinous species probably results not from succulence, but from the extreme resistance of the phycobiont (a blue-green alga) to adverse environmental conditions.

(*b*) *Inverted Thalli.* A bizarre adaptation described by Vogel (1955) from South Africa, was a *Buellia* species which partially reversed its thallus

8. Lichens of Hot Arid and Semi-arid Lands

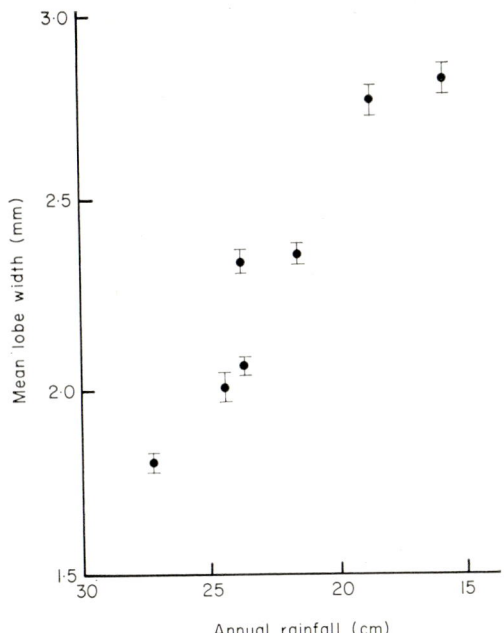

FIG. 8. The increase in width of lobe at the penultimate dichotomy of *Chondropsis semiviridis* for locations in South Australia as rainfall decreases. Scale marks represent one standard deviation each side of the mean.

structure. The thallus grows above soil level on translucent quartz pebbles, with the algal layer close against the quartz, and the medulla and cortex over it, and apothecia on the outside. The algal layer is illuminated not through the cortex as is normal, but by indirect light through the quartz pebble, getting about 10% of incident solar radiation, so reducing heat and water stress on the algal layer of the thallus.

(c) *"Fensterflechten"*. Vogel (1955) first described the "window lichens" ("lichens with periscopes" according to Follmann and Follmann-Schrag, 1964) as minute squamulose thalli of 2-mm diameter, with only the upper cortex visible, the rest being sunk into the earth. They have a thick upper cortex and algal layer, and very long (5 mm) and extensive rhizoids. Such species are known for southern Africa (Vogel, 1955) and South America (Follmann, 1965b; Follmann and Follmann-Schrag, 1964). They show the ultimate reduction of surface area, only the upper surface being exposed like a window in the soil. They are usually highly reflective, and their thick cellular upper cortex limits light and heat penetration, restricts water loss, and strongly absorbs water from the air. The extensive rhizoidal system must allow them to absorb water in appreciable quantities from the soil. All of these characteristics favour survival in a desert.

(d) *Hygrochasic thalli.* *Chondropsis semiviridis* from Australian semi-arid lands is an extreme example of the lichen variation discussed by Kappen (1974) in which thalli curl up when dry, and uncurl when wet. *Chondropsis* rolls up very tightly in a ball when dry, but when wet rolls out flat on the surface (Rogers, 1971). It has a thick upper cortex (80 μm) which is opaque when dry, and a very thick (200 μm) medulla which is white below. When rolled up, it reduces the surface area exposed to insolation, protects the algal layer behind the thick, reflective lower surface, or, should light fall on a portion of the normal upper surface, the thick opaque cortex protects it.

(e) *Erratic thalli.* A growth form which appears characteristic of arid and semi-arid areas is the erratic or unattached lichen ("Wanderflechten" of German authors). Lichens of this form are reported from every continent. From the description cited in Eversmann (1831) it appears that Pallas, a noted naturalist–explorer and prolific writer in late eighteenth century Russia, was the first to describe such forms. *Aspicilia esculenta* was first collected by him from the Kirgiz steppes. Eversmann (1831) discussed this and a number of related species, with an appendix to his paper by Nees van Esenbeck which included details of the chemistry of *Aspicilia esculenta*. Elenkin (1901b) has reviewed the early history of this species.

After examining specimens that descended in an unusually severe hail storm, Errera (1893) suggested that *Aspicilia esculenta* may have been the manna of biblical reports which rained down on the Jewish tribes wandering on the Sinai Peninsula. Analyses showing a calcium oxalate content of 60–70% (Nees van Esenbeck, 1831; Errera, 1893), however, throw doubt on this suggestion (cf. Chapter 5). Smith (1921), Harrison (1951) and Cloudsley-Thompson (1976) have discussed this subject further.

Elenkin (1901c) surveyed the erratic lichens of North Africa, the Middle East and western Asia. He recorded *Aspicilia esculenta* and *A. affinis*, each with a number of varieties, as well as *Parmelia vagans* (sub. *P. molliuscula*) and *P. ryssolea* from Russia, and *Aspicilia esculenta* from Africa.

Hale (1974) noted the wide distribution of erratic forms of *Parmelia* on desert soils, apparently referring all the unattached forms to *P. camtschadalis*, originally from Kamchatka and Nepal (Mereschkovsky, 1918). Hale reported this species as present in Russia, North America, Australia and Africa. Other writers, however, record other species from those areas: Looman (1964a) reported the North American form as *P. chlorochroa*; Rogers and Lange (1971) noted *P. australiensis* and *P. convoluta* in

Australia; Follmann (1967a), in his study of the erratic lichens of Chile, reported *P. vagans*; and Mattick (1970) reported *P. convoluta* from Namibia. Follmann (1967b), in his study of Chilean fog oases, discussed two other erratic species, *Roccella cervicornis* and *Tornabenia ephebaea*. Under similar climatic conditions in Baja California, Rundel et al. (1972) found unattached colonies of *Desmazieria pulchribarbara*.

It is apparent that many of the erratic forms are simply detached fragments of normally attached species (Smith, 1921). For example, Magnusson (1940) noted that the normally epiphytic *Teloschistes brevior* "descended onto the soil" in Mongolia and became erratic. *Tornabenia ephebaea*, normally epiphytic, was found unattached by Follmann (1967a) on soil and rocks in South America. *Aspicilia esculenta* is known from both erratic and attached specimens. Whether the various *Parmelia* species discussed above are obligately erratic or are detached portions of thalli which also grow attached to substrates is uncertain. However, *Chondropsis semiviridis* which is widespread in semi-arid regions of Australia (Rogers, 1971) is obligately erratic. It is devoid of all organs of attachment, and is found only on soil surfaces. Other obligately erratic forms appear to be *Roccella cervicornis* (Follmann, 1967a) and *Desmazieria pulchribarbara* (Rundel et al., 1972).

Erratic forms are not confined to semi-arid areas, as Smith (1921) recorded their presence in England, and Meyer (1825) described their formation in Germany. They are, however, especially well developed in semi-desert areas. The erratics are usually of such a form that they can be blown about, and because they are so convoluted or are filamentous, photosynthetic surfaces are always exposed. Perhaps the most striking erratic form is the Australian *Chondropsis semiviridis*. This species is not convoluted, but rather rolls up neatly into a ball when dry, such that when wetted it unrolls onto the soil, virtually always with its photosynthetic surface exposed to the sun. The adaptive significance of the erratic form is possibly that it allows the colonization of soil surfaces which are too unstable to support attached forms.

3. Environmental Modification and Taxonomy

The taxonomy of desert lichens, which are essentially crustose species, is bedevilled by the description of many narrowly defined species. This is partly the result of poor collections from those areas, and partly a lack of understanding of deserts by some taxonomists dealing with material sent to them.

Collections of lichens from deserts have usually been made incidentally by those interested in phanerogams—or even by non-botanists. The lichens collected on one major Australian desert expedition of last century were collected by the expedition's anthropologist! This has often resulted in fragmentary collections, unrepresentative of the areas. Terricolous material has usually been little collected, or returned as a disintegrated dust heap. Saxicolous materials collected have been those specimens on pebbles small enough to take intact, or on parts of outcrops easily broken off: both producing atypical samples. Because the collectors' main interests have been in other subjects, those lichen collections made have usually been few and show an incomplete range of the variation to be found in the field. The need for extensive and deliberate collecting of lichens from desert regions is no less acute today. The need for taxonomists to study material in the field, especially in totally unfamiliar environments, cannot be stressed too much. W. A. Weber, one of the few lichenologists to have collected extensively in deserts himself, discusses these problems in Chapter 2. Modifications caused by the desert environment have also been noted by Galun (1963, 1970), especially in respect of the astonishing plasticity of soil lichens, and these have been discussed briefly by Poelt (1974).

B. *Physiological*

The two prime physiological stress factors acting on organisms in a hot desert are heat and drought. The effect of low rainfall is reinforced by the intensity of incoming solar radiation, which rapidly evaporates water from surfaces, and tends to heat the thallus.

1. Water Relations

The water relations of lichens have been the subject of a number of studies. The more significant studies include those of Jumelle (1892), Goebel (1926), Kolumbe (1927), Stocker (1927), Smyth (1934) and Butin (1954). Smith (1962) and Blum (1974) have produced extensive reviews on the topic. Essentially, it has been demonstrated that lichens absorb water over the whole thallus, the thallus acting like a hygrophilic gel. Water content varies from 1–15% of the dry weight under drought conditions, up to 350% when saturated, with the homiomerous (gelatinous) lichens reported to have a saturated water content up to 36 times their dry weight.

Barkman (1958) calculated that lichen thalli have a water potential of -300 to -1000 atm ($-30,300$ to $-101,000$ kPa), probably due largely to the walls of the hyphae and gelatinous algal walls, not to the cytoplasmic contents of the cells. Water can therefore be absorbed directly from the air as well as from dew and rain. Since the humidity is commonly quite high in arid lands in the early morning hours (dew often forming), it is likely that water absorbed this way is ecologically significant.

Very little physiological study was actually carried out on desert prior to the important investigations of Lange and his co-workers (Lange, 1969a,b; Lange and Bertsch, 1965; Lange et al., 1970a,b). This group studied a variety of fruticose and crustose species from the Negev. At about the same time Rogers (1971) made an eco-physiological study of *Chondropsis semiviridis*, a foliose species from the Australian desert. Lange and Bertsch (1965) showed that *Ramalina maciformis* could absorb about 40% of its saturation water content from the air. At that water content, carbon dioxide uptake reached 70% of the level at optimal water content (about 80% of saturation content). Below 20% of saturation content respiration exceeded photosynthesis. Lange et al. (1970a) demonstrated that *R. maciformis* could absorb sufficient water from an atmosphere with a relative humidity of 80% to reach the photosynthetic compensation point. Such humidities are common occurrences overnight in arid areas, as the temperature falls rapidly at night. They also showed that short-term drying did not affect photosynthesis, but that after longer terms of dehydration inhibition was proportional to the duration of drought. Even after 51 weeks with a water content of 1% that of saturation, *R. maciformis* regained its initial photosynthetic capacity. They concluded therefore that damage by drought was improbable in the natural habitat.

The study also showed that at low temperatures (2°C) the light saturation intensity was 20,000 lx, rising to 48,500 lx (equivalent to full sun) at 20°C. At 48,500 lx, the optimal temperature for photosynthesis was 20°C. This figure is lower than the optimal temperature recorded by Rogers (1971) for *Chondropsis semiviridis* (25–30°C), but is still a relatively high temperature in comparison with other lichens (Table VI). While a temperature optimum for assimilation of 20°C is low for desert plants as noted by Lange (1969a), it does not suggest that *Ramalina maciformis* is in this respect specially adapted for desert existence, at least not more so than are other lichens.

In a remarkable field study, Lange et al. (1970a) confirmed their hypothesis concerning *R. maciformis*: they demonstrated (Fig. 9) that an early morning dew provided sufficient moisture to allow some 3 h of photosynthesis after dawn. This burst of photosynthesis was more than adequate to compensate for the respiratory losses during the night. They found that 1·32 mg CO_2 (g dry weight)$^{-1}$ day^{-1} was incorporated, that 0·78 mg CO_2 g^{-1} day^{-1} was lost by respiration, leaving a net gain of 0·54 mg CO_2 g^{-1} day^{-1} or, 0·146 mg C g^{-1} day^{-1}. The precise amount while varying through the year was sufficient in their estimation to allow a 5–10% growth rate per year. *Teloschistes lacunosus* behaved in an essentially similar manner.

TABLE VI.

Optimal temperatures for photosynthesis.

Species	Reference	Optimal Temperature	Habitat
Platismatia glauca	Stalfelt (1939)	1°C (19°C)	Alpine
Parmelia corei	Lange (1953)	5°C	Antarctica
Hypogymnia intestiniformis	Lange (1953)	8°C	Alpine
Letharia vulpina	Lange (1953)	8°C	Alpine
Cladonia foliacea	Lange (1953)	8°C	Alpine
Stereocaulon alpinum	Lange (1953)	8°C	Alpine
Cetraria islandica	Stalfelt (1939)	3–14° (18°C)	Alpine
Lasallia pustulata	Stalfelt (1939)	11°C (15°C)	Subalpine
Cladonia elongata	Lange (1953)	12°C	Alpine
C. "sylvatica"	Stalfelt (1939)	14°C (18°C)	Cool temperate
Usnea "dasypoga"	Stalfelt (1939)	14°C (18°C)	Cool temperate
Cora pavonia	Lange (1953)	15°C	Tropical
Cladonia rangiferina	Bliss and Hadley (1964)	15°C	Alpine
Cetraria islandica	Bliss and Hadley (1964)	15–20°C	Alpine
C. nivalis	Bliss and Hadley (1964)	15–20°C	Alpine
Ramalina farinacea	Stalfelt (1939)	18°C (22°C)	Temperate
R. fraxinea	Stalfelt (1939)	19° (19°C)	Temperate
Parmelia pachyderma	Lange (1953)	20°C	Tropical
P. magna	Lange (1953)	20°C	Tropical
Ramalina maciformis	Lange (1969a)	20°C	Hot desert
Diploschistes scruposus	R. W. Rogers (unpublished)	20–25°C	Hot desert
Chondropsis semiviridis	Rogers (1971)	25–30°C	Hot desert

The optima, which are arranged in ascending order, are generally in accord with the habitat from which the lichen was taken.

Lange et al. (1970b) have extended the study to cover crustose and foliose species (*Caloplaca ehrenbergii, C. aurantia, Lecanora farinosa, Xanthoria isidioidea, Squamarina* cf. *crassa, Diploschistes steppicus*) and an endolithic species as well. Like the fruticose species, it was found that these were all able to photosynthesize after sunrise following dewfall. These species were also able to absorb sufficient moisture from the air to permit a short period of photosynthesis after sunrise even if no dew formed. If the thalli were wet during the day at normal temperatures (36–40°C), respiratory losses occurred (Fig. 10). This loss was explained in terms of high respiration at high temperatures, and was taken to indicate

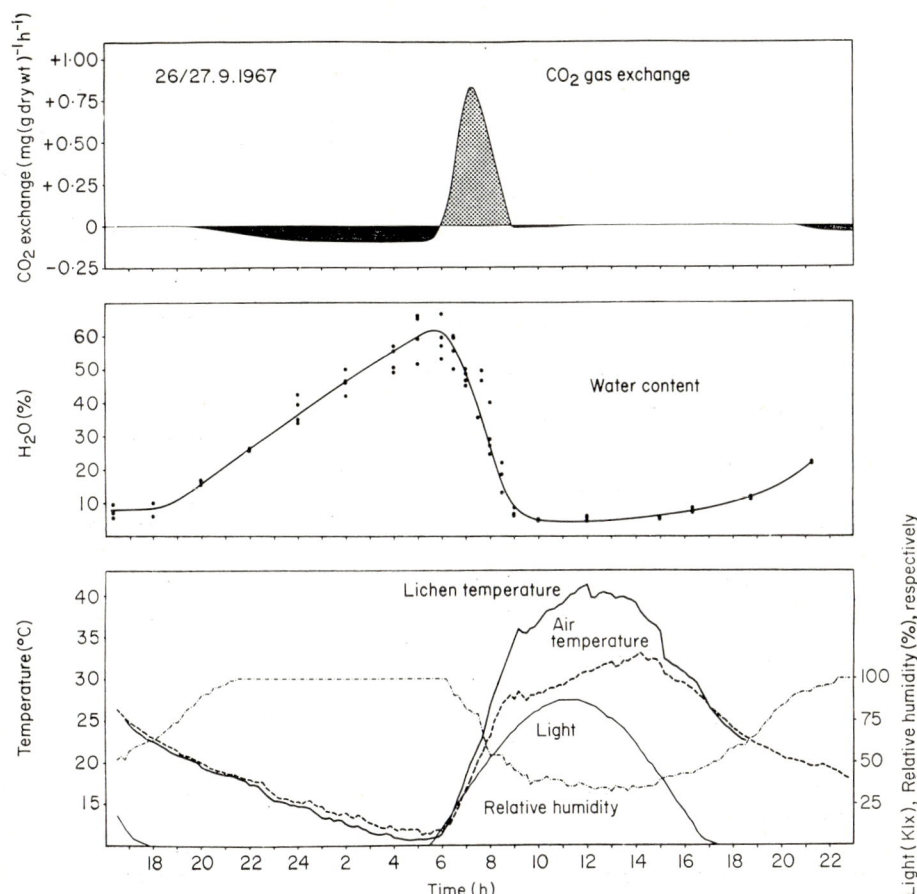

Fig. 9. Carbon dioxide exchange (top) and water content (middle) of *Ramalina maciformis* in the Negev desert. Lower graph shows thallus temperature, light intensity and relative humidity. Note the burst of photosynthesis after sunrise as the thallus dries out. (Reproduced from Lange *et al.*, 1970a, by permission.)

adaptation to low temperatures, since no such loss could be detected if the thalli were kept at a relatively low temperature (19°C).

In the eco-physiological studies of *Chondropsis semiviridis* in Australia (Rogers, 1971), high humidity was suggested as a factor limiting the distribution of the species, as the thallus did not unroll and expose the photosynthetic surface unless liquid water in the form of dew or rain was applied. Apart from this qualification, the findings of that study agreed with those of Lange and his co-workers in terms of rapid response to water, and in the finding that the respiratory surge following wetting was not of great

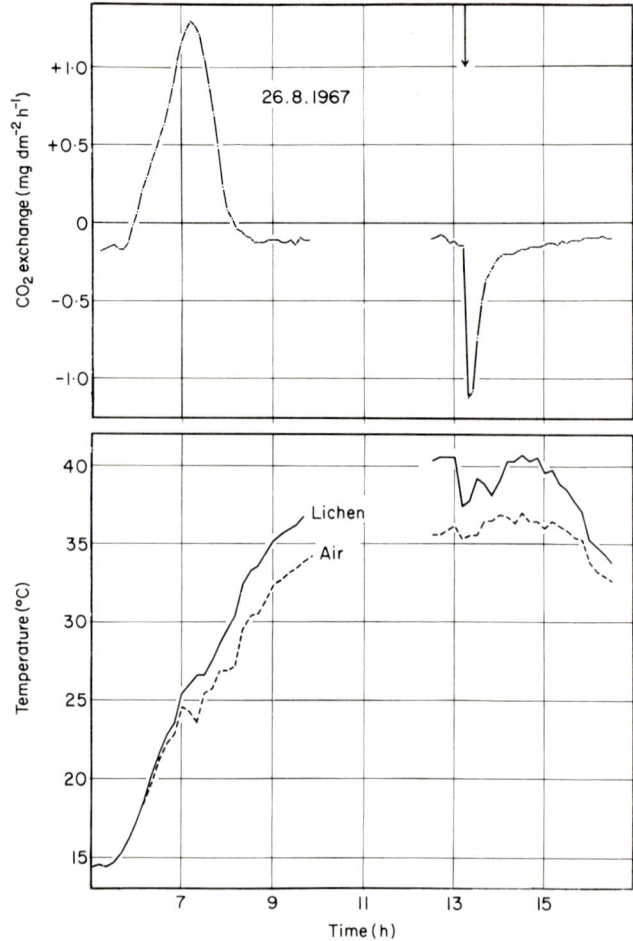

FIG. 10. Carbon dioxide exchange (top) of an endolithic lichen in the Negev desert, showing the respiratory surge caused when the thallus was wet in full sun (arrowed). The lower graphs show air and thallus temperatures. (Reproduced from Lange et al., 1970b, by permission.)

ecological significance. Like *Ramalina maciformis*, *Chondropsis semiviridis* survived prolonged drought apparently without permanent damage.

It appears, therefore, that desert lichens are adapted to use dews to provide short bursts of photosynthesis in the cool of the morning, and that frequency of dew formation is likely to be a significant factor in determining their distribution. Crustose species on soil surfaces may also avail themselves of the upward movement of water in desert soils at night, demonstrated by Rose (1968). It is possible that under desert conditions dew could form within the soil even if it did not form on the surface. This would help explain the extensive development of crustose lichens on desert soil surfaces.

Because of the varying conditions of light and temperature under which optimal water content of thalli for carbon assimilation has been determined, it is not possible to compare results from study to study. Something of a debate has developed as to whether the optimum is saturation level, or lower. Ellee (1939), Smyth (1934) and Lange (1969a) all found saturation conditions optimal, whereas Jumelle (1892), Stocker (1927), Ried (1960) and Kershaw (1972) found optima at lower levels. The precise optimum will vary according to prevailing conditions, lower optima being expected of low light intensities because while light may be the limiting factor in photosynthesis, it will not greatly influence respiration rate. It appears that xeric species under identical conditions do have lower optimal water contents for assimilation. Kershaw and Harris (1971) showed a cline across England in *Parmelia caperata*, a lichen of more mesic conditions, from an optimum at 88% of saturation water content in wet regions in Devon to only 28% of saturation in drier regions of Norfolk. Kershaw (1972) demonstrated this relationship for a number of species in Canada, *Xanthoria fallax* showing a marked shift in optima depending on its origin (Fig. 11). It is likely, therefore, that desert species will reach optimal assimilation rates at lower water contents than will temperate species.

2. Temperature

Lange (1953) reviewed the effect of high temperatures on lichens, and tested a great number of species after exposure to a 30-min period at various temperatures. As a rule, lichens are not easily damaged by high temperature if air-dry, but are sensitive when wet. Photosynthesis is more sensitive to high temperature than is respiration. Lange noted a general tendency for the temperature resistance of lichens to correspond with the conditions likely to prevail in their environment.

Lange (1969a) found that after 30 min at 60°C, air-dry *Ramalina maciformis* thalli were virtually unaffected. Similarly, Rogers (1971) found that *Chondropsis semiviridis* thalli were resistant to high temperatures (60°C) when dry, but when wet a temperature of 40°C for 30 min caused severe damage. Temperatures in excess of 60°C do occur on desert soils (Rogers, 1971), so it appears that while the heat resistance of dry thalli is usually sufficient to withstand the conditions encountered, some heat damage may occur. In some semi-desert areas where summer rains are not infrequent it is possible that wet thalli will reach quite high tempera-

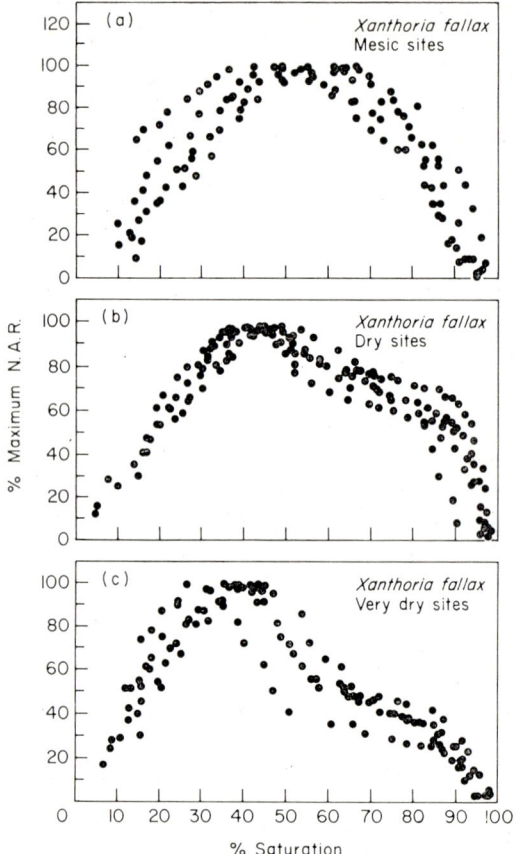

FIG. 11. Relationship between net assimilation rate (N.A.R.) and thallus saturation for *Xanthoria fallax* from various sites in Canada. Note the reduced saturation level at which material from various sites reaches maximal assimilation. (Reproduced from Kershaw, 1972, by permission of the National Research Council of Canada.)

tures and so suffer severe damage. Rogers suggested that this controlled the distribution of *Chondropsis* in eastern Australia, as *Chondropsis* only occurred in desert areas within which winter rainfall predominated.

It is apparent that lichens from hot deserts are better adapted to high temperature than other lichens. Lange (1969a) observed that the optimal temperature of 20°C for assimilation by *Ramalina maciformis* was low for a hot-desert plant, and may represent an adaptation to allow maximal photosynthesis in the cool of the morning. While this is so, an optimum of 20°C is a high temperature optimum amongst the lichens studied (Table

VI). The hot-desert lichens studied do have higher temperature optima for photosynthesis than other lichens. From Table VI it is apparent that species from warm areas tend to show higher optimal temperatures than those from cold areas, an apparent adaptation to environmental conditions. It is notable that with the exception of an unnamed lichen species from Florida (Gannutz, 1968) which has its optimum in the range 20–32°C, hot-desert lichens have the highest optimal temperatures recorded.

IV. Factors Controlling Distribution

It is apparent from the preceding discussion on the morphological and physiological response of lichens to aridity, that high temperatures and drought are important factors in their evolution. These factors are also important considerations in explaining the distribution of lichens within desert regions.

The assertion that lichens are able to survive a far longer drought than they meet in nature (Lange, 1953) was based on measurements of respiration. It is apparent that studies based on photosynthetic response show a lesser drought resistance. However, it is not so much drought resistance as the total number of hours during which a lichen may photosynthesize that will determine whether or not it survives. Studies by Lange et al. (1970a,b) and Rogers (1971) show how small is the margin by which a desert lichen survives. A reduction of a few hours per year in photosynthesis—a few less occasions when the thallus is wet—and respiration might well exceed photosynthesis. Thus, it is not unexpected that lichen distribution so closely parallels rainfall patterns, especially on the dry edge of the distribution range.

It is likely that inability to compete with other organisms, especially for light, limits the development of desert lichens in areas of high rainfall, although they would be expected to occur in dry, exposed sites in such places. Thus, the distribution of desert lichens is largely explained in terms of rainfall and its seasonal distribution.

However, not only must climate be suitable for a lichen to develop, but so must the substrate (Brodo, 1974). Rogers (1972a) has documented the distribution of soil-surface lichens on a variety of arid zone soils. The frequencies of 32 species were examined with respect to soil pH, extractable soil calcium and sodium, and soil-surface type. Some species were restricted to the acid-neutral soils (pH 6·4–7·2) while others were restricted to alkaline soils, but most occurred across a wide range of soil pH, six species occurring across the entire range surveyed (pH 6·4–8·8). The relationship between soil calcium and species range is similar to that between pH and species range, except that no species is confined to soils with high ($> 6·4$ mmol/g soil) extractable calcium. Studies on the relationship between extractable soil sodium and species range produced the surprising result that while *Collema coccophorum* was most frequent on the more saline soils ($> 10^3$ μmol Na/g soil), most species occurred over a wide range of sodium concentrations. It was found that the lichens were most common on soils with a fine texture which either crusted or was hard-setting. Lichens were rare on self-mulching clays, sands and gravels. Galun (1963) observed that the pattern of soil-surface lichen distribution in the Negev was in accord with the soil pattern of the area.

Saxicolous lichens are often quite substrate specific. Galun (1963) found that where flint and chalk stones occurred together each had distinctive lichen communities, although the environment was otherwise identical. She observed, however, that *Buellia subalbula* var. *fuscocapitellata* occurred on both substrates, although it formed only small thalli 2–3 cm in diameter on flint, but thalli as large as 10 cm in diameter on limestone.

It is also true that corticolous and lignicolous species in deserts may be substrate specific; for example Darrow (1950) found that *Caloplaca astrosanguinea* was confined to the bark of *Prosopis juliflora* var. *velutina* in the desert shrublands of Arizona. Such specificity is probably uncommon, most species apparently growing on a number of substrates, but more frequently on some than others. It is, however, the rarity of bark and wood surfaces in some deserts that prevents the development of corticolous and lignicolous species in them—Galun and Reichert (1960) found only one corticolous lichen in the Negev, *Caloplaca luteoalba* growing on *Acacia tortilis*.

In broad terms, lichens and phanerogams are distributed in response to climatic factors, so it is not surprising to find that within deserts they often show similar distribution patterns. This similarity has been referred to by Reichert (1937b,c), Faurel *et al.* (1953), Looman (1964b), Schubert and Klement (1971) and Rogers (1972b). Faurel *et al.* (1953) and Rogers (1972b) suggested that the lichen flora conveniently divided arid lands into two portions, depending on the severity of aridity.

The similarity in distribution patterns can be judged by comparing the distribution of *Kochia sedifolia*, a desert perennial shrub, with that of *Diploschistes ocellatus* (Fig. 12). This pattern will, however, only hold where the lichen and phanerogam are subject to the same stress. In the Negev, for instance, both soil lichens and phanerogams are distributed as the soils, both being sensitive to soil variations. However, rock-inhabiting lichens are sensitive to the nature of the rock, which the phanerogams are not. Hence, various lichen communities occur within the same phanerogamic community when flint and chalk pebbles are mixed (Galun, 1963).

Fog oases are a striking departure from the general principle. Because of differences in water relations, phanerogams absorb water almost exclusively

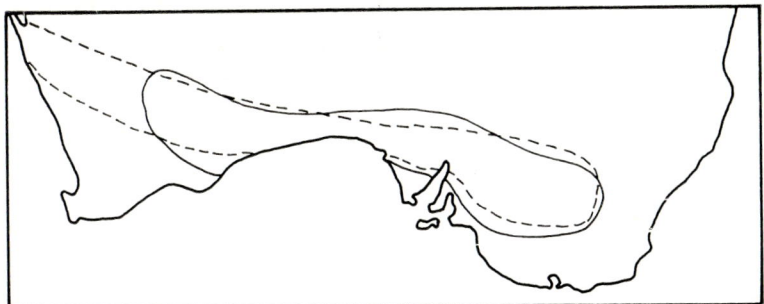

FIG. 12. Distribution of the shrub *Kochia sedifolia* (based on data of Hall et al., 1964) (solid line) and *Diploschistes ocellatus* (broken line) in southern Australia.

from the soil, whereas lichens absorb water quite readily from the air. Regions with low rainfall but where fogs or high humidity are common would thus be dry for phanerogams, but humid for lichens. There are reports of such locations in South America (Follmann, 1967b; Thomson and Iltis, 1968; Follmann and Redón, 1972), North America (Rundel et al., 1972) and South Africa (Mattick, 1970). In all these areas the desert sweeps down to tropical coasts, and fogs frequently roll in over them from the sea.

The lichen growth in these areas is quite unlike that found in other desert regions, genera such as *Usnea, Ramalina, Heterodermia* and *Arthonia* dominating. The species list for the Cerro Moreno region in Chile (Follmann, 1967b) reads like a list from a rainforest. The affinity of the flora is in fact with the wet tropics. The floras reported from Peru by Thomson and Iltis (1968) and Baja California by Rundel et al. (1972) were essentially of terrestrial species which formed extensive mats over the soil. Those discussed by Follmann (1967b) and Follmann and Redón (1972) from subtropical Chile are more diverse, having corticolous and foliicolous, as well as saxicolous and terricolous, species. In this region even the spines of arborescent cacti are festooned with *Usnea* and other fruticose genera. Further regions where fog oases are likely to be found, therefore, include the coast of Mauritania in North Africa and the coast of north-western Australia in the vicinity of Exmouth Gulf.

The uniformity of desert lichen floras around the world relative to the diversity of phanerogamic floras, and the relative lack of endemics, has caused comment (Thomson, 1961; Weber, 1962; Looman, 1964b; Rogers and Lange, 1972). Thomson (1961) observed that vegetative reproduction would limit endemism, but would cause any new mutation that did arise to develop a uniform population rapidly, as back-crossing may not occur. This apparently had occurred, but only rarely, in *Physcia* in North America.

It has long been held that the desert environment provides a strong stimulus to evolutionary change and to speciation. Stebbins (1952) accounted for this by hypothesizing a series of minor climatic changes which alternately isolated small populations allowing them to develop separately, then brought them together so they hybridized. To explain the distribution of cosmopolitan desert lichens it is necessary to assume that recombination occurs at least rarely. This assumption is compatible with the evidence for a slow rate of evolution. Hence ecological sifting of available gene combinations is probably more important than the evolution of new combinations for the development of desert lichen communities.

V. Lichens in Arid Ecosystems

In arid lands lichens are ecologically significant: their resistance to extremes of heat and drought enables them to occupy habitats unavailable to phanerogams, and nowhere is this more striking than on the soil surface. The ability of crustose, squamulose and sometimes foliose lichens to completely cover the soil surface in desert regions has an important impact on soil stability, hydrology and fertility.

Erosion of desert soils by wind and water is a critical problem. However, lichen crusts retard such erosion to a marked extent, their thalli preventing raindrop impact on to the soil, and their rhizoids binding the usually friable desert soil particles. Cameron and Blank (1966) found lichen crusts capping uneroded soil pedestals up to 10 cm higher than the surrounding eroded areas. Weber (1962, 1967) also noticed the ability of lichens to retard soil erosion. Examination of quadrats used to study revegetation of bare soil on the Koonamore Vegetation Reserve in South Australia showed, however, that lichen crusts only develop on areas of soil that are protected from erosive forces (Rogers, 1974). On that reserve a shrub layer of *Atriplex vesicaria* or *A. stipitata* appears on eroded areas before lichen crusts form. It appears that the first stage in stabilization of the soil surface is protection from wind erosion, the surface being then bound by fragile algal filaments, the lichens growing only slowly on this temporarily stabilized surface. It is apparent on these quadrats that once the shrub cover dies and decays, a well developed lichen crust is able to resist erosive forces for a considerable time.

Lichen crusts have a marked effect on the hydrology of arid lands. The rate of infiltration of water into a sandy soil at Koonamore was 7 cm h^{-1} in an area with a lichen crust, but almost 14 cm h^{-1} in a nearby area on a similar soil without a lichen crust (Hall and Specht, personal communication). Crusting of soil surfaces is generally expected to reduce infiltration and thus increase run-off (Jackson, 1958), so increasing the likelihood of erosion by flowing water. Soil-surface lichen crusts are, however, markedly vesicular, frequently having a microtopographic amplitude of 1–2 cm. Such a microtopography would tend to retain a great deal of the water which could not infiltrate into the soil, and would so reduce flow rates as to limit any erosion that might have been caused.

Lichen-stabilized soil crusts contribute significantly to the fertility of desert soils and, presumably, to growth of vascular plants. Shields (1957) and Shields *et al*. (1957) showed the nitrogen content of soils with a lichen crust to be 2–7 times higher than soils without a lichen crust. Similarly, they found the organic carbon levels in lichen-stabilized soils to be higher than that in other desert soils. These differences were presumably due to biological fixation by both free-living and lichenized algae in the soil surface. The only records of nitrogen fixation by desert lichens are those of Rogers *et al*. (1966). However, it is to be expected that any lichen containing an heterocystous blue-green alga will be capable of nitrogen fixation.

It is striking how few seedlings germinate in lichen-encrusted areas after rains compared with nearby areas without lichens. It is possible that this is simply due to the physical nature of the soil surface, but it is also possible that lichen acids are involved. Rondon (1966) and Pyatt (1967) have shown that lichen extracts inhibit seed germination, and Huneck and Schreiber (1972) have shown that at low concentrations some lichen substances promote growth of vascular plants. Other organisms may also be affected: the antibacterial effects of lichen acids have been discussed by Stoll *et al*. (1947), and Malicki (1967) showed that whereas usnic acid inhibited bacterial decomposition of cellulose, it did not inhibit nitrogen fixation by *Azotobacter*. While such effects may not be significant in most ecosystems, in a desert where the entire soil surface is carpeted with lichens these factors deserve examination.

Lichens on soil and rock surfaces undoubtedly provide habitats for a complex microfauna, of the type described by Wood (1970) from arid soils in Australia. There is little sign that insects browse on desert lichens in Australia, and no record of such from other arid lands. The impact of lichens on animal communities must, therefore, generally be indirect through their impact on the other plants.

VI. The Impact of Man

The impact of man on lichens in deserts and semi-deserts has been chiefly through the introduction of grazing herds. There can be no doubt that the soil-surface lichens have been destroyed or damaged over an area measured in millions of square kilometres.

The extent of damage has been partly documented by Rogers (1972) for Australia. From this study it is apparent that in large areas of Australia there has been an almost complete destruction of the lichen crusts on the soil following low intensity grazing by sheep. This has occurred in little more than 100 years. Damage in the Old World must be very much more extensive indeed. Detailed studies by Rogers and Lange (1971) around sheep-watering places in arid South Australia showed that different species respond differently to trampling by sheep. While all are reduced in frequency, the reduction is less marked in *Collema* and *Endocarpon* than in any other genera. *Lecidea decipiens* and *Heppia lutosa* showed reduced frequencies at distances as great as 800 m from watering places (Fig. 13).

Recovery from damage caused to soil-surface lichens by sheep is slow. On the Koonamore Vegetation Reserve, some areas ungrazed by sheep since 1925 were still not showing a development of soil-surface lichens in 1969, although areas adjacent to them did.

The destruction of soil surface lichen populations by trampling animals is likely to have far-reaching effects. Once the crust is destroyed, the surface soil which contains most of the soil nutrients is easily blown away. The soil texture changes when the crust is damaged and the living conditions for the soil microflora are radically changed. This will in turn affect the growth of phanerogams especially by alteration of seed-bed characteristics.

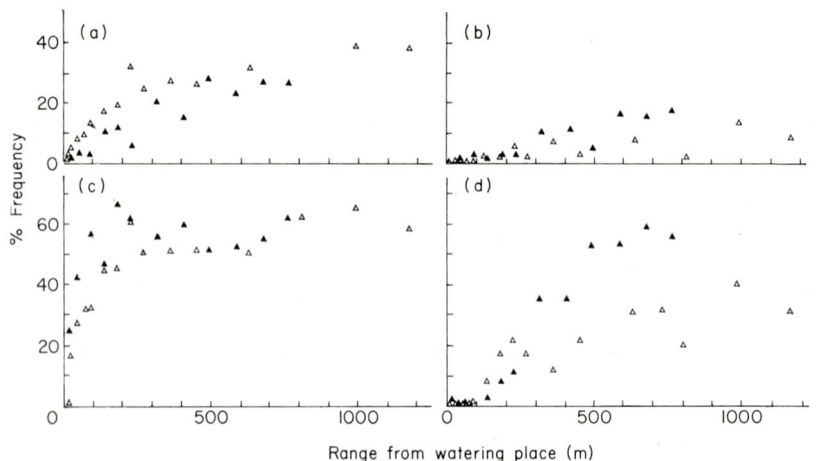

FIG. 13. Relationship between percentage frequency and distance from a sheep-watering place for soil-surface lichens in arid South Australia. (a) *Heppia lutosa*, (b) *Dermatocarpon lachneum*, (c) *Collema coccophorum*, and (d) *Lecidea decipiens*. Open triangles and solid triangles represent two nearby sites. (Reproduced from Rogers and Lange, 1971, by permission.)

It is likely, then, that by the time the soil-surface lichen crust is destroyed by trampling livestock, far-reaching and perhaps irreversible changes have been initiated in the system.

The impact of industrial activity on desert lichens is relatively limited. The principal non-pastoral industries in arid lands are mining and tourism; disturbance to lichen communities is probably due to mechanical damage caused by machinery. This situation is different from that in cold deserts (tundra) discussed by Schofield and Hamilton (1970).

Acknowledgements

Much of the information in the maps showing distribution of lichens in southern Australia was supplied by Dr G. C. Bratt, Mr R. Filson and Mr N. Sammy. The Universities of Queensland and Western Australia both provided finance to allow the collection of data on the distribution of lichens in arid south-western Australia. Dr L. Kappen kindly made available relevant parts of manuscripts before publication. Miss M. O'Sullivan and staff of the University of Queensland Biological Sciences Library provided invaluable assistance in tracing and obtaining copies of many articles.

References

Alon, G. and Galun, M. (1971). The genus *Caloplaca* in Israel. *Israel J. Bot.* **20,** 273–292.
Baker, C., Elix, J. A., Murphy, D. P. H., Kurokawa, S. and Sargent, M.V. (1973). *Parmelia reptans*, a new lichen species producing the depsidone, succinprotocetraric acid. *Aust. J. Bot.* **21,** 137–140.
Barkman, J. J. (1958). "Phytosociology and Ecology of Cryptogamic Epiphytes". van Gorcum, Assen.
Bibby, P. N. S. (1955). A remarkable lichen from arid Australia. *Muelleria* **1,** 60–61.
Bliss, L. C. and Hadley, E. B. (1964). Photosynthesis and respiration of alpine lichens. *Am. J. Bot.* **51,** 870–874.
Blum, O. (1974) ["1973"]. Water relations. *In* "The Lichens" (V. Ahmadjian and M. E. Hale, eds), pp. 371–400. Academic Press, New York and London.
Bouly de Lesdain, M. (1932). Lichens de l'état de New-Mexico. *Annls. cryptog. exot.* **5,** 89–139.
Brodo, I. (1974) ["1973"]. Substrate ecology. *In* "The Lichens" (V. Ahmadjian and M. E. Hale, eds), pp. 401–441. Academic Press, New York and London.
Butin, H. (1954). Physiologische-ökologische Untersuchungen über den Wasserhaushalt und die Photosynthese bei Flechten. *Biol. Zbl.* **73,** 459–502.

Cameron, R. E. and Blank, G. B. (1966). Desert algae: soil crusts and diaphanous substrata as algal habitats. *NASA Tech. Report* 32–971, 1–41.
Cloudsley-Thompson, J. L. (1976). "Insects and History". Wiedenfeld and Nicolson, London.
Darrow, R. A. (1950). The arboreal lichen flora of southeastern Arizona. *Am. Midl. Nat.* **43**, 484–502.
Dodge, C. W. (1966). New lichens from Chile. *Nova Hedwigia* **12**, 307–352.
Doidge, E. M. (1950). The South African fungi and lichens to the end of 1945. *Bothalia* **5**, 1–1094.
Elenkin, A. (1901a). Lichenes florae Rossiae et regionum confinium orientalum. *Trudȳ imp. S-peterb. bot. Sada* **19**, 5–52.
Elenkin, A. (1901b). Lishnaynikovaya manna (*Lichen esculentus* Pall.). *Trudȳ imp. S-peterb. bot. Sada* **19**, 53–99.
Elenkin, A. (1901c). Kochyuschie lishnayniki pustin i steppei. *Izv. imp. S-peterb. bot. Sada* **1**, 16–37; 52–71.
Ellee, O. (1939). Über die Kohlensaüreassimilation von Flechten. *Beitr. Biol. Pfl.* **26**, 250–288.
Errera, L. (1893). "Pain du ciel" provenant de Diarbekir. *Bull. Acad. r. Belg. Cl. Sci.* **26**, 83–91.
Ertl, L. (1951). Über die Lichtverhältnisse in Laubflechten. *Planta* **39**, 245–270.
Eversmann, E. (1831). In Lichenum esculentum Pallasii et species consimiles adversaria. *Nova Acta Leopoldina* **15**, 349–358.
Faurel, L., Ozenda, P. and Schotter, G. (1953). Les Lichens du Sahara Algérien. *In* "Desert Research Proceedings, an International Symposium held in Jerusalem", pp. 310–317. Israel Research Council, Jerusalem.
Filson, R. G. (1967). Supplementary descriptions for two Victorian desert lichens. *Muelleria* **1**, 197–202.
Fink, B. (1909a). Lichens of the Desert Laboratory Domain. *In* "Distribution and Movements of Desert Plants" (V. M. Spalding, ed.), pp. 24–27. Carnegie Institution of Washington, Tucson.
Fink, B. (1909b). The composition of a desert lichen flora. *Mycologia* **1**, 87–103.
Fink, B. (1935). "The Lichen Flora of the United States". University of Michigan Press, Ann Arbor.
Flagey, C. (1901). Lichenes algerienses exsiccati. *Revue Mycol.* **13**, 83–87; 107–117.
Follmann, G. (1961). Eine dornbewohnende Flechtengesellschaft der zentralchilenischen Sukkulentenformation mit kennzeichnender *Chrysothrix nolitangere* Mont. *Ber. dt. bot. Ges.* **73**, 449–462.
Follmann, G. (1962). Eine dornbewohnende Flechtengesellschaft der nordchilenischen Sukkulentenformation mit kennzeichnender *Anaptychia intricata* (Desf.) Mass. *Ber. dt. bot. Ges.* **74**, 495–510.
Follmann, G. (1965a). Eine gesteinbewohnende Flechtengesellschaft der nordchilenischen Wüstenformation mit kennzeichnender *Buellia albula* (Nyl.) Muell. Arg. *Nova Hedwigia* **10**, 243–256.
Follmann, G. (1965b). Fensterflechten in der Atacamawüste. *Naturwissenschaften* **52**, 434–435.

Follmann, G. (1967a). Chilenische Wanderflechten. *Ber. dt. bot. Ges.* **79**, 453–462.
Follmann, G. (1967b). Die Flechten der nordchilenischen Nebeloase Cerro Moreno. *Nova Hedwigia* **14**, 215–281.
Follmann, G. (1967c). Vegetationsanalytische Untersuchungen an Flechtengesellschaften zwischen Atacamawüste und Grahamland. *Ber. dt. Bot. Ges.* **80**, 199–205.
Follmann, G. and Follmann-Schrag, I-A. (1964). Planta con periscopios. Un nuevo ecotipo de vegetales encontrado en el Desierto de Atacama. *Bol. Univ. Chile* 1964, 34–39.
Follmann, G. and Redón, J. (1972). Ergänzungen zur Flechtenflora der nordchilenischen Nebeloasen Fray Jorge und Talinay. *Willdenowia* **6**, 431–460.
Galun, M. (1963). Autecological and synecological observations on lichens of the Negev, Israel. *Israel J. Bot.* **12**, 179–186.
Galun, M. (1970). "The Lichens of Israel". Israel Academy of Science and Humanities, Jerusalem.
Galun, M. and Reichert, I. (1960). A study of the lichens of the Negev. *Bull. Res. Coun. Israel* **9D**, 127–148.
Gannutz, T. P. (1968). Effects of environmental extremes on lichens. "Colloque sur les Lichens et la Symbiose Lichénique" *Mém. Soc. bot. Fr.*, pp. 169–179.
Geiger, R. (1950). "The Climate Near the Ground." Harvard University Press, Cambridge.
Goebel, K. (1926). Morphologische und biologische Studien. VII. Die Wasseraufnahme der Flechten. *Ann. Jard. bot. Buitenz.* **36**, 1–83.
Hale, M. E. (1974). "The Biology of Lichens", 2nd ed. Edward Arnold, London.
Hall, E. A. A., Specht, R. L. and Eardley, C. M. (1964). Regeneration of the vegetation on Koonamore Vegetation Reserve, 1926–1963. *Aust. J. Bot.* **12**, 205–264.
Harrison, S. G. (1951). Manna and its sources. *Kew Bull.* **5** 407–417.
Herre, A. W. C. T. (1911). The desert lichens of Reno, Nevada. *Bot. Gaz.* **51**, 286–297.
Hirst, J. M. and Hurst, G. W. (1967). Long distance spore transport. In "Airborne Microbes" (P. H. Gregory and J. L. Monteith, eds), pp. 307–344. Cambridge University Press, Cambridge.
Huneck, S. and Schreiber, K. (1972). Wachtumregulatorische Eigenschaften von Flechten- und Moss-Inhaltsstoffen. *Phytochemistry* **11**, 2429–2434.
Imshaug, H. A. (1956). Catalogue of Mexican lichens. *Revue bryol. lichén.* **25**, 321–385.
Jaccard, P. (1912). The distribution of the flora in the alpine zone. *New Phytol.* **11**, 37–50.
Jackson, E. A. (1958). Soils and hydrology at Yudnapinna Station, South Australia. *Soils Ld. Use Ser. C.S.I.R.O. Aust. Bull.* **24**, 1–66.
Jacquemin-Roussard, M. and Kilbertus, G. (1971). Quelques lichens d'Afghanistan: Ebauche écologique. *Bull. Soc. lorr. Sci.* **10**, 59–65.

Johnson, E. R. L. and Baird, A. M. (1970). Notes on the flora and vegetation of the Nullarbor Plain at Forrest, Western Australia. *J. R. Soc. West. Aust.* **53**, 56–61.
Jumelle, H. (1892). Recherches physiologiques sur les lichens. *Revue gén. Bot.* **4**, 49–64; 103–121; 305–320.
Kappen, L. (1974) ["1973"]. Response to extreme environments. *In* "The Lichens" (V. Ahmadjian and M. E. Hale, eds), pp. 311–380. Academic Press, New York and London.
Kappen, L. and Schulze, E-D. (1972). *Ramalina maciformis* (Del.) Nyl. fertile in the Western Negev, Israel. *Lichenologist* **5**, 323–325.
Keller, B. (1930). Die Erdflechten und Cyanophyceen am unteren Lauf der Volga und des Ural. *Vegetationsbilder* **20** (8), 43–48.
Kershaw, K. A. (1972). The relationship between moisture content and net assimilation rate of lichen thalli. *Can. J. Bot.* **50**, 543–555.
Kershaw, K. A. and Harris, G. P. (1971). Simulation studies and ecology. A simple defined system and model. *In* "International Symposium on Statistical Ecology" (G. P. Patil, E. C. Pielou and W. E. Waters, eds), Vol. 3, pp. 1–21. Pennsylvania State University Press, University Park.
Klement, O. (1955). Prodromus der mitteleuropäischen Flechtengesellschaften. *Repium nov. Spec. Regni Veg. Beih.* **135**, 5–194.
Klement, O. (1958). Die Stellung der Flechten in Pflanzensoziologie. *Vegetatio* **8**, 43–56.
Klement, O. (1966). Flechten aus der Mongolischen Volksrepublik. *Repium nov. Spec. Regni veg.* **72**, 98–123.
Kolumbe, E. (1927). Untersuchungen über die Wasserdampfaufnahme der Flechten. *Planta* **3**, 734–757.
Lamb, I. M. (1936). Lichens from Bahrein Island. *J. Bot., Lond.* **74**, 347–351.
Lamb, I. M. (1963). "Index Nominum Lichenum". Ronald Press, New York.
Lange, O. L. (1953). Hitze und Trockenresistenz der Flechten in Beziehung zu ihrer Verbreitung. *Flora, Jena* **140**, 39–97.
Lange, O. L. (1969a). Experimentell-ökologische Untersuchungen an Flechten der Negev-Wüste. 1. CO_2-Gaswechsel von *Ramalina maciformis* (Del.) Bory unter kontrollierten Bedingungen im Laboratorium. (National Research Council of Canada Technical Translation, 1654). *Flora, Jena* **158**, 324–359.
Lange, O. L. (1969b). Die funktionellen Anpassungen der Flechten an die ökologischen Bedingungen arider Gebiete. *Ber. dt. bot. Ges.* **82**, 3–22.
Lange, O. L. and Bertsch, A. (1965). Photosynthese der Wüstenflechte *Ramalina maciformis* nach Wasserdampfaufnahme aus der Luftraume. *Naturwissenschaften* **52**, 215–216.
Lange, O. L., Schulze, E-D. and Koch, W. (1970a). Experimentell-ökologische Untersuchungen an Flechten der Negev-Wüste. II. CO_2-Gaswechsel und Wasserhaushalt von *Ramalina maciformis* (Del.) Bory am natürlichen Standort während der sommerlichen Trockenperiode. *Flora, Jena* **159**, 38–62.
Lange, O. L., Schulze, E-D. and Koch, W. (1970b). Experimentell-ökologische Untersuchungen an Flechten der Negev-Wüste. III. CO_2-Gaswechsel und Wasserhaushalt von Krusten- und Blattflechten am natürlichen Standort während der sommerlichen Trockenperiode. *Flora, Jena* **159**, 525–538.

Looman, J. (1964a). Ecology of lichen and bryophyte communities in Saskatchewan. *Ecology* **45**, 481–491.
Looman, J. (1964b). The distribution of some lichen communities of the Prairie provinces and adjacent parts of the Great Plains. *Bryologist* **67**, 209–224.
Magnusson, A. H. (1940). Lichens from Central Asia. 1. *Rep. sci. Exp. N.W. China, XI (Bot.)* **1**, 1–167.
Magnusson, A. H. (1944). Lichens from Central Asia. 2. *Rep. sci. Exp. N.W. China, XI (Bot.)* **2**, 1–68.
Malicki, J. (1967). Wpływ kwasow porostowych na mikroorganizmy glebowe. Csez II. Wpływ wyciagow wodnych z gatunkow *Cladonia* na bakterie glebowe. *Annls Univ. Mariae Curie-Skłodowska, sect. C* **22**, 159–163.
Mattick, F. (1970). Flechtenbestände der Nebelwüste und Wanderflechten der Namib. *Namib Meer* **1**, 35–44.
Meigs, P. (1953). World distribution of arid and semi-arid homoclines. *Arid Zone Progm.* **1**, 203–210. [2 maps]
Mereschkovsky, C. (1918). Note sur une nouvelle forme de *Parmelia* vivant àlé'tat libre. *Bull. Soc. Bot. Genève. Ser. 2.* **10**, 26–34.
Meyer, G. F. W. (1825). "Die Entwickelung, Metamorphose und Fortpflanzung der Flechten". Vandenhoeck and Ruprecht, Göttingen.
Müller, J. (1880). Enumeratio Lichenum Aegyptiacorum. *Revue Mycol.* **2**, 40–44.
Müller, J. (1884a). Lichenes Palaestinenses. *Revue Mycol.* **6**, 12–15.
Müller, J. (1884b). Enumeratio Lichenum Aegyptiacorum. Supplementum Primum. *Revue Mycol.* **6**, 15–20.
Müller, J. (1893). Lichenes. *Trans. R. Soc. S. Aust.* **16**, 142–149.
Nees van Esenbeck, T. F. L. (1831). Ueber die Flechten, welche unser Freund und College, Herr Prof. Eversmann, in den Kirgischen Steppen gesammelt hat. *Nova Acta Leopoldina* **15**, 359–362.
Pallas, P. S. (1771–1776). "Reise durch Verschiedene Provinzen des Russischen Reiches". K. Akademie der Wissenschaften, St. Petersburg.
Poelt, J. (1974) ["1973"]. Systematic evaluation of morphological characters. *In* "The Lichens" (V. Ahmadjian and M. E. Hale, eds), pp. 91–115. Academic Press, New York and London.
Poelt, J. and Wirth, V. (1968). Flechten aus dem Nordöstlichen Afghanistan. Gesammelt von H. Roemer in Rahmen der Deutschen Wakhan Expedition 1964. *Mitt. bot. StSamml., Münch.* **7**, 219–261.
Pyatt, F. B. (1967). The inhibitory influence of *Peltigera canina* on the growth and germination of graminaceous seeds and the subsequent growth of seedlings. *Bryologist* **70**, 326–329.
Reichert, I. (1937a). Steppe vegetation in the light of lichen vegetation. *Proc. Linn. Soc. Lond.* **149**, 13–23.
Reichert, I. (1937b). La Libia e la sua posizione fitogeografica dal punto di vista lichenologico. *Nuovo Giorn. Bot. Ital. N.S.* **44**, 188–196.
Reichert, I. (1937c). Eine lichenogeographische Skizze Palaestinas. *Verhandl. zool.-bot. Ges. Wien.* **86–87**, 288–296.

Reichert, I. (1940). A new species of *Diploschistes* from oriental steppes and its phytogeographical significance. *Palest. J. Bot., Rehovot Ser.* **3**, 162–182.

Reimers, H. (1951a). Beiträge zur Kenntniss der Bunten Erdflechten-Gesellschaft. I. Zur Systematik und Verbreitung der Charakterflechten der Gesellschaft besonders im Harzvorland. *Ber. dt. bot. Ges.* **63**, 143–157.

Reimers, H. (1951b). Beiträge zur Kenntniss der Bunten Erdflechten-Gesellschaft. II. Allgemeine Fragen. *Ber. dt. bot. Ges.* **64**, 36–50.

Renaut, J., Marrache, P. and Trotet, G. (1969). La notion de "Spectre Biologique" adaptée aux lichens. *Mém. Soc. bot. Fr.*, **1968** (Coll. Lich.), 197–203.

Ried, A. (1960). Stoffwechsel und Verbreitungsgrenzen von Flechten. II. Wassersubmersionresistenz von Krustenflechten benachbarter Standorte. *Flora, Jena* **149**, 345–385.

Rogers, R. W. (1970). Ecology of Soil Surface Lichens in Arid South-Eastern Australia. Ph.D. thesis, University of Adelaide.

Rogers, R. W. (1971). Distribution of the lichen *Chondropsis semiviridis* in relation to its heat and drought resistance. *New Phytol.* **70**, 1069–1077.

Rogers, R. W. (1972a). Soil surface lichens in arid and sub-arid south-eastern Australia. II. Phytosociology and geographic zonation. *Aust. J. Bot.* **20**, 215–227.

Rogers, R. W. (1972b). Soil surface lichens in arid and sub-arid south-eastern Australia. III. The relationship between distribution and environment. *Aust. J. Bot.* **20**, 301–316.

Rogers, R. W. (1974). Lichens from the T. G. B. Osborn Vegetation Reserve at Koonamore in arid South Australia. *Trans. R. Soc. S. Aust.* **98**, 113–124.

Rogers, R. W. and Lange, R. T. (1971). Lichen populations on arid soil crusts around sheep watering places in South Australia. *Oikos* **22**, 93–100.

Rogers, R. W. and Lange, R. T. (1972). Soil surface lichens in arid and sub-arid south-eastern Australia. I. Introduction and floristics. *Aust. J. Bot.* **20**, 197–213.

Rogers, R. W., Lange, R. T. and Nicholas, D. J. D. (1966). Nitrogen fixation by lichens of arid soil crusts. *Nature, Lond.* **203**, 96–97.

Romano, M. (1914). Licheni della Tripolitania. *Boll. Orto bot., Napoli* **4**, 349–354.

Rondon, Y. (1966). Action inhibitrice de l'extrait du lichen *Rocella fucoides* (Dicks.) Vain. sur la germination. *Bull. Soc. bot. Fr.* **113**, 1–2.

Rose, C. W. (1968). Water transport in soil with a daily temperature wave. I. Theory and experiment. *Aust. J. Soil Res.* **6**, 31–44.

Rudolph, E. D. (1953). A contribution to the lichen flora of Arizona and New Mexico. *Ann. Mo. bot. Gdn* **40**, 63–72.

Rundel, P. W., Bowler, P. A. and Mulroy, T. W. (1972). A fog induced lichen community in north-west Baja California, with two new species of *Desmaziera*. *Bryologist* **75**, 501–508.

Santesson, R. (1942). Some lichens from Palestine and Syria. *Ark. Bot.* **30B** (5), 1–5.

Schofield, E. and Hamilton, W. L. (1970). Probable damage to tundra biota through sulphur-dioxide destruction of lichens. *Biol. Conservation* **2**, 273-280.

Schubert, R. and Klement, O. (1971). Beitrag zur Flechtenflora der Mongolischen Volksrepublik. *Repium nov. Spec. Regni veg.* **82**, 187-262.

Shields, L. M. (1957). Algal and lichen floras in relation to nitrogen content of certain volcanic and acid range soils. *Ecology* **38**, 661-663.

Shields, L. M., Mitchell, C. and Drouet, F. (1957). Alga- and lichen-stabilised surface crusts as soil nitrogen sources. *Am. J. Bot.* **44**, 489-498.

Shushan, S. and Anderson, R. A. (1969). Catalogue of the lichens of Colorado. *Bryologist* **72**, 451-483.

Smith, A. L. (1921) "Lichens". Cambridge University Press, Cambridge.

Smith, D. C. (1962). The biology of lichen thalli. *Biol. Rev.* **37**, 537-570.

Smyth, E. S. (1934). A contribution to the physiology and ecology of *Peltigera canina* and *P. polydactyla*. *Ann. Bot.* **48**, 781-818.

Stalfelt, M. G. (1939). Der Gasaustausch der Flechten. *Planta* **29**, 11-31.

Stebbins, G. L. (1952). Aridity as a stimulus to plant evolution. *Am. Nat.* **86**, 33-44.

Steiner, J. (1895). Ein Beitrag zur Flechtenflora der Sahara. *Sber. Akad. Wiss. Wien.* **104**, 383-393.

Steiner, J. (1896). Beitrag zur Flechtenflora Sudpersiens. *Sber. Akad. Wiss. Wien* **105**, 436-446.

Steiner, J. (1919). Flechten aus Transkaukasien. *Ann. Mycol.* **17**, 1-32.

Steiner, J. (1921). Lichenes aus Mesopotamien und Kurdistan sowie Syrien und Prinkipo. *Annln Naturh. Mus. Wien* **34**, 37-66.

Stocker, O. (1927). Physiologische und ökologische Untersuchungen an Laub- und Strauchflechten. *Flora Abt. A. Physiol. Biochem. (Jena)* **121**, 334-415.

Stoll, A., Brack, A. and Renz, J. (1947). Die antibakterielle Wirkung der Usninsäure auf Mykobakterien und andere Mikroorganismen. *Experientia* **3**, 115-117.

Szatala, O. (1957). Prodromus einer Flechtenflora des Irans. *Annls hist.-nat. Mus. natn. hung.* **8**, 101-154.

Thomson, J. W. (1961). Evolution in the lichen genus *Physcia*. *In* "Recent Advances in Botany", Vol. I, pp. 267-271. University of Toronto Press, Toronto.

Thomson, J. W. and Iltis, H. H. (1968). A fog induced lichen community in the coastal desert of southern Peru. *Bryologist* **71**, 31-34.

Thornthwaite, C. W. (1948). An approach toward a rational classification of climate. *Geogrl Rev.* **38**, 55-94.

Vogel, S. (1955). Niedere "Fensterpflanzen" in der Südafrikanischen Wüste. Eine ökologische Schilderung. *Beitr. Biol. Pfl.* **31**, 45-55.

Weber, W. A. (1962). Environmental modification and the taxonomy of the crustose lichens. *Svensk bot. Tidskr.* **56**, 293-333.

Weber, W. A. (1963). Lichens of the Chiricahua mountains, Arizona. *Univ. Colo. Stud. Ser. Biol.* **10**, 1-27.

Weber, W. A. (1967). Environmental modification in crustose lichens. II. Fruticose growths forms in *Aspicilia. Aquilo Ser. Bot.* **6,** 43–51.
Werner, R. G. (1954). Notes de lichénologie libano-syrienne I. *Bull. Soc. Bot. Fr.* **101,** 355–360.
Werner, R. G. (1955). Notes de lichénologie libano-syrienne II. *Bull. Soc. Bot. Fr.* **102,** 350–356.
Werner, R. G. (1956). Notes de lichénologie libano-syrienne III. *Bull. Soc. Bot. Fr.* **103,** 461–467.
Werner, R. G. (1957). Notes de lichénologie libano-syrienne IV. *Bull. Soc. Bot. Fr.* **104,** 321–326.
Werner, R. G. (1958). Notes de lichénologie libano-syrienne V. *Bull. Soc. Bot. Fr.* **105,** 238–243.
Werner, R. G. (1959). Notes de lichénologie libano-syrienne VI. *Bull. Soc. Bot. Fr.* **106,** 332–337.
Werner, R. G. (1963). Notes de lichénologie libano-syrienne VII. *Bull. Soc. Bot. Fr.* **110,** 311–315.
Werner, R. G. (1966). Notes de lichénologie libano-syrienne VIII. et égyptienne. *Bull. Soc. Bot. Fr.* **113,** 74–83.
Wetmore, C. M. (1970). The lichen family Heppiaceae in North America. *Ann. Mo. Bot. Gdn* **57,** 158–209.
Willis, J. H. (1951). Botany of the Russel Grimwade Expedition. *Mem. natn Mus. Vict.* **17,** 33–64.
Wood, T. G. (1970). Micro-arthropods from soils of the arid zone in southern Australia. *Search* **1,** 75–76.
Zahlbruckner, A. (1922–1940). "Catalogus Lichenum Universalis". Borntraeger, Leipzig.
Zahlbruckner, A. (1925). Chilenische Flechten, gesammelt von C. Skottsberg. *Acta Horti gothoburg.* **2,** 1–26.
Zahlbruckner, A. (1926). Afrikanische Flechten (Lichenes). *Bot. Jb.* **60,** 76–101.
Zukal, H. (1895). Morphologische und biologische Untersuchungen über die Flechten. *Sber. Akad. Wiss. Wien* **104,** 529–574; **105,** 197–264.

9 | Lichens of Man-made Substrates

FRANK H. BRIGHTMAN and
MARK R. D. SEAWARD

I. Introduction	253
II. Materials Derived from Animal Sources	255
A. Bone	255
B. Leather	256
C. Hair and Wool	257
D. Miscellaneous	258
III. Materials Derived from Plant Sources	258
A. Wood	258
B. Cork	264
C. Miscellaneous	265
IV. Wholly Synthetic Organic Materials	266
V. Materials Derived from Mineral Sources	266
A. Inorganic Building Materials	266
B. Glass	277
C. Non-ferrous Metals	282
D. Iron	283
VI. Discussion	286
References	290

I. Introduction

From the earliest times, when primitive man was beginning to make his mark on his surroundings, human activities have had profound effects on the abundance and distribution of lichens. In the temperate zones of the earth, the opening up and clearance of primeval deciduous forest drastically altered the lichen flora, dispersing it over wide areas, but in some circumstances apparently diversifying it. It is well known that secondary woodland and plantations are comparatively poor in corticolous species, whereas old large trees in parkland often support an epiphytic lichen flora which is quite rich, sometimes perhaps surprisingly so, but in any case very different from that found in those few relatively undisturbed semi-natural areas that remain. In general, moisture-loving species diminished in abundance and additional habitats became available to more xerophytic species. The hypertrophication of the environment which accompanies the establishment of human communities and is promoted also by the practice of animal husbandry had analogous effects on the saxicolous lichen flora. More and more habitats became available to species that previously had been largely restricted to the sea coast, and a characteristic community of lichens dependent upon the availability of relatively high levels of concentration of mineral salts spread inland and became ubiquitous in all areas settled by man.

These processes diminished some lichen communities but led to the increase of others on all types of natural substrate. The coming of the Industrial Revolution, however, brought a new factor which has tended to overshadow them. New and rapid advances in technology were accompanied by severe atmospheric pollution which had catastrophic effects on lichens, not merely in the immediate vicinity of industrial installations, but wherever fossil fuels came into use for heating and transport. Such destruction of lichen floras by this cause is still increasing.

Widespread changes brought about by agriculture and industry have mainly occurred in the more temperate regions of the world. Much of the tropics remained relatively little affected by human activities until very recently, but few areas now enjoy this immunity. In many places increasing population and rising standards of living are resulting in the effects of atmospheric pollution being superimposed on those of forest clearance and hypertrophication and it remains to be seen to what extent tropical lichen floras will have the resilience to adapt to these stringent new conditions.

Man destroys, but he also manufactures. Interesting evidence of the adaptability of a considerable number of temperate-zone lichen species is provided by the extent to which they have been successful in colonizing man-made substrates. Some are able to establish themselves on the most unlikely and apparently inhospitable materials, as can be seen in the pages that follow, where the available rather scattered information has been brought together and summarized. Certain patterns may be discerned in these data, from which tentative conclusions are drawn in the concluding section.

This is a field in which there is wide scope for research into the environmental factors involved. Apart from some data on water relations and pH, and much more scanty information on mineral accumulation in a few species, available measurements are few, and so we have been compelled to adopt an anecdotal approach. Another limitation is that most of our examples are perforce taken from Great Britain and to a less extent Europe and the temperate zone of the northern hemisphere generally. It is clear, however, that the number of lichen species which can develop on man-made substrates is quite large, including a group referred to as omnicolous by Smith (1921), which are more or less ubiquitous, and others whose habitat requirements are more specialized and restricted.

II. Materials Derived from Animal Sources

A. Bone

The raw materials used by man in antiquity may for convenience be designated as primary raw materials, and bone was certainly one of them. It was one of the earliest materials utilized by man for making tools and *objets d'art*, and in fact artefacts made from it sometimes accompany fossil hominid remains. Of course, the use of bone for such purposes continues on a minor scale at the present day. It cannot properly be described as a man-made material, but since man soon became the most numerous of carnivorous mammals and the only one that constructs refuse tips and middens, his activities certainly increased the amount of bone available for colonization by lichens. In later times lichens growing on bones acquired a certain magical significance; for instance, they figure as one of the ingredients of the wonder-working weapon-salve. This was an ointment intended for application to a sword which had caused an injury, and it was supposed to have the effect of healing the wound that had already been made. The recipe for the salve given by Porta (1658) includes "scurf that groweth thick on a man's skull left to the open air". The "scurf" would have comprised some or all of *Lecanora dispersa, L. campestris, Caloplaca citrina, C. holocarpa, Candelariella vitellina, Buellia punctata, Rinodina subexigua, Sarcogyne regularis* and *Verrucaria nigrescens*, and perhaps also *Buellia canescens, Aspicilia calcarea* and *Lecanora muralis*. Other species very commonly found on bone are *Xanthoria parietina, Physcia adscendens, P. tenella* and *P. caesia*. Lists of lichens found on bones are given by Fries (1867), Arnold (1875), Richard (1883), Darbishire (1909) and Bouly de Lesdain (1910).

TABLE I.

Lichens recorded from leather (from Arnold, 1875; Richard, 1883; Bouly de Lesdain, 1910, 1921, 1951; Salisbury, 1953; with unpublished data from F. H. Brightman, B. J. Coppins and M. R. D. Seaward).

Aspicilia calcarea	L. dispersa
A. contorta	Lecanora muralis
Bacidia inundata	L. saligna
B. muscorum	Lecidea granulosa
B. umbrina	L. symmicta
Buellia alboatra aggr.	L. uliginosa
B. canescens	L. scabra
B. punctata	Micarea denigrata
Caloplaca aurantiaca	M. nitschkeana
C. cerina	Parmelia perlata
C. citrina	P. subaurifera
C. holocarpa	P. sulcata
C. luteoalba	Peltigera canina
C. saxicola	Physcia adscendens
C. stillicidorum	P. caesia
C. variabilis	P. leptalea
C. vitellinula	P. orbicularis
Candelaria concolor	P. tenella
Candelariella aurella	Physciopsis adglutinata
C. vitellina	Physconia farrea
Cladonia chlorophaea	P. pulverulenta
C. conista	Polyblastia vouauxii
C. fimbriata	Ramalina subfarinacea
C. furcata	Rinodina bischoffii
C. pyxidata	R. subexigua
Collema flaccidum	Thelocarpon laureri
Evernia prunastri	T. magnussonii
Hypogymnia physodes	T. olivaceum
Lecania cyrtella	Trapelia coarctata
L. erysibe	Verrucaria nigrescens
Lecanora campestris	V. viridula
L. carpinea	Xanthoria aureola
L. chlarona	X. parietina
L. chlarotera	X. polycarpa

B. Leather

Another primary raw material was the skins of animals, used to make bags and containers, cut into strips for tying things together, and with the hair left on for clothing and bedding. Owing to the perishable nature of such artefacts, direct evidence is scarce, although the pieces of skin of *Megatherium* found in caves in South America are suggestive, and specimens of pieces of strap and the like have survived from Neolithic and Iron Age times in many parts of the world. The skins of small animals such as rabbits and moles do not require much treatment before use, but the hides of larger animals readily lose their flexibility and become hard and brittle. They are therefore worked into leather by scraping and kneading processes, and in some cases by longitudinal splitting with a knife, combined with steeping in a variety of solutions, such as decoctions of the bark of trees. In general, the chemical part of these treatments increases the acidity of the final product, vide the following substrate measurements: rabbit skin, pH 6·0; kid, pH 5·3; pigskin, pH 3·6; calf, pH 3·3; and sheepskin, pH 3·0. Moulds (especially *Penicillium* spp.) sometimes grow on fairly new leather, but lichens do not. However, old leather that has been weathered often carries a rich lichen flora, as can be seen from the lists of species (incorporated into Table I) recorded from this substrate by Arnold (1875) and Bouly de Lesdain (1909, 1921). This is partly due to a marked decrease in acidity: new shoe leather, pH 3·9; new strap leather, pH 3·3; weathered shoe leather, pH 6·0; and weathered strap leather, pH 5·6.

Another factor is that weathering produces a surface texture which is ideal for lichen growth. The uniformly rough porous surface which develops increases the water absorbing and retaining power of the substrate and encourages the lodgement of lichen propagules and the attachment of the growing thallus.

C. Hair and Wool

Richard (1883) found *Physcia tenella* and *Lecanora dispersa* on some old hare's paws (*Lepus timidus*) nailed to a door at St André d'Ornay, Vendée. In England in the present century *Lecanora conizaeoides* may be found occasionally on very old corpses of grey squirrel (*Sciurus carolinensis*) in gamekeepers' larders. The fur of these animals has a pH approximating to 7·0, whereas horse hair and cow hair is more acid (pH 5·5). In any case, such substrates are of an accidental and more or less ephemeral nature. Materials manufactured from hair and wool are rather more permanent, and when left undisturbed in suitable conditions may support the growth of lichens. Felt is made by compacting fur, usually that of rabbit (*Oryctolagus cuniculus*), which also has an approximately neutral pH. *Trapelia coarctata* and "*Bacidia inundata*" (this name has been variously applied to at least six closely related species—Coppins, *in litt.*) have been found on discarded old hats and pieces of carpet underfelt. Roofing materials made from felt impregnated with bitumen and surfaced with mineral dust commonly support extensive lichen floras usually including *Hypogymnia physodes*, *Xanthoria parietina*, *Physcia* spp., *Lecanora muralis* and *L. dispersa*. Sheep's wool varies in pH from 3·8 to 5·8, but nevertheless Richard (1883) records *Buellia canescens* and *Verrucaria muralis* from wool waste. Woollen textiles are not acid, however, and their texture contributes in favourable circumstances to their ability to provide a substrate for lichens; on knitted woollens an example is *Cladonia macilenta* (Brightman, 1960); as an example on woven wool, M. R. D. Seaward (unpublished data) found an extensive growth of *Lecanora conizaeoides* on a gaberdine raincoat worn by a scarecrow.

D. Miscellaneous

Bouly de Lesdain (1951) discussed the lichens that he found over a period of nearly 50 years on man-made substrates amongst the sand dunes in the area of Dunkirk. It is apparent that people will discard anything and everything at the seaside, including amongst substances of animal origin even a material like silk, on which *Physcia orbicularis*, *Physcia adscendens*, *Caloplaca citrina*, *Lecanora dispersa*, *Buellia punctata* and "*Bacidia inundata*" have been found.

More or less indestructible rubbish continues to accumulate on our more populous shores, and it is convenient to mention here a wholly man-made rather than animal substratum, namely a plastic carton found on the beach at Gadshill in the Isle of Wight, on which *Bacidia umbrina* and *Micarea nitschkeana* were growing (Coppins, *in litt.*).

III. Materials Derived from Plant Sources

A. Wood

Wood was of course a primary raw material, but like leather and unlike bone and mineral artefacts, few specimens of objects made from it have survived from prehistoric times. In any case, the thalli of lichens growing on wood do not usually persist for long once they are dead. Richardson and Green (1965) found *Pyrenula nitida* on alder wood 2500 years old from peat at Wybunbury Moss, Cheshire, but this is an exceptional case and in their paper they discuss the very few other authentic subfossil specimens that have been reported by various authors. On the other hand, postholes are a commonplace of archaeological investigation and testify that from early times the area of substrate available to lichens adapted to growing on dead wood was significantly increased by human activity. In post-medieval times, developments in agricultural practice produced on a wide scale a demand for methods of enclosure of fields which could only partially be met by traditional methods of planting and training hedges. The invention of power saws made sawn timber cheap and plentiful, and the word "pale" acquired a new meaning. In the sense of a sawn board of wood secured vertically to a framework of wood rails, "pale" became popular in lichenological literature (e.g. Smith, 1918, 1926, *passim*). Pales differ considerably in texture from the dead wood of decorticated tree trunks and branches, partly because the tracheids and xylem vessels are cut across at various angles, but also because the latter are invariably affected to a greater or lesser extent by the activities of decay microorganisms such as bacteria, the mycelia of a wide variety of fungi, and other organisms, including myxomycetes and various invertebrates. Pales, and sawn timber generally, normally have a lower water content than ordinary dead wood.

Worked timber suitable for lichen colonization is found in a variety of structures in addition to the various kinds of wooden fencing. The cladding of boarded buildings when not regularly painted, as is often the case on farms, and roofs made of wooden shingles, frequently support a lichen flora. Many railway sleepers when sufficiently old and weathered and wooden memorials in graveyards provide substrates for some species. The importance of these habitats is apparent in areas throughout the world where trees and the corticolous substrates they provide are scarce or absent, as in hot and cold deserts, at high altitudes and on wind-swept moorlands. Interesting assemblages, for example, may be observed (although little has been published about them) on shingle roofs in boreal regions where numerous species of *Cladonia* (including *Cladina*), *Cetraria* (s.l.) etc., give an almost blanket cover to the substrate which superficially resembles that of a terricolous habitat from a lichenological point of view.

Lichens that have been recorded from worked timber are listed in Tables II, III and IV (compiled mainly from data assembled by Coppins, *in litt.*). No species appears to be entirely restricted to this substrate, although *Cyphelium notarisii* comes near to it (Brightman, 1964). *Lecanora farinaria* would qualify, if it exists; it was described from worked timber in Sussex in the early nineteenth century, but has not been seen since (Laundon, 1963). Other species known in Britain only from worked timber are marked (^{a}) in Table III. Hypertrophication of fence posts and other outdoor woodwork occurs through impregnation with dust, resulting in corresponding modifications of the flora. In Britain particularly this process has been enhanced in recent years by agricultural practices such as improved drainage, the removal of hedgerows, and the burning of

Table II.

Lichens commonly found on worked timber, at least in a large part of the British Isles. Letters of compass-bearing indicate strong distribution tendencies.

Bacidia umbrina	Lecidella elaeochroma
Bryoria fuscescens	Micarea denigrata
Buellia griseovirens	M. lignaria
B. punctata	Ochrolechia turneri
Caloplaca citrina	Parmelia caperata
C. holocarpa	P. glabratula
Candelariella vitellina	P. saxatilis
Cetraria chlorophylla	P. soredians S
Cladonia coniocraea	P. sulcata
C. macilenta	P. subaurifera
Cyphelium inquinans SE	Parmeliopsis aleurites E
Evernia prunastri	P. ambigua
Hypogymnia physodes	Pertusaria amara
H. tubulosa	P. coccodes E
Lecanora chlarona E and N	Physcia adscendens
L. chlarotera	P. orbicularis
L. confusa	P. tenella
L. conizaeoides	Platismatia glauca
L. dispersa	Pseudevernia furfuracea
L. expallens	Ramalina farinacea
L. piniperda N	Usnea. hirta NE
L. varia	U. inflata W
Lecidea granulosa	U. subfloridana
L. quernea	Xanthoria candelaria
L. scalaris	X. parietina
L. symmicta	X. polycarpa
L. turgidula N	Xylographa abietina N
L. uliginosa	X. vitiligo N and W

stubbles, which have tended to increase soil erosion, and by the widespread use of artificial fertilizers. The increasingly numerous appearances of *Lecanora muralis* on fence rails may be attributable to this cause. Another interesting and quite conclusive example is provided by O'Sullivan and Mitchell (1974) who found thalli referable to the genera *Usnea*, *Ramalina* and *Xanthoria* growing on a limited area of western red cedar wood cladding by a doorway on an otherwise lichen-free wall. The flora developed after nutrient-rich dust had accumulated sufficiently through regular daily carpet-beating by the cleaning women (see Fig. 1). Species that have been found on strongly hypertrophicated wood are marked *h* in Table IV.

TABLE III.

Lichens of rare or local occurrence on worked timber in the British Isles. Letters of compass-bearing indicate strong distribution tendencies.

Arthonia impolita SE	Cyphelium tigillare E
Bacidia chlorococca	Lecanora farinaria$^{a\,b}$ S
B. aff. circumspecta N	L. jamesii W
B. sabuletorum	L. saligna
Biatorella moriformis	L. sarcopisioides N
Bryoria lanestris NE	L. subfuscata^{a}
Buellia alboatra	L. subintricata N
Calicium abietinum S	Lecidea friesii N
C. glaucellum	L. ochrococca N
C. quercinum$^{a\,b}$ S	Lecidella pulveracea S
C. salicinum	Micarea globularis N
C. trabinellum$^{a\,b}$ NE	M. melaena
Caloplaca aurantiaca	M. nitschkeana
C. cerina	M. prasina
C. furfuracea$^{a\,b}$	M. violacea
Candelaria concolor	Mycoblastus sanguinarius
Catillaria graniformis^{a} SE	Ochrolechia parella
C. griffithii	O. yasudae
C. lightfootii N and W	Opegrapha atra
Cetraria pinastri E	Parmelia exasperata
C. sepincola	P. exasperatula
Chaenotheca chrysocephala E	P. laciniatula
C. ferruginea	P. tiliacea s. l.
C. phaeocephala^{b}	Parmeliopsis hyperopta
Chaenothecopsis debilis^{b}	Pertusaria pertusa
Cladonia bacillaris	Physcia caesia
C. fimbriata	Physconia enteroxantha
C. furcata	Ramalina calicaris
C. impexa	R. fastigiata
C. pityrea	R. fraxinea
C. pyxidata	Rinodina exigua
Cornicularia aculeata	Sphinctrina microcephala^{a} S
Cyphelium notarisii^{a} SE	Teloschistes chrysophthalmus^{b} S
C. ocellatum^{a} NE	Tornabenia atlantica^{b} S

^{a} Known only from this substrate in the British Isles.
^{b} Probably extinct on this substrate.

Lundström (1970) studied the lichens growing on exposed wooden posts and slabs that had been subjected to certain preservative treatments. He found that fruticose species such as *Alectoriae* and *Usneae* and

TABLE IV.

Lichens that are normally saxicolous but have occasionally been recorded on worked timber (h = hypertrophicated, m = maritime and u = upland habitats).

Acarospora fuscata	*Lecidea lucida*
Anaptychia fusca (m)	*L. orosthea*
Caloplaca decipiens (h)	*L. tumida*
C. saxicola (h)	*Lecidella scabra* (h)
C. verruculifera (m)	*L. subincongrua* (m)
Candelariella aurella (h)	*Parmelia conspersa* (u)
Catillaria chalybeia	*P. mougeotii* (u)
Diploschistes scruposus (h)	*P. omphalodes*
Huilia crustulata (u)	*P. verruculifera*
H. macrocarpa (u)	*Pertusaria corallina* (u)
Lecania erysibe (h)	*P. pseudocorallina* (m)
Lecanora badia	*Physcia dubia* (h)
L. campestris (h)	*Protoblastenia rupestris* (h)
L. epanora	*Ramalina siliquosa* (m)
L. intricata var. *intricata* (u)	*Rhizocarpon geographicum* (u)
L. intricata var. *soralifera* (u)	*R. obscuratum* (u)
L. muralis (h)	*R. polycarpon* (u)
L. polytropa (u)	*Rinodina subexigua* (h)
L. rupicola (m)	*Stereocaulon vesuvianum* (u)
Lecidea atrata	*Umbilicaria polyphylla*
L. erratica	*U. torrefacta*
L. leucophaea (u)	*Xanthoria aureola*

foliose species such as *Platismatia glauca* and *Hypogymnia physodes* were inhibited by preservatives containing arsenic, but that crustose species (*Lecanora* sp., *Lecidea* sp. and *Bacidia* sp.) would grow on all types of treated wood except those that had been freshly creosoted. It is obvious from recent field observations in Britain that untreated sawn timber is increasingly rarely used in the open, and the consequences for certain lichen species have been repeatedly commented upon (Coppins and James, 1974; Coppins and Lambley, 1974; Hawksworth *et al.*, 1974). Species apparently already extinct are marked ([b]) in Table III. *Cyphelium notarisii*, *Sphinctrina microcephala* and *Catillaria graniformis* are unlikely to survive much longer.

Table II lists the lichens most commonly found on worked timber in Britain. It consists mainly of common species with wide ecological amplitudes that are found on bark, wood and rock, and even in some cases on soil and peat, e.g. *Caloplaca citrina, Hypogymnia physodes, Micarea lignaria, Parmelia saxatilis, Pertusaria amara, Physcia orbicularis, Platismatia glauca* and *Xanthoria parietina*. It includes also others that are

FIG. 1. Lichen assemblage on wood cladding impregnated by nutrient-rich dust on the west side of the soil laboratory at Johnstown Castle, Co. Wexford (see O'Sullivan and Mitchell, 1974). Lichen-free area ($c.$ 50 cm in radius) surrounded by zone colonized by *Usnea* sp., *Ramalina* sp., *Xanthoria* sp. etc., varying in width from 38 to 65 cm. (Photograph by A. M. O'Sullivan.)

predominantly confined to bark and wood, e.g. *Evernia prunastri, Lecanora chlarona, L. expallens, L. piniperda, Parmelia subaurifera, Ramalina farinacea* and *Xanthoria polycarpa*. The only species listed that is found only on wood is *Xylographa abietina*. Species which show a strong tendency to regional localization in Britain are marked in the table by letters indicating the major points of the compass. Wooden fencing is now being used more widely than in the past in the extreme north, in Caithness, Orkney, Shetland and Sutherland, and this has led to increased frequencies of such species as *Brytoria fuscescens* and *Pseudevernia furfuracea* (Hawksworth *et al.*, 1974). These two species and also *Parmeliopsis aleurites* are essentially species of northern or upland areas (Hawksworth, 1972; Hawksworth and Chapman, 1971) but extend their range into south-east England mainly on fence rails. *Cyphelium inquinans* shows differing substrate preferences at the opposite ends of its range in Britain. Bailey (1974a) showed the distribution of this species to be mainly south-eastern. In the northernmost localities on his map the lichen was found on the bark of *Quercus* spp., whereas most of the records in the main area of distribution are of specimens on fence posts and rails. Since 1974 *C. inquinans* has been discovered on the bark and wood of *Pinus sylvestris* and more rarely on that of *Betula* spp. in the native pine woods of Scotland at Rothiemurchus, Abernethy, Glen Affric, Guisachan, Glen Moriston and in the Black Wood of Rannoch. It is apparently absent from worked timber in these areas.

Table III lists epiphytic species that are common and widespread over much of Britain, but that are only of sporadic occurrence on worked timber, e.g. *Calicium glaucellum, Catillaria griffithii, Micarea prasina, Parmelia exasperatula* and *Pertusaria pertusa*. It also includes species that are most typically found on terricolous substrates but which occasionally occur on rotten woodwork, e.g. *Cladonia* spp. and *Cornicularia aculeata*, and others that are rare on any substrate, e.g. *Catillaria graniformis, Chaenotheca phaeocephala, Cyphelium* spp. and *Cetraria pinastri*.

Table IV lists species that are normally saxicolous, but occasionally occur on worked timber. For some species this may occur quite frequently under special conditions, for instance, *Lecanora polytropa* and *L. intricata* in upland Britain and *Lecidella scabra* on hypertrophicated wood.

B. Cork

The cork cells which form the outer bark of trees comprise a non-living tissue when mature. Nevertheless profound changes begin to occur in the corticolous lichen flora as soon as the tree itself is dead. These changes appear to be due at first to alterations in the water relations of the bark, and ultimately to changes in the surface texture. Experiments in which discs of natural bark carrying *Lecanora conizaeoides* were embedded in sheets of cork, initially smooth but experimentally roughened in various ways, and exposed at a slight angle to the vertical over a period of 5 years, showed that the lichen spread downwards, but only became established on roughened areas. Cork floats from fishermen's nets cast up on the shore weather to a very rough texture, and Richard (1883) found *Lecanora dispersa* and *Buellia punctata* on such objects. He also found *Lecidea enteroleuca* on old bottle corks. Bouly de Lesdain (1951) records *Xanthoria parietina*, *X. polycarpa*, *Physcia adscendens*, *P. orbicularis*, *Caloplaca citrina*, *C. lithophila*, *Lecania cyrtella*, *Rinodina subexigua*, *Catillaria melanobola* and "*Bacidia inundata*" from cork debris amongst the sand dunes near Dunkirk.

Linoleum, according to the original patent by Walton in 1860, is manufactured from a mixture of powdered cork and linseed oil spread on a coarse hessian canvas, and so may be mentioned here. Bouly de Lesdain (1951) states that he found all the species recorded on cork growing on discarded linoleum, and in addition *Bacidia umbrina* and the basal squamules of a *Cladonia* sp.

C. Miscellaneous

Roofs are conveniently exposed to the sun and the rain, and at the same time are well drained; they are free from disturbance, and not liable to the accumulation of leaf litter, drifting soil and the like; they have a variety of aspects, and parts of them, for instance along ridges and by chimneys, are subject to hypertrophication by bird droppings. They thus provide many suitable habitats for lichens. Roofs with wood shingles are mentioned above, and those with inorganic coverings of various kinds are discussed in the next section. Some roofs, nowadays restricted mainly to country areas, are thatched with straw (from cereal crops) or with reed (*Phragmites australis* or *Cladium mariscus*). Lichens, particularly *Cladonia* spp. such as *C. chlorophaea*, *C. pityrea*, *C. macilenta* and *C. bacillaris*, are sometimes found on old thatch. Seaward (1977) noted in addition *Parmelia caperata*, *P. perlata*, *P. revoluta* and *Lecidea uliginosa* on the thatched roof of a cottage near Waterford, Ireland. Hawksworth *et al.* (1974) reported that *Lobaria scrobiculata* formerly occurred on thatch in Devon and Suffolk. It seems certain that lichens are much less common on thatch than they were in the past, and the reason appears to be the practice of protecting the straw or reed from the attentions of birds by covering it with galvanized wire netting. The zinc of the galvanizing is inimical to lichens (see Section V. p. 282). The records of Bouly de Lesdain (1951) show that there seems to be nothing of vegetable origin that will not ultimately support the growth of at least one species of lichen if it is left exposed long enough under suitable conditions. He found lichens on rubber (*Physcia adscendens*), cotton cloth ("*Bacidia inundata*"), and coconut shell (*Catillaria melanobola*). He draws attention to an interesting contrast between cardboard and paper on the one hand and leather on the other. He points out that only a very few species of lichen such as *Lecidea granulosa* and *L. uliginosa* are found at all commonly on cardboard and paper, whereas he was able to record over 30 species of fungus. For leather his results were almost exactly the opposite; he found only three of fungi, but over 50 lichens (see Table I).

IV. Wholly Synthetic Organic Materials

Having exploited primary raw materials to the full, and in some cases exhausted available supplies of them, man is now increasingly creating his own from chemicals of varying degrees of complexity. Fabrics woven from synthetic fibres such as nylon and terylene are now used as widely as wool, silk and vegetable fibres, but so far there are no records of lichens growing on them. So-called "plastics" are mostly high polymers, and frequently they are fabricated under very high pressures into sheeting, containers of various kinds, and other artefacts. Hitherto unknown in nature, many of them are proving so far to be non-biodegradable, and their mode of formation produces very hard and uniform surface textures. They provide therefore particularly inhospitable habitats, but nevertheless already there are indications that lichens will be able in due course to grow on some at least of these ultimate man-made substrates. For instance, *Bacidia umbrina* and *Micarea nitschkeana* have been found growing on a carton made of high-impact polystyrene on the beach at Gadshill, Isle of Wight (Coppins, *in litt.*), and several well developed thalli of *Physcia caesia* have been recorded (Earland-Bennett, personal communication) from the fibre-glass roof of a farm building in Huntingdonshire. This latter material, of which the main constituent is a wholly man-made synthetic resin, is coming increasingly into use as a building material, and its properties as a substrate for lichens would be well worth investigation.

V. Materials Derived from Mineral Sources

A. Inorganic Building Materials

A large number of lichens are adapted to colonizing weathered rock surfaces, and are therefore particularly abundant in upland and mountain areas, where there are extensive natural exposures of rock, and on rocky

9. Lichens of Man-made Substrates

shores by the sea. Elsewhere their natural habitats are restricted to localized outcrops, and in areas which have been subject to glaciation, to occasional erratic boulders which may have been carried considerable distances from their place of origin by ice action. Such erratics frequently provide interesting extensions to, or disjunct localities for, lichen distributions, as for instance when a mass of siliceous rock is found in a predominantly limestone area. Examples are the occurrence of *Lecanora badia* on a siliceous rock amongst limestone boulders by a stream at Malham in Yorkshire and "*Bacidia inundata*" (see remarks on this "species" on p. 257) on a fragment of sandstone amongst a shingle bank of flint pebbles at Dungeness in Kent, the former substrate having been transported by the Pleistocene ice-sheet and the latter by wave action of the sea.

Human activity has from early times created new rock exposures. Mining for flint had begun in southern England over 6000 years ago (Renfrew, 1974). The spoil from such workings is colonized by a sparse lichen flora in which the commonest species are *Aspicilia gibbosa*, *Lecidea erratica* and *Rhizocarpon obscuratum*. Not much later, in Mesopotamia and Egypt, the invention of metal tools made possible the quarrying of large blocks of stone for use in building. Frequently building stones were, and are, transported over considerable distances. Quarries may be regarded as artificial outcrops, and transported building stones as erratics.

However, it is found that the lichen flora of abandoned quarries is in many cases sparser, less species-rich, and different in composition from that of nearby natural outcrops and boulders. It would appear that the man-made vertical faces generated by quarrying are often not favourably aligned to the physical and biotic weathering processes that encourage lichen colonization. A similar phenomenon may often be seen in dry-stone walls constructed from locally quarried materials, which tend to bear different and sparser lichen floras.

The erection of monoliths, dolmens, chamber-graves and henges in Neolithic times frequently involved the use of stones which were not available locally. Today many of these monuments bear most remarkable lichen floras: the presence, for example, of typically maritime lichens in inland localities still requires detailed interpretation. Laundon (1966) reported the occurrence of *Ramalina siliquosa* on a prehistoric dolmen in Wiltshire which was reconstructed in the early part of this century when a lintel was re-erected. The lintel now has an abundance of this lichen on its sides, indicating that colonization had occurred in recent years. However, its status on the monument prior to reconstruction requires further investigation. He pointed out that the Wiltshire localities for *R. siliquosa* are about 80 km from the sea, and listed other English inland localities between 15 and 30 km inland from the coast. He also reported another typical maritime lichen, *Anaptychia fusca*, from Wiltshire. It has been pointed out by Seaward (1975a, 1977) that the lichen floras of Irish monuments often bear very little resemblance to the assemblages found on rocks of the same geological origin in the neighbourhood, and indeed those found on dry-stone walls made from the same stone. He recorded *R. siliquosa* from two prehistoric monuments in Ireland at distances of 28 and 35 km from the sea, and *A. fusca* on an Irish dolmen 50 km from the sea. The influence of maritime conditions on all of the above sites is very limited or non-existent; the factors responsible for the difference in status of these species in maritime and non-maritime localities remain unknown (cf. Fletcher, 1973, 1975).

Megalithic and other forms of dry-stone construction were eventually superseded by building methods involving dressed stones held together by various forms of mortar and cement. At this stage in building development the substrates created may be regarded as distinctly man-made, rather than merely man-influenced as in the case of dry-stone constructions. The stonework used for such purposes as facing buildings and constructing memorials is for the most part polished or otherwise dressed smooth. Nevertheless, under circumstances favourable to lichens, colonization takes place, especially on limestone. The ubiquitous species *Lecanora dispersa* soon more or less completely covers all exposed surfaces, and even in the centres of cities is represented by the forma *dissipata*. In fact, this latter forma and another species, *Candelariella aurella*, are responsible for much of the discoloration of light-coloured building materials, both new and freshly cleaned, in urban areas. Brightman (1965) found between 300 and 400 apothecia cm^{-2} of *L. dispersa* on the walls of the Fitzwilliam Museum in Cambridge. On older limestone structures in the city, such as Clare Bridge, he recorded species-rich closed communities of crustose species. He was also able to show that the much poorer flora of sandstone was due in part at least to the feeble buffering powers of the substrate. The sandstone parapet of King's Bridge had a very sparse flora consisting mainly of *Lecanora conizaeoides*, and a pH of 3·5. Elsewhere by the river Cam, similar sandstone at soil level, which was exposed to the buffering effects of soil water, had a pH of 7·0, and was colonized by such species as *Trapelia coarctata* and *Lecanora muralis*.

Consideration should also be given here to the effect of mortar on adjacent non-calcareous materials. *Caloplaca heppiana* was observed to spread from mortar on to brick, and on a particularly porous type of brick which was normally colonized by *Lecidea lucida*, this species was replaced by *Caloplaca citrina* when contamination by lime from mortar had occurred. Gilbert (1971) found that the mortared junctions between sandstone wall capstones could act as refugia for *Parmelia saxatilis* under conditions of fairly severe atmospheric pollution near Newcastle upon Tyne. This species is typically a lichen of acidic substrates, but under conditions of increasing pollution it prefers alkaline niches. He demonstrated by mapping particular areas of wall-top on a quadrat basis at intervals of about a year that *P. saxatilis* was retreating into such niches adjacent to the mortar between the stones. Seaward (1975b) records for another species the opposite of this phenomenon. He found that near Leeds the establishment of *Lecanora muralis* on brick (see Fig. 2) and siliceous stone walls was made possible in the first instance by the mortar. The lichen became established on this at first, and then as time proceeded the thalli spread beyond the limited mortar zone to exploit the surrounding acidic substrate. Over a 5-year period of observation in an area 8 km to the northwest of the centre of Leeds there have been dramatic changes in the developmental and distribution patterns of *L. muralis*. Rosettes, once confined to mortared areas, in the first instance spread to the adjacent millstone grit, and more recently, thalli have appeared independently on the acidic substrate only. The form of these latter thalli is noticeably different from that of the rosettes established on the mortar, but the reasons for this are not known. This diversification of substrate colonization by *L. muralis* is of particular interest because the area under discussion has shown no significant fall in sulphur dioxide pollution level in recent years.

It is well known that lichens are adversely affected by the ecological conditions in urban areas. Chief of these adverse factors is no doubt atmospheric pollution, though extremes of temperature, humidity and water supply also play a part. The decline in the lichen flora along transects running into urban centres has been investigated in cities all over the world. These studies have been directed almost exclusively to epiphytic lichens, which do not concern us here, but Gilbert (1965) recorded an approximately 30% reduction in the number of saxicolous species between the 7- and 6-mile zones from the centre of Newcastle upon Tyne (compared with an approximately 70% reduction in the number of corticolous species). The inner urban limit for the distribution of certain species is quite clearly defined, but changes fairly rapidly with time, so that the ecological factors operating in the marginal area at or immediately preceding a particular date must be critical for the lichen's performance, or even to its existence. *Lecanora muralis*, for example, has a distinctive zonation in urban areas as exemplified by its distribution in the vicinity of Leeds (Seaward, 1976, his Fig. 10). Its inner limits coincide with its distribution on asbestos-cement roofing; a more detailed pattern of its distribution according to substrate preference in part of this urban area is presented in Fig. 3 and the mean distances from the city centre in 1970 and 1975 are given in Table V. It will be seen from these data that over the 5-year

Fig. 2. Coping stones of brick wall in moderately polluted suburban area, to show the establishment loci for *Lecanora muralis* on the calcareous mortar bonding and the eventual spread of the thalli on to the acidic substrate, which at this stage is mainly colonized by *Candelariella vitellina*. (Photograph by M. R. D. Seaward.)

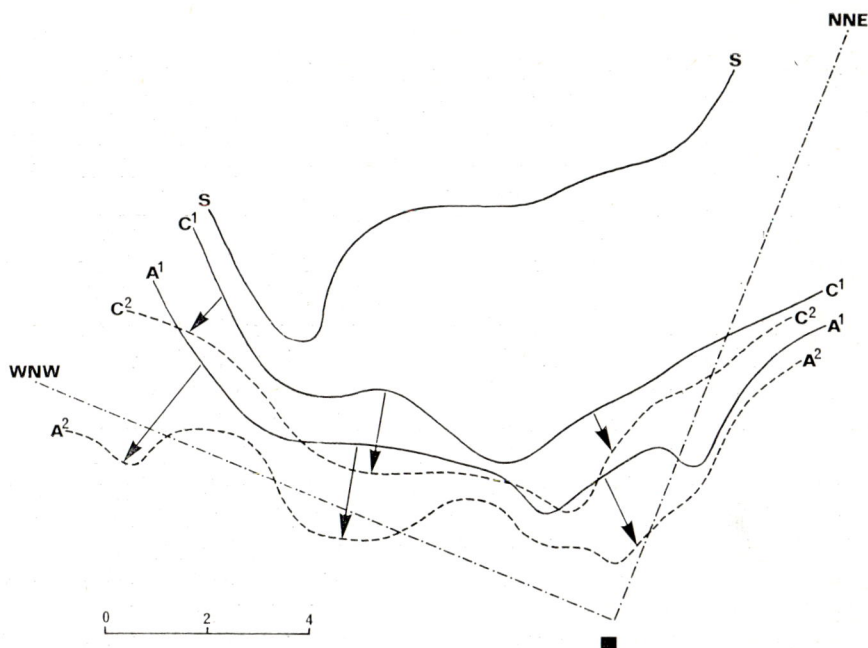

FIG. 3. The major inner limits of *Lecanora muralis* growing on asbestos-cement sheeting and tile roofs (A^1 = 1970, A^2 = 1975), on cement, concrete and mortar (C^1 = 1970, C^2 = 1975), and on siliceous wall capstones (S = 1970 and 1975) within the WNW to NNE sector of Leeds, Yorkshire.

period *L. muralis* has spread, on man-made substrates, towards the centre of the city at a rate of about 150 m per annum, but there has been no noticeable change in its distribution on siliceous substrates. These observations appear to correlate very well with trends in sulphur dioxide measurements obtained from National Survey gauges which show that levels continue to fall significantly in the central and suburban areas of the city, but less significant changes, if any, are recorded in semi-rural areas adjacent to the conurbation. It is possible from such analyses to construct a biological scale based entirely on the substrate preference limits of *L. muralis*. From this scale estimates can be made of the mean daily sulphur dioxide concentration and the rainwater pH which correspond closely with the values obtained from instruments (see Table VI).

Churchyards, especially those containing calcareous stone buildings and memorials, provide the most important habitats for lichens in many urban areas in the British Isles. According to Laundon (1970), 65% of the species of the London lichen flora are found in such habitats, and old

TABLE V.

The mean distances, in km from Leeds city centre, of the major inner limits of Lecanora muralis *on three types of substrate in 1970 and 1975.*

Substrate	1970	1975	Inward Spread (m year^{-1})
Asbestos-cement tile and sheet roofs	5·30	4·55	150
Asbestos-cement roofs, cement, concrete and mortar	6·25	5·55	140
Asbestos-cement roofs, cement, concrete, mortar and siliceous wall capstones	9·30	9·30	0

TABLE VI.

Relationship between sulphur dioxide concentration and rainfall pH with increasing distance from Leeds city centre in 1970 and 1975, as determined from a biological scale based on the ecological status of Lecanora muralis.

Distance from city centre (km)	Mean daily SO$_2$ concentration (μg m^{-3})		Rainwater pH	
	1970	1975	1970	1975
0–2·40	>240	>200	4·4–4·7	4·7–4·9
2·40–4·00	200–240	170–200	4·7–4·9	4·9–5·1
4·00–5·60	170–200	125–170	4·9–5·1	5·1–5·3
5·60–8·85	125–170	<125	5·1–5·5	5·3–5·5
8·85+	<125	<125	5·5+	5·5+

churchyards in the suburban areas support well developed communities comparable with their counterparts in rural areas. For instance, *Caloplaca heppiana* and a few other species survive in London churchyards as relicts from earlier times of more rural surroundings and lower levels of air pollution; they are absent from nearby cemeteries of more recent date. The percentage occurrence of *Caloplaca heppiana* on limestone memorials dated between 1750 and 1950 in a London churchyard shows a dramatic decline at the turn of the nineteenth century, and a total absence in the twentieth century (Laundon, 1967, his Fig. 3). Calcareous memorial stones imported into city churchyards and cemeteries in northern England bear a richer lichen flora than the local siliceous stonework, although *Caloplaca heppiana* is much less likely to occur in any abundance, or indeed at all. More recently Laundon (1976) has described a species (*Lecanactis hemispherica*) which appears to be relict in rural churchyards. It has been found only on plaster on the outside of church walls, now a rare substratum, although formerly much more common. In the same paper he comments on the distribution of *Caloplaca ruderum* which has never been found on natural rock outcrops nor on hard limestone masonry but only on the soft limestone and mortar of old church walls. Even rarer is *Geisleria jamesii* described by Swinscow (1967), and found only once on the mortar of a church wall on Brownsea Island, Hampshire.

Cement, increasingly employed for a wide variety of uses in urban areas, is particularly valuable as a substrate for lichens. Within 4 years of the erection of cement coping stones in 1967 at the entrance to Trinity and All Saints' Colleges, near Leeds, there was a coverage of approximately 80% of *Lecanora dispersa*, with some *Candelariella aurella* (Seaward, 1975b). The average count of *L. dispersa* apothecia was 346 cm^{-2} (range 146–497 cm^{-2}). Hawksworth (1969) reports an 85% coverage of *L. dispersa* and *C. aurella* on concrete walls over a 5-year period in Leicester, and Laundon (1967) notes that cement made in 1961 was well covered with *Caloplaca citrina*, *Lecanora dispersa* and *Verrucaria muralis* by 1964 at Morden, south London. He also mentions the interesting point that elsewhere in the vicinity, cement-filled holes in limestone produced by the wartime removal of iron railings, have become colonized by lichens while the surrounding harder limestone remains uncolonized.

Sewage farms, now obsolete and rapidly being dismantled and the land converted to other purposes, in the recent past provided notable habitats for lichens in urban areas. The concrete used for sprinkler and tank surrounds and for open-flow channels had both buffering and nitrophilous potentialities and well developed communities of such lichens as *Caloplaca citrina*, *C. decipiens*, *Candelariella aurella*, *C. medians*, *C. vitellina*, *Lecanora dispersa*, *L. muralis*, *Physcia caesia*, *P. orbicularis*, *Rinodina subexigua* and *Xanthoria parietina* were frequent. The concrete sides of reservoirs within urban boundaries also exhibit well developed lichen communities. These reservoirs are much frequented by birds, and their perching areas are comparable in this respect with the concrete constructions of the old sewage farms.

Asbestos-cement sheeting, used either in corrugated form or as rectangular tiles for roofing purposes, is even more favourable than ordinary cement or concrete as a lichen substrate. Brightman (1959) pointed out that not only does the buffering power of this material contribute to its suitability for colonization by lichens in urban environments, but the fibrous texture of weathered surfaces provides conditions for improved water relations and also encourages the lodgement of lichen propagules and the attachment of growing thalli. More recently the growth of lichens on asbestos-cement tiles has been exhaustively investigated and fully reported on by Seaward (1972, 1975b, 1976).

Walls built of brick frequently display an interesting mosaic of communities. The top surface and ledges (formed when as is frequently the case there is a protruding damp course of tiles or header bricks) dry rather slowly after rain and may also tend to accumulate a scanty humus deposit. Such situations are often occupied by foliose species such as *Hypogymnia physodes*, *Parmelia sulcata*, *Physcia orbicularis*, *P. adscendens* and *Xanthoria parietina*, and the basal squamules of *Cladonia fimbriata*. The bases of walls are damper than the higher parts and tend to be eutrophicated. They are often colonized by such species as *Buellia canescens* which can tolerate some degree of shading, although on parts of walls which are in permanent shadow the only lichen to be found is *Lepraria incana*. Brightman (1965) found that surface texture has a considerable influence on the lichen flora of vertical brick surfaces. In Cambridge, the most favourable brick for lichens was a sand-faced red brick with a pH of 5·8 and a water absorbing power of 12%. In addition to *Physcia* spp. and other foliose lichens, crustose species such as *Ochrolechia parella* and *Lecidea sulphurea* were common. On the other hand, ordinary red stock bricks had a more acid pH (6·0) and a lower water absorbing power (9%), and were colonized only by *Lecanora conizaeoides*. Smooth porous surfaces, provided they were moist enough, supported an abundant growth of *Lecidea lucida*. A soft red brick with pH 6·8 and water absorbing power of 17–20% was outstanding in this respect; but a yellow hand-made brick with a similar smooth surface (pH 6·6, water absorbing power 6%) could only support *L. lucida* when due to local conditions it was more or less permanently saturated. Glazed bricks have not been observed to provide a substrate for lichens, although Bouly de Lesdain (1951) mentions a few species including *Physcia adscendens*, *Lecidea umbrina*, *Lecanora dispersa* and "*Bacidia inundata*" growing on glazed earthenware and pottery, and M. R. D. Seaward (unpublished data) found *Caloplaca chalybaea*, *Candelariella medians* etc. on Roman pottery fragments at Thuburbo Majus in Tunisia.

It has been mentioned above that a number of species which succeed in habitats influenced by man were adapted originally to maritime situations. An interesting example of an upland species which is spreading under urban conditions, mainly in the London area, is *Stereocaulon pileatum* (Kershaw, 1963). Under natural conditions this lichen is a rather local species of the north and west of the British Isles. In 1955 it was found on a fence by a railway yard in Surrey, and soon after on a railway bridge in Essex. Other occurrences in suburban Essex were noted later, and then it was found in several localities in south London, some of them associated with railways. It is now a common species on the tops of brick walls. A species frequently associated with it is *Trapelia coarctata*. This is one of the three pioneer lichens which have colonized the lava rocks or the island of Surtsey off Iceland, which was formed by a volcanic eruption in 1967. Seven years after the island appeared out of the sea, *Trapelia coarctata*, *Stereocaulon vesuvianum* and *Placopsis gelida* had established themselves on the lava fields (Kristinsson, 1972).

As indicated above, the recent spread on man-made substrates of *Lecanora muralis* and *Stereocaulon pileatum* is quite well documented. Equally well known, though not yet the subject of an individual study, is the increasing frequency of *Candelariella medians*, which is rare on natural calcareous rocks but is now widespread on buildings, walls and gravestones constructed from limestone in south and east England. *Xanthoria elegans* is also becoming commoner, particularly on asbestos-cement and concrete. *Parmelia mougeotii* has been found on slate roofs and *Acarospora heppii* on concrete in rural parts of southern England (Coppins, *in litt.*), and these species may be beginning to spread similarly.

Land utilization surveys have shown that footpaths, pavements and roadways constitute up to 35% of the ground surface of urban areas. The replacement of natural soil by concrete and tarmacadam restricts plant colonization to all but a few species, and saxicolous lichens have only marginal success in exploiting such habitats. A few species are usually to be found on concrete close to perpendicular surfaces where some measure of protection is afforded against mechanical wear, and microclimatic niches exist with local water regimes and pollution levels that differ markedly from those of more exposed situations. Tarmacadam is rarely colonized in polluted atmospheres, but in suburban and rural areas it may support an interesting lichen flora, including high cover-values of *Candelariella vitellina*, *Lecanora muralis*, *L. polytropa* and, more rarely, foliose species such as *Parmelia conspersa*. Asphalt roofing materials similarly support a variety of acidophilous, mainly crustose, lichens in the less polluted rural and suburban atmospheres. *Lecanora conizaeoides* has recently been recorded from cable-lagging at Northwood Hills Station, London (Hawksworth, personal communication). Derelict areas, spoil tips and rubbish dumps are seemingly an inevitable part of the urban landscape. When active accretion of refuse ceases, some colonization by lichens takes place, especially by *Cladonia* species such as *C. subulata* and *C. chlorophaea*, the foliose species *Peltigera spuria*, and the crustose *Baeomyces rufus*. On old buildings, monuments and sculptures, lichens may develop to such an extent that they constitute a nuisance. Macmillan (1861) puts an extreme view:

> ... wherever they fasten their tiny fangs the process of disintegration commences; and though carried on slowly and imperceptibly, though ages may elapse before any apparent effects have been produced, except the increase of individuals and the more shaggy and picturesque appearance of the rocks, yet the object of that steady, ceaseless labour will one day be accomplished; and it is humiliating to the pride of man to find, that the noble piles of architecture built by him as if for eternity, though apparently as solid as the rock out of which each individual stone had been hewn, and as hard as the famous Roman cement which had resisted the utmost efforts of Goth and Vandal, must yield in the end to the slow but persevering assaults of the most diminutive and contemptible vegetables, and be brought back again by these apparently feeble agents to the bosom of nature, out of which he had reared them with such labour and skill.

In fact, structural damage caused by lichens is most unlikely (cf. Lloyd, 1972; Richardson, 1973), although they are often very conspicuous on architectural and other stonework. Lichen communities create a mosaic of colour and texture on such surfaces which in some circumstances has a considerable aesthetic appeal. On the other hand, they have undeniably detrimental effects on delicately sculptured work. Paleni and Curri (1972) have pointed out that a number of common species, for instance *Verrucaria marmorata*, speedily disfigure marble statues and carvings, and that spoilation of stonework in Venice is quite far advanced. Lichen damage, mainly by *Parmelia tinctorum*, to the sculptured panels of Barabudur, the Buddhist monument in central Java, has been reported by Seshadri and Subramanian (1949). There, *P. tinctorum* defaces the panels and obscures the fine detail; further, water-soluble lichen acids are said to be the cause of deterioration of the calcareous stonework. During the process of cleaning the sculptures mechanically, additional disfigurement results from the removal of the surface of the substrate attached to the lichens. Another example may be given from Angkor, the Khmer shrine in Cambodia, where a statue of a god of death has become so eroded by crustose lichens that it is known locally as "The Leper King", although it is not a representation of the real personage to whom this sobriquet is traditionally ascribed. The problem is worldwide; similar deterioration is occurring on Maya sculptures in Yucatan (Hale, personal communication).

The aesthetic/functional contradiction may be seen in microcosm on the gravestones in churchyards in the British Isles. Loosely attached foliose species, such as *Parmelia saxatilis*, sometimes effectively obliterate the incised lettering of epitaphs, and conservationists cannot reasonably object to their removal. On the other hand, some more closely adherent sub-squamulose or areolate species and crustose ones with varying degrees of thalline immersion in the substrate will usually leave the inscriptions quite visible and decipherable. This is the case with the thalli of *Acarospora fuscata* and *Lecanora polytropa* which tend not to grow into cracks, and also those of *L. intricata* var. *soralifera* and *Lecidea tumida* on West Yorkshire gravestones (Henderson and Seaward, 1976). Sometimes inscriptions may actually be highlighted by lichens such as *Lecidea lucida* which grow preferentially in the micro-environmental niche provided by the letters. Species such as these enhance the churchyard setting; the British Lichen Society has paid particular attention to their ecological, as well as lichenological, importance in both urban and rural areas through the work of its Conservation Committee. Conversely, newly erected buildings, especially large farm buildings constructed of asbestos-cement sheeting, are apt to offend the eye in rural parts of the British Isles, and attempts have been made to promote lichen growth by treatment of the substrate with weak suspensions of organic substances as part of a programme of landscaping.

B. Glass

H. G. Wells (1915) describes in a felicitous phrase the boundary wall of an estate as "crested with uncivil glass against the lower orders". It was from such a situation, on old pieces of broken glass on top of a wall in Marennes, that Richard (1877) recorded *Xanthoria parietina*, *Buellia canescens*, *Lecanora dispersa*, *L. crenulata*, *Lecania erysibe* and *Rinodina subexigua*. He also noted *Physcia tenella* in a similar habitat in Saint-Maixent. Later, in a study of lichen substrates (Richard, 1883), he recorded *Caloplaca cerina* from pieces of broken glass at the Château de Meudon. Crombie (1879), in a note entitled "Vitricole Lichens", remarked that "in this country we have only observed them on broken pieces of bottles on garden wall tops, chiefly in Scotland", but mentions neither species nor precise localities. In the present century there are no published records of lichens on "uncivil glass", and we only know of one locality, at Frindsbury in Kent, where *Xanthoria parietina* occurs on such a substrate (see Fig. 4).

Lichens on windows have attracted a slightly more extensive literature, and specimens are available for study. J. E. Smith's miscellaneous lichen herbarium, now in the herbarium of the British Museum (Natural History), includes four small pieces of glass from a church window near St Germans, Norfolk, collected in 1793, which bear traces of *Caloplaca heppiana*. Also from a church window, in Falsterbo, Scania, Fries (1831) reported *Xanthoria*

Fig. 4. *Xanthoria parietina* on "uncivil glass" from a wall-top at Frindsbury, Rochester, Kent; collector F. H. Brightman. (Photograph: S. Alexander.)

parietina. According to Richard (1883), Bouteille found *Buellia canescens* and *Caloplaca saxicola* on window panes at Chapelle-en-Vexin in 1873. Nylander (1879) has a number of references to lichens growing on windows in Île d'Yeu, including *Lecanora dispersa*, *Rinodina subexigua*, *Buellia alboatra*, *Caloplaca heppiana*, *Candelariella reflexa* and *Phylloporina elaeospila*. Buchet (1890) discussed the growth of lichens on stained glass church windows in Moustoir, without, however, naming the species he observed. Richard (1883) gives a considerable list of records from Noirmoutier (see Table VII), which he says "is very remarkable for the number and beauty of its vitricolour species". A final nineteenth-century record by Oehlert of *Xanthoria parietina* on a church window in Mayenne in 1896 is reported by Mellor and de Virville (1921).

Mellor (1922b) described 77 specimens of stained glass collected by F. Gandin from churches in Brittany, Normandy, Rhampogne and the Île de France in a thesis entitled *Les lichens vitricoles et la détérioration des vitraux d'église*. All of them did in fact show evidence of deterioration, and the majority bore lichens, the commonest species being *Buellia canescens*, *Pertusaria leucosora*, *Lepraria candelaris*, *Caloplaca saxicola* and *Biatorina erysiboides*. A summary of the thesis had already been published elsewhere (Mellor, 1922a), and in the previous year preliminary notes had appeared (Mellor, 1921a, b, c). Also in 1921, Mellor collaborated in a study of 17 specimens of glass collected by M. Alleaume from churches in Mayenne (Mellor and de Virville, 1921). Only three species of lichen were observed:

TABLE VII.

Some lichens recorded from glass (nomenclature modernized).

Buellia alboatra agg. A, E	*L. dispersa* A, E
B. canescens A, E	*Lepraria candelaris* E
B. stellulata C	*Ochrolechia parella* A
Caloplaca aurantiaca A	*Opegrapha saxatilis* E
C. citrina A, B	*O. saxicola* E
C. heppiana A	*Pertusaria amara* A
C. holocarpa A, B	*P. flavicans* E
C. holocarpa var. *pyrithroma* A	*P. leucosora* E
C. saxicola A, B, E	*Physcia dubia* E
C. vitellinula A	*P. lithotea* A
Candelariella aurella A	*P. orbicularis* A
Catillaria chalybeia A	*Physciopsis adglutinata* A
C. erysiboides E	*Placynthium nigrum* A
Collema crispum A	*Porina chloritica* E
Diploschistes caesioplumbeus A	*Ramalina polymorpha* E
Lecania erysibe A	*Rinodina subexigua* A, E
L. olivacella E	*Verrucaria nigrescens* A
Lecanora atra A	*Xanthoria aureola* E
L. campestris A, B, D	*X. parietina* A, E
L. crenulata A	

Sources: A—Noirmoutier, France (Richard, 1883); B—Île d'Yeu, France (Richard, 1883); C—Île de Groix, France (Richard, 1883); D—l'Eglise de Montaudin, Mayenne, France (Mellor and de Virville, 1921); E—Precise localities unknown (compiled from observations on material in the British Museum (Natural History); Mellor, 1921a,c, 1922a,b; Mellor and de Virville, 1921).

Buellia canescens, *Rinodina subexigua* and *Lecanora campestris*. Most of the specimens studied by Mellor, including the type specimen of *Caloplaca vitricola* Mellor, were deposited in the laboratory of the École Normale Supérieure (Mellor, 1924) and cannot now be traced. Nine specimens, however, are in the herbarium of the British Museum (Natural History), together with a number of others marked by her "received too late for inclusion in *Les lichens vitricoles*". One of these latter is from Eastleach Church, Lechlade, Gloucestershire, collected by J. Constance in 1923, and bears *Buellia canescens*, *Caloplaca teicholyta* and *Dirinia stenhammari*. More recently, *Buellia canescens* and *Lecanora conizaeoides* have been found on a church window at Shorncote, also in Gloucestershire (Bailey, 1974b). A list of lichens recorded from glass, not enumerated above, is given in Table VII.

Mellor (1921) had no doubt that lichen growth caused deterioration of stained glass. She states that some colours, especially amethyst, but also purple, green, blue, red and amber, are susceptible to lichen attack, but that grey tones are less so and that golden-yellow glass, in which the colour is due to silver salts, is more or less immune. This corresponds with the remarks of Buchet (1890) who says that he observed lichen growths following the pattern of the picture, but he adds that a colour which is unaffected in one window may be attacked in another. Mellor (1922b) was able to demonstrate the presence of small fragments of glass within the thallus in some of her lichens, particularly *Pertusaria leucosora*, but also occasionally in some other species. She concluded that lichens cause mechanical damage to glass which has already been altered chemically. In a subsequent publication (Mellor, 1923) and in her contribution to a symposium (Mellor, 1924), she tended to over-emphasize these deleterious effects. Other contributors to the symposium were of the opinion that the damage observed was due to chemical weathering. Heaton (1925) pointed out that the degree of deterioration was related to the composition of the glass, and Turner (1925) stated that decomposition of glass was largely dependent on the percentage of sodium in it, and that the form taken by the pitting and furrowing which resulted from chemical weathering was related to the mechanical treatment received during manufacture.

Thus it may be questioned whether lichens are a primary cause of the damage observed. This characteristically takes the form of circular pitting, and it is noteworthy that more often than not pitted glass is quite free from lichen growth, and where lichens are present, far more pits are exposed than are covered by thalli. Ivy (*Hedera helix*) often clings to pitted glass, and is sometimes accused of causing the damage, but experiments with this plant and with *Ficus repens* show that they are incapable of attaching themselves to undamaged glass and when held artificially against it their haptera stain the surface slightly but do not etch or pit it at all (F. H. Brightman, unpublished data). In this connection, the chemical data of Harvey and King (1971) are instructive (see Table VIII). The Roman glass has a composition close to that of modern glass, and is very stable. The York fourteenth-century glass, with its lower silica content, is very corroded. The history of the Winchester College chapel windows is interesting. They were originally installed in about 1393, but were "restored" in 1821. In practice, this meant that a replica was installed, and the original glass dispersed. Some of it was re-used at Ettington in Warwickshire, and in 1950 this was reclaimed and reinstalled in Thurbern's Chantry at Winchester. It was corroded, and some of it was covered with lichens, apparently mainly *Buellia canescens*. Harvey and King (1971) point out that corrosion was worst in the sample with the highest lime/silica

TABLE VIII.

Composition of some ancient glasses (from Harvey and King, 1971).

Glass Sample	% Silica	% Lime and Magnesia
Roman	69	11
York Minster	54	20
Winchester College	47	28

ratio. They go on to say that lichens aid the normal processes of weathering by retaining water in contact with the glass and rendering this water slightly acid with carbon dioxide. Furthermore, they can cause spalling and mechanical breaking away of minute flakes of glass, but it is reasonable to assert that lichens cannot gain a hold on glass unless the surface is previously damaged by purely chemical weathering. Examination of smooth glass in church windows never reveals the growth of anything other than green algae; sometimes when mixed with cobwebs it has a markedly leprose appearance, but under the microscope has never yielded anything other than *Desmococcus* (F. H. Brightman, unpublished data).

A recently published note (Green and Snelgar, 1977) adds a new dimension to the foregoing discussion, and underlines our remarks in the introduction to this review about the paucity of published information from the warmer parts of the world. Green and Snelgar report the growth of *Parmelia scabrosa* on the windscreens of derelict motor cars near Hamilton, North Island, New Zealand. The latitude of Hamilton

(37° 46'S) corresponds approximately to that of Seville in the northern hemisphere, but the climate is more equable and relative humidities are often high. The authors state that *P. scabrosa* is quite prevalent on glass, and is sometimes a nuisance on glasshouses.

C. Non-ferrous Metals

Richard (1883) reported *Psorotichia pictava, Caloplaca citrina, Candelariella aurella* and *Lecanora dispersa* growing on lead used to seal the parapets of the chateau at Fontainebleau, and *Candelariella vitellina* growing on lead at Versailles. Bouly de Lesdain (1910) mentions *Rinodina subexigua* and *Lecanora dispersa*, laconically stating the substrate to be "plomb" and giving no indication of locality other than Dunkirk. The authors have never seen any species of lichen growing on lead, in spite of taking every opportunity of examining roofing leads, the lead of stained glass windows and lead used to seal iron railing into stonework and the like, nor are they aware of any other records apart from those quoted above, either published or unpublished. It may be significant that Mellor (1922b) in her detailed account of lichens on stained glass does not mention any case of plants growing on the lead. Perhaps the lead sealing at Fontainebleau and Versailles was much decayed and overlaid by stone, grit or soil particles. The overgrowing of calcifuge species from brick on to old mortar, or conversely of calcicole species from new cement on to old brick is a fairly common phenomenon. On the other hand, when lichens are found growing on the putty of aluminium-framed greenhouses, it is only rarely that they encroach on to the metal proper even when the latter is considerably corroded and pitted. Rayner (1976) recorded *Bacidia umbrina* growing on aluminium sheet, but did not comment on the state of the surface of the metal.

Lichens have not been reported from other non-ferrous metals. Lead and aluminium have the property of forming coherent films of oxide on the surface, as do some forms of iron and steel (see next section). It is possible that in some circumstances the oxide layer develops a suitable texture to permit the attachment of lichen propagules and at the same time limits the amounts of toxic metallic ions passing into solution, thus accounting for the rare occurrences mentioned above. The available evidence about the toxicity of various metallic ions to lichens is, however, rather conflicting. Fairly high amounts of lead have been found in apparently undamaged thalli of, for instance, *Parmelia* spp. growing by motorways, and it is possible that some lichen species may accumulate aluminium. Lambinon *et al.* (1964) recorded several species of lichens from furnace slag that was comparatively rich in zinc. On the other hand, lichens have never been reported from sheet zinc, and when the metal is employed as galvanizing to put a protective coat on to iron wire, it dissolves fairly readily and destroys lichens in the track of the rain from it (Seaward, 1974). Gilbert and Wathern (1976) mention finding a lichen (*Acarospora* aff. *smaragdula*) growing on a post on an island in the Outer Hebrides where, they say, "its distribution was related to water dripping from a piece of copper wire". Yet the toxicity to lichens of rainwater falling from copper wires is a fact of common observation, for telephone wires running above clay or asbestos-cement roofs produce a clearly visible "shadow" where lichens are unable to grow, and such "shadows" are only colonized very slowly if the wires are removed. Similarly, apparently paradoxical observations in connection with iron are discussed in the next section. Clearly, further research on this subject is urgently needed, but even from the rather scanty basis provided by the evidence already available, we can conclude that there is a considerable range of response to toxic metals in different species. In cases where metals do accumulate it appears in some species at least that the process is largely passive, and considerable accumulations of metals are extracellular.

D. Iron

Iron on the other hand, in contrast to non-ferrous metals, can provide quite a favourable substrate for lichens. Schaerer (1850) mentions *Xanthoria parietina* growing on an iron cover in a public footway. Nylander (1862) recorded 12 species from iron conduits carrying water from Marly to Versailles: *Xanthoria parietina, Physcia leptalea, P. orbicularis, Caloplaca saxicola, C. cerina, Candelariella vitellina, Acarospora fuscata, Lecanora muralis, L. campestris, Aspicilia calcarea, Lecidea goniophila* and *Verrucaria nigrescens*. Arnold (1868) found eight species growing on old iron nails, hinges and the like in the Tyrol: *Xanthoria parietina, X. elegans, Physcia tenella, P. caesia, P. orbicularis, Caloplaca luteoalba, Aspicilia gibbosa* and *Rhizocarpon geographicum*. Later he published a rather similar but somewhat longer list of species growing on iron at Eichstätt, Bavaria (Arnold, 1875): *Xanthoria parietina, X. elegans, Physcia tenella, P. orbicularis, Caloplaca saxicola, C. decipiens, Candelariella vitellina, Pertusaria amara, Aspicilia calcarea, Catillaria sphaeroides* and *Verrucaria nigrescens*. Richard (1883) reported *Xanthoria parietina* from old cannon "*exfoliés par la rouille*" by the river Charente at Rochefort and from the iron bands on a millstone in the village of Miron. He also found *Lecania erisybe* on the cannon, and nine other species on old ironwork in various places: *Physcia tenella, P. orbicularis, Caloplaca citrina, C. cerina, Cande-*

lariella vitellina, C. aurella, Rinodina subexigua, Lecanora dispersa and *L. campestris.*

In this century, Bouly de Lesdain (1910) found a dozen species growing on old iron amongst the sand dunes around Dunkirk: *Xanthoria parietina, Physcia orbicularis, P. adscendens, Caloplaca citrina, C. heppiana, C. holocarpa, Candelariella vitellina, Rinodina subexigua, Lecanora campestris, L. dispersa, Lecania erysibe,* and *Bacidia inundata*; later he added two further species, *Xanthoria polycarpa* and *Verrucaria papillosa* (Bouly de Lesdain, 1914).

Since then the authors and a number of other lichenologists have recorded lichens growing on iron from localities all over the British Isles, although these later records have not been published. Crustose species found most frequently include *Lepraria incana, Caloplaca citrina, Candelariella vitellina, Lecanora dispersa, L. expallens, L. polytropa, L. conizaeoides, L. intricata, Lecidea lucida, Bacidia umbrina* and *Rinodina subexigua*. Foliose species occur less often, the commonest being *Xanthoria parietina, Physcia adscendens* and *P. orbicularis.*

Three apparently distinct types of iron substrate may be broadly distinguished: exfoliated iron, rusted iron and painted iron. The exfoliated substrate originates usually from wrought iron which has become so extensively rusted that layers of iron oxide have flaked off in sheets which remain together as long as they are undisturbed, rather like the pages of a partly open book. Richard's cannon, mentioned above, is the earliest published example, but many others are known, most often perhaps supporting *Lepraria incana* and similar powdery crustose species. The substrate referred to as rusted iron is often, more strictly, steel, which has acquired a firmly adhering thin and coherent coat of oxide. The surface is smooth but more or less finely pitted. The horizontal members of steel bridges and the horizontal supports of iron railings from which the original paint has long since flaked away may rust in this manner and often support *Lecanora polytropa* and similar crustose species. Painted iron will support lichens when the paint coating is sufficiently weathered. They do not grow on loose flaky paint, but on pitted paint surfaces with a texture similar to that of rusted iron. An interesting example is an abandoned motor car observed by one of the authors (F. H. Brightman) in a field near Glenbarr in Kintyre, Argyllshire: it was possible to establish that the vehicle, a 1960 Vauxhall Victor, had last been registered for the year ending 30 June 1969, and presumably had been left in the field since the summer of that year. When observed in the summer of 1973, there were a number of thalli of *Hypogymnia physodes, Parmelia exasperata* and *P. subaurifera* growing on the paintwork of the bonnet, roof and boot. The paint finish was intact, but weathered to a finely pitted surface.

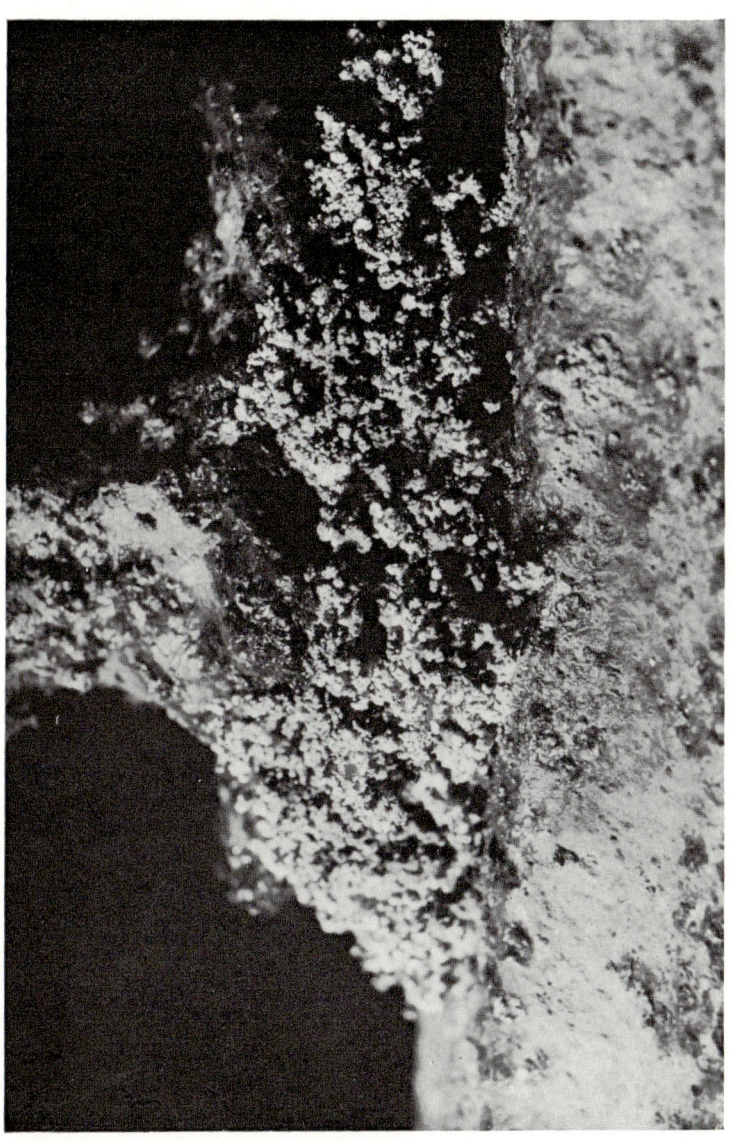

FIG. 5. *Stereocaulon pileatum* on the iron railings of a grave in Rusthall churchyard, Tunbridge Wells, Kent. (Photograph: A. G. Side.)

Although the three types of substrate seem quite distinct, there appears to be no consistent difference in the flora they support. Practically all the species mentioned above have been found at one time or another on all three. On the other hand, small-scale local differences in conditions appear to have significant effects. For instance, very well developed thalli of *Stereocaulon pileatum* are to be found on the iron railings surrounding a grave in a churchyard near Tunbridge Wells in Kent (see Fig. 5). However, the plants grow only on the upper horizontal supports in the immediate vicinity of the ornamental spikes at the top of the vertical railings, and not elsewhere.

A number of saxicolous lichens are characteristic of rocks that are rich in iron compounds, e.g. *Lecidea atrata, Rhizocarpon oederi, Acarospora sinopica* and *Lecanora epanora.* They tend to accumulate iron, and some of them (e.g. *L. atrata* and *R. oederi*) secrete it as a mixture of oxides on the surface of the thallus. Weber (1962) used the term "oxydated" for this phenomenon. Lange and Ziegler (1963) found as much as 5832 p.p.m. of iron in *L. epanora* (and/or *L. subaurea*—see Earland-Bennett, 1975) growing on slag heaps in the Harz mountains. *L. epanora* is locally frequent on ironwork surrounding graves in the churchyard at Flash in Staffordshire (Hawksworth, 1976), and *A. sinopica* has been found growing on rusty iron bolts and the like in Scotland, but the characteristically "oxydated" species have not been found on iron as distinct from iron-containing rock. Also, the species most commonly found on iron show no particular preference for naturally occurring iron-containing substrates. Certainly there is scope for experimental investigation into this problem, but the observations so far available are consistent with our general conclusion that lichens on man-made substrates are more affected by the texture than by the chemical nature of the substrate.

VI. Discussion

In the popular imagination the appearance of lichens on human artefacts such as masonry and other stonework is indicative of antiquity. It is indeed true that colonization of freshly man-made substrates is often a slow process, because in general they do not provide very hospitable habitats. In time, however, they frequently become modified by various influences in a number of ways that encourage the growth of lichens. These weathering processes are discussed further below. On the other hand, although actual measurements of growth rates are in most cases entirely lacking, lichen species which may be said to be omnicolous in Smith's (1921) sense are all of comparatively rapid growth. It seems likely also that in the long term they are unsuccessful in competition with other slower growing lichen species. This is certainly the case with *Lecanora conizaeoides*, and by these two criteria and on account of its general ubiquity in western Europe, it is omnicolous. It is interesting to note as well that a number of omnicolous species appear in James' (1973) tentative "scale of eutrophication". In order of tolerance to increasing levels of hypertrophication they are: *Buellia canescens*; *Physcia adscendens* and *Xanthoria parietina*; *Physcia orbicularis*, *Caloplaca citrina* and *Candelariella* spp. Comparatively rapid growth combined with hypertrophication tolerance enable omnicolous lichens to flourish in habitats where competition from other species is lacking.

The most important weathering processes from the point of view of the establishment and growth of lichens are hypertrophication, modifications of the surface texture and changes in water-holding capacity. These will now be considered in turn. The term hypertrophication is used here in a broad and somewhat vague sense, as is frequently the case in lichenological literature. It implies the accumulation of soluble substances in the lichen substrate, without specifically indicating their precise nature. For instance, bird-perching stones, which develop a characteristic lichen flora, are said to be eutrophicated, and the meaning of this usage is reasonably clear. However, whether it is the nitrogenous or the phosphorus-rich substances in the bird droppings, or indeed possibly some other consideration altogether, which is the key factor affecting the flora, is a completely open question. The roofs and walls of farm buildings, and walls and tree trunks beside tracks used by cattle, develop similar lichen floras. This effect can be reproduced experimentally by impregnating stone with weak suspensions of animal dung. However, it is noteworthy that the same results can be obtained with other suspensions, such as water in which potatoes or rice have been boiled. The species found on bird-perching stones and farm buildings are in general very common by the sea. Clearly, in maritime situations the balance of the ions in the moisture permeating the rock will be different again; possibly it is sodium, magnesium or chlorine that is important rather than nitrogen or phosphorus.

Changes in the pH of the substrate resulting from the absorption of soluble substances are also relevant. Lichens generally regarded to be indicators of eutrophication develop freely on the trunks of trees which normally have acid bark, when the trees are growing in the vicinity of limeworks and the pH of the bark is thereby increased. Perhaps the lowering of the pH of extremely alkaline substrates like some forms of asbestos-cement by the absorption of acid pollutants from the atmosphere should be regarded as a kind of eutrophication. In this case rainfall is important in leaching some substances from the substrate and impregnating it with others carried in solution, but frequently eutrophication is brought about by the deposition of dust. The effect of dust from limeworks has been mentioned above, and it is farmyard dust, rather than the direct splashing of liquids, which affects the roofs of farm buildings. Impregnation with dust is important in eutrophicating exposed woodwork, particularly pales, and encouraging the growth of lichens. Again, it is far far from clear which constituents of the dust are the key ones in suitably weathering the substrate.

Surface texture affects the growth of lichens probably mainly at the establishment stage by affecting the ease with which propagules are able to find lodgement. Lichens appear first at the tops of pales after a certain amount of weathering has occurred, and that this is not merely due to the horizontal aspect is shown by the fact that they develop at the same time on exposed knots in the timber while the rest of the vertical surfaces remain clear. Some weathering is necessary to open up and roughen the end grain, but the rate at which this occurs is impossible to measure, and how far it has to proceed for lichens to establish themselves is difficult to assess. Asbestos-cement tiles with a fibrous surface are more readily colonized than those with smooth surfaces. Roofing slates are only very slowly and usually poorly colonized due to their smooth surface. Frequently a roof will include a few slates with a rougher surface than normal due to a foreign inclusion in the original rock along a plane of cleavage, and these will carry a growth of lichens while all the others remain bare. Surfaces treated with paints to which sand or other particles have been added to give a rough-cast texture are rapidly colonized by lichens, whereas smooth paint surfaces are not. In fact, this has led paint manufacturers to add powerful fungicides to their products of this kind in order to prevent what they regard as the undesirable growth of lichens. The colonization of twigs by *Lecanora conizaeoides* is primarily related to the roughness, rather than the age, of the twig. The concept of roughness is a difficult one to evaluate. A surface with a system of grooves 1 μm across separated by ridges 1 μm wide, would be regarded as rougher than a surface with 1-μm grooves separated by 1 mm by most observers, but on the other hand, a system of 1-mm grooves 1 mm apart would be said to be rougher still. Another variable affecting judgement in this matter is the depth of the grooves. Quantitative measurement of texture in any real situation appears to be an impossibility.

The water-retaining power of the substrate, although relevant to some extent at the establishment stage, becomes more important when growth of the propagules commences. The higher the water-retaining capacity, the slower will the substrate dry out. Lichen thalli dry out quite rapidly, and are able to survive prolonged periods of almost complete desiccation, but are only capable of active growth for as long as their water content is near to saturation. Thus prolongation of the periods during which this is possible is advantageous. On absorption of water after a period of drought, respiration rates increase more rapidly than rates of photosynthesis, so too rapid and frequent alternations of wetting and drying are inimical to the plant. A high substrate water content can tide over short periods of drought, and benefit the lichen in this way as well. Water retaining power is linked to surface texture to some extent. Deeply textured surfaces retain a certain amount of water by capillarity, even if the water absorbing power of the material itself is not very high. This effect is relevant, for instance, with the denser types of brick, and substrates such as exfoliated iron and surface-pitted glass.

When once established, the lichens themselves contribute to the weathering processes. They cause hypertrophication, both directly and also indirectly by providing lodgement for dust particles and detritus of various kinds. They modify the surface texture; for instance, the ubiquitous *Lecanora dispersa* rapidly becomes established on marble and limestone, roughening even the smoothest surfaces, and thus making the attachment of foliose species possible. The latter modify the water relations of the substrate by retaining capillary films of moisture between the thallus and the surface. Even so, in towns the weathering of limestone, especially the purer forms, is mainly chemical, and it has been shown above that the effects of lichens on such substrates as glass have tended to be exaggerated in the past.

At the present time new kinds of man-made substrates are proliferating as never before. It is noticeable that the hard smooth surfaces of substances unknown in nature which are now coming into use for cladding buildings are not so far becoming softened and variegated with lichen growth. No doubt most architects would probably be dissatisfied if this were not the case, although some, together with their counterparts in government planning offices, are having second thoughts about the impact on the countryside of buildings that do not mellow with age. The polythene containers that now litter our beaches have not developed a lichen flora (cf. p. 266) either, presumably because of the uncompromising surface texture as well as the (apparently) non-biodegradable nature of the material. Similarly, although lichens will grow on discarded linoleum, they do not do so on polyvinyl floor covering thrown away in the same situations. In so far as the new materials are not subject to weathering processes, even such versatile plants as lichens are unable to gain a foothold, but there is no reason to believe that in the future they may not succeed in extending their range to even more unlikely substrates than some of those referred to in this chapter.

References

Arnold, F. (1868). Die Lichenen des fränkischen Juras. *Flora, Jena* **51**, 520–524.
Arnold, F. (1875). Die Lichenen des fränkischen Juras. *Flora, Jena* **58**, 524–528.
Bailey, R. H. (1974a). Distribution maps of lichens in Britain. Map 8. *Cyphelium inquinans*. *Lichenologist* **6**, 169–171.
Bailey, R. H. (1974b). Notes on Gloucestershire lichens—10. *J. Gloucesters. nat. Soc.* **25**, 374–375.
Bouly de Lesdain, M. (1909). Lichens des environs de Versailles. *Bull. Soc. bot. Fr.* **57**, 3–7.
Bouly de Lesdain, M. (1910). "Recherches sur les Lichens des environs de Dunkerque." Michel, Dunkirk.
Bouly de Lesdain, M. (1914). "Recherches sur les Lichens des environs de Dunkerque: 1er Supplément." Michel, Dunkirk.
Bouly de Lesdain, M. (1921). Lichens des environs de Versailles: 4e supplément. *Bull. Soc. bot. Fr.* **68**, 16–24.
Bouly de Lesdain, M. (1951). Remarques sur la végétation lichénique des substratum variés, disséminés dans les dunes à l'est et à l'ouest de Dunkerque, de Mardyck (Nord), à la frontière belge. *Revue bryol. lichén.* **20**, 289–296.
Brightman, F. H. (1959). Some factors influencing lichen growth in towns. *Lichenologist* **1**, 104–108.
Brightman, F. H. (1960). Field meeting at Flatford. *Lichenologist* **1**, 203–206.
Brightman, F. H. (1964). *Cyphelium notarisii* in Britain. *Lichenologist* **2**, 283–284.
Brightman, F. H. (1965). The lichens of Cambridge walls. *Nature Cambs.* **8**, 45–50.
Buchet, G. (1890). Les lichens attaquent le verre et, dans les vitraux, semblent préférer certaines couleurs. *C. r. Séanc. Soc. Biol., Sér.* 9 **2**, 13.
Coppins, B. J. and James, P. W. (1974). Distribution maps of lichens in Britain. Map 9. *Cyphelium notarisii*. *Lichenologist* **6**, 172–174.
Coppins, B. J. and Lambley, P. W. (1974). Changes in the lichen flora of the parish of Mendlesham, Suffolk, during the last fifty years. *Trans. Suffolk nat. Soc.* **16**, 319–335.
Crombie, J. M. (1879). Vitricole lichens and the Schwendenerian hypothesis. *Grevillea* **8**, 30–31.
Darbishire, O. V. (1909). "Report of the Second Norwegian Expedition in the *Fram* 1898–1902." Part 21, Lichens. Videnskabs-selskabet i Kristiania.
Earland-Bennett, P. M. (1975). *Lecanora subaurea* Zahlbr., new to the British Isles. *Lichenologist* **7**, 162–167.
Fletcher, A. (1973). The ecology of maritime (supralittoral) lichens on some rocky shores of Anglesey. *Lichenologist* **5**, 401–422.
Fletcher, A. (1975). Key for the identification of British marine and maritime lichens. I. Siliceous rocky shore species. *Lichenologist* **7**, 1–52.
Fries, E. M. (1831). "Lichenographia Europaea reformata." Berling, Lund.
Fries, Th. M. (1867). Lichenes Spitzbergenses. *K. Svenska VetenskAkad. Handl.* **7** (2), 1–53.

Gilbert, O. L. (1965). Lichens as indicators of air pollution in the Tyne Valley. *In* "Ecology and the Industrial Society" (G. T. Goodman, R. W. Edwards and J. M. Lambert, eds), pp. 35–47. Blackwell, Oxford.

Gilbert, O. L. (1971). Studies along the edge of a lichen desert. *Lichenologist* **5**, 11–17.

Gilbert, O. L. and Wathern, P. (1976). The flora of the Flannan Isles. *Trans. Proc. bot. Soc. Edinb.* **42**, 487–503.

Green, T. G. A. and Snelgar, W. P. (1977) *Parmelia scabrosa* Tayl. on glass in New Zealand. *Lichenologist* **9**, (in press).

Harvey J. H., and King, D. G. (1971). Winchester College stained glass. *Archeologia* **103**, 149–177.

Hawksworth, D. L. (1969). The lichen flora of Derbyshire. *Lichenologist* **4**, 105–193.

Hawksworth, D. L. (1972). Regional studies in *Alectoria* (Lichenes). II. The British species. *Lichenologist* **5**, 181–261.

Hawksworth, D. L. (1976). Field meeting in north Staffordshire. *Lichenologist* **8**, 189–196.

Hawksworth, D. L. and Chapman, D. S. (1971). *Pseudevernia furfuracea* (L.) Zopf and its chemical races in the British Isles. *Lichenologist* **5**, 51–58.

Hawksworth, D. L., Coppins, B. J. and Rose, F. (1974). Changes in the British lichen flora. *In* "The Changing Flora and Fauna of Britain" (D. L. Hawksworth, ed.), pp. 47–78. Academic Press, London and New York.

Heaton, N. (1925). The deterioration of stained glass. *J. Br. Soc. Master Glass Painters* **2**, 7.

Henderson, A. and Seaward, M. R. D. (1976). The lichens of Harewood. *Naturalist, Hull* **101**, 61–71.

James, P. W. (1973). The effect of air pollutants other than hydrogen fluoride and sulphur dioxide on lichens. *In* "Air Pollution and Lichens" (B. W. Ferry, M. S. Baddeley and D. L. Hawksworth, eds), pp. 143–175. Athlone Press, London.

Kershaw, K. A. (1963). Lichens. *Endeavour* **22**, 65–69.

Kristinsson, H. (1972). Studies on lichen colonization in Surtsey 1970. [*Surtsey Research Progress Report* **6**, 77.] Reykjavik.

Lambinon, J. L., Maquinay, A. and Ramaut, J. L. (1964). La teneur en zinc de quelques lichens des terrains calaminaires Belges. *Bull. Jard. bot. État Brux.* **34**, 273–282.

Lange, O. L. and Ziegler, H. (1963). Der Schwermetallgehalt von Flechten aus dem *Acarosporetum sinopicae* auf Erzschclackenhalden des Harzes. 1. Eisen und Kupfer. *Mitt. flor.-soz. ArbGemein.*, N.F. **10**, 156–183.

Laundon, J. R. (1963). The taxonomy of sterile crustaceous lichens in the British Isles. 2. Corticolous and lignicolous species. *Lichenologist* **2**, 101–151.

Laundon, J. R. (1966). Hudson's *Lichen siliquosus* from Wiltshire. *Lichenologist* **3**, 236–241.

Laundon, J. R. (1967). A study of the lichen flora of London. *Lichenologist* **3**, 277–327.

Laundon, J. R. (1970). London's lichens. *Lond. Nat.* **49**, 20–69.

Laundon, J. R. (1976). Lichens new to the British flora: 5. *Lichenologist* **8**, 139–180.

Lloyd, A. O. (1972). An approach to the testing of lichen inhibitors. *In* "Biodeterioration of Materials" (A. H. Walters and E. H. Hueck-van-der Plas, eds), vol. 2, pp. 185–191. Applied Science Publishers, London.

Lundström, H. (1970). Epixyler på impregnerade trästolpar i Bogesund. [Report no. 101.] Swedish Wood Preservation Committee, Stockholm.

Macmillan, H. (1861). "Footnotes from the Page of Nature or First Forms of Vegetation." Macmillan, Cambridge.

Mellor, E. (1921a). Sur les lichens vitricoles. *C. r. Séanc. Soc. Biol.* **84**, 650–651.

Mellor, E. (1921b). L'action mécanique des lichens dans la détérioration des vitraux d'église. *C. r. Séanc. Soc. Biol.* **85**, 634–635.

Mellor, E. (1921c). Les lichens vitricoles et leur action mécanique sur les vitraux d'église. *C. r. hebd. Séanc. Acad. Sci., Paris* **173**, 1106–1108.

Mellor, E. (1922a). Lichens vitricoles. *Revue gén. Bot.* **34**, 1–16.

Mellor, E. (1922b). Les lichens vitricoles et la détérioration des vitraux d'église. Doctorate thesis, Paris.

Mellor, E. (1923). Lichens and their action on the glass and leadings of church windows. *Nature, Lond.* **112**, 299–300.

Mellor, E. (1924). Deterioration of stained glass. *Trans. Soc. Glass Technology* **8**, 182–186.

Mellor, E. and de Virville, D. (1921). La détérioration des vitraux d'église de la Mayenne par les lichens. *Bull. Mayenne-Sci.* **1921**, 1–15.

Nylander, W. (1862). Circa Lichenes ferricola notula. *Bot. Z.* **41**, 319.

Nylander, W. (1879). Addenda nova ad lichenographiam europaeam, Continuatio secunda et tricesima. *Flora, Jena* **62**, 353–364.

O'Sullivan, A. M. and Mitchell, M. E. (1974). An unusual lichen habitat. *Ir. Nat. J.* **18**, 23–24.

Paleni, A. and Curri, S. (1972). Biological aggression of works of art in Venice. *In* "Biodeterioration of Materials" (A. H. Walters and E. H. Hueck-van-der Plas, eds), vol. 2, pp. 392–400. Applied Science Publishers, London.

Porta, G. B. della (1658). "Natural Magick . . . in twenty books . . . wherein are set forth all the Riches and Delights of the Natural Sciences." Young and Speed, London.

Rayner, R. W., ed. (1976) ["1975"]. "The Natural History of Pagham Harbour", Part II. Bognor Regis Natural History Society, Bognor Regis.

Renfrew, C., ed. (1974). "British Prehistory: A New Outline." Duckworth, London.

Richard, O.-J. (1877). "Catalogue des Lichens des Deux-Sèvres." Clouzot, Niort.

Richard, O.-J. (1883). Étude sur les substratums des lichens. *Acta Soc. linn. Bordeaux* **37**, 1–88.

Richardson, B. A. (1973). Control of biological growths. *Stone Industries* **8**, 22–26.

Richardson, D. H. S. and Green, B. H. (1965). A subfossil lichen. *Lichenologist* **3**, 89–90.

Salisbury, G. (1953). The genus *Thelocarpon* in Britain. *NWest. nat.* **24**, 66–76.
Schaerer, L. E. (1850). "Enumeratio lichenum Europaeorum." Staempfl, Bern.
Seaward, M. R. D. (1972). Aspects of urban lichen ecology. Ph.D. thesis, University of Bradford.
Seaward, M. R. D. (1974). Some observations on heavy metal toxicity and tolerance in lichens. *Lichenologist* **6**, 158–164.
Seaward, M. R. D. (1975a). Contributions to the lichen flora of south-east Ireland—I. *Proc. R. Ir. Acad.*, *B* **75**, 185–205.
Seaward, M. R. D. (1975b). Lichen flora of the West Yorkshire conurbation. *Proc. Leeds phil. lit. Soc. sci. sect.* **10**, 141–208.
Seaward, M. R. D. (1976). Performance of *Lecanora muralis* in an urban environment. *In* "Lichenology: Progress and Problems" (D. H. Brown, D. L. Hawksworth and R. H. Bailey, eds), pp. 323–357. Academic Press, London and New York.
Seaward, M. R. D. (1977). Contributions to the lichen flora of south-east Ireland—II. *Proc. R. Ir. Acad.*, *B* **77**, 119–134.
Seshadri, T. R. and Subramanian, S. S. (1949). A lichen (*Parmelia tinctorum*) on a Java monument. *J. scient. ind. Res.*, *B* **8**, 170–171.
Smith, A. L. (1918). "A Monograph of the British Lichens", Vol. 1, 2nd ed. British Museum (Natural History), London.
Smith, A. L. (1921). "Lichens". Cambridge University Press, Cambridge.
Smith, A. L. (1926). "A Monograph of the British Lichens", Vol. 2, 2nd ed. British Museum (Natural History), London.
Swinscow, T. D. V. (1967). Pyrenocarpous lichens: 12. The genus *Geisleria* Nitschke. *Lichenologist* **3**, 418–422.
Turner, W. E. S. (1925). The deterioration of stained glass. *J. Br. Soc. Master Glass Painters* **2**, 9.
Weber, W. A. (1962). Environmental modification and the taxonomy of the crustose lichens. *Svensk bot. Tidskr.* **56**, 293–333.
Wells, H. G. (1915). "Boon." Unwin, London.

10 | Lichen Communities in the British Isles: A Preliminary Conspectus

PETER W. JAMES, DAVID L. HAWKSWORTH
and FRANCIS ROSE

I. Introduction	296
II. The Phytosociological Approach	297
A. Nomenclature	298
B. Taxonomy	301
III. Epiphytic Communities	304
Calicion hyperelli	306
Cladonion coniocraeae	313
Graphidion scriptae	314
Lecanorion subfuscae	318
Lecanorion variae	319
Lobarion pulmonariae	322
Parmelion laevigatae	327
Parmelion perlatae	330
Pseudevernion furfuraceae	334
Usneion barbatae	338
Xanthorion parietinae	342
IV. Limestone Communities	349
Aspicilion calcareae	349
Xanthorion parietinae	360
V. Other Basic Rock Communities	361
VI. Siliceous Rock Communities	364
A. Shaded	365
Leprarion chlorinae	365

B. Exposed	370
Lecideion tumidae	370
Pseudevernion furfuraceae	371
Rhizocarpion alpicolae	372
Umbilicarion cylindricae	374
C. Nutrient-enriched	378
Parmelion conspersae	378
Xanthorion parietinae	382
D. Mineral-rich	382
Acarosporion sinopicae	382
E. Marine and Maritime Communities	384
F. Aquatic Communities	389
VII. Terricolous Communities	393
A. Pebbles	394
B. Basic soils	397
C. Coastal soils and dunes	400
D. Acid soils and peat	402
VIII. Summary	407
Note on the Relevé Tables	408
Acknowledgements	408
References	409

I. Introduction

Compared with investigations into the systematics and geographical distribution of lichens within the British Isles, the study of the structure and composition of the communities they comprise has been sadly neglected. The first author to attempt a detailed survey of the lichens occurring in different habitats in the British Isles was W. Watson. He examined the lichens and bryophytes to be found on sand dunes (Watson, 1918a), calcareous soil (Watson, 1918b), in freshwater (Watson, 1919), arctic-alpine vegetation (Watson, 1925), moorlands (Watson, 1932) and woodlands (Watson, 1936a, 1936b). In this period a few other detailed examinations of particular sites were also carried out, of which the most noteworthy are perhaps those concerned with the marine and maritime species at Howth (Knowles, 1913), sand dune species at Blakeney Point (McLean, 1915), heaths in Breckland (e.g. Farrow, 1916) and upland communities on Cader Idris (Evans, 1932). Data from these and other ecologically orientated lichenological studies were incorporated into Tansley's (1939) classic monograph of the vegetation of the British Isles.

In continental Europe in the early and mid-1920s, the recording of lichen communities began to enter a more quantitative phase, data from particular quadrats being compiled into tables; communities also started to be provided with latinized names (see Barkman, 1958). This, the phytosociological approach of the Zurich-Montpellier and Uppsala schools, is open to a number of criticisms (see pp. 297–298), and was not readily adopted in the British Isles; indeed, the first author to endeavour to apply continental phytosociological systems to lichen communities in this country appears to have been Laundon (1956, 1958).

Within the last two decades, ecological studies in Britain have been carried out by descriptive (e.g. Alvin, 1960; Brown and Brown, 1969; Dickinson and Thorp, 1968; Sheard, 1968; Sheard and Ferry, 1967), numerical (e.g. Fletcher, 1973a,b; Yarranton, 1967) and phytosociological (e.g. Coker, 1968; Graham, 1971; Hawksworth, 1969, 1972a, 1973a; Laundon, 1956, 1958, 1967, 1970; McVean and Ratcliffe, 1962; Rose and James, 1974) methods. Most of these investigations have, however, been restricted to particular localities or habitats and no attempt either to draw the published information together or to provide a synopsis of the major lichen communities present in the British Isles as a whole has previously been made.

Studies by the present authors, both in the British Isles and elsewhere, have led them to the view that within a single climatically uniform region, each particular substrate *tends* to assume, eventually, a characteristic and often remarkably uniform lichen vegetation under the influence of similar environmental factors. A preliminary survey of these noda in the British Isles is presented here. The volume of data currently available precludes the possibility of a definitive treatment at this time, but an attempt is made here to recognize the *major* alliances present and a few of the more distinctive associations within them.

We hope that this preliminary conspectus will both serve as a stimulus to phytosociological investigations in the British Isles and provide a framework for the future discussion of lichen communities in these islands.

II. The Phytosociological Approach

There are a number of recent reviews concerned with the theory and practice of phytosociological approaches to the study of plant communities (Barkman, 1973; Guinochet, 1973; Hawksworth, 1974; Shimwell, 1971) and these should be consulted for further information on these aspects. The fundamental criticism levelled against phytosociological methods is to question whether plant communities really exist as distinct entities. Investigation by both phytosociological and numerical methods make it clear that groupings can be recognized, but equally clear that all stands cannot be accommodated within particular units (although the majority generally can be), and may be intermediate between one or more (whether they are defined numerically or by phytosociological methods). Phytosociology, in our opinion, should aim to determine those major noda in the continuum of plant communities which are related to clearly recognizable ecological and environmental parameters, rather than to fit all stands encountered into a rigid system of too strictly defined associations. In most investigations, phytosociologically defined noda are derived essentially by floristic and intuitive methods, but noda can also be recognized as a result of numerical analyses of data obtained from randomized sampling techniques. The relative merits of these approaches have already been the subject of considerable debate, but only recently has a comparison been made between intuitive phytosociological and a number of numerical approaches (Frenkel and Harrison, 1974). These authors found that the intuitive phytosociological method was extremely valuable for the rapid reconnaissance and establishment of basic groupings, particularly in terms of the man-hours expended, although a combination of both methods was likely to prove more useful than either on its own in the production of an entirely comprehensive system.

In the case of lichen communities, the design of a randomized sampling technique or programme which will lead to the detection of the major noda presents special problems, mainly unsolved as yet, on account of the large number of variables involved affecting the communities. For this reason, intuitive methods of sampling are currently the most satisfactory means of detecting the diversity of the lichen communities in an area.

A. *Nomenclature**

Accepting that the subjective, or intuitive, method is a valid one for the recognition of major noda, the question then arises as to whether such noda should be provided with latinized names and ordered in a hierarchical system. The provision of a latinized name enables (or should enable) the concept of a particular community to be readily communicated and

---

* While this work was in press, a Code of Phytosociological Nomenclature prepared by the Nomenclature Commission of the International Society for Vegetation Science appeared (Barkman *et al.*, 1976). Many of the procedures followed in this chapter are in line with proposals in this Code, with two particularly important exceptions: (1) changing the generic stems to conform to current idiotaxonomic usage is rejected, and (2) a starting point date of 1910 is proposed whereas we followed Barkman (1958) who proposed 1922 for epiphytic communities. If this Code becomes accepted by the botanical community at large (i.e. at an International Botanical Congress) it will thus be necessary to change some of the syntaxonomic names employed in this chapter.

10. Lichen Communities in the British Isles

discussed in the literature; for reasons of precision such names require author citations and rules to control their validity and application. Although somewhat complex systems of rules have been proposed (e.g. Meijer-Drees, 1953; see Hawksworth, 1974), no internationally accepted code comparable to the International Code of Botanical Nomenclature is currently available. Thus, even within Europe, variations in the procedures followed in naming plant communities exist. One of the most important of these from the point of view of the stabilization of names is the practice of changing the names of phytosociological units as the names of the species utilized in those units are changed; Laundon (1967) did not do this, for example, while some authors have, and yet they have left author citations for the syntaxon unchanged (e.g. Kalb, 1970; Wirth, 1972). Other problems arise when authors adopt latinized names based on relevé data in earlier works where the original authors not only did not coin latinized names, but also gave no indication that they considered their communities of syntaxonomic value, as in Klement's (1955) treatment of Almborn's (1948) "communities".

In our view, a closer correspondence between the procedures of syntaxonomy and the rules of idiotaxonomy (namely the International Code of Botanical Nomenclature, Stafleu *et al.*, 1972) is required and to this end we have endeavoured to apply the following criteria in this chapter. In a few cases (see p. 353) the nomenclatural situations are so complex that we have preferred to follow continental usage even where this is almost certainly contrary to the views expressed below; it may be that a case could be made for the conservation of some extremely well established names of syntaxa.

1. Names in all languages are accepted and latinized *provided that their authors accorded them a phytosociological rank*. Thus, tables of records to show taxa associated with a species in autecological studies are rejected and syntaxa referred to by their authors as "communities", "noda" etc., are omitted unless the original author latinized them (i.e. implied they had phytosociological rank). Under this rule the "communities" of Almborn (1948) and the "noda" of Rose and James (1974) are thus considered invalid.

2. In cases where community names *not* fulfilling criterion (1) above have been taken up later and attributed to the earlier author either alone or as a "combination", such names are treated as new and attributed to the later author alone.

3. Where the name of a syntaxon fulfilling criterion (1) and based on two or more specific names is contracted by a later author, it is treated as a new name and attributed to the later author alone. This procedure seems desirable as such names are frequently open to contraction in a number of

different ways. Names of syntaxa based on either (i) a single specific epithet (e.g. *Amaretum* Almb., *Conizaeoidion* Laund.), (ii) two species of different genera, or (iii) more than two species, are treated as unacceptable here; for simplicity it might, however, be preferable to take this further and accept only names based on a single generic and a single specific name.

4. The principle of homonymy is difficult to apply in the earlier lichen phytosociological literature as author citations to syntaxa were rarely provided and the names were not latinized (e.g. Hilitzer, 1925). Some syntaxonomists have treated all such names as distinct and homonyms of the first name published. We have accepted the first publication as the acceptable one where other criteria of acceptability are complied with. Definite homonyms (i.e. cases where an author clearly intended to describe a new syntaxon, being unaware of the existence of an earlier homonym) must clearly be rejected. "Misapplied" names of syntaxa are not considered in synonymies presented here as they can have no nomenclatural significance.

5. With respect to the descriptive data accompanying names of associations, subassociations or variants acceptable under criterion (1), where such names are "nomina nuda" (i.e. lack any information as to the species present and their frequencies), they are considered invalid. Such data provided either in text or tabular form is acceptable in our view. Ideally, tables with several records, one of which is designated as the "type record", should be presented by authors describing new syntaxa today. Names of alliances and suballiances are treated as validly published if one or more validly published subordinate taxon is indicated; where no type is clearly designated, the association name based on the same species name as the higher syntaxon is treated as its holotype.

6. The correct name for an association is taken to be a combination of the earliest available acceptably published syntaxal epithet, combined or recombined where necessary, with a name derived from the modern idiotaxonomic genus in which that species epithet is currently placed; changes in *species* names are not followed in syntaxal names here. Where names are recombined, the author of the recombination is indicated and that of the basionym placed in brackets before the later author's name; the same method of citation is followed when there is a change of syntaxonomic rank even if the actual name is unaltered (as in the case of a transfer from suballiance to alliance).

Although contrary to procedures in the International Code of Botanical Nomenclature, it will be noted from the above (criterion 1) that the syntaxonomist's usual practice of accepting names in languages in addition to Latin is followed here. Whether it is advisable to continue this may be debatable, as while devalidation would very considerably reduce the number of syntaxal names to be considered by modern syntaxonomists, at the same time it would lead to changes in some of the better known and widely accepted community names.

In our opinion, the adoption of an internationally agreed set of rules of phytosociological nomenclature is urgently required, as only through this will stability and consistency in the application of names be achieved. Ideally, this matter should be considered by the XIIIth International Botanical Congress to be held in Sydney in 1981.

B. Taxonomy

The phytosociological approach, in addition to the technical nomenclatural problems outlined above, is beset by a number of taxonomic difficulties when one considers the application of existing lichen community names:

1. There is an extremely large number of names of lichen communities already published to consider. Delzenne-van Haluwyn (1976) compiled about 1700 but her listing is very far from complete and the actual number is probably considerably in excess of 2500 to judge from an unpublished catalogue of lichen syntaxal names drawn up by D. L. Hawksworth in connection with the preparation of the present contribution. Furthermore, copies of many of the earlier papers in this field which appeared in the 1920s are extremely difficult to locate; this also applies to some articles published in central and eastern Europe in the 1940s. In the case of the names of syntaxa cited here with full details of their place of publication, all have been checked by us in their original sources. It proved necessary to spend an inordinate amount of time during this study in bibliographical work.

2. In relevés a considerable number of the species recorded may present idiotaxonomic difficulties in themselves or have been incorrectly named; in some instances the names of associations have been based on misdetermined material or have had their characteristic and faithful species wrongly identified. As phytosociologists (perhaps fortunately from the standpoint of conservation) rarely collect samples of all taxa in their relevés and preserve them in herbaria, in many cases the true identity of species included cannot always be definitely ascertained by later workers.

3. A considerable number of the described associations may not have been based on representative stands, i.e. they may have been recognized on the basis of an insufficient preliminary subjective assessment of the major noda of a large geographical area. In practice, associations have generally been described from rather small geographical regions and relatively few studies have considered even whole countries, let alone major climatic zones (e.g. the Atlantic coasts of Europe). For this reason, species stated

to be characteristic of or faithful to particular associations may not really be so over the full range of the association if they are limited by factors other than those limiting the association as a whole.

4. Lichen communities of similar floristic content on different substrates have frequently been provided with different names because of this rather than their species composition; some communities with identical species compositions do, however, occur on both trees and siliceous rocks. Lastly, it has been argued that epiphytic communities in particular should not be recognized (and so not accorded names) independently from the vascular plant (tree) phytosociological taxa in which they occur; in our view, epiphytic communities merit independent recognition as they may occur (i) on the bark of different trees, (ii) in quite different higher plant syntaxa (i.e. woodland types), and (iii) are influenced by environmental factors other than those determining the presence or absence of their phorophytes. It should perhaps be noted that this is contrary to the situation with macromycetes which are often host-dependent owing to mycorrhizal or other nutritive associations and so have generally been studied within their host communities and not given latinized names (see Hawksworth, 1974, for selected references to examples of this approach); the only major departure from this approach is Darimont's (1975) important study in which numerous fungal syntaxa are proposed.

Some additional problems arise from the effects of man on lichen communities. In general, man's influence on lichen communities leads to a simplification in them; thus, in areas subjected to, for example, air pollution, agricultural sprays or trampling, species normally dominant, characteristic or faithful of an association may be rare in, or absent from it. In the case of phytosociological surveys in areas extensively influenced by man, there is consequently a tendency to recognize as distinct phytosociological taxa communities which are essentially simplified facies of those occurring in unaffected areas. If, however, a sufficiently large number of relevés were made over the whole range from the affected to unaffected areas, a gradual continuum could be found; we feel that such simplified communities should not be recognized formally. A few communities now predominating in areas affected by man, however, do merit recognition in that they can form identical communities in unaffected areas (e.g. the *Lecanoretum pityreae*, the *Xanthorion* associations of dust-enriched barks in natural forests in dry areas of Mediterranean Europe; see pp. 321 and 342, respectively).

One final factor to be borne in mind is that of the relationship between communities themselves; such relationships may be either spatial or temporal (e.g. successional). In some instances seral communities may merit phytosociological recognition where they persist for considerable periods of time, represent distinctive noda and differ markedly in species content from those to which they lead (e.g. some communities on smooth young branches and twigs). In contrast, where succession follows the pattern of a gradual migration of species into a community with little or no elimination of pioneer species, formal recognition of the essentially species-poor pioneer facies of the "community" appears superfluous.

Hierarchical classifications of varying complexity above the rank of alliance were adopted by Mattick (1951), Klement (1955) and Barkman (1958) whose systems have been followed, sometimes with modifications, by most recent lichen phytosociologists. We have not recognized any taxa above the rank of alliance here as the described higher syntaxa have been largely based on substrate and abstract species content concepts, with little reference to the climatic and environmental factors controlling their development. As it would clearly be unsatisfactory to propose a scheme of higher syntaxa on the basis of the British data alone, this aspect is not considered further here.

In the treatment for the British Isles presented in this contribution, we have tended to adopt wider concepts of some communities than previous authors, e.g. Barkman (1958); only by doing this does it seem possible to relate many of the associations to environmental factors (albeit in a qualitative way at present) and view the communities as an interlocking mosaic in which the noda occur in different habitats. Narrow concepts of associations, like very narrow species concepts in idiotaxonomy, tend to confuse rather than to clarify; we feel that it is important to endeavour to recognize the major peaks (noda) in the continuum of lichen communities but of very doubtful value to attempt to distinguish minor undulations in the topography of the cline. Our approach might thus be regarded as "synusial" in the sense of Barkman (1973), but we feel it is unnecessary to develop a system of synusial names independent of those proposed for syntaxa.

The procedure adopted in the selection of the communities to be recognized in this preliminary conspectus was consequently (1) an initial selection based on the species present (i.e. floristic), and (2) an attempt to correlate these with environmental factors in so far as these can be judged from field observations alone. The communities distinguished by this approach appear to be the most worthy of syntaxonomic recognition. Most of such communities contain both a number of lichens more or less faithful (confined) to them, and also groupings of characteristic but not strictly faithful species (i.e. those also regularly occurring in one or more other noda).

In the course of field work in the British Isles we have found that on the discovery of one or two species either characteristic of or faithful to a particular community (e.g. the *Lobarion*) it is well worth spending further time searching for additional members of that community. A knowledge of the composition of the various communities described here is consequently an aid to the making of comprehensive surveys of the species present in a particular site.

III. Epiphytic Communities

The very large number of factors determining the development of particular corticolous and lignicolous assemblages of lichens have been the subject of detailed reviews by Barkman (1958) and Brodo (1974) whose works should be consulted for further information on these aspects. Amongst the more important of these in the British Isles are: (1) degree of illumination; (2) humidity of the environment; (3) age of the bark surface; (4) degree of corrugation of the bark; (5) degree and rate of sloughing of bark; (6) continuity and age of woodland cover in a particular site; (7) inclination of surfaces; (8) aspect; (9) degree of bark leaching by rain; (10) degree of impregnation of bark with organic nutrients; (11) air pollution; (12) pollution by agricultural chemicals; (13) pH of the bark surface; (14) basic nutrient status of bark; (15) presence of tannins, betulin or resins etc., and (16) moisture-retaining and absorbing properties of the bark.

As corticolous communities occur on a living substrate they are particularly fluid and this dimension has to be borne in mind in their delimitation as transitions between noda will inevitably be of frequent recurrence.

The epiphytic lichen communities of the British Isles are arranged here in eleven alliances. These are treated alphabetically, as are the associations within them, for ease of reference. The main relationships of the alliances to each other are summarized in Fig. 1 and the following key illustrates the main floristic differences between them.

1. Old forest indicator species (see Table II; e.g. *Lobaria* spp., *Nephroma* spp., *Pachyphiale cornea*, *Pannaria* spp., *Parmeliella* spp., *Sticta* spp. and *Thelotrema lepadinum*) frequent to abundant .... *Lobarion pulmonariae* (p. 322).
Old forest indicator species absent or rare.................. 2

2. Crustose lichens dominant 3
Foliose and/or fruticose lichens dominant 6

3. Arthoniales, stalked Caliciales, Hysteriales (including Lecanactidaceae), *Lepraria* spp., or pyrenocarpous lichens dominant....... 4
Lecanorales or *Cyphelium* dominant 5

4. Arthonioid, lirelliform or pyrenocarpous lichens dominant; communities of young twigs or smooth bark often in somewhat shaded situations .... *Graphidion scriptae* (p. 314).
Caliciales, Lecanactidaceae, *Arthonia impolita* or *Lepraria* spp. dominant; communities of dry bark, bark recesses, or quickly drying lignum .... *Calicion hyperelli* (p. 306).

5. *Lecanora chlarotera*, *L. confusa*, *L. pallida*, *Lecidea symmicta* or *Lecidella elaeochroma* abundant; pioneer communities of twigs forming mosaics of small juxtaposed thalli
.... *Lecanorion subfuscae* (p. 318).
Bacidia chlorococca, *Cyphelium inquinans*, *Lecanora conizaeoides*, *L. varia* or *Lecidea scalaris* abundant to dominant; communities not confined to twigs, often in moderately polluted areas, often forming extensive pure stands and rarely mosaic-like
.... *Lecanorion variae* (p. 319).

6. *Acrocordia*, *Anaptychia*, *Buellia*, *Caloplaca*, *Gyalectina*, *Physcia*, *Physciopsis*, *Physconia*, *Teloschistes* or *Xanthoria* species, or *Parmelia acetabulum*, *P. elegantula*, *P. laciniatula* or *P. quercina* frequent to dominant; communities of nutrient-rich or nutrient-enriched barks in well lit situations .. *Xanthorion parietinae* (p. 342).
Above genera and species rare or absent; communities of mainly nutrient-poor barks 7

7. *Cladonia* species dominant; on bark in humid situations, rotting wood, stumps and tree bases .... *Cladonion coniocraeae* (p. 313).
Cladonia species not dominant; communities mainly of well lit sites... 8

8. *Ramalina* and/or *Usnea* species dominant.. *Usneion barbatae* (p. 338).
Alectoria, *Bryoria*, *Cetraria*, *Hypogymnia*, *Parmelia*, *Parmeliopsis*, *Platismatia* or *Pseudevernia* species dominant 9

9. *Parmelia caperata*, *P. perlata*, *P. reticulata*, *P. revoluta* or *P. soredians* present to dominant or co-dominant; communities of well lit moderately acid barks becoming rare in northern Endland and Scotland .... *Parmelion perlatae* (p. 330).
Alectoria, *Bryoria*, *Cetraria*, *Parmelia laevigata*, *P. taylorensis*, *Parmeliopsis*, *Platismatia* or *Pseudevernia* species dominant 10

10. *Parmelia laevigata*, and/or *P. taylorensis* present; *Cetrelia*, *Menegazzia*, *Mycoblastus*, *Ochrolechia* or *Sphaerophorus* species often present to abundant; communities of well lit extremely acid leached bark in areas of exceptionally high rainfall .. *Parmelion laevigatae* (p. 327).

Alectoria, Bryoria, Cetraria, Hypogymnia, Parmelia saxatilis, Parmeliopsis, Platismatia or *Pseudevernia* species present to dominant or co-dominant; communities of acid barks commonest in northern and upland areas of England and central and eastern Scotland with lower rainfall .. *Pseudevernion furfuraceae* (p. 334).

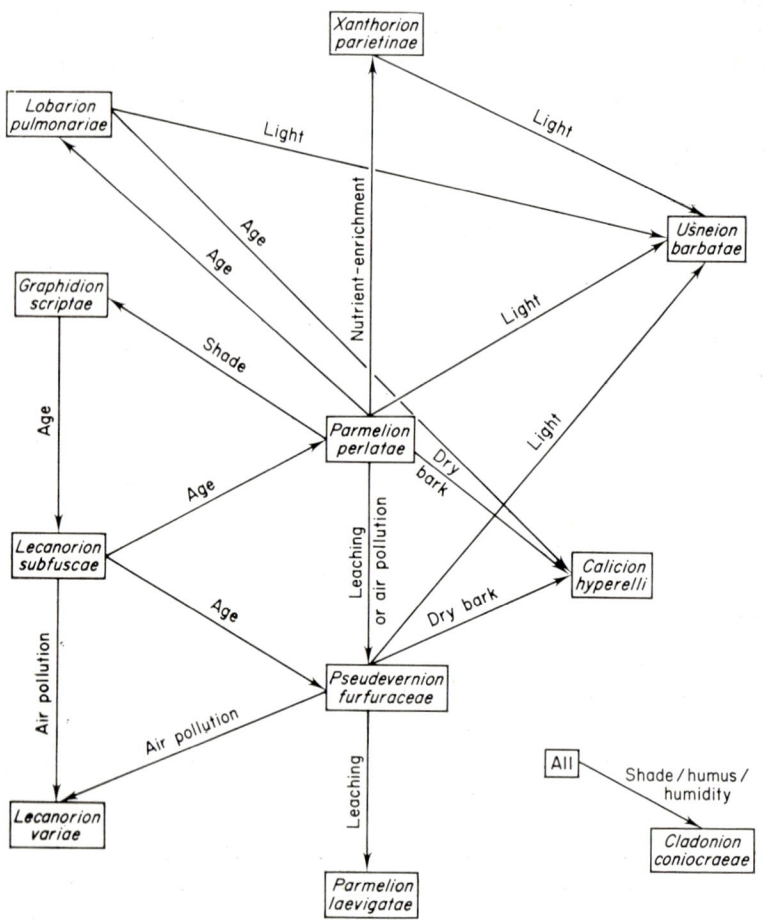

FIG. 1. Principal relationships between the epiphytic alliances present in the British Isles.

All. 1. *Calicion hyperelli*

Calicion hyperelli Čern. & Hadač, *in* Hadač, *Příroda* **36,** 254 (1944) [notes only seen]; type: *Lecanactidetum abietinae* Hil.—*Coniocybion gracilentae* Klem., *Beih. Feddes Repert.* **135,** 146 (1955).—*Leprarion* Almb., *Bot. Notiser, Suppl.* **1** (2), 33 (1948).—*Calicion viridis* Fab., *Monogr. Bot., Warsaw* **26,** 24 (1968) [as "Hadač em. Barkm."].

This large and species-diverse alliance is united by a common habitat rather than its species composition, which varies very greatly from one association to another. The alliance is characteristic of aged dry bark of deciduous trees and decorticate wood in well lit situations. The associations include many species which are strictly faithful to them; a rather unusual situation for epiphytic lichen communities. Many occur in ecological niches where they either never receive direct rain (i.e. are entirely dependent on atmospheric humidity for their moisture) or are in very exposed sites where their substrates dry out very rapidly following rain.

Arthonietum impolitae Almb.

Bot. Notiser, Suppl. **1** (2), 33 (1948).—*Impolitetum* Almb., K. svenska Vetensk-Akad. Avh. natursk. **11**, 27 (1955).

This markedly xerophilous community was treated by Barkman (1958) in a rather broader sense than is adopted here (see p. 309). We take it up for *Arthonia impolita* dominated communities on dry, well lit, often somewhat basic, barked trees in southern England. The association is rather poor in species but *Schismatomma decolorans* is a particularly common component of it. This association is not infrequent on trees in pastures where *Buellia canescens*, *B. punctata*, *Catillaria griffithii* and *Lecanora expallens* may enter into it.

Calicietum abietini Kalb

Hoppea [Denkschr. Regen. bot. Ges.] **26**, 104 (1966).

This association, characterized by stands of *Calicium glaucellum*, or rarely *C. abietinum*, is optimally developed on decorticate pine wood in the Scottish Highlands when additional members of the Caliciales enter into it (e.g. *Chaenotheca brunneola*) but is also encountered occasionally throughout the British Isles in this habitat and also on the exposed wood of a wide range of deciduous trees and, more rarely, fence-posts. *Chaenotheca trichialis* is also a locally important component of this association, but can also occur in a facies of the *Calicietum hyperelli* (see below). Many distinct noda may perhaps be recognized within the *Calicietum abietini* in the Scottish Highlands but these are too poorly understood at the present time to describe here.

Calicietum hyperelli Hil.

Spisy Přirod. Fac. Karl. Univ. **41**, 98 (1925) [as "Association à *Calicium hyperelum*" (sic)].—*Chaenothecetum melanophaeae* Barkm., Phytos. Ecol. Crypt. Ep., 356 (1958).

This association was described by Hilitzer (1925) for communities dominated by *Calicium viride* and in which *Lecanactis abietina* was often a major component. In the British Isles this generally species-poor association occurs on dry, rough or moderately smooth (e.g. *Acer*) bark of deciduous trees in lowland areas (most frequently in those subjected to moderate air pollution) and also on coniferous trees in central and eastern Scotland. Communities dominated by *Lecanactis abietina* appear distinct from the *Calicietum hyperelli* in the British Isles and that species is rarely a major component of this association here. *Lecanora expallens*, *Lecidea scalaris*, *Lepraria incana*, sometimes *Buellia schaereri* (central and eastern Scotland), *Catillaria griffithii* and *Schismatomma decolorans*, enter this association but it most commonly consists of almost pure stands of *Calicium salicinum*, *C. viride*, *Chaenotheca ferruginea* and *Lepraria candelaris*. *Chaenotheca ferruginea* predominates in polluted areas whilst *Schismatomma virgineum* occurs in the driest facies of this association.

In lowland Britain the *Calicietum hyperelli* is most commonly found on the dry (usually north-east) sides of well lit deciduous trees which carry either the *Parmelion perlatae* or *Xanthorion parietinae* on their better lit and wetter sides. Allied to this association is a further easily overlooked community which may be distinct, which occurs in dry sheltered bark crevices of rather basic-barked trees (e.g. *Fraxinus*, *Salix*, *Ulmus*) particularly in south-east England and central Scotland. This community is dominated by *Chaenotheca trichialis* together with *C. hispidula*, *C. carthusiae*, *C. laevigata* and *Coniocybe sulphurea*.

Coniocybetum furfuraceae Kalb

Ber. bayer. bot. Ges. **41**, 70 (1969).

The *Coniocybetum furfuraceae*, characterized by abundant *Coniocybe furfuracea*, occurs in more shaded sites than the *Calicietum hyperelli* and when on trees is restricted to basal crevices, exposed roots etc. This association is not exclusively corticolous and also occurs on soft humus and in rock crevices where it is protected from direct rain. The *Coniocybetum furfuraceae* is mainly restricted to western and northern parts of the British Isles. When on rock this association may include elements of the *Micareetum sylvicolae* (p. 366) and its relationships to that community merit further study.

Lecanactidetum abietinae Hil.

Spisy Přírod. Fac. Karl. Univ. **41**, 94 (1925) [as "Association à *Lecanactis abietina*"].

Communities dominated by *Lecanactis abietina* occur in generally similar habitats to the *Calicietum hyperelli* but favour somewhat more shaded and humid situations, for example dry, very acid bark of *Pinus* or *Quercus* within a wood rather than isolated trees in pastures. The association can perhaps be viewed as a dry bark counterpart of the *Parmelietum laevigatae*. Where this association is optimally developed, the *L. abietina* with apothecia, *Calicium viride* is often completely absent, and while *Chaenotheca ferruginea* is often to be found in the vicinity, it tends to be on slightly less sheltered parts of the trunks and to represent a fragment of the *Calicietum hyperelli* rather than forming a part of the present association. The *Lecanactidetum abietinae* is rather poor in species in the British Isles although *Lepraria incana* and *Schismatomma decolorans* may occur in it from time to time. *Arthonia leucopellaea*, *Lecanactis amylacea* and sometimes *L. corticola* or *L. dryophila*, are rare members of this association occurring in, for example, south-western and north-eastern Scotland and the New Forest. In old woodland areas *Schismatomma niveum* is also a frequent member of this association.

Lecanactidetum premneae ass. nov. (Table I, Fig. 2)

This very distinctive association, the "*Schismatomma decolorans–Lecanactis premnea–Opegrapha lyncea* community" of Rose and James (1974), appears to ultimately become the post-climax community of very ancient (over 300 years) *Quercus* trees and can cover their entire surfaces both in open parklands and in forested areas (e.g. the New Forest) in southern Britain. It colonizes bark surfaces which have become dry and brittle with age and lost their water-holding capacity but retained a relatively high pH. Though an association clearly relict from ancient forests, it does not need a forest microclimate to survive and can occur on ancient, now isolated, trees. The association is unknown in Scotland and appears very rarely in Ireland.

The characteristic species of the *Lecanactidetum premneae* are *Lecanactis premnea*, *Opegrapha lyncea*, *O. prosodea* (extreme south of England only), *Schismatomma decolorans*, and sometimes *S. virgineum*. *Arthonia impolita* and *Buellia canescens* are sometimes present in pasture sites. Although almost entirely restricted to aged *Quercus* (though also known on aged *Fagus*), a facies of this association (lacking *O. lyncea*) occurs on extremely ancient *Taxus* trees (usually in churchyards).

This community has been little understood in the past, doubtless due to the scarcity of ancient *Quercus* trees outside the British Isles, but elements of it were included by Barkman (1958) in his rather broad concept of the

Table I.*

Lecanactidetum premneae ass. nov.

Species	\multicolumn{8}{c}{Stands}
	1	2	3	4	5	6	7	8
Arthonia impolita	3.3	–	–	–	–	1.3	1.3	–
Buellia canescens	–	–	–	–	–	+.2	3.2	–
Catillaria griffithii	+.2	–	+.2	–	–	–	–	–
Chaenotheca brunneola	–	–	–	–	–	+.0	–	–
Cladonia coniocraea	–	–	–	–	+.0	+.2	–	–
C. parasitica	–	–	–	–	+.0	–	–	–
Enterographa crassa	–	–	2.2	2.3	2.3	2.3	–	–
Haematomma ochroleucum var. *porphyrium*	–	1.2	–	–	–	–	–	–
Lecanactis abietina	–	–	+.2	–	–	–	–	–
L. premnea	1.2	3.3	4.5	3.3	2.3	–	1.2	3.3
Lecanora expallens	+.2	–	–	–	–	–	–	2.3
Lecidea granulosa	–	–	–	+.2	–	–	–	–
L. quernea	+.2	–	+.0	–	–	+.0	1.2	–
L. uliginosa	–	–	–	–	+.2	–	–	–
Lepraria candelaris	–	–	2.3	–	+.2	–	–	1.2
L. incana	+.2	–	2.1	2.2	1.2	1.2	–	–
Normandina pulchella	–	–	–	+.2	–	–	–	–
Opegrapha lyncea	–	2.3	–	1.2	2.3	1.3	+.2	2.3
O. prosodea	–	–	–	–	–	–	1.2	–
Parmelia caperata	–	–	–	–	–	+.0	–	–
P. glabratula	–	–	–	+.2	–	–	+.0	–
P. sulcata	–	–	–	1.2	–	–	–	–
Pertusaria hemisphaerica	–	–	–	–	–	–	–	1.3
P. hymenea	–	–	+.1	–	–	–	–	–
Phlyctis argena	–	–	–	–	–	–	–	+.3
Ramalina farinacea	–	–	–	+.0	–	–	–	–
Rinodina roboris	–	–	–	+.0	–	–	1.2	–
Schismatomma decolorans	4.3	3.3	+.1	3.3	2.2	3.3	4.4	1.3
Hysterium angustatum	–	–	1.1	–	–	–	–	–
Bryum capillare	–	–	–	+.0	–	–	–	–
Hypnum cupressiforme	–	–	–	+.2	+.0	+.0	–	–
Isothecium myosuroides	–	–	–	+.2	–	–	–	–
Lophocolea heterophylla	–	–	–	+.2	+.0	–	–	–
Metzgeria furcata	–	–	–	+.2	–	–	–	–
Mnium hornum	–	–	–	+.0	–	–	–	–
Orthodontium lineare	–	–	–	–	+.0	–	–	–

*See p. 408 for notes on relevé tables.

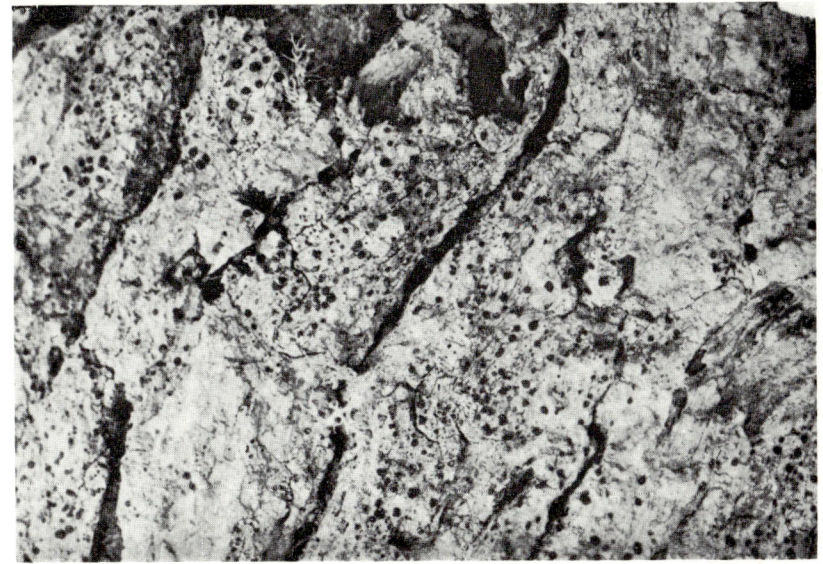

FIG. 2. *Lecanactidetum premneae* on dry bark on an ancient oak. Species present include: *Calicium viride*, *Lecanactis premnea* (predominating), *Lepraria incana*, *Ramalina farinacea* and *Schismatomma decolorans* (Hampshire: New Forest, Vinney Ridge, 1969, P. W. James).

1. Cumbria, Naworth Park (35/– – 6 – –): old *Quercus* in pasture, 1·0 m diam, incl. 90°, aspect SW, 0·5 × 0·5 m, cover 60%, 31 August 1973, F.R.
2. Hampshire, New Forest, Moyles Court (41/1 – – 0 – –): ancient *Quercus* by road, 2·5 m diam, incl. 90°, aspect N, 1·0 × 0·5 m, cover 70%, 20 September 1967, F.R.
3. Hampshire, New Forest, Eyeworth Wood (41/2 – – 1 – –): *Quercus robur*, 0·6 m diam, incl. 90°, aspect E, 1·0 × 0·5 m, cover 90%, 25 June 1968, F.R. and J. J. Barkman.
4. Kent, Lullingstone Park (51/5 – – 6 – –): ancient *Quercus* in relict valley forest, 1·6 m diam, incl. 90°, aspect S, 1·0 × 1·0 m, cover 60%, 10 January 1969, F.R.
5. Norfolk, Merton Park, Merton Oaks Wood (52/9 – – 9 – –): ancient *Quercus* in wood, 3·0 m diam, incl. 90°, aspect S, 1·0 × 1·0 m, cover 70%, 5 June 1970, F.R.; *type record*.
6. Norfolk, Kimberley Park (63/0 – – 0 – –): ancient *Quercus* by drive in parkland, 2·0 m diam, incl. 90°, aspect S, 1·0 × 1·0 m, cover not determined, 8 March 1969, F.R.
7. Sussex, Parham Park (51/0 – – 1 – –): very old *Quercus* in park, 1·5 m diam, incl. 90°, aspect N, 1·0 × 0·5 m, cover 90%, 5 November 1968, F.R.
8. Wiltshire, Longleat Park (31/8 – – 4 – –): ancient *Quercus* in open parkland, 3·0 m diam, incl. 90°, aspect S, 1·0 × 1·0 m, cover 70%, 5 June 1970, F.R.

Arthonietum impolitae treated above. It certainly occurs in France in the Fôret de Fontainebleau and Brittany.

Leprarietum candelaris (Matt.) Barkm.

Phytos. Ecol. Crypt. Ep., 355 (1958).—*Lepraria candelaris* Gesellschaft Matt., Ber. West-preuss. bot.-zool. Ver. **59**, 16 (1937).

The *Leprarietum candelaris* is dominant on the dry north and east sides of hardwood trees in forests in relatively unpolluted parts of the British Isles where the rainfall is rather low (particularly in north-east Scotland near the Moray Firth). Elsewhere in southern and eastern England it is largely confined to dry bark crevices, where it is often associated with *Schismatomma decolorans*, although it does occasionally spread from crevices to fully exposed, raised flattened bark surfaces in a few low-rainfall southern areas (e.g. extreme south Devonshire).

The *Leprarietum candelaris* is closely related to the *Calicietum hyperelli* and appears to be a more lowland counterpart of it in the British Isles.

Leprarietum incanae ass. nov. prov.

This association has a similar ecological amplitude to the *Coniocybetum furfuraceae* occurring not only on the shaded parts of trees but also on humus, on dry powdery soil in rock crevices and on rock. *Lepraria incana* often occurs in almost pure stands but *Cladonia coniocraea*, *Lecidea granulosa* and *L. uliginosa* frequently enter into the association on trees. The *Leprarietum incanae* is the most shade and air-pollution tolerant of the associations of the *Calicion hyperelli* treated here extending far into urban areas in cracks of the bark on the basal parts of deciduous trees.

Opegraphetum fuscellae Almb.

Bot. Notiser, Suppl. **1** (2), 140 (1948).—*Fuscelletum* Almb., K. svenska Vetensk-Akad. Avh. natursk. **11**, 37 (1955).

A rather species-poor community dominated by *Opegrapha vermicellifera* occurring mainly in woodlands on dry bark of deciduous trees but sometimes spreading on to adjacent *Hedera* stems or siliceous rocks. Additional important components include *O. ochrocheila*, *O. vulgata* and *Pyrenula nitida*. This association occurs on a wide range of deciduous trees, although also known from conifers, but in Britain prefers somewhat base-rich barks. Interestingly, like the *Parmelietum revolutae* (see p. 331) it is found on conifers only in areas where there is little or no air pollution.

Two facies of this community can be recognized in the British Isles: (1) the typical nodum on shaded basic barks (e.g. *Ulmus*) with *Opegrapha vermicellifera, O. ochrocheila* and *O. lichenoides,* and (2) one on shaded more acid barks (e.g. *Fagus*) where *O. niveoatra* predominates and *Pertusaria leioplaca* and *Graphis* species are characteristically present. The *Opegraphetum fuscellae,* as conceived here, has a rather southern and western distribution in the British Isles.

All. 2. *Cladonion coniocraeae*

Cladonion coniocraeae Duvign., *Bull. Soc. r. bot. Belg.* **74,** 49 (1942); type: *Cladonietum coniocraeae* Duvign.

This alliance was first described from Belgium for communities dominated by *Cladonia coniocraea* and applied to British communities by Laundon (1956, 1958, 1967) and later authors. Two associations are recognized in Britain both of which can occur on tree bases and peaty soils.

Cladonietum cenoteae Frey

Veröff. Geobot. Inst. Rubel **4,** 249 (1927).

This association, first described from Scandinavia, is used here in its original somewhat restricted sense (see below) for the *Cladonia*-rich communities which are found on very dry and acidic tree bases, dead stumps and on peaty soil under mature native pine woods in the Scottish Highlands. It is particularly well developed in the Glen More and Rothiemurchus Forests where it includes as its characteristic species the rare or local *C. botrytes, C. carneola, C. cenotea, C. deformis* and *C. nemoxyna* as well as the more widespread *C. furcata, C. gonecha, C. gracilis, C. scabriuscula* and other moorland *Cladoniae.*

Cladonietum coniocraeae Duvign.

Bull. Soc. r. bot. Belg. **74,** 49 (1942).

The *Cladonietum coniocraeae* is a well marked community which is both pollution tolerant in the British Isles and has a wide ecological amplitude, occurring on the bases of shaded trees (sometimes extending far up the trunks in ravine-like sites), rotting wood, clayey and peaty soils, and siliceous rocks in damp situations. The community, when optimally developed, is rich in species of *Cladonia,* although *C. coniocraea* is the most important; others commonly encountered include *C. chlorophaea, C. digitata, C. fimbriata, C. macilenta, C. pyxidata* s. str. and *C. squamosa.* On dry decorticate wood *C. coniocraea* may be almost entirely replaced by *C. parasitica*, and in damp woods in the south and south-west, the rather rare *C. caespiticia* also enters into the association. The most pollution-tolerant facies of this community, however, merely comprises swards of *C. coniocraea* and/or *C. macilenta*. The association intergrades with other noda containing species such as *Lecidea granulosa*, *L. uliginosa*, *Micarea prasina*, *Peltigera canina*, *P. polydactyla*, *P. praetextata* and *Sphaerophorus globosus*. Bryophytes are important components in this community, the introduced *Orthodontium lineare* (see Rose and Wallace, 1974) being particularly characteristic in the south-eastern and Midland counties. In relatively unpolluted areas *Usnea*-rich facies occur in better lit sites but these, following Barkman (1958), are placed in the *Cladonieto–Usneetum tuberculatae* here (see p. 338).

The *Cladonietum coniocraeae* has been treated as a synonym of the *Cladonietum cenoteae* by a number of lichen phytosociologists (e.g. Wilmanns, 1966) following Klement (1955), but this view was rejected for the British Isles by Hawksworth (1972a); from the information provided above it will be evident that these two associations differ fundamentally in species composition, habitat and distribution and thus clearly merit treatment as distinct associations.

Various infrassociational taxa have been described within this association (e.g. var. *macrocladonietum digitatae* Duvign., var. *microcladonietum coniocraeae* Duvign., var. *sphaerophoretum globosae* Duvign., subass. *cladonietum digitatae* Klem.) and while most can be recognized within the British Isles they appear to be of minor synsystematic value.

All. 3. *Graphidion scriptae*

Graphidion scriptae Ochsn., *Jahrb. St. Gall. naturw. Ges.* **63**, 49 (1928); type: *Pyrenuletum nitidae* Hil.—*Graphidion* Almb., *K. svenska VetenskAkad. Avh. natursk.* **11**, 21 (1955); type: *Nitidetum* Almb.

This particularly well marked alliance is characteristic of smooth bark, usually on young trees, branches or twigs, and comprises two distinct elements, (1) the pioneer community of trees, and (2) some communities persisting on mature trees with smooth bark or restricted to deep shade. The *Graphidion scriptae* is characterized by a wide range of crustose species and where macrolichens are encountered within its associations they can be viewed as pioneers of subsequent successional stages. Most of the crustose species involved are esorediate with either lirellate or pyrenocarpous ascocarps.

The alliance is entirely restricted to the bark of deciduous trees in the British Isles. While it evidently merits subdivision into several associations, few studies have been made on this alliance in Britain so far and the following treatment is consequently to be regarded as provisional.

In more humid situations, especially on inclined boughs of *Corylus* and *Salix* in sheltered sites in western Britain and western France, the alliance becomes rapidly enriched by other taxa, both of bryophytes and foliose lichens, characteristic of the *Lobarion pulmonariae* or *Parmelion laevigatae*. The commonly occurring bryophytes include species of *Ulota* (particularly *U. bruchii* and *U. crispa*, with *U. phyllantha* in maritime situations and *U. drummondii* and *U. hutchinsiae* in hyper-oceanic sites) and *Orthotrichum* (particularly *O. affine*, *O. pulchellum*, *O. striatum* and *O. tenellum*) and hepatics such as *Frullania dilatata*, *Lejeunea ulicina*, *Metzgeria furcata* and *Radula complanata*.

Communities in this alliance are generally air-pollution intolerant. In drier unpolluted areas its species may persist on the trunks of mature *Quercus* on the level surfaces between the furrows in the bark, but in high rainfall areas they will only persist on phorophytes which have smooth bark throughout their lives (e.g. *Corylus*, *Fagus*, *Ilex*, *Salix* and *Sorbus*). The alliance may also occur as species-poor variants on medium aged *Betula* in moister parts of the British Isles but it is not typically developed on that tree.

Some hyper-oceanic lichens occur within communities of this alliance (e.g. *Anthracothecium pyrenuloides*, *Arthothelium* species (excluding *A. ruanum*), *Dermatina swinscowii*, *Graphina ruiziana*, *Lecanactis homalotropum*, *Thelotrema subtile*, *Leptogium hibernicum*, *Pyrenula dermatodes*, *P. laevigata*, *Thelotrema monosporum* and *Tomasellia ischnobela*) but the synsytematics of these have not been investigated and so these are not considered in the associations recognized below. They tend to occur where *Lobarion* on trunks is rich in members of the Pannariaceae, but rather poor in *Lobaria*, *Pseudocyphellaria* and *Sticta* species, as in the Glasdrum National Nature Reserve, Argyllshire.

Arthopyrenietum punctiformis ass. nov. prov.

This name is adopted provisionally here for the pioneer community of twigs and saplings dominated by pyrenocarpous lichens with endophloeodal thalli, and includes *Arthonia aspersella*, *A. punctiformis*, *A. radiata*, *A. tumidula*, *Arthopyrenia fallax*, *A. punctiformis*, *Dermatina quercus* (second or third year oak twigs), *Opegrapha atra*, *Stenocybe pullatula* (confined to *Alnus* twigs) and *Tomasellia gelatinosa* (mainly on *Corylus*). The age of the twigs and the degrees of preliminary scarring influence the intrusion of additional hypophloeodal species which first appear around leaf and girdle scars. *Lecanora carpinea*, *L. chlarotera*, *L. confusa*, *Lecidea symmicta*, *Lecidella elaeochroma* and *Rinodina sophodes* indicate transitions to the *Lecanoretum subfuscae*, whilst in more polluted areas *Bacidia chlorococca* and *Lecanora conizaeoides* indicate transitions to the *Bacideetum chlorococcae* and *Lecanoretum pityreae*, respectively. On saplings a facies dominated by *Porina chlorotica* var. *carpinea* with much *Opegrapha atra* and *O. vulgata* and little or no *Arthonia punctiformis* is particularly distinctive; this frequently includes species of the *Lecanoretum subfuscae* (well lit sites) or *Pyrenuletum nitidae* (shaded sites) representing transitions to those associations.

Graphidetum scriptae Hil.

Spisy Přirod. Fac. Karl. Univ. **41,** 90 (1925) [as "Association à *Graphis scripta*"].
—*Graphinetum platycarpae* var. *graphinetum anguinae* D. Hawksw., *Fld Stud.* **3,** 548 (1972).

This name is taken up here for communities of moderately shaded smooth bark dominated by some of the larger lirelliform crustose lichens (e.g. *Graphina anguina*, *Graphis elegans*, *G. scripta*, *Phaeographis dendritica* and *P. lyellii*). It characteristically occupies less densely shaded sites than the *Pyrenuletum nitidae*, transitions to which are not infrequent. *Arthonia lurida* is also commonly found in this association. In its well developed form this association is essentially oceanic; only *G. scripta* of the species listed above occurs in any quantity in continental areas of Europe.

On *Ilex*, particularly in the New Forest and south-west Ireland, a species-rich variant of this association occurs which includes *Arthonia stellaris*, *Arthothelium ilicinum*, *Mycoporellum sparsellum* and *Stenocybe septata*, together with much *Thelotrema lepadinum*.

Although Barkman (1958) subsumed Hilitzer's association within the *Pyrenuletum nitidae*, Hilitzer appears to have recognized the distinctness of these noda and thus his name is adopted here.

Pertusarietum amarae Hil.

Spisy Přirod. Fac. Karl. Univ. **41,** 87 (1925) [as "Association à *Pertusaria amara*"].
—*Pertusarietum hemisphaericae* Klem., *Beih. Feddes Repert.* **135,** 140 (1955) [as "Almborn 1948"].—*Pertusarietum wulfenii* Klem., *Beih. Feddes Repert.* **135,** 134 (1955) [as "Almborn 1948"].—*Amaretum* Almb., *K. svenska VetenskAkad. Avh. natursk.* **11,** 15 (1955).

This association was used by Barkman (1958) in a rather restricted sense but has been subsequently adopted by, for example, Fabisewski (1968) who recognized three variants. The *Pertusarietum amarae* is interpreted here as comprising somewhat shade-tolerant communities, often on rather less smooth bark, in which species of *Pertusaria* predominate. The community is particularly well developed on *Fagus* in the New Forest and *Carpinus* in south-east England. *Pertusaria hymenea* and *P. pertusa* are the most important components although other species of the genus provide high cover values locally (e.g. *P. hemisphaerica, P. leioplaca*). In old forests *Haematomma elatinum, Lecidea cinnabarina* and *Thelotrema lepadinum* also enter into it and the lirelliform species generally present (other than *Graphis elegans*) become replaced by verrucose-fertile and verrucose-sorediate taxa (characterized by *Pertusaria*).

The *Pertusarietum amarae* shows some tendency to intergrade with the *Pyrenuletum nitidae* and could be viewed alternatively as an extreme facies of that association.

Pyrenuletum nitidae Hil.

Spisy Přírod. Fac. Karl. Univ. **41,** 91 (1925) [as "Association à *Pyrenula nitida*"]. —*Lecanoretum glabratae* Klem., *Beih. Feddes Repert.* **135,** 132 (1955) [as "Almborn 1948"].—*Nitidetum* Almb., *K. svenska VetenskAkad. Avh. natursk.* **11,** 21 (1955).—*Porinetum carpineae* Barkm., *Phytos. Ecol. Crypt. Ep.*, 382 (1958) [as "*nov. ass. prov.*"].

The *Pyrenuletum nitidae*, principally characterized by extensive mosaics of *Enterographa crassa* and/or *Pyrenula nitida* and *P. nitidella*, occurs on a wide range of smooth-barked deciduous trees when growing in deep shade. Many species may accompany it, of which *Arthonia lurida, A. radiata, A. spadicea, A. tumidula, Graphis scripta, Opegrapha atra, O. viridis, O. vulgata, Pertusaria leioplaca* and *Phaeographis dendritica* merit particular note; some of these species may be very important components of this association locally. Further east in Europe (e.g. eastern Denmark) this association is important on *Fagus* in old dry forests; *Opegrapha viridis* usually occurs in such sites but *Enterographa crassa* is normally absent.

According to Barkman (1958) the *Opegraphetum herpeticae* Almb. (syn. *Rufescentetum* Almb.) is closely allied to the *Pyrenuletum nitidae*. While the components of this association occur in the British Isles they do not appear to constitute a well defined association here; *Opegrapha rufescens*, for example, is essentially a species of aged dry bark underhangs of *Fraxinus*, more rarely *Quercus*, and is perhaps better placed with other associations characteristic of this habitat (see pp. 308-309). The *Arthonietum luridae* Kalb, described from Germany, seems only doubtfully distinct from our concept of the *Pyrenuletum nitidae*. The *Porinetum carpineae* of Barkman (1958) appears to be merely an early stage of the *Pyrenuletum nitidae* and so is not recognized separately here; this facies, not usually found on young twigs, includes *Microthelia micula, Porina chlorotica* var. *carpinea* and *P. leptalea.*

All. 4. *Lecanorion subfuscae*

Lecanorion subfuscae Ochsn., *Jahrb. St. Gall. naturw. Ges.* **63**, 50 (1928); type: *Lecanoretum subfuscae* Hil.—*Lecanorion carpineae* Barkm., *Phytos. Ecol. Crypt. Ep.*, 390 (1958); type: *Lecanoretum carpineae montanum* Barkm.—*Olivaceion* Laund., *Lond. Nat.* **37**, 72 (1958) [as *nom. nov.* for "*Lecanoretum subfuscae* Ochsn. p.p.*"*].

The nomenclature of this alliance is extremely complex. Barkman (1958) rejected Ochsner's name as highly ambiguous because of the confusion which surrounds the name *Lecanora subfusca* but, as is the case with the *Usneion barbatae* discussed below (p. 338), since there can be little doubt that the composition of the alliance corresponds to Barkman's *Lecanorion carpineae*, Barkman's terminology is rejected here as it has been by, for example, Kalb (1970) and Hawksworth (1972). For further information on the nomenclature of this alliance, Kalb's (1970) detailed discussion should be consulted.

The *Lecanorion subfuscae* is essentially a pioneer community of *well lit* twigs and young trees characterized by species of *Lecanora* (particularly the *L. subfusca* species complex) and other hyperphloeodal crustose species forming intricate mosaics. A single association is recognized within it in the British Isles.

Lecanoretum subfuscae Hil.

Spisy Přírod. Fac. Karl. Univ. **41**, 84 (1925) [as "Association à *Lecanora subfusca*"].—*Lecideeto parasemo-Phlyctideetum* Hil., *Spisy Přírod. Fac. Karl. Univ.* **41**, 89 (1925) [as "Association à *Lecidea parasema* et *Phlyctis*"].—*Lecanoretum allophanae* Duvign., *Bull. Soc. r. bot. Belg.* **74**, 39 (1942) [as "*nom. nov.*"].—*Lecanoretum carpineae atlanticum* Barkm., *Phytos. Ecol. Crypt. Ep.*, 392 (1958).

This association, which is widespread through the British Isles, is particularly rich in species, the most important components being *Lecanora chlarotera*, *L. chlarona*, *L. pallida* and *Lecidella elaeochroma*. Other species frequently present include *Arthonia radiata*, *Buellia griseovirens*, *Graphis scripta*, *Lecanora carpinea*, *L. confusa*, *L. expallens*, *L. intumescens*, *Lecidea symmicta*, *Opegrapha atra*, *Parmelia exasperata*, *P. subaurifera*, *Pertusaria leioplaca*, *Phlyctis argena*, and *Rinodina sophodes*, together with the mosses *Ulota bruchii* (near the coast) and *U. phyllantha*. In slightly more shaded situations *Catillaria lightfootii*, *Haematomma elatinum*, *Lecanora jamesii*, *Lecidea tenebricosa* and *Phlyctis agelaea* enter into the association in the south and west. In southern Britain the association eventually gives way to the *Parmelietum revolutae* on deciduous trees as the macrolichens in that association tend to grow over the mosaics of crustose species characteristic of the *Lecanoretum subfuscae*. With increasing shade, in contrast, the *Lecanoretum subfuscae* merges into the *Graphidetum scriptae* to such an extent that stands which are difficult to assign conclusively to one or the other association are not uncommon. On *Prunus spinosa* and *Salix repens* in some coastal situations (e.g. Dungeness, Kent and Berry Head, South Devonshire), however, the association includes *Lecanora chlarotera*, *L. confusa*, *Lecidea symmicta* and *Lecidella elaeochroma* but is atypical in an abundance of *Caloplaca cerina;* this distinctive facies also occasionally includes *Ramalina baltica* indicating an affinity with the *Ramalinetum fastigiatae*.

Several allied associations may well occur in other parts of Europe but are in need of further investigation as Barkman's (1958) treatment of them is evidently unsatisfactory. The *Lecanoretum laevis* Barkm., in which *Lecanora chlarotera* is replaced by *L. laevis* and *Lecidella elaeochroma* replaced by *Lecidea euphorea*, is certainly distinct being characteristic of warmer areas and starting to appear in southern France; the *Lecanoretum laevis* is also a pioneer twig community and has a wide geographical range extending into, for example, the forests of the Azores and West Pakistan. Particularly in the more southerly parts of Europe (e.g. southern and western France), *Teloschistes chrysophthalmus*, *T. flavicans* and *Usnea flammea* appear to enter into communities which might be considered a regional variant of the *Lecanoretum subfuscae*.

All. 5. *Lecanorion variae*

Lecanorion variae Barkm., *Phytos. Ecol. Crypt. Ep.*, 362 (1958); type: *Psoretum ostreatae* Hil.—*Conizaeoidion* Laund., *in* Kettering and District Field Club, *First Fifty Years*, 92 (1956) [*nom. illegit.*]

This alliance is employed here for four associations, three of which are particularly tolerant of sulphur dioxide air pollution. These do not appear to be merely species-poor facies of other more pollution-sensitive assemblages of species as they occur from time to time in relatively unpolluted parts of the British Isles, particularly on lignum and the acid bark of coniferous trees.

The name of this alliance is misleading as it was based on the assumption that *Lecanora conizaeoides* (syn. *L. pityrea*) was simply a form of *L. varia*. Despite some recent Czechoslovakian studies implying the contrary (for details see Hawksworth, 1973b), this conclusion is taxonomically unacceptable as these taxa differ chemically, morphologically and in their respective distributions.

The *Caloplacetum phloginae*, placed in this alliance with some hesitation by Barkman (1958), is subsumed under the *Physcietum ascendentis* of the *Xanthorion* here (see p. 344).

Bacidietum chlorococcae LeBlanc

Can. J. Bot. **41**, 615 (1963) [as "Union à *Bacidia chlorococca*"].—Kalb, *Hoppea* [*Denkschr. Regensb. bot. Ges.*] **26**, 110 (1966) [as "*ass. nov.*"].

Although frequently overlooked, *Bacidia chlorococca* is widespread and very tolerant of air pollution (see Ahti and Vitikainen, 1974) entering far into urban areas in *Desmococcus* (*Pleurococcus* auct.)-dominated swards on deciduous trees whose barks have been impregnated with soot. This species-poor association often includes only *Bacidia chlorococca* and *Desmococcus* and appears even more pollution tolerant than the *Lecanoretum pityreae*. The *Bacidietum chlorococcae* tends to prefer somewhat more sheltered sites than the *Lecanoretum pityreae*, occurring, for example, on twigs and young branches of trees, and *Calluna* and *Ulex* stems; intermediates between the two associations are not uncommon, particularly on *Calluna*, where *Buellia pulverea* and *Micarea nitschkeana* may also occur.

Cyphelietum inquinantis Kalb

Diss. Bot., Lehre **9**, 34 (1970) [as "*Cyphelium*"].

This recently described association is characterized mainly by the abundance of *Cyphelium inquinans*, *Lecanora varia* and *Parmeliopsis ambigua*. It is possible that the *L. varia* reported by Kalb (1970) included *L. conizaeoides;* both these species occur in this association in the British Isles although not throughout its whole range. The *Cyphelietum inquinantis* is essentially a somewhat continental association which consequently has an eastern distribution in the British Isles. In lowland England it occurs mainly on decorticate horizontal rails, or on fence-posts, while in the Scottish Highlands (probably its original habitat), it is found occasionally on bark and decorticate pine wood in old pine forests where *Xylographa abietina* may be present in it as it is in the Alps. In south-east England *Parmeliopsis aleurites* is an additional component and elements of the *Pseudevernietum furfuraceae* may also enter into the association.

The *Cyphelietum inquinantis* has strong affinities with the *Calicion hyperelli* and is placed here, following Kalb (1970), because of the presence of *Lecanora conizaeoides* and (or) *L. varia*. The relationships of this association with the *Lecanoretum variae* Frey require investigation.

It is possible that the very rare eastern *Cyphelium notarisii* communities in Britain are a facies of this association (or perhaps the little understood *Cyphelietum tigillaris* Klem.) but too few data are available to make any reliable assessment as to their syntaxonomic status in Britain at the present time. As an example of the composition of this facies the following relevé may be cited: *Cyphelium notarisii* (2.3), *C. inquinans* (2.3), *Lecanora conizaeoides* (2.3), *L. varia* (1.3), *Ramalina fastigiata* (1.2) and *Xanthoria polycarpa* (2.2) [E. Sussex, Winchelsea Beach, Rye Harbour (51/9– – 1 – –): wooden post 250 m from the sea, incl. 90°, aspect 225°, 0·2 × 0·2 m, 17 April 1976, F.R.].

Lecanoretum pityreae Barkm.

Phytos. Ecol. Crypt. Ep., 363 (1958).

This is an extremely uniform association which characteristically includes only a single species, *Lecanora conizaeoides*, forming extensive pure continuous swards on well lit deciduous trees which can cover them from base to uppermost twigs. It is, however, optimally developed only where mean winter sulphur dioxide levels are in the range 55–150 µg m^{-3}, and in this range commonly also occurs on sandstones (particularly in the southern Pennines) and peaty moorland soils. Where air pollution levels are in the lower part of this range the association is also occasionally found on coniferous trees. The *Lecanoretum pityreae* appears to be pollution tolerant or a pollution exploiter rather than one needing pollution to survive as it does occur in parts of Britain almost entirely free of any air pollution; interestingly, in such areas it is almost entirely restricted to decorticate wood (especially fence-posts) and either twigs or acid-barked trees (e.g. birch and coniferous trees) and usually found in sites frequented by visitors from parts of the country where it is common. The association appears to have poor competitive ability where there is little air pollution and fails to encroach on communities already established.

The *Lecanoretum pityreae* is the commonest or only association on trees over large tracts of the British Isles extending northwards from the Thames valley, through the Midlands, and into northern Lancashire in the west and Northumberland in the east. More detailed information as to the behaviour of *Lecanora conizaeoides* in Britain is included in Hawksworth *et al.* (1973, 1974). In damper sites, the association frequently intergrades with the *Bacidietum chlorococcae* on twigs, *Calluna* stems etc.

Few other species ever form important components of this association. The most frequently encountered is perhaps *Lepraria incana*, especially in bark crevices or near the bases of trees, although plants of *Hypogymnia physodes*, *Parmelia saxatilis* and *Parmeliopsis ambigua*, for example, start to occur with increasing frequency towards the lower end of its sulphur dioxide optimal concentration range; these latter plants represent intergrades with the species-poor pollution variant of the *Pseudevernietum furfuraceae* mentioned below (p. 337). Where bark is nutrient enriched, particularly by soot, *Desmococcus vulgaris* also enters into this association. The *Lecanoretum pityreae* is often attacked by *Athelia arachnoidea* (Berk.) Jül. which can form large pale lesions in it.

Lecideetum ostreatae (Hil.) comb. nov.

Psoretum ostreatae Hil., *Spisy Přírod. Fac. Karl. Univ.* **41,** 99 (1925) [as "Association à *Psora ostreata*"].—*Lecideetum ostreatae* Schulz [Schulz-Korth], *Beih. Feddes Repert.* **67,** 48 (1931) [as "*Lecidea ostreata*-Ass."].—*Lecideion ostreatae* Laund., *in* Kettering and District Field Club, *First Fifty Years*, 94 (1956) [as "*Lecidion*"].—*Lecideetum scalaris* Kalb, *Diss. Bot.*, Lehre **9,** 37 (1970) [as "Hilitzer 1925"].

This air pollution-tolerant association is mainly confined (in moderately polluted areas) to the bark of deciduous trees in sheltered, well lit sites, but, like the *Lecanoretum pityreae*, will also occur on coniferous trees at lower sulphur dioxide levels. Although an essentially corticolous community it is also occasionally encountered on decorticate wood (particularly fence-posts and tree stumps), sandstones and brick in moderately polluted parts of the British Isles. *Lecidea scalaris* predominates in the association and is sometimes the only species present. *Cladonia coniocraea*, *C. fimbriata*, *Lecanora conizaeoides* and *Lecidea granulosa* may also occur but generally have low cover values. The *Lecideetum ostreatae* also occurs on charred trunks; *Lecidea scalaris* often fruits in this habitat and *Lecidea friesii* and *Toninia carodocensis* may also be components of this specialized facies.

All. 6. *Lobarion pulmonariae*

Lobarion pulmonariae Ochsn., *Jahrb. St. Gall. naturw. Ges.* **63,** 64 (1928); type: *Lobarietum pulmonariae* Hil. [syn. *Nephrometum laevigatae* Barkm.].

The *Lobarion pulmonariae* (Fig. 3) is composed mainly of large foliose lichens and robust bryophytes and appears to be the natural forest climax community on mature hardwood trees with barks of pH 5·0–6·0 in western Europe outside areas with Mediterranean climates. It is now very much fragmented in distribution due to the felling and management of primeval forests, drainage and various forms of pollution. In drier areas it tends to be confined either to sheltered glades in more open forests where there is more light, or to the upper boughs of trees. Outside western

Fig. 3. *Lobarion pulmonariae* on oak. Predominant species: *Lobaria amplissima* (with cephalodia) and *L. pulmonaria* (Argyllshire: Ardnamurchan Peninsula, Camasine, 1967, F. Rose).

Scotland and Brittany it is largely restricted to mature or older tree trunks but in these two regions it will occur on relatively young trees and even old *Corylus* bushes. In parts of Scotland with high humidities it is even able to colonize planted trees or avenues while elsewhere it has acquired an essentially relict status; i.e. become confined to primary woodland relics with old trees. The *Lobarion* is rarely well developed in coppice woodlands unless numerous old standard trees are present but persists in many sites in England on sheltered ancient parkland *Fraxinus*, *Quercus* and *Ulmus* trees. In drier areas (e.g. Sussex, north-east England, eastern Scotland) it is usually best developed in forest relics in sheltered humid valley floor sites or on trees close to rivers in gorges.

Today the alliance is unable to spread in drier districts into woodlands less than 200 years old unless these adjoin, or have adjoined, ancient woodlands from which dispersal could take place when the trees reached sufficient maturity. The ability to colonize younger trees and exist in more shaded sites in wetter climatic zones may be due to (1) a higher annual growth rate (longer growing season) of the macrolichens concerned in the constantly humid environments, and (2) photosynthesis being able to occur for longer periods in moist conditions and thus compensating for lower light intensities.

There is little doubt, to judge from literature sources and field evidence from relict sites, that this alliance was formerly quite general as the epiphytic community of the mature trees of lowland and lower montane forests of western Europe outside areas of intense Mediterranean summer drought (even in these areas it does, however, seem to have existed in sheltered, locally more humid, sites). In Mediterranean forest areas it is normally replaced by communities of the *Xanthorion* alliance. It may always, as is the case at the present time, have been only well developed (outside more oceanic areas) on better lit trees at the edges of glades. In the central European mountains and the Pyrenees, the *Lobarion* is also developed locally in open forest stands of coniferous trees (particularly *Abies pectinata* which has a less acid and more water- and nutrient-retentive bark than most other conifers). Evidence for the former paramount importance of the *Lobarion* in the European hardwood forests is provided by (1) the large number of species present in it, most of which today show highly disjunct and presumably fragmented distributions, (2) the relatively high constancy of many of these species over the whole geographical range of the alliance (which is by no means purely oceanic), (3) the number of faithful species it contains (i.e. species only present in communities of this alliance), and (4) its occurrence in forest situations on a very wide range of deciduous trees (only on *Alnus* which has a very acid bark, pH 4·5, is it really rare).

The synsystematics of this alliance are extremely complex. Several distinct associations occur within it but the delimitation and nomenclature of these are currently unclear. As we propose to deal with the *Lobarion* in some detail in a future publication only a brief discussion of the principal noda of the alliance seen in Britain is included here. Attention is, however, drawn to the detailed discussions of it in the southern Black Forest and Canary Islands by Wirth (1968) and Klement (1965), respectively.

The "typical" western European nodum of this alliance, termed the *Nephrometum lusitanicae* by Barkman (1958), comprises a large number of species, the most constantly occurring of which are listed in Table II. A particular feature of the nodum is the abundance of bryophytes which form a substrate on which the macrolichens grow. Even within the western European range of this nodum, many stands attributable to it lack many, or even all, of the macrolichens characteristic of it as a result of the effects of man. Such species-poor communities probably always occurred on more shaded trees, and as a pre-climax sere to the full species-rich nodum; these may be considered a "pre-*Lobarion*" community and usually comprise the species indicated by ^{c} in Table II. In many parts of southern England, and also in Normandy and Picardy in northern France, the *Lobarion* is in large measure found only as this pre-*Lobarion* nodum. Even

Table II.

Some components of the Lobarion pulmonariae *in the typical western European facies of the alliance* (= Nephrometum lusitanicae *Barkm.*).

Arthonia didyma[a c]	*Parmeliella atlantica*
Acrocordia gemmata[a c]	*P. coralloides* aggr.[a]
Bacidia affinis[a]	*P. plumbea*[a]
B. biatorina[a]	*Peltigera collina*[a]
Biatorella ochrophora[a]	*P. horizontalis*[a]
Catillaria atropurpurea[a c]	*P. praetextata*[b]
C. sphaeroides[a]	*Pertusaria hemisphaerica*[c]
Dimerella lutea[a c]	*P. hymenea*[c]
Evernia prunastri[b]	*P. pertusa*
Haematomma elatinum	*P. velata*[a c]
Lecanora quercicola[a]	*Porina coralloidea*[a]
Lecidea cinnabarina	*P. hibernica*[a]
Leptogium lichenoides[b]	*P. leptalea*
L. teretiusculum	*Ramalina farinacea*[b]
Lithographa dendrographa[c]	*Rinodina roboris*[b]
Lobaria amplissima[a]	*Sticta limbata*[a]
L. laetevirens[a]	*S. sylvatica*[a]
L. pulmonaria[a]	*Thelopsis rubella*[a c]
L. scrobiculata[a]	*Thelotrema lepadinum*[c]
Nephroma laevigatum[a]	
Normandina pulchella[b c]	Bryophytes
Opegrapha sorediifera[a c]	*Antitrichia curtipendula*[a]
Pachyphiale cornea[a c]	*Camptothecium sericeum*[b c]
Pannaria mediterranea[a]	*Frullania fragillifolia*[a]
P. pityrea[a]	*F. tamarisci*[b c]
P. rubiginosa[a]	*Isothecium myosuroides*[a c]
Parmelia crinita[a c]	*Neckera complanata*[b c]
P. glabratula[b]	*N. pumila*[b c]
P. reddenda[a c]	*Orthotrichum lyellii*[b c]
P. revoluta[b c]	*Pterogonium gracile*[a]
P. saxatilis[b]	*Zygodon baumgartneri*[a c]

[a] = species more or less faithful to this nodum.
[b] = companion species occurring with a high constancy.
[c] = species also of the "pre-*Lobarion*" pioneer community (see p. 324).

in the New Forest (see Rose and James, 1974) and parts of south-west England the pre-*Lobarion* is much more frequent than noda closely comparable to the *Nephrometum lusitanicae*.

In central, eastern and southern Scotland and parts of northern England, away from the more humid western and southern counties, the *Lobarion* tends to be relatively species poor. Often only *Lobaria pulmonaria* (and sometimes *L. scrobiculata*) represent that genus, *Sticta* species are usually absent, and the cyanophilous species may be represented only by *Parmeliella plumbea* (or *Pannaria rubiginosa* in north-east Scotland, e.g. Darnaway Forest). The same pattern of impoverishment appears to occur in this alliance in the forests of north, central and eastern Europe (e.g. Denmark, Poland).

Conversely, in south-west England, North Wales and, more particularly, western Scotland, Brittany and the French Pyrenees, additional species absent further east become locally common (e.g. *Lopadium pezizoides*, *Parmeliella atlantica*, *Pseudocyphellaria crocata*, *P. intricata*, *Sticta canariensis* (*S. dufourii* morphotype), *S. fuliginosa*; in a few areas of western and southern England and Wales, also *Rinodina isidioides*). An especially species-rich facies of this alliance occurs in moist valley-bottom woods in sheltered lowland situations in western Scotland, particularly on inclined boughs of old *Corylus* or *Salix atrocinerea* in wet carr woodland; this includes fewer crustose lichens but all four British *Lobaria* species (all fertile in western Scotland; all but *L. laetevirens* rarely so elsewhere), *Pannaria*, *Parmeliella* and *Pseudocyphellaria* species and, occasionally, also *Heterodermia obscurata*. In such constantly wet conditions *Cetrelia olivetorum* s.l., *Menegazzia terebrata*, *Parmelia endochlora*, *P. laevigata* and *P. taylorensis* (all species usually found only in the *Parmelion laevigatae*, see p. 327) may also occur, together with *Leptogium* species (particularly *L. burgessii*), *Parmelia sinuosa* (on twigs) and *Ulota* species. This nodum, which is very distinctive when well developed, shows affinities with both the *Parmelion laevigatae* (although on far less acid bark) and, through *Lecanora jamesii* communities, to the *Graphidion scriptae* (generally of smooth-barked twigs and small branches). In such sites in western Scotland and south-west Ireland, small twigs and branches often support *Graphidion* with *Lecanactis homalotropum* and *Thelotrema subtile* (see p. 315).

A euoceanic moss-dominated facies of the *Lobarion* in south-west Ireland characteristically includes *Leptogium brebissonii*, *L. burgessii*, *Lobaria pulmonaria*, *Porina hibernica*, *Pseudocyphellaria lacerata* and *Sticta canariensis* (the *S. dufourii* morphotype occurring in deep shade).

In western Ireland (particularly in the Burren, Co. Clare; e.g. Poulavallan, Den of Clab) a facies of the *Lobarion* occurs on *Corylus* in relict *Corylus* woodlands of the limestone pavements. In this community *Lobaria* species are now lacking (currently restricted to mixed *Quercus* forest areas in Ireland) but *Pannaria*, *Parmeliella* and *Sticta* species are abundant and *Leptogium burgessii*, *Normandina pulchella* and *Usnea* species occur on larger stems. *Thelotrema subtile* is present in these communities in Connemara and south-west Ireland.

Fraxinus–Ulmus forest, especially when on basic soils over limestone, basic volcanic rocks, or on rich alluvial soils, supports a further distinctive community of the *Lobarion* in which, in addition to the taxa marked [a] in Table II, there is a particularly rich development of species of *Pannaria* (including *P. sampaiana*), *Parmeliella*, *Collema* (*C. fasciculare, C. furfuraceum, C. nigricans* and *C. subflaccidum*) and *Leptogium* (*L. azureum, L. burgessii, L. cyanescens, L. hibernicum* and *L. saturninum*); *Lithographa dendrographa* may also belong here. This community is now well developed only in a few areas of western Scotland where it occurs both in open forest and on isolated well lit old trees (e.g. Loch Melfort; Ellary Woods, Loch Caolisport; Rassal Wood, Kishorn; Mull; Sunart; Loch Arkaig); fragments do, however, persist in southern (e.g. New Forest, Cranborne Chase) and western England, Wales, the Lake District (e.g. Gowbarrow) and parts of western Ireland. Also, as members of this community there may be the very rare British species *Arctomia delicatula* (euoceanic), *Collema occultatum* and *Pannaria ignobilis* (a valley species of the central Highlands, often found on trees which have *Bryoria capillaris* on their twigs).

All. 7. *Parmelion laevigatae*

Parmelion laevigatae all. nov.; type: *Parmelietum laevigatae* P. James *et al.* (monotype).

This new alliance contains a single association:

Parmelietum laevigatae ass. nov. (Table III, Fig. 4)

This is a well marked community characteristic and general on the western side of the British Isles, including parts of south-west Ireland, and is also present in western Brittany. It is largely restricted to *Betula* and *Quercus* (although it can be encountered on *Alnus* and mossy rocks) and confined to exposed, often upland, woodland sites with high rainfall (127–229 cm year^{-1}) and with at least 180 wet days year^{-1}. The only sites for this community outside the 180 wet-day isoline (see Ratcliffe, 1968; Coppins, 1976) are in north Pembrokeshire but, were more meteorological data available, we suspect they also would be within this isoline. The *Parmelietum laevigatae* is essentially an association of well lit hardwood forests, and appears to be correlated with exposure to heavy rain which leaches the upper horizons of the bark. The pH of bark supporting this community is

FIG. 4. *Parmelietum laevigatae* on mossy boulder. Predominant species: *Parmelia laevigata* and *Sphaerophorus globosus* (Argyllshire: Salen, 1967, P. W. James).

---

1. Argyllshire, Loch Sunart, Camasine (17/7 - - 6 - -): *Quercus* bough, 0·2 m diam, ± horizontal, 1 × 0·2 m, cover 90%, September 1970, F.R.
2. Gwynedd (Merioneth), Coed Ganllwyd, below Rhaedr-ddu (23/7 - - 2 - -): 3 adjacent *Betula pubescens* trunks in wood, *c.* 15 cm diam, incl. 70°, aspect WSW, 1 × 0·4 m, cover 75%, 1 April 1975, F.R.
3. Gwynedd (Merioneth), Coed Maen Ymenyn (23/848355): siliceous rock boulder, incl. 80°, aspect SE, 1 × 1 m, cover 70%, 27 March 1975, F.R.
4. Cumbria, Borrowdale, Castle Crag Woods (35/2 - - 1 - -): *Quercus petraea*, 1 m diam, incl. 85°, aspect SE, 1 × 0·5 m, cover 100%, 27 July 1971, F.R.; *type record*.
5. Cumbria, Eskdale, Dalegarth Woods (34/2 - - 9 - -): *Quercus* in wooded ghyll below Stanley Force, 0·8 m diam, incl. 90°, aspect N, 1 × 1 m, cover 90%, 25 July 1970, F.R.
6. Cumbria, Borrowdale, Thorneythwaite Wood, (35/2 - - 1 - -): *Quercus petraea* in woodland, 0·6 m diam, incl. 90°, 1 × 0·5 m, cover 70%, 25 July 1971, F.R.
7. Cumbria, Borrowdale, Seatoller, Low Stile Wood (35/2 - - 1 - -): *Quercus* (?) in woodland, 0·5 m diam, incl. 90°, aspect SW, 1 × 0·5 m, cover 90%, 26 July 1971, F.R.
8. Cumbria, Borrowdale, Seatoller, Low Stile Wood (35/2 - - 1 - -): *Quercus petraea* in open woodland, 0·5 m diam, incl. 90°, aspect SE, 0·5 × 0·5 m, cover n.a., 27 July 1971, F.R.

Table III.
Parmelietum laevigatae ass. nov.

Species	\multicolumn{8}{c}{Stands}
	1	2	3	4	5	6	7	8
Cetrelia olivetorum s.l.	–	–	–	–	1.2	–	+.2	–
Cladonia chlorophaea	1.2	+.2	–	+.2	–	–	–	–
C. coniocraea	–	–	–	2.2	+.2	–	2.2	–
C. ochrochlora	1.2	–	–	–	–	–	–	–
Cornicularia aculeata	–	–	+.2	–	–	–	–	–
Evernia prunastri	–	–	–	–	–	–	+.0	–
Hypogymnia physodes	–	+.2	–	1.2	+.0	–	1.2	1.2
H. tubulosa	1.2	+.0	–	–	–	–	–	–
Menegazzia terebrata	2.3	1.2	–	–	1.3	–	–	–
Micarea sp.	–	–	–	–	–	–	–	3.4
Mycoblastus sanguinarius	–	–	–	1.3	–	–	–	1.3
Ochrolechia androgyna	–	+.2	–	3.2	3.3	–	4.4	2.3
O. tartarea	–	–	–	2.4	–	–	–	–
Parmelia crinita	2.2	–	–	–	–	–	–	–
P. glabratula	–	–	–	–	–	–	+.0	+.2
P. laevigata	2.2	4.3	–	3.2	+.2	3.3	–	1.3
P. saxatilis	–	1.3	+.2	–	–	–	1.2	2.2
P. taylorensis	2.2	–	2.2	–	–	2.3	+.2	2.3
Pertusaria amara	–	–	–	–	–	–	–	+.2
Platismatia glauca	2.3	1.3	+.2	1.2	1.2	–	–	–
Sphaerophorus globosus	–	+.2	+.2	+.0	–	–	–	–
Usnea inflata	–	1.2	–	–	–	–	–	–
U. subfloridana	2.2	–	–	–	+.2	–	–	–
Barbilophozia attenuata	–	–	–	–	2.3	–	–	–
B. floerkii	–	–	2.3	–	–	–	–	–
Bazzania trilobata	–	–	–	–	1.3	–	–	–
Dicranum fuscescens	–	–	–	+.2	–	–	–	–
D. scoparium	–	+.2	–	1.2	–	–	1.2	–
Diplophyllum albicans	–	–	3.3	–	–	–	–	–
Frullania tamarisci	–	–	–	–	–	–	1.2	–
Hypnum cupressiforme	3.3	–	–	–	–	–	2.2	+.2
Isothecium myosuroides	–	–	1.2	–	–	–	2.3	–
Lepidozia reptans	–	–	–	–	1.2	–	–	–
Lophozia ventricosa	–	–	–	+.2	–	–	–	–
Plagiochila punctata	–	–	–	–	2.3	–	–	–
Polytrichum formosum	–	–	+.0	–	–	–	–	–
Rhacomitrium heterostichum	–	–	1.2	–	–	–	–	–
Scapania gracilis	–	–	1.2	+.2	2.3	–	–	–
Algal crust	–	–	–	3.3	–	–	–	–

in the range pH 3·75–4·60, contrasting with the *Parmelion perlatae* and *Lobarion pulmonariae* with bark pH values almost always over pH 5·0 and, in the case of the latter, sometimes over pH 6·0.

The *Parmelietum laevigatae* is particularly characteristic of the following areas in Britain:

1. High-level oak woods round the edges of Bodmin Moor, Cornwall, at 650–750 ft (*c.* 220–230 m),
2. Exposed upland oak woods of Dartmoor, Devonshire (e.g. Black Tor Copse, Wistmans Wood), at 1200–1450 ft (*c.* 360–440 m),
3. Upper parts of valley oak woods on Exmoor and the Quantock Hills, Somerset,
4. Oak–birch woodland in high rainfall areas of the Lake District,
5. Upland oak woods in western Wales, at 600–1500 ft (*c.* 200–450 m), and
6. Generally in more exposed oak–birch woodlands in western Scotland, at 150–1200 ft (*c.* 50–360 m), as far north as Wester Ross.

Altitude appears to be less important than exposure to rain-bearing winds in determining its occurrence. Oak woods below the indicated levels, if sufficiently undisturbed, support the *Lobarion pulmonariae*. Transitions between these communities are very abrupt in southern Britain (e.g. on Dartmoor and Exmoor), but much more gradual in western Scotland.

Fragmentary forms of this association occur in valley bottoms throughout western Britain on the naturally more acid bark of *Alnus*. This association is primarily characterized by *Mycoblastus sanguinarius*, *Ochrolechia androgyna*, *O. tartarea*, *Parmelia laevigata* and *P. taylorensis*. Other species largely faithful to it include *Bryoria smithii*, *Cetrelia olivetorum* s.l., *Menegazzia terebrata*, *Parmelia endochlora* and *Pertusaria ophthalmiza* (in Scotland). *Sphaerophorus* species are frequently important components, *Cladonia* species are often present, and the community is rich in calcifuge bryophytes, particularly *Dicranum scoparium*, *Hypnum cupressiforme* var. *filiforme*, *Isothecium myosuroides*, *Plagiochila punctata*, *P. spinulosa* and *Scapania gracilis*.

Where this association occurs on mossy rocks and boulders it is usually in those upland woods where it is also present on the trees.

All. 8. *Parmelion perlatae*

Parmelion perlatae nom. nov.—*Trichoterion* Laund., *Lond. Nat.* **37**, 73 (1958) [*nom. illegit.*].—*Parmelion caperatae* Barkm., *Phytos. Ecol. Crypt. Ep.*, 450 (1959) [as "suballiance"]; type: *Parmelietum revolutae* Klem.—Non *Parmelion caperatae* Felf., *Acta geobot. Hung.* **4**, 55 (1941); type: *Parmelietum caperatae* Felf.—

? Non *Parmelion perlatae* Follm., *Ber. dtsch. bot. Ges.* **80,** 201 (1967) [*nom. nud.*]; type not stated.

This alliance, in its various facies, is the characteristic community of well lit mature deciduous trees in areas of little or no air pollution in lowland Britain. In north-eastern England, central Wales and central and eastern Scotland the alliance is replaced by the *Pseudevernion furfuraceae* as the dominant epiphytic community of well lit trees.

The *Parmelion perlatae* is characterized by the abundance of *Parmelia caperata* and *P. perlata*. A single very variable association within it is recognized here as present in the British Isles.

Parmelietum revolutae Klem. (Table IV)

Beih. Feddes Repert. **135,** 163 (1955) [as "Almborn 1948"].—*Parmelietum revolutae* var. *parmeliosum laetevirentis* Barkm., *Phytos. Ecol. Crypt. Ep.*, 452 (1958) [*nom. superfl.*].—*Parmelietum revolutae* var. *caperatosum* Barkm., *Phytos. Ecol. Crypt. Ep.*, 454 (1958).—*Parmelietum cervicornis* Duvign., *Bull. Soc. r. bot. Belg.* **74,** 47 (1942) [*nom. illegit.*].—*Parmelietum subauriferae* Duvign., *Bull. Soc. r. bot. Belg.* **74,** 47 (1942).—*Parmelietum trichotero-scortea* Barkm., *Phytos. Ecol. Crypt. Ep.*, 450 (1958).—Non *Parmelietum caperatae* Felf., *Acta geobot. Hung.* **4,** 55 (1941).

This association is best developed on the trunks and ascending boughs of *Fraxinus, Quercus, Larix* (in Ireland) and other rough-barked trees of 0·3–1·0 m diam in parklands, by minor roads, in pastures and more open woodlands; it is restricted to upper well lit parts of trees in dense woodlands where it may occur above a zone of either *Lobarion* or *Graphidion*. The association has a very wide phorophyte range, however, and will also occur on the smooth bark of *Fagus, Acer* and *Ulmus* species where the bark has not been unduly enriched by dust or animal matter. Sulphur dioxide pollution and agricultural chemicals impoverish it to varying degrees. With increasing bark acidity it tends to grade into the species-poor facies of the *Pseudevernietum furfuraceae* (see p. 335), whilst with decreasing bark acidity the *Buellietum punctiformis* or algal-dominated communities may be produced. Although not usually encountered on coniferous trees, in the most unpolluted parts of southern and western Britain (particularly near the sea), it is occasionally found on mature well lit trunks of *Picea abies, Pinus sylvestris* and *Larix*. Most pH values from bark beneath this association fall within the range pH 5·0–5·5.

The *Parmelietum revolutae* is best developed in southern and western lowland areas, occurring today in the eastern and northern parts of East Anglia, throughout the counties south of the Thames, west Midlands, Welsh lowlands (especially near the coast), Lake District, lowland coastal

Table IV.

Parmelietum revolutae Klem.

Species	_ Stands _							
	1	2	3	4	5	6	7	8
Candelariella vitellina	–	–	–	2.3	–	–	–	–
Catillaria griffithii	–	–	–	–	+.2	–	–	–
Evernia prunastri	–	–	2.3	–	1.2	1.2	1.2	2.2
Hypogymnia physodes	–	–	+.0	–	+.1	–	2.2	1.2
Lecanora chlarotera	–	–	–	–	+.2	+.2	1.2	–
L. conizaeoides	–	2.2	–	–	–	–	–	–
L. expallens	–	1.2	–	3.2	–	+.2	3.2	–
Lecidea quernea	–	–	+.0	–	4.3	3.3	–	–
Lecidella elaeochroma	–	–	–	–	–	+.0	–	1.2
Lepraria candelaris	–	–	–	–	1.2	–	–	–
L. incana	+.2	–	–	–	–	+.0	–	–
Ochrolechia yasudae	–	–	–	–	2.2	–	2.4	–
Parmelia caperata	2.3	1.2	1.3	3.3	1.2	2.3	1.4	1.2
P. glabratula	+.2	1.2	2.3	1.3	–	+.2	2.2	–
P. perlata	3.4	–	–	–	–	–	–	2.3
P. reticulata	–	+.2	–	–	1.2	–	–	–
P. revoluta	–	1.2	1.3	–	–	–	–	–
P. saxatilis	3.3	–	1.3	–	–	–	–	–
P. soredians	–	2.3	–	–	–	–	–	–
P. subrudecta	–	1.2	–	–	–	2.2	1.3	+.0
P. sulcata	–	3.2	2.3	1.3	–	4.3	1.2	+.0
Pertusaria albescens	–	–	–	+.1	–	–	–	–
var. *corallina*	–	–	–	3.2	–	–	–	–
P. amara	1.3	–	–	–	3.2	1.3	+.2	–
P. pertusa	1.2	1.3	–	–	1.2	2.3	–	–
Phlyctis argena	–	1.2	–	–	–	1.3	–	–
Ramalina farinacea	+.0	2.2	+.0	–	1.1	+.0	+.2	+.0
R. fastigiata	–	–	–	–	–	+.0	–	–
Hysterium angustatum	–	–	–	–	–	+.0	–	–
Cololejeunea minutissimus	–	–	–	–	–	–	–	1.2
Frullania dilatata	–	–	–	–	–	+.2	–	3.3
Hypnum cupressiforme	1.2	–	–	+.2	1.3	1.3	–	4.3
Metzgeria furcata	–	–	–	–	–	–	–	1.2
M. fruticulosa	–	–	–	–	–	–	–	+.0
Ulota phyllantha	–	–	–	–	–	–	–	2.2
Zygodon viridissimus	–	–	–	–	–	–	–	+.2

western Scotland north to Skye and Applecross, and parts of Angus, Fife and Perthshire. Its replacement in the north and east of the British Isles (and also in the central Welsh uplands) by the *Pseudevernietum furfuraceae* as the main community on well lit hardwood trees appears to be due to climatic factors, though whether low winter temperatures or lack of summer sunshine and warmth are the limiting factors is unclear. In high-rainfall upland areas the *Parmelietum revolutae* gives way to the *Parmelietum laevigatae*.

The association similarly gives way to the *Pseudevernietum furfuraceae* in northern continental areas (e.g. Denmark, north Germany), but is very general in France north of the Mediterranean area.

The species most commonly encountered in this association are *Catillaria griffithii, Evernia prunastri, Hypogymnia physodes, Lecanora chlarotera, L. expallens, Lecidea quernea, Ochrolechia yasudae, Parmelia caperata, P. glabratula* ssp. *glabratula, P. perlata, P. saxatilis, P. subaurifera, P. subrudecta, P. sulcata, Pertusaria albescens, P. amara, P. hemisphaerica, P. pertusa, Phlyctis argena* and *Ramalina farinacea*. Additional important components in the south and/or east include *Parmelia borreri, P. reticulata, P. soredians, Pertusaria coccodes, P. flavida* and *Rinodina roboris*. *Parmelia perlata* is abundant only where air pollution levels are extremely low (e.g. relevés from Slapton, S. Devon, in Hawksworth, 1972a). Bryophytes are generally poorly represented in this association, the most frequently

---

1. Cumbria, Borrowdale, Manesty Park (35/2 - - 1 - -): *Quercus* at edge of wood, 0·8 m diam, incl. 90°, aspect SW, 0·5 × 0·5 m, cover 75%, 4 June 1969, F.R., D.L.H. and B.J.C.
2. E. Kent, Hothfield Park (51/9 - - 4 - -): old *Quercus*, 1·0 m diam, incl. 90°, asp. S, 1·0 × 1·0 m, cover 60%, 7 April 1969, F.R.
3. E. Kent, near Ashford, Willesborough Lees (61/0 - - 4 - -): *Quercus* in open pasture, 1·0 m diam, incl. 90°, aspect SW, 1·0 × 1·0 m, cover 70%, 24 September, 1968, F.R.
4. E. Kent, Maidstone, Mote Park (51/7 - - 5 - -): *Fraxinus*, 0·8 m diam, incl. 90°, aspect not indicated, 0·5 × 0·5 m, cover 80%, 10 September 1968, F.R.
5. E. Kent, Otham, Gore Court (51/7 - - 5 - -): *Fraxinus* in parkland, 1·0 m diam, incl. 90°, aspect SE, 2·0 × 0·5 m, cover 90%, 6 September 1968, F.R.
6. W. Kent, Plaxtol (51/6 - - 5 - -): *Fraxinus* in valley, 0·6 m diam, incl. 90°, aspect S, 1·0 × 2·0 m, cover 90%, September 1967, F.R.
7. Surrey, Burstow, Westlands Farm (51/3 - - 4 - -): *Quercus robur*, 0·7 m diam, incl. 90°, aspect S, 1·0 × 0·5 m, cover 80%, July 1968, F.R.
8. E. Sussex, Fairlight undercliff (51/8 - - 1 - -): *Quercus*, 0·3 m diam, incl. not indicated, aspect S, 1·0 × 0·2 m, cover 90%, 1 March 1969, F.R. and P.W.J.

encountered being *Dicranoweissia cirrata*, *Hypnum cupressiforme* and *Orthotrichum lyellii*.

The *Parmelietum revolutae* is particularly rich in species and many others, in addition to those mentioned above, occur from time to time in varying amounts. Nevertheless, only two species approach being strictly faithful to it in Britain: *Parmelia soredians* (which also rarely occurs on decorticate wood and stonework) and *P. reticulata* (most frequent in the east and south-east). It is perhaps unfortunate that the name of this association is based on that of *P. revoluta* as that species is not particularly common in it, tending to be most frequent in either slightly shaded sites or on the smoother and less water-retentive bark of *Alnus* and *Fagus* in southern woodlands. In Kent and East Anglia *P. acetabulum*, more characteristically a species of the *Xanthorion*, also occurs in the *Parmelietum revolutae* as it does in parts of Belgium and France; this phenomenon may be correlated with climatic factors approaching those optimal for this particular species (i.e. generally drier and warmer summers).

The original record of the *Parmelietum caperatae* Felf., described from Hungary, is even more transitional between the *P. revolutae* and the *Xanthorion* as it included not only *Parmelia acetabulum* but also *Anaptychia ciliaris*, *Physconia pulverulenta* and *Xanthoria parietina*. Felföldy's name is thus not taken up for the *Parmelietum revolutae* here; his community may be referred to the *Parmelietum acetabulae* var. *parmeliosum caperatae* Ochsn. (Ochsner, 1928: 62).

All. 9. *Pseudevernion furfuraceae*

Pseudevernion furfuraceae (Barkm.) comb. nov.—*Parmelion furfuraceae* Barkm., *Phytos. Ecol. Crypt. Ep.*, 456 (1958) [as "suballiance"], basionym; type: *Parmelietum furfuraceae* Hil.—*Parmelion saxatilis* Barkm., *Phytos. Ecol. Crypt. Ep.*, 450 (1958); type: *Parmelietum furfuraceae* Hil.; nom. illegit. [non *Parmelion saxatilis* Klem., *Ber. bayer. bot. Ges.* **28**, 257 (1950); type: *Parmelietum conspersae* Klem.].—*Physodion* Waldh., *K. svenska VetenskAkad. Avh. natur.* **4**, 90 (1944); type: "*Parmelia physodis*-förbundet DR. 1942".—*Parmeliopsidion ambiguae* Barkm., *Phytos. Ecol. Crypt. Ep.*, 466 (1958) [as "suballiance"]; type: *Parmeliopsidetum ambiguae* "Hil."—See also Follmann (1974).

This alliance, most frequently termed the "*Physodion*" by British authors since Laundon (1956), is essentially a northern, more acidic substrate-requiring counterpart of the *Parmelion perlatae* in the British Isles. It is characteristic of trees with moderately acidic barks in well lit situations and is widespread throughout large areas of northern England, central Wales and central and eastern Scotland. Under pollution stress, where bark tends to become somewhat acidified, it occurs further south as species-poor communities. While in Scotland it is widespread on coniferous and deciduous trees, in areas not subject to pollution stress in southern England it is mainly restricted to coniferous trees, birch and fence-posts. These factors suggest it tends to prefer more acidic substrates than the *Parmelion perlatae*. It largely replaces the *Parmelion perlatae* in lowland Scandinavia east of western Norway.

The *Pseudevernion furfuraceae* is not exclusively corticolous, however, and has similar associations and compositions when growing on siliceous rocks. It is found throughout the British Isles on acidic rocks wherever these occur. The characteristic species of the alliance are *Bryoria fuscescens*, *Cetraria chlorophylla*, *Hypogymnia physodes*, *H. tubulosa*, *Ochrolechia androgyna*, *Parmelia saxatilis*, *P. sulcata*, *Parmeliopsis ambigua*, *Platismatia glauca* and *Pseudevernia furfuracea*. Two associations in this alliance may be recognized in the British Isles at the present time.

Pseudevernietum furfuraceae (Hil.) Kalb (Table V)

Diss. Bot., Lehre **9,** 59 (1970).—*Parmelietum furfuraceae* Hil., *Spisy Přírod. Fac. Karl. Univ.* **41,** 122 (1925) [as "Association à *Parmelia furfuracea*"].— *Parmelietum furfuraceae-physodes* Frey & Ochsn., *Arvernia* **2,** 78 (1926).—*Parmelietum saxatilis* Hil., *op. cit.* **41,** 143 (1925) [as "Association à *Parmelia saxatilis*"].—*Parmelietum sulcatae* Hil., *op. cit.* **41,** 151 (1925) [as "Association à *Parmelia sulcata*"].—*Hypogymnio physodis–Parmelietum saxatilis* Wirth, *Diss. Bot., Lehre* **17,** 211 (1972) [as "(Hil. 1927) nom. nov."].—*Physodeto-sulcatetum* DR., *Svensk bot. Tidskr.* **39,** 148 (1945).—*Physodetum* Almb., *K. svenska VetenskAkad. Avh. natur.* **11,** 39 (1955).—*Parmeliopsidetum ambiguae* subass. *platismatietosum glaucae* Kalb, *Hoppea* [*Denkschr. Regensb. bot. Ges.*] **30,** 84 (1972).—See also Follmann (1974).

This association is very variable and some authors have recognized numerous subassociations and variants within it (e.g. Barkman, 1958; Fabizewski, 1968; Kalb, 1970); Hawksworth (1969) noted the presence of several of these in Derbyshire. Essentially it has a composition as for the alliance but can be simplified under air pollution stress to *Hypogymnia physodes–Parmelia saxatilis–Platismatia glauca* communities (sometimes with *Pseudevernia furfuracea*) such as are particularly widespread and luxuriant on trees and siliceous rocks over large areas of central and northern England. *Bryoria fuscescens* is particularly common in the north and in upland areas of the south-west (e.g. Dartmoor). Facies rich in *Parmeliopsis ambigua* are also not uncommon under moderate pollution stress. A valuable résumé of the syntaxonomy of this association and its composition is provided by Follmann (1974).

In the south, *Evernia prunastri*, *Lecanora pallida*, *Pertusaria amara* and

Table V.

Pseudevernietum furfuraceae (Hil.) Kalb.

Species	1	2	3	4	Stand 5	6	7	8	9	10
Bryoria fuscescens	4	6	6	4	9	7	4	7	–	–
Calicium viride	–	–	–	+	–	–	–	–	–	–
Cetraria chlorophylla	+	–	–	7	–	4	–	+	–	–
Evernia prunastri	+	–	–	+	–	–	–	–	–	–
Huilia macrocarpa	–	–	–	–	–	–	–	+	–	–
Hypogymnia physodes	4	6	6	2	+	4	3	+	5	2
H. tubulosa	+	–	–	–	–	+	+	–	–	–
Lecanora conizaeoides	–	–	–	–	–	+	+	–	7	–
L. expallens	+	–	–	+	+	–	–	–	–	–
L. intricata var. soralifera	–	–	–	–	–	–	+	–	2	–
L. scalaris	–	–	–	+	–	–	–	–	–	–
Lepraria incana	4	–	–	–	–	–	–	–	–	–
Mycoblastus sanguinarius	–	–	–	–	–	–	+	–	–	–
Ochrolechia androgyna	–	–	–	+	–	–	–	–	–	–
O. turneri	6	–	4	+	–	–	–	–	–	–
Parmelia glabratula	–	–	–	+	–	–	–	–	–	–
P. saxatilis	+	–	–	+	4	–	6	7	5	7
P. sulcata	–	–	–	3	+	–	–	–	–	–
Parmeliopsis hyperopta	–	–	+	–	–	–	–	–	–	–
P. ambigua	–	–	4	–	+	–	–	–	–	–
Phlyctis argena	–	–	–	4	–	–	–	–	–	–
Platismatia glauca	–	–	4	4	4	+	4	–	–	6
Pseudevernia furfuracea	–	5	+	+	+	–	+	–	8	7
Usnea hirta	–	4	–	–	–	–	–	–	–	–
U. subfloridana	+	–	+	+	–	+	–	–	–	–

1. E. Inverness, Rothiemurchus Forest, Loch an Eilein (27/895077): *Larix europaea*, incl. 79°, aspect 130°, pH 4·02, 10 × 10 cm, cover 58%, 3 August 1968, D.L.H.
2. E. Inverness, N. of Croftmore, near Boat of Garten (28/945175): palings by roadside, incl. 0°, aspect horizontal, pH 4·38, 10 × 10 cm, cover 80%, 4 August 1968, D.L.H.
3. E. Inverness, Glen More Forest (28/985078): *Pinus sylvestris* ssp. *scotica*, incl. 73°, aspect 270°, pH 4·02, 10 × 10 cm, cover 81%, 4 August 1968, D.L.H.
4. E. Inverness, near Croftmore, near Boat of Garten (28/934154): *Betula* sp., incl. 71°, aspect 225°, pH 4·67, 10 × 10 cm, cover 72%, 4 August 1968, D.L.H,

P. pertusa are often also found in this association in *Quercus* forests on very acid soils where they may give high cover values; this community perhaps constitutes a distinct subassociation but is in need of further investigation. Bark samples under this association indicate that it is optimally developed in the pH range 3·0–4·0. For details of substrate preference see under the alliance above.

Parmeliopsidetum ambiguae Frey

Ver. naturf. Ges. Basel **35**, 319 (1923).—*Parmeliopsidetum* DR., *Svensk bot. Tidskr.* **39**, 148 (1945).

This community is rather poorly represented in the British Isles where it is found mainly on decorticate coniferous wood in central and eastern Scotland. It is also sometimes encountered on conifers and birch trees themselves, however, and is clearly closely allied to the *Pseudevernietum furfuraceae* from which it differs in the abundance of *Parmeliopsis aleurites*, *P. ambigua* and *P. hyperopta*. It should be noted that this community is quite distinct from the *Parmeliopsis ambigua*-rich facies of the *Pseudevernietum furfuraceae* sometimes developed under pollution stress (p. 335).

At first we were inclined not to recognize this syntaxon as distinct in Britain but an examination of the data of Frey (1923) and Barkman's observations (1958) shows that the *Parmeliopsis*-rich communities of the Scottish Highlands are most probably species-poor variants of this association. Additional characteristic species of this essentially subboreal association now rarely found in Britain include *Cetraria juniperina, C. pinastri* and *C. sepincola*. In northern Europe this association appears to be

---

5. Lanarkshire, between Coulter and Biggar (36/0 – – 3 – –): *Fraxinus excelsior*, incl. 90°, aspect 205°, pH 4·77, 10 × 10 cm, cover 90%, 29 July 1968, D.L.H.
6. Peeblesshire, NE of Dolphinton (36/1 – – 4 – –): *Fagus sylvatica*, incl. 82°, aspect 180°, pH 5·25, 10 × 10 cm, cover 60%, 29 July 1968, D.L.H.
7. Derbyshire, Birchover, Rowtor Rocks (43/235622): millstone grit rocks, incl. 60°, aspect 225°, 10 × 10 cm, cover 70%, 28 August 1967, D.L.H.
8. Sutherland, Halladale, near Achiemore (29/8 – – 5 – –): Old Red sandstone outcrop, incl. 82°, aspect 270°, 10 × 10 cm, cover 80%, 12 August 1968, D.L.H.
9. Derbyshire, Chatsworth, near Beeley Lodge (43/265684): millstone grit wall, incl. 0°, aspect horizontal, 10 × 10 cm, cover 80%, 4 April 1967, D.L.H.
10. Derbyshire, Holloway, Dethick Lea Hall Farm (43/335575): millstone grit wall, incl. 0°, aspect horizontal, 10 × 10 cm, cover 100%, 15 March 1967, D.L.H.

particularly frequent on birch twigs, and communities with *Cetraria chlorophylla*, *C. sepincola* and *Parmelia septentrionalis* on birch twigs in the Scottish Highlands are thus placed here.

The synsystematics of the *Parmeliopsidetum ambiguae*, discussed in some detail by Barkman (1954), remain poorly understood. A considerable number of taxa at subassociation and variant ranks have been described in it which require re-evaluation on a European scale.

All. 10. *Usneion barbatae*

Usneion barbatae Ochsn., *Jahrb. St. Gall. naturw. Ges.* **63,** 68 (1928); type: *Usneetum barbatae* Ochsn.—*Usneion dasypogae* Barkm., *Phytos. Ecol. Crypt. Ep.*, 475 (1958); type: *Usneetum dasypogae* Frey [= *Letharietum divaricatae* Frey]. —*Usneion florido-ceratinae* Barkm., *Phytos. Ecol. Crypt. Ep.*, 470 (1958); type not designated.

Communities dominated by species of *Usnea*, characteristic of acidic barks and usually in very well lit situations are placed in this alliance. The synsystematics of this alliance in Europe are, however, in a confused state mainly as a result of the currently unsatisfactory species concepts in many groups of *Usnea* species. Five associations can be distinguished in the British Isles, but the names of four may require some revision when the identity of some *Usnea* species used to characterize alliances in continental Europe becomes firmly established.

All associations of this alliance are very sensitive to air pollution and so have more restricted distributions in the British Isles than they did in the early parts of last century.

Cladonieto-Usneetum tuberculatae Barkm.

Phytos. Ecol. Crypt. Ep., 471 (1958).

This association is treated here in the sense of Hawksworth (1972a) to include communities allied to the *Usneetum subfloridanae* in more shaded habitats where *Cladonia* species (e.g. *C. coccifera*, *C. pyxidata*, *C. squamosa*) and bryophytes form major parts of the stands. Also sometimes encountered on acid mossy rocks, this association may perhaps be viewed as an intermediate between the *U. subfloridanae* and the *Cladonietum coniocraeae*. *Usnea flammea*, *U. fragilescens* and *U. inflata* (syn. *U. intexta*) are to be found in this association in more upland areas in addition to *U. subfloridana*.

Ramalinetum fastigiatae Duvign.

Bull. Soc. r. bot. Belg. **74**, 42 (1942).—See Barkman (1958) for lists of probable synonyms.

The *Ramalinetum fastigiatae* (Fig. 5) is a variable association characterized by the dominance of *Ramalina* species (particularly *R. baltica*, *R. calicaris*, *R. duriaei*, *R. farinacea*, *R. fastigiata*, *R. fraxinea*, and rarely *R. pollinaria*). It is widespread and common, favouring better lit, more exposed trees and is particularly common on inclined branches and twigs. Two subassociations and three variants (one unnamed) were accepted by Barkman (1958), all of which occur in the British Isles, but they do not appear to merit separation as transitions are commonly encountered.

FIG. 5. *Ramalinetum fastigiatae* on nutrient-rich bark. Predominant species: *Evernia prunastri*, *Physcia tenella*, *Ramalina farinacea*, *R. fastigiata*, *R. fraxinea* and *Xanthoria parietina* (W. Inverness: Invergordon, 1976, R. O. Millar).

Although this association has generally been placed in the *Xanthorion* we suspect that it may be more closely allied to the *Usneion* as it parallels the *Usneetum articulato-floridae* but is found in somewhat more nutrient-rich sites. Some tendency to intergrade with the *Usneetum subfloridanae* also supports its positioning here.

Usneetum articulato-floridae var. *ceratinae* D. Hawksw.

Fld Stud. **3,** 543 (1972).

This often spectacular community (Fig. 6), which characteristically occurs in very well lit situations and is optimally developed on the uppermost sloping or horizontal boughs of trees, has a markedly southern and south-western distribution in the British Isles. The most important species of the variant are *Usnea articulata*, *U. ceratina*, *U. florida* and *U. rubiginea*, although *U. inflata* (syn. *U. intexta*) and *U. subfloridana* are often also present.

In the southern counties of England a facies dominated by *U. ceratina* and *U. inflata* occurs not uncommonly on almost vertical well lit trunks of *Fagus* and *Quercus* in glades; *U. articulata* is generally absent in this facies although *U. rubiginea* is often present. This nodum may in future merit separation as a distinct association.

The "typical" variant of this association, as interpreted by Barkman (1958), is still present in Brittany and includes *Heterodermia leucomelos*, *Pseudocyphellaria aurata* and *Teloschistes flavicans*. These species are now very rare in Britain, perhaps prefer more nutrient-rich barks, and seem better placed in a separate association, the *Teloschistetum flavicantis* (p. 348), which is always poor in *Usnea* species. The relationships between this community and both the *Usneetum florido-neglectae* Bibinger and the *Usneetum rubicundae* Barkm., both unrecognized in Britain, merit further study.

Usneetum filipendulae ass. nov. prov.

This poorly understood association, which is not validated here in the absence of detailed records, is characteristic of ancient coniferous forests in Scotland and dominated by *Usnea filipendula* and *U. fibrillosa*. *Alectoria sarmentosa*, *Bryoria capillaris*, and sometimes *U. hirta*, may be further important components of this association. It should be noted that this community may perhaps be in reality a species-poor variant of the central and northern European montane *Usneetum barbatae* Ochsn. (for which the later name *Letharietum divaricatae* Frey was employed, perhaps unnecessarily, by Barkman, 1958).

FIG. 6. *Usneetum articulato-floridae* var. *ceratinae* on *Salix* in marsh. Species present include *Evernia prunastri*, *Hypogymnia physodes*, *Parmelia perlata*, *P. sulcata* and *Usnea ceratina* (S. Devon: Slapton, Duck Marsh, 1975, F. S. Dobson).

Usneetum subfloridanae D. Hawksw.

Fld Stud. **3**, 543 (1972).

This is the most widespread association of the *Usneion* in the British Isles but tends to be optimally developed in the south and west. The *Usneetum subfloridanae* is particularly luxuriant on smaller horizontal branches or twigs in very well lit sites and appears to prefer slightly less acid bark than the *Usneetum articulato-floridae* var. *ceratinae*. *Usnea subfloridana* (and/or *U. florida* in the south) predominates in this association although *Ramalina calicaris* and *U. fulvoreagens* (in south-west England) are locally important. Facies rich in *R. calicaris* sometimes show a tendency to intergrade with the *Ramalinetum fastigiatae*. In particularly wet habitats *U. extensa* is also to be found in this association.

The *Usneetum sufloridanae* is able to overgrow pioneer communities of twigs such as the *Arthopyrenietum punctiformis* and *Lecanoretum subfuscae*. It is not to be confused with the *Pseudevernietum furfuraceae* subass. *usneetosum subfloridanae* Kalb (1972) which includes *Bryoria fuscescens* and prefers more acid bark.

All. 11. *Xanthorion parietinae*

Xanthorion parietinae Ochsn., *Jahrb. St. Gall. naturw. Ges.* **63**, 53 (1928); type: *Physcietum ascendentis* Frey & Ochsn.—*Parmelion caperatae* Felf., *Acta geobot. Hung.* **4**, 48 (1941); type: *Parmelietum caperatae* Felf.—*Buellion canescentis* Barkm., *Phytos. Ecol. Crypt. Ep.*, 400 (1958); type: *Ramalinetum duriae* ("Duvign.") Barkm.—? *Teloschistidion chrysophthalmi* Follm., *Ber. dtsch. bot. Ges.* **80**, 202 (1967); type: *Teloschistetum chrysophthalmi* Follm. [*nom. nud.*; ? non Ochsn., 1934.].

This alliance is used here for both corticolous and saxicolous communities. On trees it is characterized either by species of *Buellia, Caloplaca, Physcia, Physconia, Ramalina, Teloschistes* or *Xanthoria*, or occasionally *Anaptychia ciliaris, Bacidia rubella, Parmelia acetabulum, P. quercina, P. laciniatula* or *Candelaria concolor*. It occurs on nutrient-rich, nutrient-enriched or hypertrophicated barks and rocks, usually in well lit situations, and most frequently in sites frequented by birds, farm animals or alongside dusty roads. This alliance is widespread throughout the British Isles especially in the drier south and eastern lowlands. On trees it becomes increasingly dominant southwards in Europe as the Mediterranean is approached where it is the predominant alliance, even in old lowland relict forests. The non-corticolous associations of this alliance in Britain are discussed under the pertinent substrates on pp. 360–361 and 382.

The *Ramalinetum fastigiatae*, placed in this alliance by most previous authors (e.g. Barkman, 1958), is referred to the *Usneion barbatae* here (see p. 339) and is thus not treated below.

Buellietum punctiformis Barkm.

Phytos. Ecol. Crypt. Ep., 405 (1958).

A frequently species-poor association comprising mosaics of *Buellia punctata*, *B. canescens* (sometimes dominant), *Candelariella reflexa*, *C. vitellina*, *Lecania cyrtella*, *Lecanora chlarotera*, *L. sambuci*, *Lecidella elaeochroma*, *Lecidea quernea* and often also with *Xanthoria* species. This community is characteristic of nutrient-enriched or hypertrophicated barks and particularly common in areas of moderate sulphur dioxide and inorganic fertilizer pollution where other associations are not able to develop. It becomes rarer in the extreme north of England and Scotland.

Gyalectinetum carneoluteae D. Hawksw.

Fld Stud. **3**, 545 (1972).

A shade-loving community of nutrient-rich barked trees of the extreme south and south-west England characterized by an abundance of *Gyalectina carneolutea*. Other important components of this community are *Bacidia phacodes*, *B. rubella*, *Lithographa dendrographa* and, sometimes, *Opegrapha prosodea*. Better lit facies show some intergradation with the *Buellietum punctiformis*, whilst more shaded ones include elements of the *Pyrenuletum nitidae*. The presence of *Lithographa dendrographa* may indicate that this association has some affinity to the *Lobarion pulmonariae* (see p. 327).

Parmelietum carporrhizantis Crespo

An. Inst. bot. A. J. Cavanillo **32**, 191 (1975).

Parmelia quercina (syn. *P. carporrhizans*)-dominated communities occur in Britain on very well lit and slightly nutrient-enriched parts of trees, most frequently the upper branches. This is largely a southern European association, however, and is only found in the British Isles with any frequency in the low-rainfall and high-sunshine coastal parts of south Devonshire and Dorset where *P. borreri* and *P. pastillifera* are also characteristic of it. Crespo (1975) recognized two variants and one subassociation (the *parmelietosum endochlorae*). His first variant, with the differential species *P. caperata* and *P. soredians*, is the type represented in the British Isles.

The *Parmelietum carporrhizantis* has many features in common with the *Parmelion perlatae*, but as it prefers more nutrient-rich sites, it is referred to the *Xanthorion parietinae*.

Parmelietum elegantulae Klem.

Beih. Feddes Repert. **135**, 154 (1955) [as "Almborn 1948"].—*Parmelietum laciniatulae* Barkm., *Phytos. Ecol. Crypt. Ep.*, 446 (1958) [as "(Almborn) Klement 1955"].

This association, distinguished by the frequency of *Parmelia elegantula*, *P. laciniatula* and often *Ochrolechia yasudae*, shows strong affinities with both the *P. acetabulum* facies of the *Physcietum ascendentis* and the *Parmelietum revolutae*. The *Parmelietum elegantulae* is particularly well developed in sites affected by man and appears to require a somewhat lower bark pH than the *Physcietum ascendentis*. It is because of this difference in requirements that it is not subsumed under the *Physcietum ascendentis* here. *Physcia*, *Physconia* and *Xanthoria* are of relatively minor importance in this association as compared to the *Physcietum ascendentis*.

Physcietum ascendentis Frey & Ochsn. (Table VI)

Arvernia **2**, 82 (1926).—*Parmelietum acetabulae* Ochsn., *Jahrb. St. Gall. naturw. Ges.* **63**, 60 (1928).—*Teloschistetum chrysophthalmae* Ochsn., *Revue bryol. lichén.* **7**, 85 (1934).—*Xanthorietum candelariae* Frey, *Ergebn. wiss. Unters. Schweiz. NatnParks, n.f.* **3**, 476 (1952).—*Arthopyrenietum gemmatae* Barkm., *Phytos. Ecol. Crypt. Ep.*, 400 (1958).—*Caloplacetum phloginae* Barkm., *Phytos. Ecol. Crypt. Ep.*, 369 (1958).—See Barkman (1958) for further probable synonyms of these names.

The *Physcietum ascendentis* is treated in a rather broad sense here to include the majority of *Physcia*, *Physconia* and *Xanthoria* dominated communities of nutrient-rich bark in the British Isles. The numerous associations and subassociations recognized for these by continental authors appear to be of limited syntaxonomic importance here. The association is primarily distinguished by high frequencies of *Physcia* (particularly *P. adscendens*, *P. aipolia*, *P. orbicularis*, *P. tenella* and sometimes *P. tribacia*), *Physconia* (*P. enteroxantha*, *P. farrea*, *P. grisea* and *P. pulverulenta*) and *Xanthoria* (*X. candelaria*, *X. fallax*, *X. parietina* and *X. polycarpa*). The association is often very rich in species with 30 or more not infrequently present on single, mature, well lit, nutrient-enriched trees in areas of little air pollution. The predominant species may vary from tree to tree even when these are in close proximity and so this appears to be of rather limited syntaxonomic value. Communities falling within the concept of the *Parmelietum acetabulae* (including *Anaptychia ciliaris*, *Parmelia acetabulum*, *P. exasperatula* or *P. tiliacea*) represent an eastern facies (Table VI) of the association in Britain and are particularly well developed on well lit mature tree trunks in open parkland sites; this might be interpreted as a distinct variant or subassociation but, as it could also be viewed as the species-rich or optimal facies of a single association which becomes simplified both northwards and westwards in response to climatic stresses progressively eliminating species, it is not treated separately here. The abundance of *Ramalina* species in the association appears to be related to the degree of exposure, these becoming most frequent in well ventilated sites.

Some communities of this association in central and eastern Europe may have some affinities with the *Parmelietum revolutae* (see p. 331). Although primarily corticolous, species-poor communities referable to this association are occasionally encountered on nutrient-enriched calcareous rocks where they can, when optimally developed, include even *Anaptychia ciliaris*.

A community of nutrient-rich *Ulmus* wood tracks dominated by *Bacidia incompta* but including *Caloplaca luteoalba* is tentatively mentioned here but may merit association rank.

Physcietum caesiae Mot.

See p. 360 for details of nomenclature and composition.

Although primarily saxicolous, this association is sometimes well developed on trees or timber heavily impregnated with alkaline dust, for example near limestone-crushing plants and cement works (see Gilbert, 1976). In such situations the corticolous community includes species normally found in it when it occurs on calcareous substrates (e.g. *Lecanora muralis*, *Physcia caesia*).

Physciopsidetum elaeinae (Barkm.) comb. nov.

Physcietum elaeinae Barkm., *Phytos. Ecol. Crypt. Ep.*, 414 (1958).

This association, closely allied to the *Buellietum punctiformis*, is encountered on highly hypertrophicated bark, particularly of *Sambucus*. The variant *buelliosum canescentis* Barkm. is the facies represented in the British Isles and is distinguished by high cover values of *Physciopsis adglutinata* (and sometimes *Physcia orbicularis* and *P. tribacia*). The *Physciopsidetum elaeinae* tends to prefer slightly more shaded habitats and smoother bark than the *Physcietum ascendentis*. Barkman (1958) considered this to be primarily a

TABLE VI.
Physcietum ascendentis Frey & Ochsn. ("eastern facies").

Species	Stands							
	1	2	3	4	5	6	7	8
Acrocordia gemmata	–	–	+.2	–	–	+.0	–	–
Anaptychia ciliaris	–	3.4	2.3	1.2	–	3.3	2.2	2.3
Arthonia impolita	–	–	+.0	–	–	–	–	–
Buellia canescens	+.2	+.2	+.0	1.2	+.2	+.2	1.2	1.2
B. punctata	1.2	+.1	+.0	+.1	+.2	+.0	+.0	+.0
Calicium viride	–	–	–	–	–	+.0	2.2	–
Candelariella vitellina	–	–	–	+.1	+.0	–	–	–
Catillaria griffithii	+.2	–	+.2	+.1	–	–	–	–
Evernia prunastri	–	–	+.0	+.2	–	–	–	–
Gyalecta flotowii	–	+.0	–	–	–	–	–	–
Haematomma ochroleucum var. *porphyrium*	–	–	–	–	–	–	+.0	–
Lecanora chlarotera	–	–	+.0	–	–	–	–	+.0
L. conizaeoides	2.2	–	–	+.0	–	–	–	–
L. dispersa	–	–	+.0	–	–	–	–	+.0
L. expallens	2.2	–	+.0	1.2	2.3	+.2	2.2	+.0
Lecidea quernea	–	–	+.0	+.2	–	–	–	–
Lecidella elaeochroma	–	–	+.0	–	–	–	–	+.0
Lepraria candelaris	–	–	+.0	–	–	–	–	–
L. incana	–	–	–	–	–	+.0	+.0	–
Ochrolechia turneri	–	–	+.0	–	–	–	–	3.2
O. yasudae	–	–	+.0	2.3	–	–	+.0	–
Opegrapha varia s.l.	–	+.0	–	–	–	+.0	–	–
Parmelia acetabulum	2.2	2.5	2.3	+.2	5.5	+.0	1.2	+.0
P. caperata	–	–	1.3	–	–	–	–	+.2
P. elegantula	–	–	+.0	+.0	–	–	–	–
P. glabratula	–	–	+.0	–	–	–	–	–
P. laciniatula	–	–	+.2	–	–	–	–	–
P. perlata	–	–	2.2	–	–	–	–	1.3
P. saxatilis	–	–	+.0	–	–	–	+.0	–
P. subaurifera	–	–	+.0	–	–	–	–	1.3
P. subrudecta	+.0	–	+.2	+.2	2.2	+.0	+.0	+.2
P. sulcata	–	–	4.3	1.3	1.2	+.0	1.2	+.0
P. tiliacea s.l.	–	–	1.3	–	–	–	–	–
Pertusaria albescens	–	–	+.0	–	–	–	+.0	–
var. *corallina*	–	–	–	2.2	–	1.2	–	–
P. amara	–	–	+.0	1.3	–	–	–	+.0
P. coccodes	–	1.2	–	–	–	–	–	2.3
P. hymenea	–	–	+.0	–	–	–	–	–
P. pertusa	–	–	+.0	1.3	–	–	–	+.0

TABLE VI—continued

Species	Stands							
	1	2	3	4	5	6	7	8
Physcia adscendens	+.0	1.2	+.2	1.2	+.1	–	–	2.3
P. aipolia	–	–	2.2	2.2	–	1.2	–	1.2
P. leptalea	–	–	+.3	–	–	–	–	–
P. orbicularis	1.2	+.1	+.0	+.2	+.1	+.0	–	–
P. tenella	+.2	–	–	–	1.3	–	–	–
Physciopsis adglutinata	–	–	+.0	–	–	–	–	–
Physconia enteroxantha	–	–	–	–	–	+.0	–	–
P. grisea	+.0	2.2	+.0	+.0	+.2	4.3	–	–
P. pulverulenta	–	–	+.2	1.2	–	–	1.2	1.2
Ramalina baltica	–	–	+.0	–	–	+.0	–	–
R. farinacea	–	–	+.0	+.0	–	–	+.0	+.0
R. fastigiata	–	–	+.0	–	–	–	–	+.0
Rinodina roboris	–	–	–	–	–	–	–	2.3
Schismatomma decolorans	–	–	–	–	–	–	+.0	+.0
Xanthoria parietina	–	2.3	1.3	1.2	+.0	+.2	–	1.3
Camptothecium sericeum	–	1.4	+.2	1.3	–	–	2.4	–
Hypnum cupressiforme	–	–	–	2.3	1.2	+.2	–	–
Leucodon sciuroides	–	–	–	–	–	2.3	–	–
Orthotrichum diaphanum	+.0	–	–	–	–	–	–	–
Porella platyphylla	–	–	–	1.3	–	–	–	–
Tortula laevipila	–	+.1	+.2	1.2	–	–	+.0	–
Zygodon viridissimus	–	1.2	–	–	–	–	–	–

1. Hertfordshire, North Mymms Park (52/2 – – 0 – –): *Ulmus procera* in open park, 1·5 m diam, incl. 85°, aspect W, 1·0 × 0·5 m, cover 30%, 26 October 1968, F.R.
2. E. Kent, Bourne Park (61/1 – – 5 – –): *Ulmus procera*, 0·8 m diam, incl. 90°, aspect SW, 0·3 × 0·3 m, cover 70%, 16 March 1967, F.R.
3. E. Kent, Denton Court Park (61/2 – – 4 – –): *Fraxinus*, 0·8 m diam, incl. 90°, aspect S, 1·0 × 0·3 m, cover not indicated, 8 June 1968, F.R.
4. E. Kent, NE of "Tudor House" roadhouse E of Bearsted (51/8 – – 5 – –): *Fraxinus* in parkland, 1·5 m diam, incl. 90°, aspect S, 2·0 × 1·0 m, cover 75%, 4 September 1968, F.R.
5. E. Kent, Maidstone, Mote Park (51/7 – – 5 – –): *Acer campestre*, 0·6 m diam, incl. 90°, aspect SW, 1·0 × 0·5 m, cover not indicated, 10 September 1968, F.R.
6. W. Norfolk, Hilborough Park (53/8 – – 0 – –): *Fraxinus* in parkland, 0·7 m diam, incl. 90°, aspect SW, 0·5 × 0·5 m, cover 80%, 5 June 1970, F.R.
7. W. Suffolk, Ickworth Park (52/8 – – 6 – –): *Acer campestre* in open parkland, 0·6 m diam, incl. 90°, aspect N, 1·0 × 0·5 m, cover 60%, 28 September 1968, F.R.
8. E. Sussex, Udimore (51/8 – – 1 – –): *Acer campestre*, 0·6 m diam, incl. 90°, aspect S, 1·0 × 0·5 m, cover not indicated, 15 April 1968, F.R.

Mediterranean association and thus it is not surprising to note that, in the British Isles, it appears to be most frequent in southern England. Although Barkman regarded *Physcia clementei* as an important component of this association, communities including that species in the British Isles seem sufficiently distinctive to warrant treatment as a distinct association, the *Teloschistetum flavicantis*.

Teloschistetum flavicantis ass. nov. (Table VII)

This previously unrecognized association recalls, and may eventually prove to be identical to, Barkman's (1958) unnamed southern Atlantic and montane-Mediterranean variant of the *Physcietum ascendentis* subass. *physciosum leptaleae* Klem. which he considered to occur in Brittany, near Fontainebleau, Provence and the Central Atlas Mountains. The association is closely allied to the *Physcietum ascendentis* but is differentiated primarily by the abundance of *Teloschistes* species (*T. chrysophthalmus* or *T. villosus* in France; usually only *T. flavicans* in Britain). *Physcia clementei*, *P.*

TABLE VII.

Teloschistetum flavicantis ass. nov.

Species	Stands 1	2	Species	Stands 1	2
Anaptychia ciliaris	–	+	*Parmelia subaurifera*	2	+
A. fusca	+	–	*P. subrudecta*	–	+
Evernia prunastri	2	–	*P. sulcata*	3	4
Ochrolechia parella	+	2	*Ramalina farinacea*	3	3
O. yasudae	4	–	*R. fastigiata*	3	–
Parmelia caperata	3	3	*Teloschistes flavicans*	4	5
P. perlata	3	4	Bryophyta	–	2

Additional species present in this community in the type locality but outside the areas of the quadrats were: *Buellia canescens, Lecanora chlarotera, L. expallens, Lecidea quernea, Lepraria candelaris, Opegrapha atra, Phlyctis argena, Physcia aipolia, P. adscendens, P. clementei, P. leptalea, P. tenella, P. tribacia, P. tribacioides, Physconia pulverulenta, Physciopsis adglutinata, Ramalina baltica, R. calicaris* and *Xanthoria parietina*. A colour photograph of this community in the type locality appeared on the front cover of *Environment and Change* **2** (6), February 1974.

1. S. Devon, Torcross, Widdicombe House (20/811417): mature *Acer pseudoplatanus* in pasture, 30 cm diam, incl. 85°, aspect 290°, 20 × 20 cm, cover 87%, 11 August 1973, D.L.H.
2. S. Devon, Torcross, Widdicombe House (20/811417): mature *Acer pseudoplatanus* in pasture, 40 cm diam, incl. 87°, aspect 265°, 20 × 20 cm, cover 90%, 11 August 1973, D.L.H.; *type record*.

leptalea and *P. tribacioides*, which can form extensive stands in the sunniest and driest parts of southern England, are very closely associated with it. This community appears to have been formerly widespread in southern England but is now largely restricted to the south-west (where it is now much rarer than it was last century) from Dorset to Cornwall and is more rarely found in Pembrokeshire (S. Wales). The twig facies (with *T. chrysophthalmus*) is still common in the Mediterranean area and in southwest France (from southern Brittany southwards), where it is largely found on twigs in very well lit situations. In south-west England today it occurs both on twigs and on well lit, well ventilated nutrient-enriched tree trunks, but *T. chrysophthalmus* seems now to be extinct in Britain outside the Channel Islands (Hawksworth *et al.*, 1974).

This community is not identical to the *Teloschistetum chrysophthalmae* Ochsn. (see p. 344), the type record of which lacked any *Teloschistes* species and is referable to the *Physcietum ascendentis*.

IV. Limestone Communities

Hard limestones (e.g. Carboniferous and Devonian) support several distinct lichen communities in the British Isles. Most of these are able to spread on to softer calcareous rocks and a wide range of basic man-made substrates (e.g. asbestos-cement, concrete, basic brickwork and mortar), but then occur essentially as species-poor variants.

The communities developed on coastal limestones are essentially species-poor facies of associations of inland limestone rocks and so are not recognized separately here. Some elements of the *Caloplacetum marinae* and *Verrucarietum maurae*, normally well developed on siliceous rocks, may be encountered on hard coastal limestones, e.g. marine *Caloplaca*, *Verrucaria* and rarely *Lichina* species (see Fletcher, 1975b). *Arthopyrenia halodytes* can also occur on softer limestones and chalk as well as its more usual habitat of barnacles and other intertidal mollusc shells.

Species on soil and humus in crevices, soil and turf around limestone, chalk, pebbles etc., are treated separately on pp. 393–407, while intermediate communities on other basic rocks are discussed on pp. 361–364.

All. 12. *Aspicilion calcareae*

Aspicilion calcareae (Alberts.) comb. nov.—*Lecanorion calcareae* Alberts., *Acta phytogeogr. suec.* **20**, 34 (1946), basionym.—*Caloplacion decipientis* Klem., *Ber. bayer. bot. Ges.* **28**, 263 (1950).—*Gyalection cupularis* Matt., *Bot. Jb.* **75**, 420

(1951) [nom. nud.].—*Lecanactinion stenhammari* Matt., *Bot. Jb.* **75**, 415 (1951) [nom. nud.].—*Collemation tunaeformis* Degel. (1950) *fide* Degel., *Symb. bot. upsal.* **13** (2), 127 (1954).—*Caloplacion pyraceae* Klem., *Beih. Feddes Repert.* **135**, 72 (1955).—*Collemion rupestris* Klem., *Beih. Feddes Repert.* **135**, 89 (1955).—*Lecanorion galactinae* Laund., *in* Kettering and District Field Club, *First Fifty Years*, 95 (1956).—*Lecanorion dispersae* Laund., *Lichenologist* **3**, 294 (1967).

Although numerous syntaxa have been described for lichen communities on limestones mainly dominated by various *Caloplaca*, *Collema*, *Lecanora* and *Verrucaria* species, in view of the number of species which are, or are sometimes, common to all of them it seems preferable to recognize a restricted number of associations within a single alliance*. Diversity in this alliance is very wide and many of the species indicated in Table VIII as present in it predominate from time to time. To recognize all such facies would in our view be an unwarranted inflation of lichen syntaxonomy, particularly as a very wide range of variation can be seen to occur in apparently uniform habitats with similar moisture and light regimes at a single site; this diversity, basically one of varying local dominants, may well arise from a combination of four factors: (1) the first species that

TABLE VIII.

Lichens characteristic of limestone in the British Isles and the associations in which they usually occur.

Acarospora glaucocarpa[e]	*C. decipiens*[a]
A. heppii[e]	*C. heppiana*[a]
A. macrospora[e]	*C. holocarpa*[a f]
Acrocordia conoidea[b e]	*C. lactea*[a]
A. salweyi[e]	*C. ruderum*[a]
Arthopyrenia saxicola[a]	*C. saxicola*[c]
Aspicilia calcarea[a e]	*C. teicholyta*[a]
A. contorta[a]	*C. tetrasticha*[a c e]
A. prevostii[e]	*C. variabilis*[a e]
Bacidia cuprea[b]	*C. velana*[a]
B. sabuletorum[e]	*Candelariella aurella*[a]
Buellia canescens[f]	*C. medians*[a f]
B. epipolea[a]	*Catillaria lenticularis*[a e]
Caloplaca aurantia[a]	*Clathroporina calcarea*[b]
C. chalybaea[e]	*Collema auriculatum*[e]
C. cirrochroa[b]	*C. crispum*[e]
C. citrina[a e f]	*C. cristatum*[a e]

* Apart from the *Physcietum caesiae* discussed on p. 360.

TABLE VIII—*continued*

C. multipartitum[a,e]
C. polycarpon[a,e]
C. tenax[a,e]
C. tuniforme[a,e]
C. undulatum[a]
Dermatocarpon miniatum[b,e]
Diploschistes gypsaceus[b]
Dirina repanda[c]
D. stenhammari[c]
Encephalographa cerebrina?[a]
Gyalecta jenensis[b]
Ionaspis epulotica[b]
Lecania erysibe[a,e]
Lecanora crenulata[a]
L. dispersa[a]
L. muralis[f]
Lecidea jurana[a,b]
Lecidella stigmatea[a,e,f]
Lempholemma botryosum[e]
Lepraria crassissima[d]
L. incana[d]
L. sp. (bright green)[d]
Leproplaca chrysodeta[d]
L. xantholyta[d]
Leptogium lichenoides[e]
L. plicatile[e]
Opegrapha calcarea[c]
O. chevallieri[a]
O. mougeotii[c]
O. persoonii[a]
O. saxatilis[b]
O. saxicola[b]
Petractis clausa[b]
Physcia adscendens[f]
P. caesia[f]
P. dubia[f]
P. nigricans[f]
P. orbicularis[f]
P. tenella[f]
P. wainioi[f]

Physconia enteroxantha[b]
P. grisea[f]
P. pulverulenta[f]
Placynthium nigrum[e]
P. subradiatum[a]
P. tremniacum[b]
Polyblastia albida[a]
P. cupularis[a,e]
P. schraderi[a]
Porina chlorotica var. *persicina*[b]
Protoblastenia immersa[a,e]
P. incrustans[e]
P. metzleri[e]
P. monticola[a]
P. rupestris[a,b,e]
Psorotichia schaereri[e]
Rhizocarpon umbilicatum[a]
Rinodina bischoffii[a,e]
Sarcogyne regularis[a]
Solenopsora candicans[a,e]
Staurothele caesia[a]
S. hymenogonia[e]
S. rupifraga[e]
S. succedens[e]
Synalissa symphorea[e]
Thelidium decipiens[a,b,e]
T. incavatum[a]
T. papulare[e]
T. pyrenophorum[e]
Toninia aromatica[e]
Verrucaria coerulea[a,e]
V. dufourii[a,e]
V. glaucina[a,e]
V. hochstetteri[a]
V. muralis[a]
V. nigrescens[a]
V. sphinctrina[a]
V. viridula[a,e]
Xanthoria aureola[f]
X. parietina[f]

---

[a] *Caloplacetum heppianae* (p. 353)
[b] *Gyalectetum jenensis* (p. 355)
[c] *Dirinetum stenhammariae* (p. 356)
[d] *Leproplacetum chrysodetae* (p. 357)
[e] *Placynthietum nigri* (p. 357)
[f] *Physcietum caesiae* (p. 360)

chances to establish itself at a site, (2) successional and competition effects between the species present, (3) the softness (e.g. the calcium–silica ratio) and texture of the rock, and (4) the chemical composition of the rock (e.g. magnesium–calcium ratios).

The associations accepted within this alliance here all occupy distinctive microhabitats (Fig. 7) as well as having characteristic floristic compositions

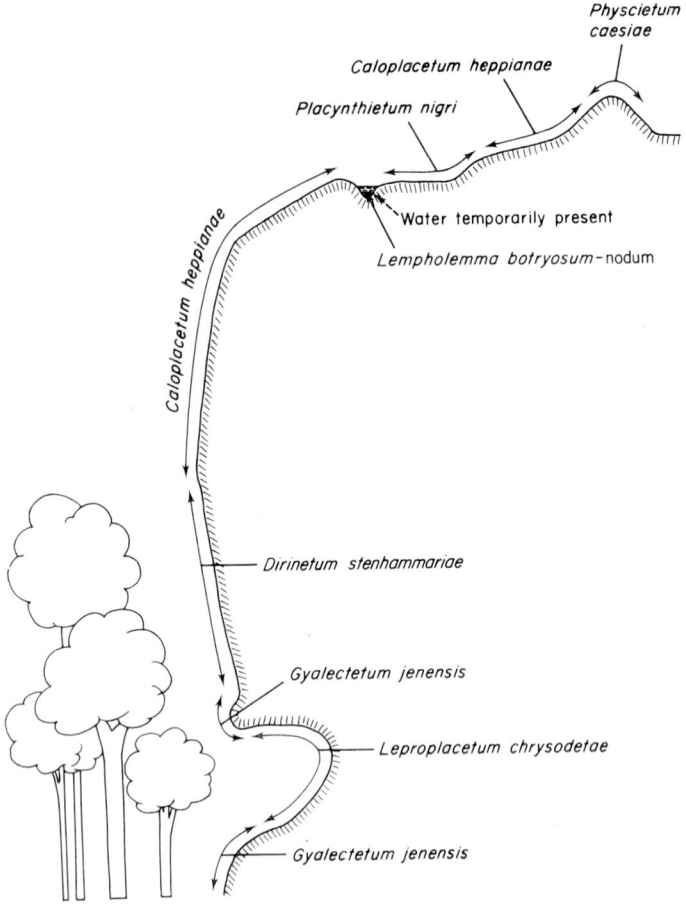

Fig. 7. Diagrammatic section through a limestone bluff to illustrate the microhabitats occupied by the principal lichen communities on limestone.

and so appear to be of greater syntaxonomic importance than communities in similar microhabitats differing only in dominant species. It must, however, be stressed that of the lichen communities treated in this chapter, those which occur on limestones are currently the least understood from the phytosociological standpoint. For this reason it is possible that future studies may be able to justify a larger number of associations than is accepted here.

The correct nomenclature of this alliance, like that of two of the main associations treated below, is not entirely clear at the present time. For this reason no attempt has been made to typify the names in the rank of alliance cited as synonyms above.

Caloplacetum heppianae DR.

Svensk. växtsociol. Sällsk. Handl. **2,** 47 (1925) [as "*Caloplaca heppiana*-Ass."].

"Synonyms"*: *Aspicilietum calcareae* "(DR.)" Klem.; *Aspicilietum contortae* "(Kaiser)" Klem.; *Caloplacetum aurantiae* Klem.; *Caloplacetum citrinae* "(Gallé)" Beschel; *Caloplacetum murorum* "(DR.)" Kaiser; *Caloplacetum variabilis* Klem.; *Lecanoretum calcareae* Alberts.; ? *Lecanoretum campestris* Massé; *Lecanoretum dispersae* Beschel.

This association, which further study may show does indeed merit subdivision as has been the practice to various degrees of most phytosociologists since Du Rietz (1925) and Kaiser (1926), is essentially that of exposed well lit dry limestones. It is characterized by *Aspicilia calcarea*, foveolate *Verrucaria* species (e.g. *V. hochstetteri*, *V. muralis* and particularly *V. sphinctrina*) and, to a lesser extent, species of *Polyblastia* (e.g. *P. albida*, *P. cupularis* and *P. dermatodes*) and *Thelidium* (*T. decipiens* and *T. incavatum*), but has many facies of which those dominated by one or more placodioid, orange *Caloplaca* species (e.g. *C. aurantia* and *C. heppiana*, both becoming rare in Scotland) are the most spectacular (Fig. 8) and so have most commonly been treated as distinct associations. Facies in which *C. teicholyta* (eastern) and *Solenopsora candicans* (western) are generally abundant also occur, particularly in the south. In the driest sites *Caloplaca aurantia* and *C. heppiana* become rarer and the frequency of *Rhizocarpon umbilicatum* increases; *Caloplaca variabilis*, *Polyblastia deminuta*, *Protoblastenia incrustans* and foveolate *Verrucaria* species are generally present in such reduced communities. Many lichens which occur in this association are, however, small and easily overlooked (particularly the pyrenocarpous species; see Table VIII).

In particularly high-rainfall areas with low sunshine, especially on

---

* The nomenclature of this and the *Placynthietum nigri* (p. 357) remain confused and require a very detailed investigation, particularly in view of numerous non-latinized names in the literature (e.g. Kaiser, 1926). Rather than take up unfamiliar names we have employed the earliest familiar name in this case and retained an incorrect but widely used name in that of the *Placynthietum nigri*. Synonyms indicated in both cases are preliminary listings as the citations and the validity of many of these names are in need of a more critical study.

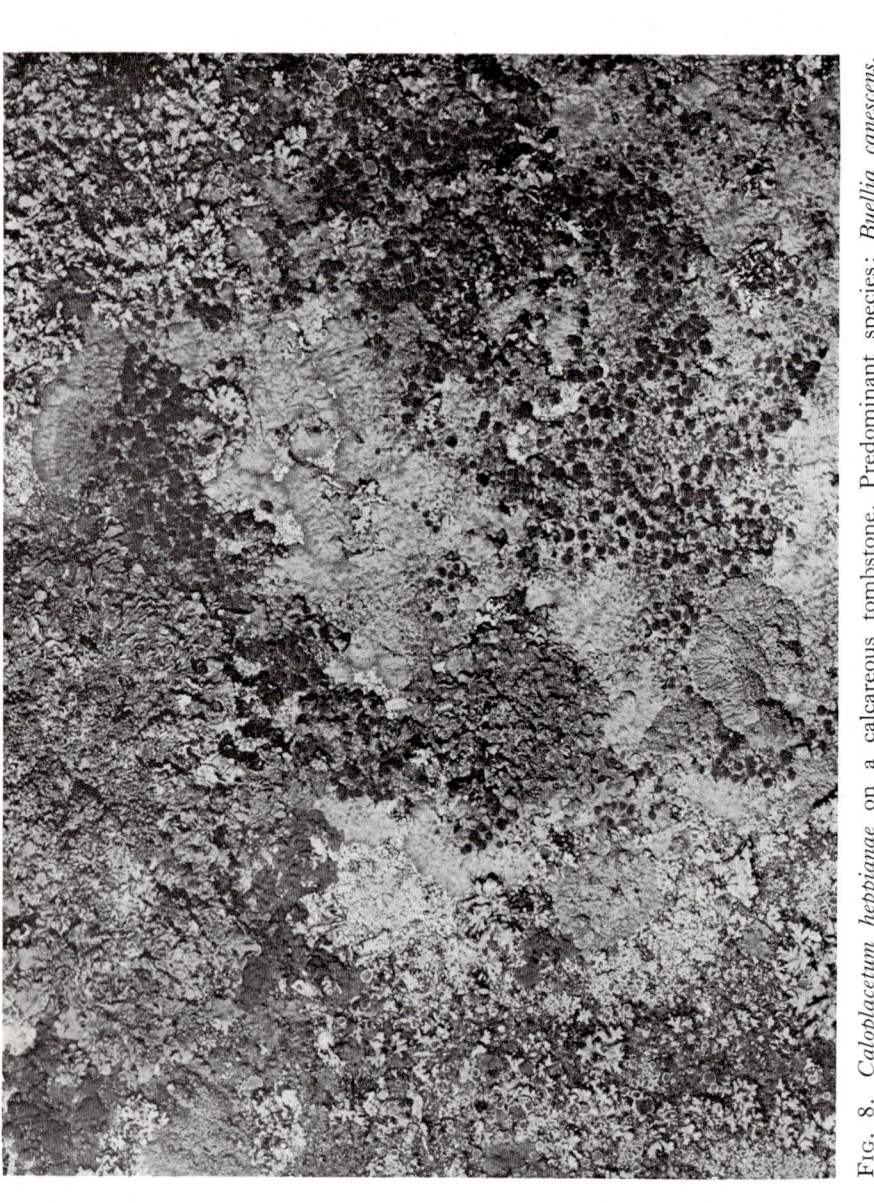

FIG. 8. *Caloplacetum heppianae* on a calcareous tombstone. Predominant species: *Buellia canescens, Caloplaca aurantia, C. saxicola, Lecanora dispersa, Physcia adscendens, P. orbicularis* and *Xanthoria aureola* (Surrey: Cobham churchyard, 1977, F. S. Dobson).

Cambrian limestones, a moss-rich facies poor in *Caloplaca* species is developed, including *C. stillicidiorum* (on mosses), *Collema multipartitum, Lecidea templetonii, Protoblastenia incrustans, Solorina saccata* and *Squamarina crassa*.

The species-poor, air pollution-tolerant *Lecanora dispersa*-dominated facies, which is very abundant on concrete, is viewed here as a pioneer phase of this association; that it now often lacks placodioid *Caloplaca* species in urban areas is almost certainly due to pollution effects (see Laundon, 1967). *Candelariella aurella–L. dispersa* communities are very rapid colonizers of fresh concrete and gravestones, and can provide very high cover values within about 5 years (Hawksworth, 1969; Seaward, 1975). On mortar in eastern Scotland *Caloplaca decipiens* and *Xanthoria elegans* (?*X. resendei*) are particularly frequent.

Temporal relationships within this association are evidently complex and have been investigated in Britain only by Syers (1964) who demonstrated, for example, that the foveolate *Verrucaria* element of this community might be a pioneer stage in that it could be overgrown by placodioid *Caloplaca* species or crustose species with epilithic thalli.

Some relevé data for this association in the British Isles are included in Laundon (1967) and Hawksworth (1969) but extensive surveys of it on naturally occurring limestones have not been carried out. In areas of the British Isles that lack naturally occurring limestones, the *Caloplacetum heppianae* is widespread on artificial substrates, particularly on well lit vertical calcareous gravestones and walls in churchyards.

Gyalectetum jenensis Klem.

Beih. Feddes Repert. **135,** 76 (1955).—*Gyalectetum cupularis* Matt., Bot. Jb. **75,** 420 (1951) [as "Gesellschaft von *Gyalecta cupularis*"], nom. nud.—*Lecideetum juranae* Klem., Beih. Feddes Repert. **135,** 75 (1955) [as "(Kaiser 1926) Klement"]. —*Caloplacetum cirrochroae* Poelt ex Klem., Beih. Feddes Repert. **135,** 88 (1955) [as "Poelt (*in litt.* 1952)"], nom. inval. [no frequency values].—*Caloplacetum cirrochroae* Poelt ex Klem. ex Wirth, Diss. Bot., Lehre **17,** 258 (1972) [table with frequency values].

A shade and prolonged moisture-loving community which tends to prefer damp limestone where, for example, water also runs over or drips on to it during periods of heavy rain or where it is shaded by trees and herbage. Although *Gyalecta jenensis* is often dominant in this association, it may sometimes be replaced in this role by *Petractis clausa*. Other species characteristic and often extremely well developed in this community include *Acrocordia* (*Arthopyrenia*) *conoidea, Caloplaca cirrochroa, Diploschistes gypsaceus, Lecidea jurana, Opegrapha saxatilis, O. saxicola, Porina chlorotica* var. *persicina* and *Protoblastenia metzleri*. *Clathroporina calcarea*, an often overlooked but locally abundant species in the British Isles, is probably faithful to this association (communities including it are particularly well developed in the limestone dales of the Peak District, and at Symond's Yat, Ross-on-Wye, but it also occurs on mortar in churchyards in south-west England).

Klement (1955) included *Leproplaca xantholyta* as a characteristic species of his *Caloplacetum cirrochroae* but no *Lepraria* species were cited by him or Wirth (1972); this thus appears to be a more moisture-requiring community than the *Leproplacetum chrysodetae*. It is consequently placed as a synonym of the *Gyalectetum jenensis* rather than of the *Leproplacetum* here.

Wirth (1972) also mentions *Dermatocarpon miniatum* and this is also usually encountered in the *Gyalectetum jenensis* in the British Isles although it can also be well developed in slightly shaded facies of the moisture-requiring *Placynthietum nigri*. The *Opegraphetum saxicolae* Mot. (Motyka, 1926, p. 224) may perhaps be the earliest name for this association but is not taken up here as *Opegrapha saxicola* is rarely dominant in the British Isles.

Dirinetum stenhammariae (DR.) comb. nov.

Lecanactinetum stenhammariae DR., *Svensk. växtsociol. Sällsk. Handl.* **2**, 47 (1925) [as "*Lecanactis stenhammari* Ass."].

The *Dirina stenhammari*-dominated communities on limestone have many lichens in common with the *Caloplacetum heppianae* but tend to be rather poor in placodioid *Caloplaca* species although *C. tetrasticha* (including *C. ochracea*) may be abundant. The *Dirinetum stenhammariae*, first recorded in Britain by Hawksworth (1973a), occupies a distinctive ecological niche on both naturally occurring limestones and calcareous church walls which are slightly shaded, usually north-facing, vertical or almost vertical, and receive little or no direct rainfall. Rather flaky surfaces are perhaps also preferred. In its environmental requirements this association is not unlike the corticolous *Lecanactidetum premneae* (p. 309). In southern England this association occurs far from natural limestone outcrops on church walls. Other species encountered in this community in the British Isles include *Dirina repanda*, *Lecanactis grumulosa*, *Opegrapha calcarea*, *O. mougeotii* and *O. subelevata*. This association, which clearly has Mediterranean or Lusitanian affinities, is, however, generally rather poor in species in the British Isles.

Leproplacetum chrysodetae ass. nov. (Table IX)

This previously unrecognized association is restricted to moderately shaded dry underhangs, recesses, cave entrances and sheltered sides of mortar-stone walls where the substrate is never directly wetted by rain or water run-off; presumably the component species obtain their moisture from the atmosphere (which is often particularly humid in such habitats). The abundance of *Lepraria crassissima*, *L. incana*, *Leproplaca chrysodeta* and *L. xantholyta* characterize this association although the frequencies of these four species may vary markedly from recess to recess and all may not be present in each site. A further species encountered in this community is an apparently undescribed bright green *Lepraria* which is widespread in this habitat in the British Isles; this lichen is probably faithful to the *Leproplacetum chrysodetae*.

As the specialized niche occupied by this association (Fig. 7) is unfavourable to most species of its alliance, this community tends to be rather poor in species although components of the *Gyalectetum jenensis* may occur in it occasionally at low cover values.

The *Leproplaca*-absent but *Lepraria*-rich *Arthopyrenietum conoideae* Codr. has some affinity with the *Leproplacetum chrysodetae* but appears somewhat intermediate between the present association and the *Gyalectetum jenensis*; for this reason it is not treated as identical to either here. In Britain *Acrocordia conoidea–Lepraria* communities occur where stands of the *Gyalectetum jenensis* and *Leproplacetum chrysodetae* adjoin one another and appear to be of little syntaxonomic importance.

Placynthietum nigri Klem.*

Beih. Feddes Repert. **135**, 89 (1955) [as "(DR. 1925) Klem."].—*Placynthieto-Verrucarietum nigrescentis* DR., Svensk. växtsociol. Sällsk. Handl. **2**, 44 (1925) [as "*Placynthium nigrum-Verrucaria nigrescens*-Assoziation"].

"Synonyms"*: *Collematetum crispi-Verrucarietum muralis* Gallé; *Collematetum pulposi* Kaiser; *Collematetum multipartitis* DR.; *Collematetum tunaeformis* Alberts.

This association, under a variety of names, has often been treated as forming an alliance distinct from the *Aspicilion calcareae* (by, for example, Klement, 1955; Degelius, 1954; Hawksworth, 1969) but, in view of the large number of species it has in common with that alliance, it seems preferable to regard it as a distinct association in the same alliance as that

---

* See footnote on p. 353. For other synonyms or possible synonyms see Kaiser (1926) and Klement (1955).

comprising the *Aspicilia calcarea-* and placodioid *Caloplaca*-containing communities (i.e. the *Aspicilion calcareae*).

The *Placynthietum nigri* prefers damper situations than the *Caloplacetum*

TABLE IX.

Leproplacetum chrysodetae ass. nov.

Species	Stand					
	1	2	3	4	5	6
Acrocordia conoidea	2	–	–	–	–	–
Caloplaca cirrochroa	–	–	–	–	–	1.–
C. heppiana	2	–	–	–	2	–
C. saxicola	–	2	–	–	–	–
Catillaria lenticularis	2	–	–	–	–	–
Clathroporina calcarea	2	4	3	3	–	–
Dirina stenhammari	–	–	4	–	–	–
Gyalecta jenensis	–	–	–	–	–	1.–
Lecanora dispersa	–	2	–	–	–	–
Lepraria crassissima	4	+	5	–	8	2.2
L. incana	–	–	–	–	–	0.2
L. sp. (bright green)	–	–	–	–	–	1.1
Leproplaca chrysodeta	–	6	5	+	–	–
L. xantholyta	6	7	7	7	–	2.2
Porina chlorotica var. *persicina*	–	–	–	–	5	2.2
Protoblastenia rupestris	–	–	–	–	2	–
Verrucaria muralis	2	–	–	3	–	–
V. nigrescens	2	–	–	–	–	–
Bryophytes	–	–	–	–	–	1.1

1. Derbyshire, Lathkill Dale (43/165660): Carboniferous limestone underhang, incl. 87°, aspect 30°, 10 × 10 cm, cover 100%, 25 April 1975, D.L.H.
2. Derbyshire, Lathkill Dale (43/160661): Carboniferous limestone underhang, incl. 101°, aspect 120°, 10 × 10 cm, cover 100%, 25 April 1975, D.L.H.
3. Derbyshire, Lathkill Dale (43/160661): Carboniferous limestone underhang, incl. 95°, aspect 55°, 10 × 10 cm, cover 100%, 25 April 1975, D.L.H.; *type record*.
4. Derbyshire, Lathkill Dale (43/160661): Carboniferous limestone underhang, incl. 92°, aspect 50°, 10 × 10 cm, cover 80%, 25 April 1975, D.L.H.
5. Derbyshire, Lathkill Dale (43/165660): Carboniferous limestone underhang, incl. 90°, aspect 120°, 20 × 10 cm, cover 100%, 25 April 1975, D.L.H.
6. West Yorkshire, Settle, Malham Cove (44/903678): Carboniferous limestone underhang, incl. 90°, aspect 270°, 1·0 × 2·0 m, cover 65%, 1970, P.W.J.

heppianae but, like it, requires light although it is rather more tolerant of slight shading. This association thus tends to occur in rather less quickly draining microhabitats, such as damp vertical faces, undulating limestone surfaces and slopes where water run-off keeps them moist. Some facies of the association are rich in bryophytes whilst others may have the large thalli of *Dermatocarpon miniatum* (see also p. 356) abundant, but it is essentially characterized by the abundance of cyanophilous lichens and shade-loving pyrenocarpous species (see Fig. 9). *Placynthium nigrum* is

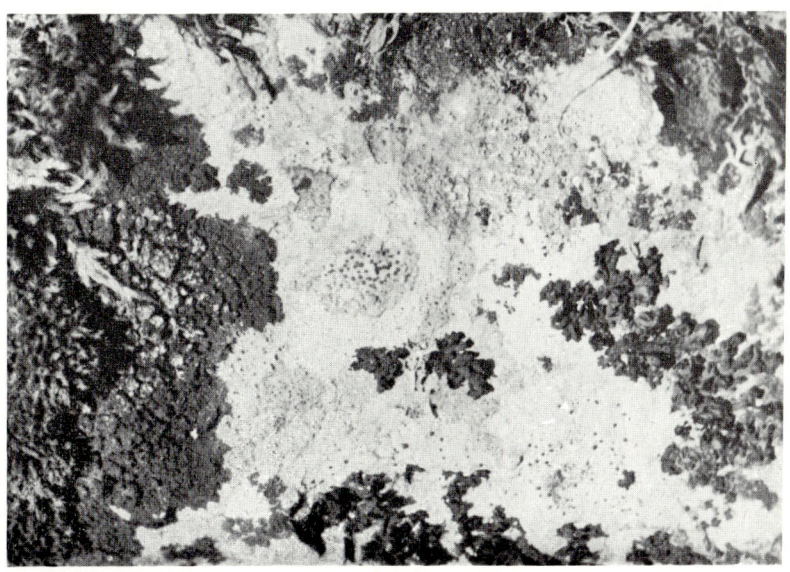

FIG. 9. *Placynthietum nigri* on limestone. Species present include: *Collema multipartitum, Placynthium lismorense, P. nigrum, Protoblastenia rupestris, Verrucaria dufourii, V. hochstetteri,* and *V. sphinctrina* (Argyllshire: Island of Lismore, 1971, P. W. James).

almost invariably present and often luxuriantly developed in this community; one or more *Collema* species also occur of which the most commonly encountered are *C. auriculatum, C. crispum* and *C. cristatum,* although other members of this genus may be abundant in it in some localities (e.g. *C. multipartitum, C. polycarpon, C. tuniforme* and *C. undulatum*); *Leptogium lichenoides*, saxicolous morphotypes of *L. teretiusculum* as well as *Psorotichia schaereri,* are other regularly occurring cyanophilous species. The rare *Synalissa symphorea* is also probably faithful to this association. Of the pyrenocarpous species, *Thelidium decipiens, T.*

papulare, *T. pyrenophorum*, *Verrucaria coerulea* and *V. dufourii* are particularly characteristic of the association. A facies not unlike the *Toninietum candidae* Kaiser but in which *Toninia candida* is replaced by *T. aromatica* occurs but scarcely merits recognition at the association level as it is rather poorly delimited. On mossy limestone rocks in woodland, a facies with abundant *Collema auriculatum*, *C. crispum*, *Cladonia pocillum*, *Polyblastia tristicula* and *Squamarina crassa* can also be recognized. For further species encountered in this species-rich association see Table VIII.

Like the *Caloplacetum heppianae*, the *Placynthietum nigri* is able to extend from natural limestone outcrops on to limestone tombstones and calcareous walls. In churchyards its different ecological requirements can easily be seen by, for example, comparing the lichen communities on the driest and dampest walls of the church, or those on vertical tombstones and around the basal parts of horizontal tops of chest-tombs. Similar contrasts can be seen in comparing limestone boulders in turf and wide expanses of limestone rock faces. The *Placynthietum nigri* is conspicuously rarer in churchyards in the low-rainfall areas of East Anglia than in south and west England.

Allied to this community is one sometimes encountered in depressions in limestone which often hold water for considerable periods of time and in which *Lempholemma botryosum* predominates; this may represent a distinct association but as it is in need of further study it is not formally recognized here but rather treated as a "nodum" (see Fig. 7). Free-living *Nostoc* is often conspicuous in such hollows when it is or has just been raining.

All. 13. *Xanthorion parietinae*

Xanthorion parietinae Ochsn.—see p. 342 for nomenclature.

Physcietum ascendentis Frey & Ochsn.

See p. 344 for nomenclature.

A community not unlike the corticolous association (see p. 344) occasionally occurs on nutrient-enriched calcareous rocks and walls, particularly in farmyards. In this habitat *Buellia canescens*, *Physcia tribacia* and *Ramalina duriaei* are particularly characteristic of it.

Physcietum caesiae Mot.

Bull. int. Acad. pol. Sci. Lett. [Cracovie], sér. B, **1924**, 843 (1925).

This association is widespread on nutrient-enriched sites (e.g. birds' perching stones) but, while it is particularly frequent on limestones (especially the horizontal faces on the tops of gravestones and concrete posts), it is less specific to rock type than nutrient availability. In moderately polluted areas the association is especially well developed on asbestos-cement and can extend far into urban areas as a species-poor facies in which *Lecanora muralis* often predominates. Although primarily a community of calcareous substrates, it is also to be encountered on siliceous rocks enriched by bird droppings and, more rarely, on manure-enriched wood near farmyards and limestone dust-enriched bark near limestone-crushing plants and quarries (see p. 345).

The *Physcietum caesiae* is characterized by high frequencies of *Physcia* species (e.g. *P. adscendens*, *P. caesia*, *P. orbicularis*, *P. wainioi*), *Physconia* (e.g. *P. grisea*) and *Xanthoria* (*X. aureola* and/or *X. parietina*); in eastern Britain *Physcia dubia* and *P. nigricans* also enter into this association and the predominance of *Xanthoria aureola* should be noted. *Caloplaca citrina*, *Candelariella medians* and *Lecanora muralis* are amongst the numerous other species frequently found in this community (see also Table VIII). Relevé data from stands of this association in London are included in Laundon (1967).

It is probable that this association has a number of synonyms (e.g. *Caloplacetum elegantis* Mot., *Physcietum dubiae* R. Sant., *P. teretiusculae* Hil., *Xanthorietum aureolae* Beschel) but the status and application of these names has not been studied by us. The *Physcietum caesiae* has clear affinities with the *Physcietum ascendentis* (see p. 344) on the basis of its species composition but characterizes particularly nutrient-enriched sites where fresh bird droppings are often present. There is also a strong affinity with the *Candelarielletum corallizae* (see p. 378).

V. Other Basic Rock Communities

In addition to species-poor facies of essentially limestone associations occurring on calcareous man-made substrates and softer calcareous rocks and basic sandstones, two specialized unnamed communities of basic rocks occur in the British Isles and these are discussed here. Maritime and nutrient-enriched siliceous rocks, which are basic due to environmental factors, are treated separately on pp. 384–389 and pp. 378–382, respectively.

Montane epidiorite nodum

An extremely specialized but rare assemblage of lichens is associated with epidiorite and mica schist rocks on several of the higher mountains in

Scotland (e.g. Ben Hope, Ben Lawers, Caenlochan and Glen Clova). The community spreads from rock on to adjacent rock debris, bryophytes and mica schist-rich soils and thus can be treated as both terricolous and saxicolous. Detailed lists of the species of this nodum on Ben Lawers have been provided by James (1965). On Ben Hope the communities are similar but *Pannaria hookeri* is dominant on rock. Many of the lichens in this nodum are extremely rare and restricted to it in the British Isles. The following may be mentioned as characteristic: *Belonia russula, Biatorella fossarum, Caloplaca stillicidiorum* (on mosses), *Euopsis granatina, Gyalidea fritzei, Pannaria hookeri, Pertusaria glomerata, Placynthium dolichoterum, P. pannariellum, Polyblastia cupularis, P. inumbrata, P. scotinospora, P. theleodes, Protoblastenia siebenhaarina, Sporopodium fuscoluteum, Thelopsis melathelia* and *Thyrea radiata*.

This nodum is apparently distinct from the *Aspicilietum verrucosae* Frey (see Klement, 1955) although it has a number of species in common with it. Its syntaxonomic position is in need of further study but it has clear affinities with some communities of alpine habitats in Norway of which it probably represents a species-poor facies.

Rhizocarpon petraeum nodum (Table X)

Siliceous rocks at the margins of limestone outcrops support a very characteristic assemblage of species (Table X) whose presence is indicative of the site being somewhat calcareous. This is essentially a transitional nodum occurring chiefly in highland Britain and in which "substrate-switch" species are an important component. Some species of this community (e.g. *Caloplaca flavovirescens, Buellia alboatra, Rhizocarpon petraeum*) occur, for example, on granite stones in a stone and calcareous mortar wall in lowland Britain. Species found in this nodum include *Bacidia sabuletorum, B. umbrina, Caloplaca aurantiaca, Candelariella vitellina, Lecanora muralis, Lecidea illita, L. speirea, L. umbonata, Lecidella stigmatea, Lithographa tesserata, Polyblastia theleodes, Protoblastenia monticola, P. rupestris* (not usually as "var. *calva*"), *Rhizocarpon petraeum* and *Thelidium pyrenophorum*. This community is not accorded the status of an association here as it is essentially transitional between limestone and siliceous rock associations and is usually very limited in extent.

Ability to recognize this nodum is of considerable importance to ecologists as it enables base-rich rocks in essentially siliceous rock areas to be recognized and thus indicates sites of potential importance for their phanerogamic floras.

Table X.

Rhizocarpon petraeum nodum.

Species	Stand 1	2	3	4	5	6
Acarospora glaucocarpa	–	–	–	1.2	–	4.0
Bacidia sabuletorum	+.2	+.2	–	–	–	+.2
B. umbrina	+.2	1.3	–	+.2	+.2	1.2
Buellia aethalea	–	+.2	–	–	–	–
Caloplaca aurantiaca	–	1.2	1.3	+.2	–	–
C. citrina	+.0	+.2	–	+.0	–	–
Candelariella vitellina	–	1.2	+.2	1.3	1.3	1.3
Catillaria lenticularis	1.2	+.2	1.2	–	+.0	+.2
Clathroporina calcarea	–	+.2	–	–	–	–
Collema cristatum	+.0	+.2	–	–	+.2	–
C. polycarpon	–	+.0	–	–	+.2	–
Gyalecta jenensis	–	–	+.2	–	–	1.2
Huilia albocaerulescens	–	+.2	–	–	+.2	–
Ionaspis epulotica	–	–	–	+.0	–	–
Lecanora crenulata	–	–	+.2	–	–	–
L. dispersa	–	+.2	1.2	–	+.2	+.2
L. muralis	–	–	–	1.2	1.3	–
Lecidea illita	–	1.2	–	–	1.2	–
L. lithophila	–	–	1.2	–	+.2	–
L. pelobotryon	+.2	+.0	–	–	+.2	–
L. speirea	2.2	–	2.3	–	1.3	–
L. umbonata	–	–	–	1.2	–	–
Lecidella scabra	–	+.0	–	–	–	–
L. stigmatea	1.2	1.3	1.2	–	+.2	1.3
Lithographa tesserata	–	–	+.2	3.3	1.2	–
Pertusaria amara	–	–	–	+.2	–	–
P. pseudocorallina	–	–	–	+.2	–	–
Placynthium lismorense	–	+.2	–	–	–	–
Polyblastia scotinospora	–	+.2	+.0	–	–	–
P. theleodes	+.2	1.2	+.2	–	+.2	–
Protoblastenia monticola	1.2	1.3	+.2	+.2	+.0	+.2
P. rupestris	1.2	2.3	1.3	2.3	3.2	1.3
Rhizocarpon obscuratum	1.2	–	+.2	+.2	–	–
R. petraeum	2.3	1.2	+.2	1.3	+.0	3.2
Thelidium decipiens	+.2	–	–	1.2	–	+.2
T. pyrenophorum	–	–	1.2	–	1.3	–
Trapelia coarctata	–	–	+.2	–	+.2	–
Verrucaria coerulea	+.2	1.3	1.2	–	+.2	2.2

TABLE X—continued

Species	Stand					
	1	2	3	4	5	6
V. glaucina	–	+.0	–	–	–	–
V. hochstetteri	–	+.0	–	–	–	+.0
V. muralis	+.2	+.2	+.2	–	+.2	+.2
V. sphinctrina	1.2	+.2	+.2	+.2	–	1.2
V. viridula	+.2	–	+.0	+.0	+.2	+.2

1. Sutherland, Inchnadamph, lower slopes of Ben More Assynt (29/286205): transition between acid rock and limestone, exposed site, incl. ±30°, aspect 180°, 0·5 × 0·5 m, cover 75%, 16 June 1961, P.W.J.
2. Argyllshire, Island of Lismore, Port Ramsey (17/892462): transition between acid rock and Dalradian limestone, exposed site, incl. ± 10°, aspect 270°, 1 × 1 m, cover 80%, 12 June 1972, P.W.J.
3. Perthshire, Kinloch Rannoch, lower slopes of Schiehallion above Lochan an Dairn (27/719568): incl. ± 30°, aspect 45°, 1 × 1 m, 14 August 1970, P.W.J.
4. S. Aberdeenshire, Braemar, above Croft of Muicken (37/143903): exposed site, incl. ± 20°, aspect 315°, 1 × 1 m, 16 August 1968, P.W.J.
5. Mid Ebudes, Island of Mull, Ross of Mull, Carsaig, near Nuns' Pass (17/525205): incl. ± 45°, aspect 225°, 1 × 1 m, 10 August 1969, P.W.J.
6. West Yorkshire, Settle, Malham Tarn, Great Close (44/903678): small outcrop in limestone, incl. ± 30°, aspect 225°, 0·5 × 0·5 m, 16 August 1967, P.W.J.

The *Rhizocarpetum concentricae* Mot. (Motyka, 1926, pp. 219, 226) may be synonymous with this nodum but this was poorly circumscribed by Motyka and included the non-British *Lecidea niveoatra* as a main component.

VI. Siliceous Rock Communities

Communities of lichens on siliceous rocks have, with the notable exception of those on rocky shores (pp. 384–389), received scant attention in the British Isles. Siliceous rocks do, however, support more lichen species than any other major habitat in the British Isles. While many communities meriting phytosociological recognition occur, problems of identification of crustose lichens and the relatively little survey work undertaken mean that the treatment presented here is particularly tentative.

Factors controlling the development of these communities parallel those listed for epiphytic communities (see p. 304) with the addition of (1) roughness and fine texture, (2) hardness and friability, and (3) heavy metal contents.

In order to draw similar alliances and associations together, the siliceous rock communities have been grouped into six broad habitat categories: (A) shaded, (B) exposed, (C) nutrient-enriched, (D) mineral-rich, (E) marine and maritime, and (F) aquatic.

A. Shaded

All. 14. *Leprarion chlorinae*

Leprarion chlorinae Šmarda & Hadač, *in* Klika & Hadač, *Příroda, Brno* **36**, 253 (1944).—*Cystocoleion nigri* Wirth, *Diss. Bot., Lehre* **17**, 105 (1972).

An alliance of shaded underhangs and recesses in hard siliceous rocks which are sheltered from direct rain; the acid rock counterpart of communities in dry bark recesses (pp. 306–313) and similar situations on limestone (pp. 349–360). Crustose and leprose species predominate in this alliance and many of these are abundantly sorediate; indeed a high proportion are unknown fertile.

A comprehensive key to the lichens of shaded acid rock crevices and underhangs in Britain is provided by James (1970) and includes information on the distribution of the species.

Some species-poor communities referable to this alliance are also able to occur on upturned tree roots, soil, pebbles in roadside cuttings or in old acid woodlands, as well as in the dry overhangs and recesses in rock of which they are characteristic. The essentially corticolous *Coniocybetum furfuraceae* (p. 308) is also able to spread on to rock and dry soil in suitably shaded sites and form intergrades with both the *Lecideetum lucidae* and the *L. sylvicolae*. The broad ecological amplitude of the *Leprarietum incanae* (p. 312) includes siliceous rocks on which it can form transitions to the *Lecideetum lucidae* and *Racodietum rupestris* in particular.

The *Sclerophytetum circumscriptae* of recesses in maritime rocks, which occupies comparable but more base-rich habitats than the *Leprarion chlorinae*, is discussed separately below (pp. 387–388), while the heavy-metal requiring *Lecanoretum epanorae*, which also has an affinity for dry crevices, is discussed on p. 383.

Mention should also be made here of a community of uncertain syntaxonomic position which is currently poorly known and thus not recognized formally here: the *Lecidea phaeops–L. taylori* nodum. This is essentially a community of damp overhangs or exposed rocks within sheltered woodland sites to which *Gyalidea hyalinescens* may also be faithful; that it requires damp rocks indicates that it should be placed in an alliance other than the *Leprarion chlorinae* and may perhaps be close to the *Aspicilietum lacustris* (p. 391).

Lecideetum lucidae (Schade) Wirth

Diss. Bot., Lehre **17**, 11 (1972).—*Biatoretum lucidae* Schade, *Beih. Feddes Repert.* **76**, 20 (1934).—*Biatoreto-Chaenothecetum* Schade, *Ber. dtsch. bot. Ges.* **41**, (58) (1924).

A species-poor and very widespread association of dry recesses in siliceous rocks, also encountered on stonework, gravestones and brick, and often consisting only of *Lecidea lucida* although *Lepraria incana* may be locally important. It is pollution-tolerant and particularly common in moderately polluted areas on shaded brickwork. This association shows some tendency to merge into the *Micareetum sylvicolae* (p. 366) and is also allied to both the *Coniocybetum furfuraceae* (p. 308) and the *Leprarietum incanae* (p. 312), both of which can also occur on rock.

Lecideetum orostheae (Hil.) Wirth

Diss. Bot., Lehre **17**, 122 (1972).—*Lecanoretum orostheae* Hil., *Čas. nár. Mus.* **1927,** 9 and 15 (1927) [as "Association à *Lecanora orosthea*"].

This association prefers somewhat better lit sites than other members of this alliance and as treated here tends also to be relatively species-rich. It is particularly well developed on the andesitic rocks of the Welsh border counties. *Haematomma ochroleucum* (including var. *porphyrium*), *Lecanora gangaleoides*, *L. subcarnea*, *Lecanactis dilleniana* and *Lecidea orosthea* are particularly characteristic of this community. Other species commonly encountered in it are *Catillaria chalybeia*, *Huilia albocaerulescens*, *Lecanora atra*, *L. grumosa*, *L. rupicola*, *Opegrapha saxatilis*, *O. saxicola*, *Parmelia glabratula* ssp. *fuliginosa* and *Rhizocarpon geographicum* s.l.

Leprarietum chlorinae Schade

Ber. dtsch. bot. Ges. **41**, (57) (1924).

This association is rather rare in Britain and largely confined to the central highlands of Scotland. It may be viewed as an upland counterpart of the *Lecideetum lucidae* and rarely includes species other than *Lepraria chlorina*.

Micareetum sylvicolae ass. nov. (Table XI)

This previously unrecognized association is closely allied to the *Lecideetum lucidae*, of which we were first inclined to treat it as a facies, but appears sufficiently distinct in both species composition and habitat, preferring

Table XI.

Micareetum sylvicolae ass. nov.

Species	Stands					
	1	2	3	4	5	6
Coniocybe furfuracea	–	–	–	–	1.2	–
Cystocoleus niger	1.2	–	–	–	–	1.3
Haematomma ochroleucum	–	1.2	–	+.0	–	–
Lecidea lucida	2.3	3.4	2.3	1.2	2.3	1.2
Lepraria incana	1.2	1.2	1.3	2.3	1.2	1.3
L. membranacea	–	–	–	+.0	1.2	–
Micarea clavulifera	–	1.2	–	–	–	–
M. polioides	–	–	–	–	2.3	–
M. semipallens	–	–	2.3	–	–	1.3
M. sylvicola	2.3	1.2	1.2	3.4	1.2	2.3
Opegrapha gyrocarpa	–	+.0	–	1.2	–	+.2
O. zonata	+.0	+.2	+.0	–	+.0	–
Porina chlorotica	+.0	–	–	–	–	+.2
P. lectissima	–	–	–	1.2	–	–
Racodium rupestre	–	1.2	–	–	–	–

1. Radnorshire, Rhayader, Glam lyn (22/940693): rock outcrop in sheltered old woodland, incl. 60°, aspect 235°, 1·0 × 1·0 m, cover 70%, 16 April 1976, P.W.J.
2. Island of Mull, Tobermory, Aros Woods, Sput Dùbh (17/509545): rock outcrop in dense old *Fagus* wood, incl. 90°, aspect 135°, 1·0 × 1·0 m, cover 60%, 17 October 1970, P.W.J.
3. Somerset, Porlock, Horner Combe, Cloutsham (31/898430): rock outcrop in woods, incl. 60°, aspect 270°, 0·5 × 0·5 m, cover 95%, 18 July 1969, P.W.J.
4. W. Ross, Upper Loch Torridon, Torridon House (18/868575): rock outcrop by woodland path, incl. 100°, aspect 235°, 1·0 × 1·0 m, cover 90%, 12 August 1966, P.W.J.; *type record*.
5. Merionethshire, Dolgellau, Aber Gwynant, near Kings Youth Hostel (23/683161): dry boulders in earth bank by stream, incl. 80°, aspect 270°, 1·0 × 1·0 m, cover 85%, 8 March 1961, P.W.J.
6. Merionethshire, Maentwrog, Llyn Mair (23/647416): sheltered boulder in old wood, incl. 90°, aspect 235°, 1·0 × 1·0 m, cover 80%, 16 April 1965, P.W.J.

particularly humid recesses (e.g. in woods), to merit separate treatment. Characteristic species of the *Micareetum sylvicolae* are *Lecidea lucida*, *Micarea clavulifera*, *M. polioides*, *M. semipallens*, *M. sylvicola* and *Lepraria incana*. It should be noted that the *Micarea sylvicola* species complex is poorly understood in Britain and in need of critical study.

Opegraphetum horistico-gyrocarpae Wirth

Herzogia **1**, 195 (1969).—*Opegrapha zonata* Soz. Degel., *Uppsal. Univ. Årsskr.* **1939** (1), 88 (1939).—*Opegraphetum zonatae* Wirth, *Herzogia* **1**, 195 (1969) [as "Degel. em."].

An association of continuously shaded rock underhangs and recesses in humid situations but preferring somewhat drier niches than the *Racodietum rupestris*. *Opegrapha gyrocarpa* and *O. zonata* are the most characteristic species of the association but may be joined by *O. lithyrga, O. saxigena, Porina chlorotica* and *P. lectissima*; the latter two species sometimes having high cover values. Two rare species apparently faithful to the *Opegraphetum horistico-gyrocarpae* in the British Isles are *Enterographa hutchinsiae* and *Rinodina oxydata*.

Racodietum rupestris Schade (Table XII)

Ber. dtsch. bot. Ges. **41**, (52) (1924) [as "*Rhacodietum*"], *Beih. Feddes Repert.* **76**, 16 (1934) [as "*Racodietum rupestris*"].—*Coenogonio-Racodietum rupestris* Schade, *Beih. bot. Cbl.* **49**, 436 (1932).—*Pertusario-Racodietum rupestris* Tobol., *Bull. Soc. Amis Sci. Lett. Poznan, sér.* D, **2**, 47 (1961).—*Cystocoleo-Racodietum rupestris* Kalb, *Diss. Bot.*, Lehre **9**, 93 (1970).

This distinctive association consists primarily of felted mats of *Cystocoleus niger* and/or *Racodium rupestre* mixed with varying amounts of *Lepraria incana* and/or *L. membranacea*. The *Racodietum rupestris* is confined to vertical or almost vertical hard siliceous rock faces in shaded humid situations and recesses which are not subject to direct rain. A considerable number of species occasionally enter into this association but generally at low frequencies; these include members of the *Opegraphetum horistico-gyrocarpae*, free-living *Trentepohlia* species and shade-loving bryophytes (e.g. *Dicranella heteromalla, Diplophyllum albicans, Sphenobilus minutus*). Although essentially saxicolous this association can spread on to adjacent compacted earth; in somewhat drier sites *Coniocybe furfuracea* may also

---

1. Yorkshire, near Goathland, Beck Hole Gorge (45/825028): fine grained sandstone, incl. 90°, aspect 0°, 20 × 20 cm, cover 95%, 29 May 1969, D.L.H.
2. Yorkshire, near Goathland, Mallyan Spout Gorge (45/824009): fine grained sandstone, incl. 90°, aspect 225°, 20 × 30 cm, cover 95%, 29 May 1969, D.L.H.
3. Yorkshire, near Goathland, Mallyan Spout Gorge (45/824009): fine grained sandstone, incl. 90°, aspect 180°, 25 × 30 cm, cover 80%, 29 May 1969, D.L.H.

TABLE XII.

Racodietum rupestris Schade.

Species	Stands							
	1	2	3	4	5	6	7	8
Baeomyces rufus	–	3	–	–	–	–	–	–
Cladonia coniocraea	–	–	–	–	–	–	–	–
C. macilenta	–	–	–	–	–	3	+	–
C. squamosa	–	–	–	–	–	+	–	–
Cystocoleus niger	8	8	4	–	–	5	8	8
Diploschistes scruposus	–	–	7	–	–	–	–	–
Fuscidea cyathoides	–	–	–	–	–	4	–	–
Huilia albocaerulescens	–	–	–	–	–	4	–	4
Hypogymnia physodes	–	+	–	–	–	–	–	–
Lecidea tumida	–	–	–	–	–	–	4	–
Lepraria incana	–	–	–	5	6	–	2	–
L. membranacea	7	–	4	6	4	3	–	4
Ochrolechia androgyna	–	–	–	–	–	3	–	–
Parmelia saxatilis	–	–	–	–	–	5	–	–
Pertusaria corallina	–	–	–	–	–	3	–	–
Racodium rupestre	–	–	–	7	7	–	–	–
Sphaerophorus globosus	–	–	–	–	–	+	–	–
Trapelia coarctata	–	3	–	–	–	–	–	–
Usnea subfloridana	–	–	–	–	–	+	–	–
Conocephalum conicum	–	3	–	–	–	–	–	–
Diplophyllum albicans	–	–	–	3	3	3	–	–
Frullania dilatata	–	–	–	–	–	3	–	–
Hypnum cupressiforme	–	–	4	–	–	+	–	–
Rhacomitrium heterostichum	–	–	–	–	–	–	–	5
Scapania umbrosa	–	3	–	–	–	+	–	–

4. Yorkshire, near Goathland, Mallyan Spout Gorge (45/824009): sandstone, incl. 90°, aspect 350°, 15 × 20 cm, cover 80%, 29 May 1969, D.L.H.
5. Durham, near Cotherstone, High Shipley Wood (45/014204): fine grained sandstone, incl. 120°, aspect 205°, 20 × 20 cm, cover 80%, 30 May 1969, D.L.H.
6. Cumberland, Buttermere, Scales Wood (35/17 – 16 –): Borrowdale volcanic rocks, incl. 96°, aspect 15°, 10 × 10 cm, cover 80%, 4 June 1969, D.L.H.
7. Derbyshire, Holloway, Lea Hurst (43/320560): millstone grit, incl. 95°, aspect 90°, 20 × 20 cm, cover 95%, 5 August 1967, D.L.H.
8. Cumberland, Keswick, Castle Head Wood (35/269226): Borrowdale volcanic rocks, incl. 88°, aspect 250·, 20 × 15 cm, cover 80%, 3 June 1969, D.L.H.

be present in it producing stands transitional to the mainly corticolous *Coniocybetum furfuraceae* (p. 308).

Mention should also be made here of a particularly interesting community of vertical rock faces in old coniferous woodland which is encountered rarely in central Scotland (e.g. Glen Strathfarrar) and comprises *Arthonia arthonioides*, *Cystocoleus niger*, *Diploschistes scruposus* and *Haematomma ochroleucum*. The syntaxonomy of this community requires further study but it may prove to represent an undescribed association.

B. Exposed

All. 15. *Lecideion tumidae*

Lecideion tumidae Wirth, *Diss. Bot.*, *Lehre* **17,** 131 (1972).

This species-diverse alliance was described by Wirth (1972) to accommodate communities on exposed siliceous rocks, boulder scree and heathland pebbles in inland sites. The communities are dominated by crustose lichens of which *Fuscidea cyathoides*, *Huilia crustulata*, *Lecidea lithophila*, *L. tumida*, and *Rhizocarpon geographicum* aggr. are particularly characteristic. *Stereocaulon dactylophyllum* and *S. evolutum* are also important species in some associations of the alliance of which three are here accepted as British. The alliance, poorly understood in this country, appears to have very close affinities with species-poor facies of both the *Parmelietum glomelliferae* and the *Umbilicarion cylindricae*; the relationship of the *Pertusarietum corallinae* in particular merits detailed investigation.

Huilietum crustulatae (Klem.) comb. nov.

Lecideetum crustulatae Klem., *Ber. bayer. bot. Ges.* **28,** 254 (1950) [as "*Lecanoretum coarctatae* Duvign. 1939 p.p."], basionym.

This is primarily an association of small, more or less firmly anchored pebbles or stones in open tracts of heathland, but it can also occur on larger outcrops and boulders. The characteristic species are *Baeomyces rufus*, *Huilia macrocarpa*, *H. crustulata*, *Lecanora polytropa*, *Lecidea erratica*, *L. plana* (chiefly in Scotland), *L. tumida*, *Rhizocarpon obscuratum* and sometimes *Trapelia coarctata* s.s. Adjacent terricolous communities are frequently rich in the more widespread species of *Cladonia* as well as *Lecidea granulosa* and *L. uliginosa*. Communities dominated by *Baeomyces rufus* and *Trapelia coarctata* s. lat. on shaded rocks and stones are treated here as a species-poor facies of this widespread moorland association which requires particularly humid sites.

Lecideetum lithophilae Wirth

Herzogia **1,** 202 (1969).

This association is essentially one of sunny hillside rock outcrops and screes as well as siliceous walls in well lit situations in western areas of Britain, but is also sometimes encountered in a fragmentary state on acid gravestones in, for example, the least polluted areas of southern England. The *Lecideetum lithophilae* is commonly rich in species, the most frequent of which are *Buellia aethalea*, *Huilia albocaerulescens*, *H. macrocarpa*, *Lecanora intricata* (including var. *soralifera*), *Lecidea fuscoatra*, *L. lithophila*, *L. pantherina*, *L. tumida* and *Rhizocarpon geographicum* aggr.

Lecidea leucophaea, *L. pelobotryon*, *Placopsis gelida*, *Pilophorus strumaticus*, *Trapelia coarctata* s.l., *T. moorei* and *Stereocaulon pileatum* are often associates in a wetter facies of this association, while in nutrient-rich sites *Acarospora fuscata* and *Candelariella vitellina* are often also present. The presence of *Stereocaulon* species characteristic of sites rich in heavy metals, including *S. delisei*, *S. dactylophyllum* and *S. pileatum*, suggests that this association has affinities with the *Acarosporion sinopicae* (p. 382).

Pertusarietum corallinae Frey

Mitt. naturf. Ges. Bern **6,** 163 (1922) [as "*Pertusaria corallina*-Ass."].

This is a widespread species-rich community of well lit siliceous rock outcrops, boulders and walls in sites which are not nutrient-enriched. Important component species of this association in the British Isles include *Cladonia coccifera*, *C. squamosa*, *Diploschistes scruposus*, *Fuscidea cyathoides*, *Lecanora badia*, *L. polytropa*, *Lecidea leucophaea*, *L. tumida*, *Parmelia glabratula* ssp. *fuliginosa*, *P. saxatilis*, *Pertusaria corallina*, *P. dealbata*, *P. lactea*, *P. pseudocorallina*, *Rhizocarpon geographicum*, *R. lecanorinum*, *Schaereria cinereorufa* and *Umbilicaria polyphylla*.

The *Pertusarietum corallinae* tends to prefer moister sites than the *Umbilicarietum cylindricae* and nutrient-poorer sites than the *Lecanoretum sordidae*, preferring, for example, vertical rather than horizontal rock surfaces and walls.

All. 16. *Pseudevernion furfuraceae*

Pseudevernion furfuraceae (Barkm.) P. James *et al.*; see p. 334 for nomenclature.

Pseudevernietum furfuraceae (Hil.) Kalb

See p. 335 for nomenclature and composition.

This primarily corticolous association is widespread as a species-poor variant (see p. 335) on hard acid siliceous rocks and walls subject to moderate air pollution. The *Pseudevernietum furfuraceae* also occurs in the British Isles on walls and rock debris; in this case the association appears to represent a seral stage giving way to associations of the *Lecideion tumidae* or the *Umbilicarietum cylindricae*.

All. 17. *Rhizocarpion alpicolae*

Rhizocarpion alpicolae Frey, Ber. Geobot. Inst. Rübel **1932**, 46 (1933).

The *Rhizocarpion alpicolae* is an arctic-alpine alliance of nutrient-poor siliceous rocks. Due to the relatively mild oceanic climatic conditions, the alliance seems to be poorly represented in the British Isles and mainly confined to the summits of mountains in Scotland and northern England. The two associations recognized here are characterized by a variety of crustose species of which *Fuscidea kochiana* and *Rhizocarpon alpicola* are the respective diagnostic components. Crustose lichen elements of the *Umbilicarion cylindricae* are almost invariably present and macrolichens of the same alliance may also be represented.

This alliance is currently poorly understood in Britain due mainly to the lack of information on the lichen flora of the summits of many Scottish mountains. The following two associations are tentatively delimited here as a guide to further study of this interesting alliance.

Fuscideetum kochianae (Wirth) comb. nov.

Lecideetum kochianae Wirth, Diss. Bot., Lehre **17**, 190 (1972) [as "Ullrich et Wirth 1969 em. et nom. nov."]; basionym.—*Lecideetum kochiano-aggregatilis* Ullrich & Wirth, *Herzogia* **1**, 199 (1969) [non Tobol. 1961].

The name *Fuscideetum kochianae* is applied to an association dominated by *Fuscidea kochiana* and including *Lecanora atra*, *L. intricata* (including var. *soralifera*), *L. polytropa*, *Lecidea leucophaea*, *L. lithophila*, *L. pantherina*, *L. tenebrosa*, *Parmelia incurva*, *Pertusaria corallina*, *P. pseudocorallina* and *Rhizocarpon geographicum*. Species of *Umbilicaria*, such as *U. polyphylla*, *U. polyrrhiza* and *U. torrefacta*, together with *Lasallia pustulata*, are sometimes also represented but are confined to the apices of slightly nutrient-enriched erratics.

This association is characteristic of very hard siliceous rocks, especially quartzite, granite and gritstones. *Fuscidea kochiana* is usually abundant, the associated species often forming scattered mosaics between the thalli of this dominant species. The *Fuscideetum kochianae* occurs in moderately exposed to relatively sheltered and shaded situations and is particularly well developed on rocks and boulders on the upper slopes of Penyghent and Ingleborough, Yorkshire, and parts of Cairngorm, Scotland.

The *Fuscideetum kochianae* has particularly close affinities with the *Parmelietum omphalodis*.

Rhizocarpetum alpicolae Frey

Verh. naturf. Ges. Basel **35,** 310 (1923).—*Rhizocarpetum geographicae* Frey, *Veröff. Geobot. Inst. Rübel* **4,** 230 (1927) [as "*Rhizocarpon geographicum*-Ass."].

The arctic-alpine *Rhizocarpetum alpicolae* is rare in the British Isles but is locally widespread on rock outcrops and large boulders on the summits of mountains of the Cairngorm range. The association, which is characteristic of rocks rich in silica but very poor in nutrients and calcium, appears to represent a "climax" stage following the *Umbilicarietum cylindricae* but, unlike that association, is dominated by crustose species rather than macrolichens.

The *Rhizocarpetum alpicolae* is best developed on the south- to east-facing vertical sides of rock outcrops which are partially buried in snow for 2 or 3 months. From a distance these sites appear noticeably yellow to pale green in colour due to an abundance of *Lecanora intricata, L. polytropa, Rhizocarpon alpicola* and *R. geographicum* aggr. The mosses *Andraea alpina, A. nivalis* (very rare), *Grimmia doniana, Rhacomitrium heterostichum* var. *gracilescens,* and *R. lanuginosum* may enter into the association. Additional lichens characteristic of the association are *Lecidea atrata, L. confluens* aggr., *L. lapicida, L. pantherina, L. tumida* and *Pseudephebe pubescens.* The rare *Lecanora leptacina* (on mosses) and *Ophniospora atrata* may also be present. Elements of the *Umbilicarietum cylindricae* often also occur, particularly *Cornicularia normoerica, Haematomma ventosum, Lecanora badia, Lecidea tenebrosa, Parmelia alpicola, P. incurva, P. omphalodes, Umbilicaria cylindrica, U. hyperborea, U. polyphylla* and *U. torrefacta.*

In wetter sites characteristic of late snow lie, *Rhizocarpon badioatrum* enters as an important species of the association. On Cairngorm mountain itself, *Lecidea pycnocarpa* is abundant on small quartzite pebbles in shallow hollows in wind-eroded plateau sites; *Cetraria delisei, C. nivalis* and, more rarely, *Cornicularia divergens* may utilize these pebbles for anchorage.

All. 18. *Umbilicarion cylindricae*

Umbilicarion cylindricae Frey, Ber. Geobot. Inst. Rübel **1932**, 40 (1933).—*Umbilicarion hirsutae* Follm., Hess. flor. Briefe **2**, 25 (1973).

This alliance is characterized by many species which are strictly montane in the British Isles, including *Cetraria commixta, C. hepatizon, Cornicularia normoerica, Hypogymnia intestiniformis, Parmelia alpicola, P. stygia, Platismatia norvegica, Pseudephebe pubescens* and species of *Umbilicaria*, notably *U. crustulosa* (Langdale Pikes, Lake District only), *U. cylindrica, U. hyperborea, U. polyphylla, U. polyrrhiza, U. proboscidea* and *U. torrefacta*; *U. deusta* occurs in deep hollows and is characteristic of a moist facies of this alliance.

Many crustose species enter the associations of this alliance of which *Fuscidea cyathoides, F. tenebrica, Haematomma ventosum, Lecanora intricata, L. polytropa, Lecidea pantherina* aggr., *L. sulphurea, Rhizocarpon geographicum* aggr., *R. lecanorinum, R. obscuratum* and *R. polycarpon* are particularly characteristic. The alliance tends to occur on nutrient-poor rather than nutrient-enriched siliceous rocks.

The two associations recognized here are distinguished by both their habitats and species composition.

Parmelietum omphalodis DR.

Akad. Abhandl. Uppsala **1921**, 164 (1921) [not seen]; see Delzenne-van Haluwyn (1976) for later usages.

The *Parmelietum omphalodis* has a strong affinity with the saxicolous facies of the *Pseudevernietum furfuraceae* but is richer in species and differs in the abundance of *Parmelia omphalodes* (see Fig. 10) and species of *Cladonia*, especially *C. cervicornis, C. chlorophaea* aggr., *C. coccifera, C. crispata* var. *cetrariiformis, C. furcata, C. gracilis, C. squamosa, C. subcervicornis* and, occasionally, *C. uncialis* ssp. *dicraea*. Other species characteristic of the association, which is widely distributed in the British Isles on nutrient-poor siliceous rock outcrops, boulders and scree, include *Bryoria bicolor, B. fuscescens, Cetraria chlorophylla, Haematomma ventosum, Hypogymnia physodes, H. tubulosa, Lecanora polytropa, Lecidea tenebrosa, Lepraria neglecta, Ochrolechia androgyna, O. tartarea, Parmelia glabratula* ssp. *fuliginosa, P. saxatilis, Pertusaria corallina, P. lactea, P. monogona* (rare), *P. pseudocorallina, Platismatia glauca, Rhizocarpon hochstetteri, Sphaerophorus globosus* and *Usnea flammea*.

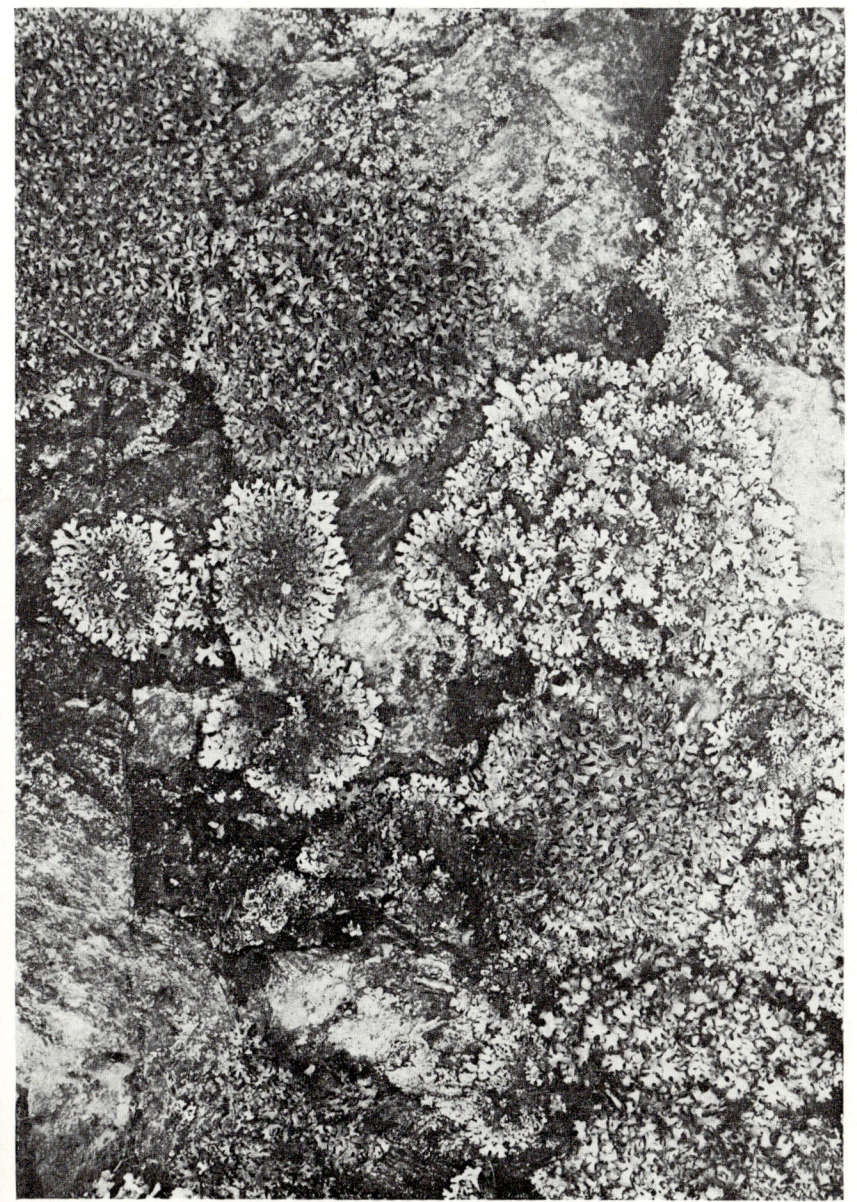

FIG. 10. *Parmelietum omphalodis* on siliceous rock; the *Parmelia*-rich facies. Predominant species: *Parmelia omphalodes*, *P. saxatilis* and *Trapelia coarctata* s.l. (Anglesey: 1972, R. O. Millar).

The association is often rich in bryophytes of which *Andraea rothii*, *A. rupestris* (syn. *A. petrophila*), *Cynodontium polycarpum* (rare), *Hypnum cupressiforme*, *Isothecium myosuroides*, *Orthotrichum anomalum*, *Rhacomitrium fasciculare* and *R. heterostichum* may be considered as characteristic. The *Parmelietum omphalodis* occurs at lower altitudes and latitudes than the *Umbilicarietum cylindricae* often occurring even in or near coastal areas in western parts of the British Isles. At higher altitudes it tends to grade into the *Umbilicarietum cylindricae* and species such as *Cornicularia normoerica*, *Rhizocarpon lecanorinum*, *R. polycarpon*, *Sphaerophorus fragilis*, *Umbilicaria polyphylla* and *U. polyrrhiza* become more abundant. As a rule the *Parmelietum omphalodis* prefers more sheltered sites than the *Umbilicarietum cylindricae*.

Klement (1955) treated Du Rietz's name *Parmelion omphalodis* as a synonym of the *Umbilicarion cylindricae* reserving a homonym of the latter name for a different, more upland, community including, for example, *Hypogymnia intestiniformis*. The usage adopted here follows that of Wirth (1972).

Umbilicarietum cylindricae (Frey) Frey

Ber. Geobot. Inst. Rübel **1932**, 40 (1933).—*Gyrophoretum cylindricae* Frey, Mitt. naturf. Ges. Bern **6**, 168 (1922) [as "*Gyrophora cylindrica*-Ass."].—*Umbilicarietum deustae* Frey, Ber. Geobot. Inst. Rübel **1932**, 49 (1933).

The *Umbilicarietum cylindricae* is a more upland and more light-demanding counterpart of the *Parmelietum omphalodis* characterized by species typical of the alliance (p. 374). Both associations may occur in very close proximity to one another and the relative abundance of the characteristic species may vary considerably in different sites depending on both geographical location and altitude (particularly as its important components are limited climatically in the British Isles). This association is optimally developed on the uppermost slopes of mountains in the Scottish Highlands where it represents a seral stage leading to the *Rhizocarpetum alpicolae*, an association with a very limited distribution in the British Isles.

The synsystematics of the *Umbilicarietum cylindricae* are poorly understood in Britain but our observations suggest that they are very complex. Preliminary studies indicate that there are several distinctive noda (indicated below), some of which may be referable to the *Parmelietum omphalodis*, and others of which may well prove to merit recognition as separate associations when thoroughly investigated.

1. Dominated by *Fuscidea tenebrica* with *F. cyathoides* and *Rhizocarpon*

geographicum as associated species, and characteristic of sunny rocks in exposed mountainous situations which are particularly abundant in the Lake District and western Scotland.

2. A shady and rather dry facies dominated by *Fuscidea taeniarum* known in Britain only from Scotland where it is rare and always found on acid, nutrient-poor rock.

3. Dominated by *Fuscidea cyathoides* but including *Lecidea leucophaea, L. lithophila, L. pantherina, Pertusaria corallina* and *Rhizocarpon geographicum* aggr. (often *R. riparium*) as characteristic associates. This nodum tends to occur at lower altitudes and latitudes than the preceding two noda and is particularly well developed in the Southern Pennines and Peak District on millstone grit rocks. This nodum is close to the *Parmelietum omphalodis* but is placed here tentatively as it sometimes includes *Umbilicaria polyphylla, U. polyrrhiza* or *U. torrefacta*.

4. A facies on mineral-rich rocks often in rather moist situations as in shallow valleys or near mountain lakes dominated by *Lecidea lithophila* but including *L. lapicida* and *L. leucophaea*; *Alectoria nigricans* and *Cornicularia aculeata* are frequent associates on mosses in such communities. This facies has a clear affinity with the *Acarosporetum sinopicae* (p. 383) but is poor in species restricted to mineral-rich sites while including many mineral-tolerant ones.

5. A sun-loving markedly western community particularly well developed in the Rhinog mountains in Wales dominated by *Lecanora mauroides* and *Rhizocarpon geographicum* but including *Lecidea pantherina* and *L. tenebrosa* at high frequencies.

6. The *Umbilicaria deusta*-dominated nodum treated as a distinct association by many authors (the *Umbilicarietum deustae* Frey, see above) which is well developed by the sides of upland mountain streams.

7. The very high altitude *Umbilicaria hyperborea* nodum.

8. The *Umbilicaria crustulosa*-dominated nodum of almost vertical south-facing, hard volcanic rocks found on the Langdale Pikes in the Lake District (see Brightman, 1962, for lists of associates).

The "typical" *Umbilicarietum cylindricae* is taken here to include high frequencies of *Umbilicaria cylindrica, U. polyphylla, U. polyrrhiza, U. proboscidea* and/or *U. torrefacta*. This nodum has a particular preference for well lit rather coarse-grained rocks in mountainous areas (particularly granite) with *Cornicularia normoerica, Pseudephebe pubescens* and the other macrolichens mentioned as characteristic of the alliance on p. 374 occurring in it. Associated bryophytes include *Andraea rupestris, Grimmia doniana, G. trichophylla, Gymnomitrion* spp., *Hedwigia ciliata, Ptychomitriun polyphyllum, Rhacomitrium fasciculare* and *R. heterostichum* var. *heterostichum*.

C. Nutrient-enriched

All. 19. *Parmelion conspersae*

Parmelion conspersae Hadač, *in* Klika & Hadač, *Příroda, Brno* **36**, 254 (1944), emend. Wirth, *Diss. Bot., Lehre* **17**, 131 (1972).—*Parmelion saxatilis* Klem., *Ber. bayer. bot. Ges.* **28**, 257 (1950).—*Acarosporion fuscatae* Klem., *Ber. bayer. bot. Ges.* **28**, 257 (1950).

The *Parmelion conspersae* comprises associations of well lit and slightly to markedly nutrient-enriched siliceous rocks. It can be considered as the acid rock counterpart of the *Xanthorion parietinae*, differing from that alliance in too many species to be treated as part of the same alliance. Some elements of the *Parmelion conspersae* are well represented in the terrestrial zone in coastal areas (Fletcher, 1973b) and there is thus some affinity with the *Ramalinetum scopularis* (see p. 386).

Candelarielletum corallizae Massé

Vegetatio **12**, 173 (1964).

Essentially an association of nutrient-enriched siliceous rocks such as birds' perching stones, the *Candelarielletum corallizae* is frequent in suitable sites in both coastal and upland parts of the British Isles. The characteristic species of this community include *Acarospora fuscata*, *Anaptychia fusca*, *Aspicilia caesiocinerea*, *Buellia canescens*, *B. punctata*, *Candelariella coralliza*, *C. vitellina*, *Lecanora muralis*, *Physcia caesia*, *P. dubia* (especially on gravestones in eastern and central England), *P. tribacia*, *P. wainioi*, *Ramalina polymorpha*, *R. subfarinacea*, *Rinodina subexigua*, *Xanthoria candelaria* and *X. parietina*. On the sea-shore, additional species entering the association include *Aspicilia leprosescens*, *Caloplaca* sp. (undescribed), *C. verruculifera*, *Lecanora poliophaea* and *Parmelia britannica* as well as the three rare species *Caloplaca scopularis*, *Candelariella arctica* and *Lecanora straminea*. *Lecanora fugiens* and the rarer and related *L. andrewii* may also belong here; *Physcia subobscura* and *Verrucaria fusconigrescens* may be locally abundant also.

In our view the relationship between the *Candelarielletum corallizae* and the *Physcietum caesiae* (see p. 360) is in need of a critical re-investigation. Further study may also indicate that the sea-shore facies of this association is better subsumed under the *Ramalinetum scopularis*.

Lecanoretum sordidae Hil. (Table XIII)

Čas. národ. Mus. **1923**, 4 (1924) [as "asociace *Lecanora sordida*"].—*Lecanoretum rupicolae* Wirth, *Diss. Bot., Lehre* **17**, 166 (1972) [as "Hilitzer 1925"].

Although the *Lecanoretum sordidae* is accepted here in accordance with Wirth (1972), we are not entirely convinced that it is sufficiently distinct from the *Parmelietum glomelliferae* to merit a permanent status. The major difference between these two communities is that the *Lecanoretum sordidae* is richer, in terms of both species number and cover, in crustose lichens. Species which are characteristic of the association include *Acarospora fuscata*, *Buellia stellulata*, *Candelariella vitellina*, *Diploschistes scruposus*, *Huilia albocaerulescens*, *Lasallia pustulata*, *Lecanora atra*, *L. badia*, *L. grumosa* (chiefly confined to Scotland), *L. polytropa*, *L. rupicola*, *Lecidea insularis* (parasitic on *Lecanora rupicola*), *L. sulphurea*, *L. tenebrosa*, *Ochrolechia parella*, *Parmelia conspersa*, *P. glabratula* ssp. *fuliginosa*,

TABLE XIII.

Lecanoretum sordidae Hil.

Species	Stands				
	1	2	3	4	5
Acarospora fuscata	–	1.2	–	–	1.2
Aspicilia caesiocinerea	+.1	1.2	+.2	1.2	+.2
A. sp.	+.2	–	+.0	–	–
Bacidia umbrina	–	–	1.2	–	–
Buellia stellulata	1.2	1.2	2.3	+.0	1.2
Candelariella vitellina	1.2	1.2	–	2.3	1.2
Catillaria chalybeia	1.2	–	–	–	1.2
Diploschistes scruposus	1.2	–	–	1.2	–
Lasallia pustulata	–	–	1.2	+.2	–
Lecanora atra	1.2	1.2	1.2	+.1	1.4
L. badia	1.2	1.3	2.3	–	2.2
L. gangaleoides	1.3	+.0	1.2	–	–
L. grumosa	–	–	–	–	2.3
L. intricata	+.2	–	–	–	+.0
L. polytropa	1.4	1.2	+.2	1.2	1.4
L. rupicola	3.3	2.3	2.2	1.3	2.3
Lecidea orosthea	–	+.0	–	1.2	–
L. sulphurea	+.2	1.2	–	2.3	1.3
L. tumida	1.4	1.2	1.2	+.2	1.2
Ochrolechia parella	–	–	1.2	–	1.2

TABLE XIII—continued

Species	Stands				
	1	2	3	4	5
Parmelia glabratula ssp. *fuliginosa*	+.2	1.2	1.3	–	+.0
P. conspersa	–	–	–	2.3	3.2
P. verruculifera	–	1.2	–	1.3	+.0
Pertusaria pseudocorallina	1.2	–	3.2	1.2	1.2
Rhizocarpon geographicum	+.2	–	1.2	+.0	–
R. obscuratum	–	–	–	+.2	+.0
R. viridiatrum	1.2	+.0	1.2	2.3	–

1. Radnor, New Radnor, Stanner Rocks (32/262584): rock outcrop (andesite), incl. 45°, aspect SW, 1·0 × 1·0 m, cover 95%, 11 April 1976, P.W.J. and R. Woods.
2. Shropshire, Church Stoke, Roundton (32/291948): rock outcrop (andesite), incl. 30°, aspect S, 1·0 × 1·0 m, cover 98%, 13 April 1976, P.W.J.
3. Radnor, Llandrindod Wells, above Shakey Bridge, Cefnllys (32/087614): rock outcrop, slight shade, incl. 20°, aspect W, 1·0 × 1·0 m, cover 90%, 11 April 1976, P.W.J. and R. Woods.
4. Montgomeryshire, Welshpool, Criggion, Breidden Hill (33/29 – 14 –): large boulder on sunny hillside, incl. *c*. 20°, aspect NW, 1·0 × 1·0 m, cover 100%, 12 April 1963, P.W.J.
5. Kirkcudbrightshire, Dalbeattie, Moyl Peninsula (25/830526): large boulder in sunny situation: incl. 10°, aspect S, 1·5 × 1·5 m, cover 80%, 25 April 1976, P.W.J. and P. Topham.

P. verruculifera, *Pertusaria flavicans*, *P. pseudocorallina*, *Rhizocarpon geographicum* aggr., *R. viridiatrum* and *Rinodina atrocinerea*. In inland sites in the British Isles the *Lecanoretum sordidae* is well developed on andesitic outcrops in the West Midlands.

In addition to the strong affinity of this association with the *Parmelietum glomelliferae*, it also appears to be allied to a crustose lichen-dominated facies of the *Ramalinetum scopularis* on coastal rocks where, in addition to the species listed above, the following exclusively maritime species occur in it: *Diploschistes caesioplumbeus*, *Lecidella subincongrua*, *Pertusaria ceuthocarpoides* and *Rhizocarpon constrictum*.

Parmelietum glomelliferae Hil. (Table XIV)

Čas. národ. Mus. **1923**, 7 (1924) [as "*Parmelia glomellifera* . . . asociaci"].—*Umbilicarietum pustulatae* Hil., *Preslia* **3**, 16 (1925) [as "Asociace *Umbilicaria pustulata*"].—*Lasallietum pustulatae* (Hil.) Wirth, *Diss. Bot.*, Lehre **17**, 152 (1972).—*Parmelietum isidiotylae* Frey, *in* Frey & Ochsner, *Arvernia* **2**, 68 (1926).

—*Parmelietum conspersae* Hil., *Preslia* **3**, 16 (1925) [as "asociace *Parmelia conspersa*"].

The *Parmelietum glomelliferae* is treated here in a broad sense to include communities of nutrient-enriched rocks dominated by species of *Parmelia* and *Umbilicaria* s.l. The *Parmelietum conspersae* and *Umbilicarietum pustulatae* have frequently been maintained as distinct entities by continental authors but as they commonly appear to intergrade in the British

TABLE XIV.

Parmelietum glomelliferae Hil.

Species	Stands				
	1	2	3	4	5
Acarospora fuscata	2	3	2	1	–
Candelariella vitellina	2	2	4	–	2
Cladonia coccifera	–	1	–	2	–
Fuscidea cyathoides	2	3	–	4	2
Lasallia pustulata	–	8	–	–	–
Lecanora polytropa	2	2	4	+	2
L. rupicola	–	–	2	–	1
Lecidea sulphurea	–	2	2	–	1
Parmelia conspersa	6	4	5	4	7
P. glabratula ssp. *fuliginosa*	–	1	–	–	3
P. loxodes	6	–	3	3	5
P. mougeotii	–	–	6	–	3
P. omphalodes	4	1	–	2	4
P. saxatilis	2	–	–	3	3
Rhizocarpon geographicum aggr.	2	2	1	2	2
Trapelia coarctata	2	–	2	3	3
Umbilicaria polyrrhiza	–	–	–	5	–
Xanthoria candelaria	–	3	–	–	–

1. South Devonshire, Dartmoor, Harford (20/642601): granite blocks in pasture, ± level, 10 × 10 cm, cover 95%, 3 April 1969, D.L.H.
2. South Devonshire, Dartmoor, near Wistman's Wood (20/613768): granite boulder used as bird perch, uneven, 20 × 20 cm, cover 90%, 15 September 1969, D.L.H.
3. South Devonshire, Dartmoor, Rippon Tor (20/748757): granite boulder in clatter, incl. 160°, aspect 45°, 20 × 20 cm, cover 80%, 30 August 1976, D.L.H.
4. South Devonshire, Dartmoor, Sharp Tor (20/686731): granite boulder in heathland, ± level, 20 × 20 cm, cover 75%, 29 July 1974, D.L.H.
5. South Devonshire, Dartmoor, Crockern Tor (20/616757): granite boulder in clatter near summit, incl. 140°, aspect 165°, 20 × 20 cm, cover 85%, 15 September 1969, D.L.H.

Isles they are treated as a single association here and subsumed under the earliest available name we have been able to locate—the *Parmelietum glomelliferae*.

The characteristic species of this community in the British Isles are *Acarospora fuscata*, *Candelariella vitellina*, *Cladonia coccifera*, *C. pityrea*, *C. squamosa*, *Lasallia pustulata*, *Lecanora polytropa*, *Parmelia britannica*, *P. conspersa*, *P. disjuncta*, *P. loxodes*, *P. mougeotii*, *P. saxatilis*, *P. sulcata*, *P. verruculifera*, *Trapelia ornata*, *Umbilicaria deusta* and *U. polyrrhiza*. A very large number of other species may occasionally enter the association, particularly elements of the *Parmelietum omphalodis* and *Pseudevernietum furfuraceae*; affinities with the *Ramalinetum scopularis* (p. 386) are also marked. The relationship of the present association to the *Lecanoretum sordidae* is discussed under the latter community (p. 380).

All. 20. *Xanthorion parietinae*

Xanthorion parietinae Ochsn. See p. 342 for nomenclature.

Physcietum caesiae Mot.

See p. 360 for composition and nomenclature.

This association, characteristic of extremely nutrient-enriched sites, can occur on suitably modified siliceous rocks, especially near colonial nesting sites of birds, as well as on limestones, man-made substrates and, rarely, trees.

D. Mineral-rich

All. 21. *Acarosporion sinopicae*

Acarosporion sinopicae Wirth, *Diss. Bot.*, Lehre **17,** 131 (1972); type: *Acarosporetum sinopicae* Hil.

This distinctive alliance was introduced by Wirth (1972) for associations restricted to rocks rich in heavy metals. The two associations recognized here are rather specialized and frequently contain several species restricted to metal-rich rocks. In Britain they occur (1) on spoil tips from old mine workings for copper, lead, silver and associated metals, (2) on walls contaminated by heavy metals, principally lead derived from car exhausts or particulate fall-out from smelters, (3) on naturally occurring metal-rich rock outcrops, and (4) occasionally in the vicinity of, or even on, rusted iron rails, worked lead (as in stained-glass windows), or in areas affected by water run-off from corrugated iron roofing. The thalli of several crustose species characteristic of this alliance are often wholly or partially rust red in colour ("oxydated"; see Chapter 2).

A review of heavy metal tolerance in lichens is provided by James (1973).

Acarosporetum sinopicae Hil.

Čas. národ. Mus. **1923**, 8 (1924) [as "*Acarospora sinopica*-asociace"].

Characteristic species of this association in Britain include *Acarospora sinopica, A. smaragdula* (including var. *lesdainii*), *Bacidia umbrina, Candelariella vitellina, Diploschistes scruposus, Huilia macrocarpa* (as "f. *oxydata*"), *Lecanora intricata* var. *soralifera, L. polytropa, L. subaurea, Lecidea atrata, L. lapicida* (as "f. *oxydata*"), *L. pantherina, L. silacea* (confined to upland sites), *Rhizocarpon obscuratum, R. oederi, Stereocaulon nanodes, S. pileatum* and *Toninia leucophaeopsis* (upland sites only). Other species of *Stereocaulon* (*S. dactylophyllum, S. delisei, S. evolutum* and *S. vesuvianum*) may also be present. In areas subject to moderate to severe sulphur dioxide pollution *Lecanora conizaeoides* enters this community, sometimes becoming an important component of it; in such sites the association becomes species-poor.

Lecanora subaurea is a very significant member of this association in the southern Pennines (Earland-Bennett, 1975).

Lecanoretum epanorae Wirth

Diss. Bot., Lehre **17**, 173 (1972).

This association is closely allied to the *Acarosporetum sinopicae* but tends to prefer more sheltered, dry situations, being found particularly in dry crevices of mineral-rich acid rocks or on the more sheltered sides of old walls. The community, which is rather species-poor, tends to be dominated by *Lecanora epanora*. Other commonly associated species are *Lepraria incana* aggr., *L. membranacea* aggr. and *Rhizocarpon obscuratum*. Details of the ecology and British distribution of *Lecanora epanora* are given by Earland-Bennett (1975). The occasional presence of *Lecanora subaurea* in the *Lecanoretum epanorae* is indicative of the close relationship of this association with the *Acarosporetum sinopicae*.

Hilitzer (1924) regarded *Lecanora epanora* as an important component of the *Acarosporetum sinopicae* but his material was most probably *L. subaurea*.

E. Marine and Maritime Communities

The marine and maritime lichen vegetation of rocky shores of the British Isles has recently been the subject of considerable attention (Fletcher, 1973a, 1975a,b, 1976). On the basis of Fletcher's careful analyses on the ecology and physiology of sea-shore lichens it seems desirable to recognize four associations for maritime lichen communities in Britain. Three of these, broadly corresponding to the zones defined by Fletcher (1973a,b), are the littoral (black) zone (*Verrucarietum maurae*), the mesic- and sub-mesic-supralittoral (orange) zone (*Caloplacetum marinae*), and the xeric-supralittoral (grey) zone (*Ramalinetum scopularis*); the drier more sheltered aspects of the latter zone support the fourth and very distinctive association described here as the *Sclerophytetum circumscriptae*.

Approximately 65 lichen species are more or less exclusively maritime in Britain, the greatest concentration of these ecologically restricted species occurring in the littoral zone and decreasing further up the shore to very few in the terrestrial zone. The terrestrial region above the xeric-supralittoral zone supports an assemblage of diverse elements of basically non-maritime affinities described elsewhere in this contribution; the species occurring there are more or less maritime influence-tolerant or show non-maritime preferences although some species of the *Ramalinetum scopularis* can occur far inland in Britain (p. 386). Some of the nutritional aspects underlying lichen zonation on rocky shores are discussed by Fletcher (1976).

Communities developed on coastal limestones, dunes, and shingle are treated on pp. 349, 400–402 and 394–395, respectively.

The associations recognized in this section are not referred to alliances as their affinities with other communities seem remote; all may merit placing in distinct alliances.

Caloplacetum marinae DR.

Svensk. växtsociol. Sällsk. Handl. **2**, 50 (1925) [as "*Caloplaca marina*-Ass."].

The *Caloplacetum marinae* is the characteristic association of the sub-mesic- and mesic-supralittoral zones on rocky shores around the British Isles. As with the *Ramalinetum scopularis* which occurs higher up the shore, the component species, though rarely, if ever, submerged in sea-water, do nevertheless appear to have a strong requirement for sea-water spray and are related to a moisture source of neutral or alkaline pH. The association is dominated by *Caloplaca* species (Fig. 11), particularly *C. marina*, *C. microthallina*, *C. thallincola* and, in more nutrient-enriched sites,

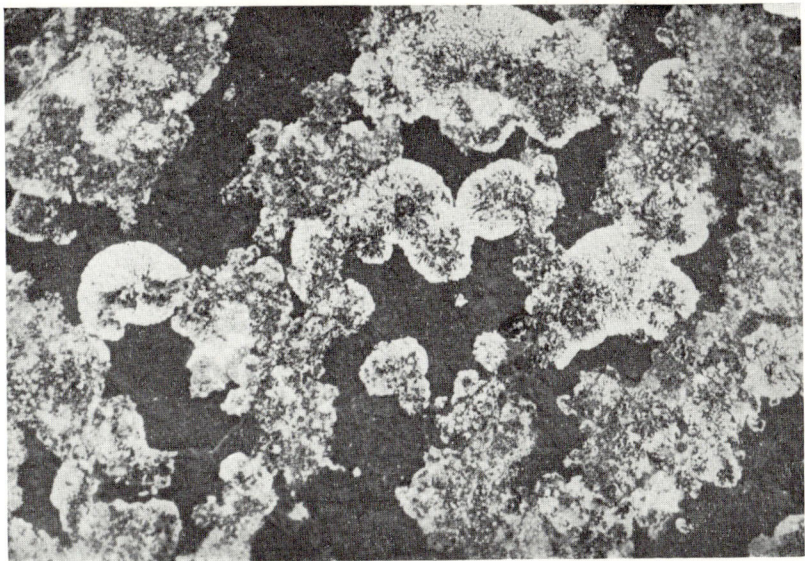

Fig. 11. *Caloplacetum marinae* on maritime siliceous rocks. Predominant species: *Caloplaca marina*, *C. thallincola* and *Verrucaria maura* (Mid-Ebudes: Island of Mull, Ross of Mull, Carsaig Bay. 1972, P. W. James).

C. verruculifera. *Lecanora actophila* and *L. helicopis* are faithful to the association in the British Isles, the former tending to prefer more sunny, well lit situations. *Verrucaria maura* is often present (e.g. Fig. 11), especially on more sheltered shores where it is often overgrown by *Caloplaca thallincola*. In northern and some western areas *Arthonia phaeobaea* and *Lecania aipospila* also form part of this association and may become locally abundant. *Catillaria chalybeia* and *Lecania erysibe* are often common although these species, unlike the others mentioned above, are by no means restricted to coastal rocks. *Lichina confinis*, occurring in the lowest part of the mesic-supralittoral zone, forms a link with the *Verrucarietum maurae*; similarly *Xanthoria parietina*, characteristic of the submesic-supralittoral, interconnects with the *Ramalinetum scopularis* association.

A more nutrient-enriched facies of this association, commonly including *Aspicilia leprosescens*, *Caloplaced verruculifera*, *Lecanora poliophaea*, *Rinodina subexigua* and the alga *Prasiola quadrata* occurs on birds' perching rocks along the shore or near nesting sites. The rare *Caloplaca scopularis*, *Candelariella arctica* and *Lecanora straminea* may also occur in this community. The relationship between this facies and the *Candelarielletum corallizae* is noted on p. 378.

Ramalinetum scopularis Klem.

Beih. Feddes Repert. **135,** 68 (1955) [as "(DR. 1925) Klem."].—*Ramalina scopularis-Anaptychia fusca* Ass. DR., *Svensk bot. Tidskr.* **19,** 333 (1925), *nom. inval.*— *Ramalina scopularis-Lecanora atra-Rhizocarpon constrictum*-Ass. DR., *Svensk bot. Tidskr.* **19,** 333 (1925), *nom. inval.*—*Ramalinetum siliquosae* Follm., *Phillipia* **2,** 8 (1973) [as "(DR.)"].—*Lecanoretum atrae* Massé, *Revue bryol. lichén.* **34,** 889 (1966).

The characteristic association of the xeric-supralittoral (grey) zone of rocky shores in the British Isles is the *Ramalinetum scopularis* (Fig. 12). A species-poor facies of this association may also occasionally be encountered on walls of churches, ancient monuments and stone walls or rock outcrops in inland sites subject to maritime influence, as for instance throughout most of Devonshire and on Avebury Circle and Stonehenge in Wiltshire. This community is exceptionally rich in species (see Fletcher, 1975a) and occupies that part of the sea-shore where moisture derived from the sea and soil from the terrestrial environment are minimal. Amongst the characteristic species of this community are *Acarospora atrata**, *Anaptychia fusca**, *A. mamillata**, *Buellia canescens*, *B. stellulata*, *Caloplaca ferruginea*,

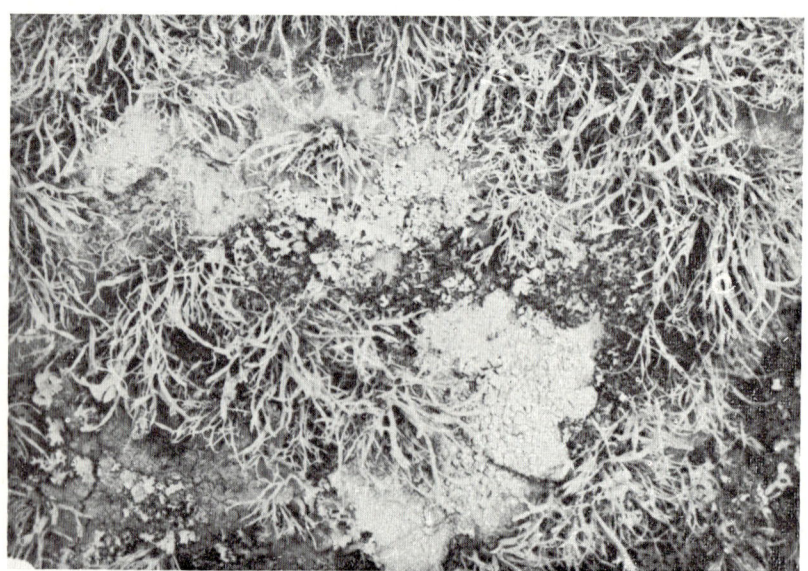

FIG. 12. *Ramalinetum scopularis* on maritime rocks siliceous. Species present include *Anaptychia fusca*, *Ochrolechia parella*, *Parmelia reticulata*, *P. saxatilis* and *Ramalina siliquosa* (Mid-Ebudes: Island of Iona, 1972, P. W. James).

* Species predominantly maritime in the British Isles.

Diploschistes caesioplumbeus*, *Huilia albocaerulescens*, *Lecanora atra*, *L. fugiens**, *L. gangaleoides*, *Lecidea diducens* (on quartzite), *L. sulphurea*, *L. tumida*, *Lecidella subincongrua**, *Parmelia delisei**, *P. glabratula* ssp. *fuliginosa*, *P. pulla**, *P. loxodes**, *P. verruculifera* (syn. *P. isidiotyla*), *Pertusaria pseudocorallina*, *Physcia subobscura**, *Ramalina cuspidata**, *R. siliquosa**, *R. subfarinacea**, *Rhizocarpon constrictum**, *R. geographicum* aggr., *R. obscuratum*, *Rinodina atrocinerea*, *R. luridescens**, *Verrucaria fusconigrescens** and *Xanthoria parietina*. This association is, however, very variable and other species may be very important components of it locally, for example, *Buellia aethalea*, *B. subdisciformis**, *B. verruculosa*, *Fuscidea cyathoides*, *F. tenebrica*, *Lecanora rupicola*, *Lecidea tenebrosa*, *Pertusaria chiodectonoides*, *P. ceuthocarpoides**, *P. monogona* and *Rinodina confragosa**.

A wetter and more sheltered facies of the *Ramalinetum scopularis* which is optimally developed on friable rocks includes *Acrocordia salweyi*, *Bacidia scopulicola**, *Enterographa hutchinsiae*, *Lecania rupicola**, *Lecanora dispersa*, *Lecidella subincongrua**, *Opegrapha confluens*, *O. gyrocarpa*, *O. lithyrga*, *O. zonata*, *Porina chlorotica*, *P. curnowii**, *Solenopsora holophaea**, *S. vulturiensis**, *Toninia aromatica*, *T. mesoidea**, *Verrucaria internigrescens** and *V. prominula**.

Species characteristic of the *Lobarion*, for example *Lobaria laetevirens*, *Nephroma laevigatum*, *Pannaria microphylla*, *Parmeliella plumbea* and *Sticta canariensis* (both morphotypes) also enter this assemblage on sheltered shores, particularly in western Scotland (e.g. Balnabraid Glen, Campbeltown, Kintyre; near Newton Stewart, Kirkcudbrightshire). Further studies may well show that this facies merits the status of a distinct association.

On rocky coasts, dry sheltered overhangs within the *Ramalinetum scopularis* support the *Sclerophytetum circumscriptae*, discussed below.

Sclerophytetum circumscriptae ass. nov. (Table XV)

The *Sclerophytetum circumscriptae* is a very distinctive association characteristic of dry, often shaded, recesses in siliceous maritime rocks which are not subject to direct rainfall. This association, which is largely restricted to the south and west coasts of the British Isles, includes many rare species often at the northern edges of their distributional ranges. The most characteristic species of this community are *Caloplaca arnoldii*, *C. littorea*, *Arthonia lobata* aggr., *Catillaria littorella*, *Dirina stenhammari*, *Lecanactis dilleniana*, *L. monstrosa*, *Lecanora tenera*, *Opegrapha cesareensis* and *Sclerophyton circumscriptum*. *Lecanora praepostera*, *Rinodina subglaucescens*,

* See footnote on previous page.

TABLE XV.

Sclerophytetum circumscriptae ass. nov.

Species	Stands	
	1	2
Arthonia lobata aggr.	1.–	–
Buellia leptoclinoides	–	+
Chiodecton petraeum	–	2.2
Lecanactis monstrosa	2.2	–
Lecanora atra	+.1	–
L. gangaleoides	1.–	+
L. praepostera	2.1	3.3
L. tenera	–	3.3
Lecidea orosthea	–	1.2
Pertusaria flavicans	+	–
Ramalina siliquosa	1.1	+.1
Roccella fuciformis	–	1.+
R. phycopsis	–	2.1
Sclerophyton circumscriptum	3.3	2.2

1. Isles of Scilly, St Mary: sheltered dry acid rock, incl. 90°, aspect 315°, 0·5 × 0·5 m, 1968, P.W.J.
2. Channel Islands, Alderney, La Roche peninsula: acid rock, aspect 45°, 1 × 2 m, cover 60%, April, 1975, P.W.J.; *type record*.

Roccella fuciformis and *R. phycopsis* are present as far north as the island of Skomer, Pembrokeshire, whereas *Buellia leptoclinoides, Chiodecton myrticola* and *C. petraeum* are restricted to south-west England and the Channel Islands. *Arthonia atlantica* is confined to western Ireland. Further notes on the ranges of many of the species in this association within the British Isles are included in James (1970).

It is of interest to note that in south-west Brittany, northern Spain and Portugal, this association is not confined to crevices but is also able to extend on to exposed rock faces.

Verrucarietum maurae DR.

Svensk. växtsociol. Sällsk. Handl. **2**, 51 (1925) [as "*Verrucaria maura*-Ass."].— *Verrucaria maura-Lichina confinis*-Association DR., *Beih. bot. Cbl.* **49**, 81 (1932)

The *Verrucarietum maurae* is the association of the littoral (black) zone on all siliceous rocky coasts of Britain. The 15 species which comprise it are more or less regularly inundated by sea-water on sheltered shores or lie within the splash zone on more exposed coasts; only *Lichina pygmaea* is regularly submerged by the tide. Although considerable patterns of zonation may occur within the *Verrucarietum maurae* these are often obscured due to variations in shoreline topography and the density of algal cover. This results in the upper and lower elements of this association becoming spatially close as on respectively drier exposed and wetter, seaweed sheltered surfaces in juxtaposition. The *Verrucarietum maurae* is dominated by species of *Verrucaria* the identification of which still presents considerable problems (Fletcher, 1975a). *Verrucaria maura* generally delimits the upper part of the zone and extends into the *Caloplacetum marinae*; on some shores this species is associated with *V. amphibia*. The maximum immersion time for any species in the association is that of 52% of one year recorded for *V. striatula* and *V. mucosa* (Fletcher, 1973a); these two species constitute the lowest extension of the zone. Intermediate species are *Arthopyrenia halodytes*, *Arthopyrenia* sp., *Verrucaria degelii* (northern Scotland only), *V. ditmarsica*, *V. erichsenii*, *V. microspora* auct. and *V. sandstedei*. *Arthopyrenia halodytes* exhibits a very wide range of substrate tolerance, occurring on acid or limestone rocks, on a wide range of shells of molluscs and on soft chalk (as in south-east England).

F. Aquatic Communities

The lichen flora of siliceous rocks in, or at the margins of, lakes and streams includes a small but distinctive assemblage of species, many of which are almost exclusive to the habitat. The horizontal zonation patterns formed by particular lichen associations within or below the splash zone are comparable with those occurring on coastal rocks (see Santesson, 1939; Wirth, 1972). Recently the zonation formed has proved of value even in the determination of river channel capacity (Gregory, 1976). The ecology and phytosociology of the freshwater communities in the British Isles have received little study although the taxonomy of the aquatic species of *Verrucaria* and *Polyblastia*, two of the most difficult genera represented, has been critically reviewed by Swinscow (1968, 1971). According to the information available there appear to be at least four associations which merit consideration. As with the case of the marine and maritime lichen communities, no attempt has been made here to assign the associations of aquatic habitats to alliances.

Some of the ecological factors which account for the delimited distribution of aquatic lichens have been discussed by Ried (1960a,b). Probably the frequency and sum extent of submersion are two important controlling factors for the survival of most species; only *Verrucaria aquatilis*, *V.*

kernstockii and *V. silacea* (the siliceous morphotype of *V. elaeomelaena* according to Wirth, 1972) require prolonged submersion. Several species, for instance, *Porina ahlesiana*, *Verrucaria aethiobola*, *V. hydrela* and *V. margacea*, are markedly tolerant of low illumination and may, in sheltered consistently humid sites, extend to damp rocks some distance from the margins of lakes and streams. Alternatively, a facies of the *Opegraphetum horistico-gyrocarpae*, including *Enterographa hutchinsiae*, *Opegrapha lithyrga*, *O. zonata* and *Porina lectissima*, as well as *Catillaria chalybeia* and *Porina chlorotica*, two species with an extremely wide ecological amplitude, may in similar shaded conditions enter into characteristic aquatic associations.

Both the pH and mineral content of the water may exert an important influence on the composition of the lichen flora. It is noteworthy that intermittently inundated siliceous rocks at the margins of lakes which are fed by streams flowing over basic rocks may develop species-poor facies of the *Gyalectetum jenensis* or the *Rhizocarpon petraeum* nodum (p. 362), including such species as either *Gyalecta jenensis*, *Lecidella stigmatea*, *Placynthium nigrum*, *Polyblastia scotinospora*, *P. theleodes*, *Lecidea speirea*, *Protoblastenia rupestris* or *Rhizocarpon petraeum*, respectively. As with coastal limestone rocks, there are few lichens characteristic of semi-inundated limestones or other types of base-rich rock: in Britain only *Verrucaria elaeomelaena* and the rare species *Placynthium tantaleum* and *Staurothele succedens* seem to belong exclusively to this category. As a rule, semi-inundated limestone supports either moisture-tolerant associations of the *Aspicilion calcareae* or a facies of the communities of dry limestones tolerant of some submersion of which *Thelidium decipiens* can be cited as a notable example.

The communities of lakes and streams seem to be particularly vulnerable to contamination by inorganic fertilizers which find their way into aquatic ecosystems by run-off and seepage from treated agricultural land. The consequent effect is one of more or less intensive hypertrophication which encourages the rapid blanket colonization of swards of blue-green and green algae which rapidly smother the pre-existing lichen communities, nearly all of which are dominated by species with crustose thalli and are unable to compete under such adverse conditions. Damage to aquatic lichen communities in Britain has been accentuated noticeably in the last 5 years, particularly in Wales, south-west Scotland and south-west England, areas where there has been a rigorously applied policy for the extensive reclamation of heathland and moorland for agriculture and forestry; many upland streams have become wholly or partly contaminated due to this cause. This aspect of freshwater pollution is in need of urgent study. To a lesser degree, the lowering or raising of lake levels and the diversion of water from streams and rivers for hydro-electric schemes and the formation of reservoirs has had a notable impact on some of the more interesting aquatic lichen communities.

Aspicilietum lacustris Wirth

Diss. Bot., Lehre **17**, 223 (1972).—*Aspicilietum lacustris* Frey, *in* Frey and Ochsner, *Arvernia* **2**, 67 (1926), ? *nom. inval.* as labelled "une fragment".

In the British Isles, the *Aspicilietum lacustris* is the widespread association of the splash zone of lakes and streams subject also to periodic inundation. The characteristic species in well lit sites are *Aspicilia lacustris, Dermatocarpon fluviatile, D. meiophyllum, Rhizocarpon laevatum, Staurothele fissa* and *Verrucaria aethiobola*, often in association with *Huilia albocaerulescens, H. macrocarpa, Lecidea tumida* and *Rhizocarpon obscuratum* in less frequently inundated habitats. With an increase in the degree of shading, a nodum also including *Bacidia inundata, Catillaria chalybeia, Aspicilia laevata* and *Verrucaria praetermissa* is established; this assemblage of species is perhaps worthy of recognition at association rank. In very shaded and more humid situations the association grades into the *Verrucarietum siliceae* marked by increased frequencies of *Verrucaria hydrela* and *V. margacea*. The *Physcietum caesiae* (p. 360) is not uncommon on the tops of boulders in streams used as bird perches.

Several other noda are discernible in the *Aspicilietum lacustris*, for example, that occurring on very shaded mica schist rocks in small streams in western Scotland and Ireland including *Microglaena larbalestieri* and *Porina guentheri* var. *grandis*; the more widely distributed *Thelidium pyrenophorum* is also often present in this nodum.

Another important nodum mainly confined to Ireland but also represented as a species-poor facies in western Scotland is that characterized by the presence of *Huilia hydrophila*, and which may correspond to the subunion *Haplocarpon hydrophilum* of Wirth (1972). Additional characteristic species of this nodum are *Porina guentheri* var. *lucens, P. interjungens* (rare), *Porocyphus kenmorensis* and the ubiquitous *Aspicilia lacustris, Catillaria chalybeia, Huilia albocaerulescens, H. macrocarpa,* and *Rhizocarpon* species.

The *Lecidea phaeops–L. taylori* community discussed on p. 365 might also be placed near here. Most of its characteristic species are markedly euoceanic; the previously "endemic" species *Lecidea phaeops* and *L. taylori* have recently been collected in the Açores (Faial and Santa Maria) where they are dominant in moist woodlands and semi-inundated river valleys similar to those in which they occur in Britain.

Ephebetum lanatae Frey

Mitt. naturf. Ges. Bern **6**, 172 (1922) [as "*Ephebe lanata*-Ass."].

The *Ephebetum lanatae* is a rather local community of lake and stream margins as well as more or less persistent seepage tracks on acid, nutrient-impoverished siliceous rocks. In Britain the association is largely confined to the north and west of the country and is characterized by the presence of *Aspicilia gibbosa* aggr., *A. lacustris*, *Ephebe lanata*, *Rhizocarpon geminatum* (rare), *R. obscuratum* (morphotype), *Scytonema* sp., *Stereocaulon pileatum*, *Stigonema* sp., *Trapelia involuta* and often various Cyanophyceae; *Trapelia moorei* may also be locally abundant as, for instance, on the island of Mull. Cyanophilic algae and the bryophytes *Bryum alpinum* and *Rhacomitrium protensum* are additional important components of the association. When the bryophyte cover is well developed, *Massalongia carnosa* and, more rarely, *Polychidium muscicola*, enter the association.

Ionaspidetum suaveolentis Frey

Mitt. naturf. Ges. Bern **6**, 170 (1922) [as "*Jonaspis suaveolens*-Ass."].—*Ionaspidetum odorae* Wirth, Diss. Bot., Lehre **17**, 223 (1972) [as "(Frey 1922) *nom. nov.*"].

A rare and very restricted association in the British Isles, the *Ionaspidetum suaveolentis* includes *Ionaspis suaveolens*, *Placynthium flabellosum* and *Polyblastia cruenta* as its characteristic species. The rare species *Dermatocarpon rivulorum*, *Gyalidea fritzei*, *Polyblastia quartzina* and *Thelidium fumidum* may also belong here. More ubiquitous species recorded are *Aspicilia lacustris*, *Bacidia inundata*, *Catillaria chalybeia*, *Dermatocarpon fluviatile*, *Rhizocarpon laevatum*, *Staurothele fissa* and *Verrucaria margacea*, although these are often present only in small quantities and in direct competition with numerous species of blue-green algae. *Thelidium aeneovinosa* cited by Wirth (1972) as a characteristic species in the association on the Continent has only once been recorded from the British Isles (Durham). The association is best developed in higher mountain areas of Scotland, particularly in the Cairngorm mountains where it seldom occurs below an altitude of 910 m. At lower altitudes the community grades into the *Aspicilietum lacustris* in which *Placynthium pannariellum* and *Polyblastia cruenta*, at least above 200 m, may also be present as subsidiary species.

Verrucarietum siliceae Wirth & Ullrich

In Wirth, Diss. Bot., Lehre **17**, 219 (1972).

This association, the *Verrucarietum siliceae*, comprises communities growing on siliceous rocks in streams and lakes where there is a moderate to rapid water flow, sufficient light and lack of mud and silt. Although the association is frequently submerged throughout the year it may also be found in shaded stream beds which dry out for relatively short periods of the year. *Verrucaria aquatilis*, *V. kernstockii* and *V. silicea* are characteristic of the *Verrucarietum siliceae*. According to Wirth (1972), *V. silicea* is the siliceous rock morphotype of *V. elaeomelaena* which occurs in similar situations on submerged limestones. Various free-living algae of which *Hildenbrandtia* and *Lemanea* merit particular mention are often present as is the moss *Fontinalis antipyretica*. The rare *Collema fluviatile* is faithful to the association in Britain.

A nodum characterized by *Bacidia inundata*, *Staurothele fissa*, *Verrucaria latebrosa* and *V. margacea*, not uncommon in western areas of the British Isles, is also referred to the *Verrucarietum siliceae*. This may be identical to the *Verrucarietum laevato-denudatae* Wirth.

VII. Terricolous Communities

In general, terricolous lichens present much greater problems of phytosociological delimitation than communities characteristic of other substrates, because lichens on soil usually form an intimate and integral part of already designated higher plant associations; they do not form the more or less exclusive stands seen in the case of most corticolous and saxicolous lichen communities. In the case of terricolous communities, lichenologists have often adopted classificatory schemes which pay only minor attention to the vascular plants involved (e.g. Klement, 1955). While this approach can be sympathized with, as those describing higher plant syntaxa have all too frequently ignored lichens in their surveys, in our view an entirely acceptable system for terricolous communities must take a proper account of both phanerogam and cryptogam components.

An outline of the higher plant syntaxa which may be recognized in Britain is included in Shimwell (1971). Those present in Britain as a whole were summarized by Tansley (1939) although he did not adopt the nomenclatural system used by continental workers, and those in Scotland are reviewed in some detail in the work edited by Burnett (1964). A survey of the plant associations of lowland Britain is currently in progress but results from it will not be generally available for some time.

For simplicity, we have separated the terricolous lichen communities into the following four categories, those of pebbles, basic soils, coastal soils and acid soils or peat.

A. Pebbles

Lichens on pebbles may be sporadic within vascular plant-dominated communities or form extensive stands in unstable pebble-dominated situations. Extremely mobile pebbles are scarcely colonized by lichens, perhaps largely due to abrasion of any incipient thalli. Three pebble communities are particularly distinctive in the British Isles and are recognized as associations here as they can occur over large areas with few or no flowering plants.

Huilietum crustulatae (Klem.) P. James *et al.*

This association is discussed under saxicolous communities above (p. 370), as it can also occur on larger rocks and boulders.

Lecideetum erraticae ass. nov. (Table XVI)

This association is particularly well developed on pebbles at Dungeness in East Kent, an area which has a noteworthy facies of the *Lecanoretum subfuscae* on *Sarothamnus* and *Prunus spinosa* (see p. 319) and also *Cladonia*-dominated heathland-like communities on decaying *Sarothamnus*. The characteristic species of the *Lecideetum erraticae* are *Buellia aspersa**, *B. aethalea*, *Lecidea erratica* and *L. tumida*. In a slightly shaded facies, an undetermined *Aspicilia* becomes dominant. Other species occasionally found in this association are *Catillaria chalybeia*, *Buellia verruculosa*, *Lecanora dispersa*, *L. polytropa*, *Rhizocarpon constrictum*, *R. obscuratum* and *Verrucaria nigrescens*. This association is optimally developed on pebbles in small declivities in shingle or where there is partial shading from adjacent scrub. The species present in this community seem able to withstand periodic disturbance; pebbles with the *Aspicilia* mentioned above frequently have their upper and lower sides colonized by this species.

Although best developed at Dungeness, this association has also been recorded in a more fragmentary state from the Chesil Beach, Dorset; Orford Ness, Suffolk; Start Bay, Devon; Pevensey Beach and Pagham, Sussex (Rayner, 1976); and Ballantrae shingle beach, Ayrshire. At Slapton the association occurs on more stabilized shingle, particularly bare patches exposed within the lichen-dominated *Cladonietum alcicornis* (p. 401), and *Buellia punctata* (saxicolous morphotype) is also a rare component there (Hawksworth, 1972a).

* This is a small sterile sorediate species resembling a slightly brown-grey form of *Lecidea tumida* and, as in that species, has a delicately fimbriate dark prothallus; gyrophoric acid is present in the medulla and in the internal part of the soralia.

TABLE XVI.

Lecideetum erraticae ass. nov.

Species	Stands					
	1	2	3	4	5	6
Aspicilia sp.	+.2	–	3.3	4.3	–	–
Buellia aethalea	2.3	1.2	1.3	1.2	1.2	2.3
B. aspersa	1.2	3.3	2.2	1.2	2.3	1.3
B. stellulata	+.0	–	–	–	+.2	–
B. verruculosa	–	+.0	–	–	+.2	1.2
Catillaria chalybeia	+.0	–	–	–	–	–
Lecanora dispersa	–	+.0	–	–	–	+.2
L. polytropa	–	–	1.1	–	–	–
Lecidea erratica	3.3	1.2	2.2	+.0	1.3	1.3
L. tumida	2.3	2.3	1.3	1.3	2.2	1.2
Rhizocarpon constrictum	–	1.1	–	–	–	2.2
R. obscuratum var. reductum	1.2	1.3	1.3	+.2	1.3	1.3
Verrucaria nigrescens	–	–	1.1	–	–	+.0

1. East Kent, Denge Beach, Open Pits (61/073186): pebbles (exposed), level, 1·0 × 1·0 m, cover 40%, 16 July 1976, P.W.J.
2. East Kent, Denge Beach, Open Pits (61/077184): pebbles (exposed), incl. 15°, 1·0 × 1·0 m, cover 55%, 16 July 1976, P.W.J.
3. East Kent, Denge Beach, Open Pits (61/063173): pebbles (shaded), incl. 10°, 1·0 × 1·0 m, cover 35%, 16 July 1976, P.W.J.; *type record*.
4. East Kent, Lydd Ranges (61/015182): pebbles, incl. 25°, 1·0 × 1·0 m, cover 45%, 28 June 1968, P.W.J.
5. Dorset, Isle of Portland, Chesil Beach, Fortuneswell (30/675745): pebbles, level, 1·0 × 1·0 m, 16 April 1970, P.W.J.
6. East Suffolk, Orford, Orford Ness (62/448494): pebbles, incl. 5°, 1·0 × 1·0 m, 25 June 1964, P.W.J.

Lecideetum watsoniae ass. nov. (Table XVII)

Unlike the other two pebble associations recognized here, this association is characteristic of basic situations, particularly calcareous (chalk) slopes of south-east England. The predominant species of the *Lecideetum watsonii* on chalk nodules are *Lecidea watsonii*, *Protoblastenia immersa*, *P. metzleri*, *P. monticola*, *Sarcogyne regularis*, *Staurothele hymenogonia*, *Thelidium decipiens*, *T. incavatum*, *Verrucaria hochstetteri*, *V. muralis*, *V. mutabilis* and *V. viridula*. On flints within the same association *Aspicilia calcarea*, *A. contorta*, *Caloplaca citrina*, *C. holocarpa*, *Lecanora dispersa*, *Lecidella stigmatea*, *Protoblastenia rupestris*, *Verrucaria nigrescens* and *V. viridula*

TABLE XVII.
Lecideetum watsoniae ass. nov.

Species	Stands				
	1	2	3	4	5
Bacidia muscorum	+.0	–	+.0	–	–
B. sabuletorum	+.2	+.0	+.0	+.0	–
Caloplaca citrina	–	–	+.2	–	–
Collema auriculatum	–	+.0	+.0	–	–
C. tenax	+.2	–	–	+.0	–
Dermatocarpon hepaticum	+.0	–	–	–	–
Lecidea watsonii	1.2	+.2	+.2	+.0	+.2
Lecidella stigmatea	+.0	+.2	–	–	–
Leptogium schraderi	–	–	+.0	–	–
L. cf. *subtile*	–	+.0	–	–	–
Petractis clausa	–	+.0	–	+.0	+.0
Physcia adscendens	–	+.0	–	–	–
Protoblastenia immersa	+.2	–	+.2	–	–
P. metzleri	–	+.0	+.2	–	–
P. monticola	–	+.0	–	–	–
P. rupestris	–	+.2	–	+.0	–
Sarcogyne regularis	+.2	–	–	–	–
Staurothele hymenogonia	+.2	1.2	+.2	+.0	–
Thelidium decipiens	+.2	–	–	+.2	+.2
T. microcarpum	–	–	–	1.2	1.3
Verrucaria hochstetteri	+.2	1.3	1.3	+.0	+.0
V. muralis	1.3	+.2	+.2	+.0	+.0
V. mutabilis	–	–	–	+.2	+.0
V. nigrescens	+.2	+.2	+.0	+.2	+.2
V. viridula	+.2	+.2	+.2	+.0	–
Algae	+.0	+.0	–	+.2	1.2
Bryophytes	1.2	+.2	+.2	+.0	+.0
Phanerogams	+.2	+.2	+.0	+.0	–

1. Surrey, Dorking, Box Hill (51/184525): exposed chalk/flint pebbles, incl. 10°, aspect 180°, cover 65%, 0·5 × 0·5 m, 16 October 1968, P.W.J.
2. West Norfolk, Thetford, Thetford Heath (52/845796): recently exposed chalk/flint pebbles, incl. 5°, aspect 180°, cover 75%, 0·5 × 0·5 m, 12 December 1965, P.W.J.
3. West Sussex, Midhurst, Heyshott (41/902167): chalk/flint nodules in rabbit warren, incl. 35°, aspect 225°, cover 55%, 0·5 × 0·5 m, 6 September 1963, P.W.J.
4. West Sussex, Midhurst, Heyshott (41/902167): chalk/flint nodules in deep shade of *Fagus*, incl. 35°, aspect 225°, cover 70%, 1·0 × 1·0 m, 6 September 1975, P.W.J.
5. Hertfordshire, Tring, Coombe Hill (42/896106): chalk/flint nodules in shade of *Corylus*, incl. 50°, 1·0 × 1·0 m, 3 March 1962, P.W.J. and T. D. V. Swinscow.

can occur. More locally *Petractis clausa* and *Polyblastia dermatodes* on chalk, and *Ochrolechia parella*, *Physcia adscendens*, *P. caesia*, *P. tenella*, *Xanthoria aureola*, *X. elegans* and *X. parietina* on flint may enter the community. A few primarily soil-loving species may also be present, for example *Bacidia sabuletorum*, *Collema tenax*, *Dermatocarpon hepaticum*, *Leptogium lichenoides* and *L. sinuatum*. The chalk nodule–flint communities form a part of those widespread on basic soils in areas such as the Breckland of East Anglia and the Kent and Sussex Downs, and species found in these (see pp. 397–400) may also occur in them occasionally.

The *Lecideetum watsoniae* is optimally developed on dry eroded south-facing hillsides in the downland areas of south-east England. It may have a somewhat transitory nature to judge from its tendency to occur in areas of relatively recent disturbance, as near old rabbit burrows where a fresh supply of chalk nodules suitable for colonization has been disinterred.

A somewhat shaded facies of this association is occasionally encountered, most frequently under *Fagus*, in which *Acrocordia monensis*, *Thelidium macrocarpum* and *Verrucaria muralis* on chalk, and *V. mutabilis* and *V. viridula* on flints become dominant; the moss *Seligeria paucifolia* is also often represented in such communities.

B. Basic Soils

The lichen communities encountered on soil in the immediate vicinity of basic rock outcrops are most appropriately treated as an integral part of those occurring on the rocks themselves (pp. 349–364). For successful colonization in such sites the terricolous lichens require open and sufficiently stable habitats, for example on closely cropped exposed hillsides (particularly those with high rabbit populations) and more especially pockets of soil between and on the outcrops themselves. In such habitats the vascular plant flora is either reduced in height by grazing, or by the overall thinness, and frequently by the dryness of the soil. Some lichens generally treated as terricolous are characteristic of bryophyte-rich communities and are themselves predominantly bryophilous (muscicolous) rather than strictly terricolous.

The ubiquitous indicator lichens of basic soils are *Bacidia sabuletorum*, *Cladonia pocillum*, *Collema tenax* (often a primary colonizer of recently disturbed soils) and *Polyblastia tristicula*; a variety of unicellular and filamentous blue-green algae (Cyanophyceae) are also generally present. Additional species which are frequently present include *Bacidia muscorum*, *Caloplaca citrina* (terricolous morphotypes), *Cladonia rangiformis*, *Collema crispum*, *Dermatocarpon hepaticum*, *Leptogium lichenoides*, *L. sinuatum*, *Microglaena muscorum*, *Placidiopsis custnanii* (in Scotland), *Polyblastia gelatinosa*, *Squamarina crassa*, *Toninia coeruleonigricans* and *T. lobulata*. Of the rarer or more local species to be found on such soils are *Bacidia herbarum*, *Buellia asterella* (syn. *B. epigaea* auct. angl.), *Fulgensia* sp., *Lecidea decipiens*, *Polyblastia agraria*, *P. wheldonii*, *Squamarina lentigera* and *Verrucaria psammophila*.

In the Breckland of East Anglia, the basic soils support particularly interesting assemblages of species, including *Buellia asterella*, *Lecidea decipiens*, *Fulgensia* sp. and *Squamarina lentigera* (Table XVIII). These communities may well correspond to the *Fulgensietum fulgentis* Gams recognized by many central European authors, but they are in need of further study. *Fulgensietum fulgentis* is a widespread continental community, especially in central and southern France.

TABLE XVIII.

Lichen-rich community of the Breckland, East Anglia (aff. Fulgensietum fulgentis *Gams).*

Species	Stands					
	1	2	3	4	5	6
Bacidia muscorum	–	2	1	–	–	–
Buellia asterella	2	2	2	2	–	–
Cladonia foliacea	4	4	2	–	–	–
C. furcata	–	1	–	2	2	–
C. pocillum	–	–	–	2	1	–
C. rangiformis	–	–	–	–	–	5
Collema tenax	1	–	–	–	–	–
Cornicularia aculeata	1	1	–	–	–	–
Dermatocarpon hepaticum	2	2	2	2	–	–
Diploschistes scruposus var. *bryophilus*	1	2	3	–	–	2
Fulgensia aff. *fulgens*	–	3	–	–	–	–
Lecanora dispersa	–	–	–	–	1	–
Lecidea decipiens	6	2	2	–	–	–
Peltigera canina	–	–	–	–	–	2
P. rufescens	–	2	+	–	1	–
Protoblastenia rupestris	–	–	+	–	–	–
Sarcogyne regularis	–	2	+	–	–	–
Squamarina lentigera	2	+	2	–	1	1
Toninia coeruleonigricans	4	4	3	4	4	2
Verrucaria hochstetteri	–	–	1	–	–	–
V. muralis	2	2	–	–	–	–
V. nigrescens	1	–	+	–	1	–
Bryum capillare	–	–	–	–	2	2
Camptothecium lutescens	–	–	+	3	4	–

TABLE XVIII—*continued*

Species	Stands					
	1	2	3	4	5	6
Ditrichum flexicaule	2	4	+	–	–	–
Encalypta streptocarpa	2	4	4	–	–	–
Rhytidium rugosum	–	–	–	–	2	–
Astragalus danicus	3	3	–	–	–	–
Botrychium lunaria	–	–	+	–	–	–
Carex arenaria	–	–	2	–	–	–
C. ericetorum	–	–	+	–	–	–
Carlina vulgaris	3	2	1	–	1	–
Centaurium erythraea	–	+	1	–	–	–
Erigeron acer	2	2	–	–	–	–
Euphrasia officinalis aggr.	2	2	–	–	–	–
Festuca ovina	7	5	3	7	6	4
Galium verum	–	2	–	–	–	–
Gentianella amarella	+	–	–	–	–	–
Koeleria gracilis	4	4	2	3	6	2
Leontodon taraxacoides	2	4	1	–	–	–
Linum catharticum	–	3	2	2	–	–
Lotus corniculatus	–	2	–	–	–	–
Medicago lupulina	–	–	–	+	–	–
Ononis repens	+	–	–	–	–	–
Ornithopus perpusillus	–	3	–	–	–	–
Pilosella officinarum	5	6	2	5	–	5
Pinus sylvestris	+	–	–	–	–	–
Prunella vulgaris	–	–	1	–	–	–
Sedum acre	–	–	–	–	–	2
Senecio jacobaea	1	–	+	–	1	–
Taraxacum laevigatum	–	–	+	–	2	3
Thymus spp.	7	–	2	3	4	5

1. West Suffolk, Lakenheath Warren (52/750805): 1·0 × 1·0 m, cover 90%, 1973, P. W. Lambley.
2. West Suffolk, Lakenheath Warren (52/750805): 1·0 × 1·0 m, cover 90%, 1973, P. W. Lambley.
3. West Suffolk, Lakenheath Warren (52/750805): 1·0 × 1·0 m, cover 100%, 1973, P. W. Lambley.
4. West Suffolk, Deadman's Grave (52/779748): 1·0 × 1·0 m, cover 100%, 1973, P. W. Lambley.
5. West Suffolk, Thetford Warren (52/849796): 1·0 × 1·0 m, cover 100%, 1974, P. W. Lambley.
6. West Norfolk, Weeting Heath (52/757878): 1·0 × 1·0 m, cover 100%, 1973, P. W. Lambley.

Future investigations may indicate that specific habitat preferences shown by particular species are related to the soil chemistry and composition as well as to those climatic factors determining which communities occur on calcareous rocks. *Bacidia sabuletorum* and *Dermatocarpon hepaticum*, for example, both have a wide ecological amplitude and can be both shade- and moisture-tolerant; hence their frequent occurrence with the *Gyalectetum jenensis*. As in the case with dune systems, the amount of calcium carbonate in the soil determines the abundance and species diversity of lichen-dominated communities in calcareous sites. Soils with 100 mg Ca/100 g and a pH usually above 6·2 are the first on which true lichen calcicoles are able to survive and where *Bacidia sabuletorum, Cladonia pocillum, C. rangiformis* and *Collema tenax* (often in abundance) occur. With increasing pH additional species start to appear but it is only at the highest pH levels (to pH 8·2) where lichens such as *Fulgensia* sp., *Lecidea decipiens* and *Squamarina crassa* are found.

There is little evidence to suggest that certain basiphilous lichens can grow equally well in sites where magnesium replaces calcium as the predominant cation. The lichen flora of the magnesium-rich serpentine outcrops in Shetland, as at Baltasound on Unst, for example, have little in common with that of limestone areas in Sutherland.

Whilst many of the species mentioned above can be expected to occur in most limestone areas of the British Isles, several others appear only with increasing altitude. *Aspicilia verrucosa, Caloplaca stillicidiorum, Lecidea templetonii, Solorina saccata* and *S. spongiosa*, for example, are predominantly montane in England, although they are to be found at sea-level in Scotland where they enter the coastal *Dryas octopetala–Carex rupestris* nodum of McVean and Ratcliffe (1962). A few species are characteristic of basic soils between outcrops of limestones and epidiorites in the Highlands of Scotland; the communities on rocks of the latter type have already been referred to (p. 361). Many of the species in such sites are very rare in the British Isles and the communities they form may be fragments of those more widely distributed in comparable habitats in western Scandinavia; the most important species of these soils in Scotland are *Biatorella fossarum, Collema ceraniscum, Dermatocarpon cinereum, Gyalecta geoica, Lopadium fecundum, Microglaena sphinctrinoides, Polyblastia sendtneri, Sagiolechia rhexoblephara* and *Thelopsis melathelia*.

C. Coastal Soils and Dunes

Very few lichens are restricted to coastal soils in the British Isles and, as with most other terrestrial communities, these form an integral but often minor part of vascular plant-dominated communities. Most lichens found in such situations require open habitats with bare but generally more or less stable soil; exceptions are provided by *Cladonia furcata* and *C. rangiformis* whose erect and relatively fast growing thalli are able to compete in more extensive developments of higher plants.

In coastal sites, terricolous lichens are often most frequent on windswept cliff tops, eroded banks or on thin layers of soil associated with rock outcrops. Small rosette-forming vascular plants such as *Plantago coronopus*, *Pilosella officinarum* and *Rumex acetosella* are commonly present in such sites when the lichens can become almost dominant; when this occurs they can be termed the *Cladonietum alcicornis* Klem. (see Hawksworth, 1972*a*). *Cladonia* species comprise the most important element in such communities and include *C. conistea*, *C. cervicornis* (including *C. verticillata*), *C. chlorophaea*, *C. foliacea*, *C. nylanderi* (rare), *C. pityrea* and *C. pyxidata*, in addition to the ubiquitous *C. furcata* and *C. rangiformis*; *Peltigera* species are additional important components locally.

On bare soil in exposed situations *Lecidea wallrothii* and *Solenopsora vulturiensis* occur; the latter is also encountered in very sheltered sites or on damp rocks in overhangs. *Lepraria* species (common), small subcrustose *Leptogium* species and *Micarea subviridescens* are also sometimes common in sheltered sites but this shade-loving community is poorly understood. Loose friable soil of banks may support *Moelleropsis nebulosa*, *Vezdaea aestivalis* and *V. leprosa*, and where the soil has become compacted and is more or less basic, pyrenocarpous species may also colonize it (e.g. *Acrocordia salweyi*, *Verrucaria hochstetteri*, *V. muralis*); this latter type of soil is perhaps not unlike the compacted mortar of old walls in some respects, on which such species normally occur.

The lichens in sand-dune systems have been studied in Britain by several workers since the pioneer investigations of Watson (1918a) which were primarily concerned with those in western Britain, emphasis being placed on Braunton Burrows, North Devon. Of the more recent investigations, those of Alvin (1960) on Studland Heath, Dorset, Brown and Brown (1969) on Blakeney Point, Norfolk, and Prince (1974) on the Sands of Forvie, Aberdeenshire, merit particular mention. The extent to which lichens form an important component of the vegetation on sand dunes depends on a variety of factors, such as the stability of the sand, its moisture-retentive properties, the frequency and permanency of dewfall, and the humus and calcium carbonate contents. The calcium carbonate in dunes is responsible for some of our most interesting dune lichen vegetation which includes many calcicolous species; this is derived mainly from the accumulation of mollusc shells but can be supplemented by run-off and erosion of adjacent calcareous outcrops (e.g. Broadhaven, Pembrokeshire). The most consolidated soils of basic dunes have lichen assemblages similar to those characteristic of limestone crevices (p. 397) and the Breckland (p. 398) in areas where the vascular plants permit their development. The most highly calcareous region in a dune system often corresponds to the transitional zone between the unstable white dunes and the stabilized grey dunes, as in older dunes there is a leaching of the calcium carbonate, a lowering of the pH and an accumulation of surface humus; all these factors favouring the establishment of the acid heathland community types are discussed below (p. 403).

Most species found on calcareous dunes are also known in other lowland calcareous sites; *Squamarina crassa* f. *pseudocrassa*, however, is noteworthy in being exclusively coastal. *Bacidia sabuletorum, Cladonia rangiformis, C. foliacea, C. pocillum, Diploschistes scruposus* var. *bryophilus, Leptogium lichenoides, L. sinuatum* and *Peltigera rufescens* are indicators of very low but perceptible concentrations of calcium carbonate in the dune sand. With increasing pH and calcium carbonate levels, species such as *Dermatocarpon hepaticum, Polyblastia gelatinosa, P. tristicula, P. wheldonii* (rare) and *Toninia coeruleonigricans* appear. *Arthopyrenia subareniseda* and *Polyblastia agraria* are so far confined to this habitat in Britain. In areas where the soil is extremely calcareous, *Fulgensia* sp., *Lecidea decipiens, Placynthium nigrum, Squamarina crassa, Toninia aromatica* and *T. lobulata* may be found.

D. Acid Soils and Peat

Most acid heathlands in Britain are dominated by a single species, *Calluna vulgaris*, which often forms more or less extensive tiered stands over large areas. Gimingham (*in* Burnett, 1964) notes that since dominance by this species is established over such a wide area, there is in consequence a considerable floristic diversity in the communities dominated by *Calluna* in the British Isles; the term "*Callunetum*" thus conveys little ecological information unless qualified in some way. *Calluna* may occur in almost pure stands or form integrated associations with other species indicative of particular habitats, for example *Arctostaphylos uva-ursi* (wetter facies or higher altitudes), *Empetrum nigrum* (high montane but descending to sea-level in Shetland), *Erica cinerea* (dry oceanic moorland), *E. tetralix* (wet moorland), *Vaccinium myrtillus* (dry lowland heaths or heathland under trees) and *V. uliginosum* (high montane heathland). The age of the community and frequency of burning are of considerable importance in relation not only to the diversity of the vascular plant flora but also to that of the cryptogams, including lichens. Open communities show a sequence of lichen development after burning; *Lecidea granulosa, L. uliginosa* and sometimes *Baeomyces rufus* are amongst the first species to appear, but in the following 5–6 years a more species-rich community forms. Eventually the *Calluna* becomes too dense and/or tall, and lichens are gradually excluded from the community, often forming aberrant morphotypes difficult to determine before they are finally lost. In the case of old *Calluna* bushes, the larger *Cladonia* species (e.g. *C. arbuscula*, *C. impexa*) and others (e.g. *C. floerkeana*) can often be seen to colonize the central open dying parts. The various effects of man on acid heathlands are discussed further by Hawksworth *et al*. (1974).

The *Cladonia*-dominated communities of rotting logs, rotting tree stumps and tree bases not uncommonly spread on to adjacent soil provided that this has a very high humus content. These communities, which cannot be regarded as strictly terricolous or corticolous and are treated under the *Cladonion coniocraeae* here (pp. 313–314), include those of the floor of mature pine woods in the Scottish Highlands with several very rare *Cladonia* species (e.g. *C. carneola*, *C. botrytes* and *C. cenotea*).

1. Lowland Heaths

Lowland heathland communities in dry sandy situations support a ubiquitous assemblage of species, most of which belong to the genus *Cladonia* (see Fig. 13): *C. arbuscula*, *C. bacillaris* (rare), *C. cervicornis* (incl. *C. verticillata*), *C. chlorophaea* (all chemotypes), *C. coccifera*, *C. coniocraea*, *C. crispata* var. *cetrariiformis*, *C. floerkeana*, *C. furcata*, *C. glauca*, *C. gracilis*, *C. gonecha* (rare), *C. impexa*, *C. pityrea*, *C. polydactyla*, *C. squamosa* (including var. *allosquamosa*), *C. subulata*, *C. tenuis* and *C. uncialis* ssp. *dicraea*. Additional important species include *Baeomyces roseus*, *B. rufus*, *Cornicularia aculeata*, *C. muricata*, *Icmadophila ericetorum* (rare), *Lecidea granulosa*, *L. oligotropha* (rare), *L. uliginosa* and *Pycnothelia papillaria*. Small siliceous pebbles in lowland heaths often support fragments of the *Huilietum crustulatae* (p. 370).

A particularly wet facies of the above assemblage, whilst retaining *Cladonia crispata*, *C. squamosa* and *C. uncialis* ssp. *dicraea*, may also be partially dominated by *C. strepsilis*; the other species indicated above then tend to be restricted to the crowns of tussocks or other raised areas. This is a particular feature of dying tussocks of *Carex paniculata* and *Molinia coerulea* in bogs where lichens are otherwise absent. Another facies of particular interest is the occurrence of *Cetraria islandica* on lowland heaths of East Yorkshire, Lincolnshire and Norfolk; old records suggest that this species was formerly common in such areas. This facies is almost certainly allied to the richer but comparable communities occurring near sea-level in the Netherlands and west Jutland, Denmark.

Although most lichens in heaths disappear with the increasing density

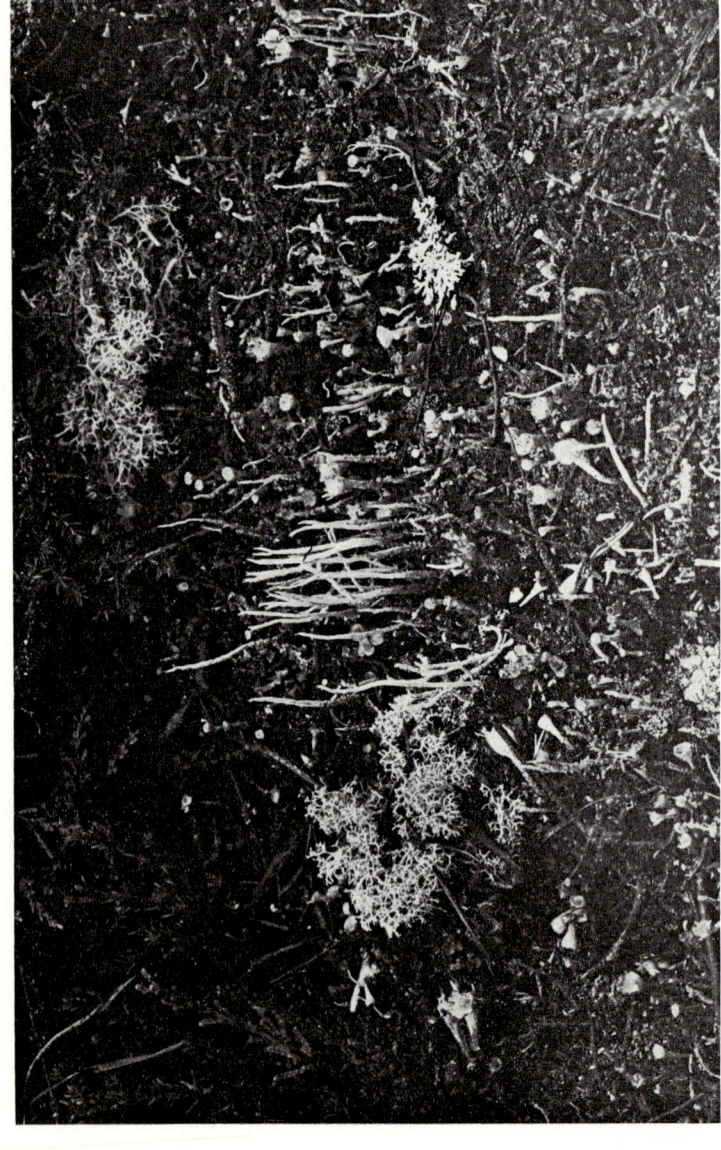

Fig. 13. Lowland heath community dominated by *Cladonia* species in the centre of an aged *Calluna* tussock. Species present include: *Cladonia chlorophaea, C. furcata, C. impexa* and *Hypogymnia physodes* (the latter on a *Calluna* stem) (Surrey: Wisley, Ockham Common, 1977, F. S. Dobson).

of the vascular plant vegetation (and sometimes bryophyte cover), *Cladonia arbuscula*, *C. impexa*, *C. tenuis* and, to a lesser extent, *C. furcata* and *C. gracilis* may persist. This tendency is almost certainly due to their upward growth (towards the light) and ability to decay from the base upwards as they grow. Furthermore, these species have a relatively rapid growth rate amongst the lichens (see Chapter 3) and so are more able to compete with the less vigorously growing vascular plants.

2. Blanket Peat

Blanket peat bogs, which cover large tracts of countryside in northern and western Britain, support lichen communities generally similar to those of the more lowland heathlands discussed above but occasionally have species which also occur in upland heaths. As peat becomes exposed either through drainage, river action, peat-cutting by man or erosion, lichen-dominated stands develop on the lips or "hags" which remain. Of particular importance in such sites are *Cladonia coccifera*, *C. bellidiflora*, *C. cervicornis*, *C. gonecha*, *C. polydactyla*, *C. squamosa*, *Icmadophila ericetorum* and *Pycnothelia papillaria*, but several other noteworthy species occur in this type of habitat, for example *Coriscium viride* (with *Thelocarpon epibolum*), *Cladonia fragilissima* (rare oceanic) and *Lecidea glaucolepidea*.

3. Upland Heaths

Detailed surveys of the floristics of upland heaths in the Scottish Highlands have been prepared by McVean and Ratcliffe (1962) and McVean (*in* Burnett, 1964). These authors presented data on the lichens in many of the communities they recognized and in some cases used them in the delimitation of the syntaxa themselves. The lichen-rich heaths as interpreted by McVean and Ratcliffe fall into two main categories, the dwarf shrub heaths and the moss heaths. The former include noda such as the *Arctoeto-Callunetum* of the northern highlands, the *Cladineto-Callunetum* which occurs at high altitudes and includes species such as *Alectoria sarmentosa* ssp. *vexillifera* and *Cetraria nivalis*, and the *Cladonia*-rich *Cladineto-Vaccinetum* which is central and eastern as well as montane in Scotland. Most facies of the moss heaths are dominated by *Rhacomitrium lanuginosum*, but a few associated with patches of late snow-lie lack this species and are dominated by the diverse *Gymnomitrium concinnatum–Salix herbacea* associations. The lichen flora of the mountain tops in the Scottish Highlands is still relatively poorly understood both from the taxonomic and ecological standpoint; indeed many upland areas have not been studied by lichenologists at all in recent decades. For these reasons we have not attempted here to relate our observations to the communities distinguished by McVean and Ratcliffe (1962).

Some of the rarest and most interesting species in Britain occur in upland heaths, such as those on the summit peaks of the Cairngorm mountains. The lichen-rich communities in montane situations in Scotland appear in general, however, to be species-poor variants or relics of more widespread Scandinavian (particularly Norwegian) communities. The most important lichen species in these communities in Britain are mainly concentrated in sites which are relatively dry and dominated by *Salix herbacea*; wet sites become bryophyte dominated. The low-growing, prostrate, woody stems of *S. herbacea* which shed their leaves in winter provide adequate anchorage for many macrolichens and soil-stabilizing species. The lichen flora developed might perhaps be considered as an oligotrophic counterpart of the lichen assemblage associated with the base-rich mica schist soils of the Ben Lawers range (see p. 362) and includes: *Baeomyces roseus*, *B. rufus*, *Catillaria contristans*, *Cladonia luteoalba*, *Lecanora epibryon*, *Lecidea assimilata*, *L. caesioatra*, *L. granulosa*, *L. stenotera*, *Micarea melaena*, *M. turfosa*, *Ochrolechia androgyna*, *O. frigida*, *O. geminipara*, *O. tartarea*, *Pertusaria oculata*, *Porina mammillosa*, *Thamnolia vermicularis* (var. *subuliformis*), *Toninia havaasii*, *T. tristis*, *T. squalescens* and *T. squalida*.

In bare areas adjacent to *Salix herbacea* (see Fig. 14), colonies of *Cetraria delisei*, *Cornicularia aculeata*, *C. muricata*, *Stereocaulon saxatile*, and, very rarely, *C. divergens* are developed. These species, together with *Alectoria ochroleuca** (very rare), *A. nigricans*, *A. sarmentosa* ssp. *vexillifera*, *Cetraria ericetorum* (rare), *C. islandica*, *C. nivalis*, *Pertusaria dactylina*, *Thamnolia vermicularis* and *Cladonia* species [especially *C. bellidiflora*, *C. coccifera*, *C. deformis* (rare), *C. gonecha*, *C. impexa*, *C. gracilis*, *C. tenuis* and *C. uncialis*] occur in *Rhacomitrium lanuginosum*-dominated areas and all have the ability of either growing on or over this moss, or of vertical growth which enables them to compete successfully with it. The very rare *Nephroma arcticum* and *Platismatia norvegica* occur in the *Polygoneto-Rhacomitretum lanuginosi* and *Arctoeto-Callunetum* respectively of McVean and Ratcliffe (1962).

While emphasis has been placed on the lichen-rich communities of mountain tops in the above, it is important to note that many of the same species also form an integral part of the subalpine *Calluna–Vaccinium uliginosum–Empetrum nigrum* communities which often lie on more inclined adjacent mountain slopes; in such situations the abundance of

---

* It should be noted that, as pointed out by Hawksworth (1972b), the numerous references to this species by McVean and Ratcliffe (1962) refer to *Alectoria sarmentosa* ssp. *vexillifera*.

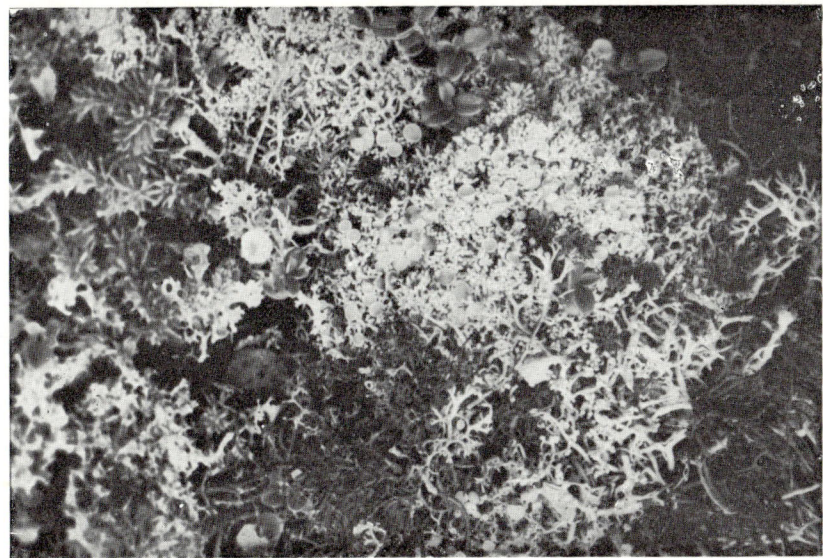

Fig. 14. Upland heath community. Species present include: *Cetraria nivalis, Cladonia bellidiflora, C. coccifera, C. crispata, C. impexa, C. uncialis* ssp. *dicraea, Empetrum nigrum, Ochrolechia frigida* (predominant) and *Vaccinium vitis-idaea* (Cairngorm mountains, Cairn Gorm summit plateau, 1974, P. W. James).

Alectoria sarmentosa ssp. *vexillifera, Cetraria nivalis, C. islandica* and species of *Cladonia* subgen. *Cladina* is particularly striking.

Although the optimal development of upland heaths in the British Isles occurs in the Scottish Highlands, comparable but species-poor communities are to be found on the highest mountains in England (e.g. Cheviot Hills, Lake District) and Wales (e.g. Snowdonia).

VIII. Summary

Following a brief review of previous studies on lichen communities in the British Isles, the value of the phytosociological approach to these is discussed as are the difficulties inherent in such methods. Some of the nomenclatural problems arising in the naming of plant communities are discussed and proposals are made for changes in order both to overcome these and to bring the nomenclatural practice for syntaxa more in line with that for plant taxa (idiotaxa). The major part of the chapter is concerned with a survey of the principal lichen communities in the British Isles. In the absence of a great deal more field work, an entirely definitive classification cannot be presented, but it is hoped that this preliminary conspectus will serve both as a framework for the future naming of lichen communities in Britain and as a stimulus to further work in this field. Topics requiring more detailed investigation are emphasized throughout.

The epiphytic communities are grouped in eleven alliances, those on limestone in two, and those on siliceous rocks in eight. Some associations, however, have not been referred to any alliances, as in the case of those which occur on moderately basic rock and in marine, maritime and aquatic habitats. Terricolous communities remain mostly unnamed here as they appear to be most appropriately described together with the vascular plants with which they occur. Two new alliance names are introduced (*Parmelion laevigatae*, *P. perlatae*) and three new combinations made at this rank in accordance with changes in the generic names of the species on which their names were based. Eleven new associations are also described and five new combinations made at that rank. A key to the epiphytic alliances is included as are tables of relevé data for most of the main associations treated and synopses of the synonymy of the community names employed.

Note on the Relevé Tables

The relevé data are expressed in the tables of this chapter by one of two methods depending on the recorder: either by the $+\to 10$ Domin scale or the Braun-Blanquet $+\to 5$ scale, in which case a sociability score on the $0\to 5$ scale is appended after a full-stop. Detailed information on the meaning of these units is included in Shimwell (1971). Data on inclinations, aspects, altitude etc. are included wherever available. Angles of inclination are measured from the horizontal, the recorder facing the substrate (i.e. $90°$ = vertical, $100°$ = inclined away from the recorder at $10°$).

Acknowledgements

We are very grateful to Mr B. J. Coppins for his constructive comments on our manuscript, to Drs O. Wilmanns and V. Wirth for the loan of copies of several important early publications on lichen phytosociology, and to Mr P. W. Lambley for making available to us his unpublished relevés from the Breckland.

References

Ahti, T. and Vitikainen, O. (1974). *Bacidia chlorococca*, a common toxitolerant lichen in Finland. *Mem. Soc. Fauna Flora fenn.* **49**, 95–100.
Almborn, O. (1948). Distribution and ecology of some south Scandinavian lichens. *Bot. Notiser, Suppl.* **1** (2), 1–252.
Alvin, K. L. (1960). Observations on the lichen ecology of South Haven Peninsula, Studland Heath, Dorset. *J. Ecol.* **48**, 331–339.
Barkman, J. J. (1954). Zur Kenntnis einiger *Usneion*-Assoziationen in Europa. *Vegetatio* **4**, 309–333.
Barkman, J. J. (1958). "Phytosociology and Ecology of Cryptogamic Epiphytes". Van Gorcum, Assen.
Barkman, J. J. (1973). Synusial approaches to classification. *In* "Ordination and Classification of Communities" (R. H. Whittaker, ed.), pp. 435–491. W. Junk, The Hague.
Barkman, J. J., Moravec, J. and Rauschert, S. (1976). Code of Phytosociological Nomenclature. *Vegetatio* **32**, 131–185.
Brightman, F. H. (1962). Field meeting at Arnside. *Lichenologist* **2**, 97–100.
Brodo, I. M. (1974) ["1973"]. Substrate ecology. *In* "The Lichens" (V. Ahmadjian and M. E. Hale, eds), pp. 401–441. Academic Press, New York and London.
Brown, D. H. and Brown, R. M. (1969). Lichen communities at Blakeney Point, Norfolk. *Trans. Norfolk Norwich Nat. Soc.* **21**, 235–250.
Burnett, J. H., ed. (1964). "The Vegetation of Scotland". Oliver and Boyd, Edinburgh and London.
Coker, P. D. (1968). The epiphytic communities of the Ruislip district. *J. Ruislip Distr. nat. Hist. Soc.* **17**, 26–37.
Coppins, B. J. (1976). Distribution patterns shown by epiphytic lichens in the British Isles. *In* "Lichenology: Progress and Problems" (D. H. Brown, D. L. Hawksworth and R. H. Bailey, eds), pp. 249–278. Academic Press, London and New York.
Crespo, A. (1975). Vegetación liquenica epifita de los pisos mediterráneo de Meseta y Montaño Ibero-atlantico de la Sierra de Guadarrama. *An. Inst. bot. A. J. Cavanillo* **32**, 185–197.
Darimont, F. (1975) ["1973"]. "Recherches Mycosociologiques dans les Forêts de Haute Belgique". 2 vols. Institut Royal des Sciences Naturelles de Belgique, Brussels.
Degelius, G. (1954). The lichen genus *Collema* in Europe. *Symb. bot. upsal.* **13** (2), 1–499.
Delzenne-van Haluwyn, C. (1976). "Bibliographia phytosociologica syntaxonomica" (R. Tüxen, ed.). Supplement I. "Bibliographia societatum lichenorum". J. Cramer, Vaduz.
Dickinson, C. H. and Thorp, T. K. (1968). Epiphytic lichens on *Corylus avellana* in The Burren, County Clare. *Lichenologist* **4**, 66–72.
Du Rietz, G. E. (1925). Götlandische Vegetationsstudien. *Svensk. växtsociol. Sällsk. Handl.* **2**, 1–65.

Earland-Bennett, P. M. (1975). *Lecanora subaurea* Zahlbr., new to the British Isles. *Lichenologist* **7**, 162–167.

Evans, E. P. (1932). Cader Idris: a study of certain plant communities in south-west Merionethshire. *J. Ecol.* **20**, 1–52.

Fabiszewski, J. (1968). Porosty Śnieżnika Kłodzkiego i gór Bialskich. *Monographiae bot.* **26**, 1–116.

Farrow, E. P. (1916). On the ecology of the vegetation of Breckland. I. General description of Breckland and its vegetation. *J. Ecol.* **3**, 211–228.

Fletcher, A. (1973a). The ecology of marine (littoral) lichens on some rocky shores of Anglesey. *Lichenologist* **5**, 368–400.

Fletcher, A. (1973b). The ecology of maritime (supralittoral) lichens on some rocky shores of Anglesey. *Lichenologist* **5**, 401–422.

Fletcher, A. (1975a). Key for the identification of British marine and maritime lichens. I. Siliceous rocky shore species. *Lichenologist* **7**, 1–52.

Fletcher, A. (1975b). Key for the identification of British marine and maritime lichens. II. Calcareous and terricolous species. *Lichenologist* **7**, 73–115.

Fletcher, A. (1976). Nutritional aspects of marine and maritime lichen ecology. *In* "Lichenology: Progress and Problems" (D. H. Brown, D. L. Hawksworth and R. H. Bailey, eds), pp. 359–384. Academic Press, London and New York.

Follmann, G. (1974). Nordhessische Flechtengesellschaften. II. Das *Pseudevernietum furfuraceae* (Hil.) Ochsn. *Hess. flor. Briefe* **23**, 40–47.

Frenkel, R. E. and Harrison, C. M. (1974). An assessment of the usefulness of phytosociological and numerical classificatory methods for the community biogeographer. *J. Biogeogr.* **1**, 27–56.

Frey, E. (1923). Die Berücksichtigung der Lichenen in der soziologischen Pflanzengeographie, speziell in den Alpen. *Ver. naturf. Ges. Basel* **35**, 303–320.

Gilbert, O. L. (1976). An alkaline dust effect on epiphytic lichens. *Lichenologist* **8**, 173–178.

Graham, G. G. (1971). Phytosociological studies of relict woodlands in the north-east of England. M.Sc. thesis, University of Durham.

Gregory, K. J. (1976). Lichens and the determination of river channel capacity. *Earth Surface Processes* **1**, 273–285.

Guinochet, M. (1973). Phytosociologie et systématique. *In* "Taxonomy and Ecology" (V. H. Heywood, ed.), pp. 121–140. Academic Press, London and New York.

Hawksworth, D. L. (1969). The lichen flora of Derbyshire. *Lichenologist* **4**, 105–193.

Hawksworth, D. L. (1972a). The natural history of Slapton Ley Nature Reserve. IV. Lichens. *Fld Stud.* **3**, 535–578.

Hawksworth, D. L. (1972b). Regional studies in *Alectoria* (Lichenes). II. The British species. *Lichenologist* **5**, 181–261.

Hawksworth, D. L. (1973a). The lichen flora and vegetation of Berry Head, South Devonshire. *Trans. Torquay nat. Hist. Soc.* **16**, 55–66.

Hawksworth, D. L. (1973b). Ecological factors and species delimitation in the lichens. *In* "Taxonomy and Ecology" (V. H. Heywood, ed.), pp. 31–69. Academic Press, London and New York.

Hawksworth, D. L. (1974). "Mycologist's Handbook". Commonwealth Mycological Institute, Kew.
Hawksworth, D. L., Rose, F. and Coppins, B. J. (1973). Changes in the lichen flora of England and Wales attributable to pollution of the air by sulphur dioxide. In "Air Pollution and Lichens" (B. W. Ferry, M. S. Baddeley and D. L. Hawksworth, eds), pp. 330–367. Athlone Press, London.
Hawksworth, D. L., Coppins, B. J. and Rose, F. (1974). Changes in the British lichen flora. In "The Changing Flora and Fauna of Britain" (D. L. Hawksworth, ed.), pp. 47–78. Academic Press, London and New York.
Hilitzer, A. (1924). Příspěvky k lisejnikům šumavy a Pošumaví. Čas. národ. Mus. **1923**, 1–14. [Reprint only seen.]
Hilitzer, A. (1925). Étude sur la végétation épiphyte de la Bohême. Spisy Fac. Sci. Univ. Charles **41**, 1–202.
James, P. W. (1965). Field meeting in Scotland. Lichenologist **3**, 155–172.
James, P. W. (1970). The lichen flora of shaded acid rock crevices and overhangs in Britain. Lichenologist **4**, 309–322.
James, P. W. (1973). The effect of air pollutants other than hydrogen fluoride and sulphur dioxide on lichens. In "Air Pollution and Lichens" (B. W. Ferry, M. S. Baddeley and D. L. Hawksworth, eds), pp. 143–175. Athlone Press, London.
Kaiser, E. (1926). Die Pflanzenwelt des Hennebergisch-Fränkischen Muschelkalk-Gebietes. Beih. Rep. spec. nov. regni veg. **44**, 1–280.
Kalb, K. (1970). Flechtengesellschaften der vorderen Ötztaler Alpen. Diss. Bot., Lehre **9**, 1–118.
Kalb, K. (1972). Rindenbewohnende Flechtengesellschaften im Nürnberger Reichswald. II. Hoppea [Denkschr. Regensb. bot. Ges.] **30**, 73–91.
Klement, O. (1955). Prodromus der mitteleuropäischen Flechtengesellschaften. Beih. Rep. spec. nov. regni veg. **135**, 5–194.
Klement, O. (1965). Zur Kenntnis der Flechtenvegetation der Kanarischen Inseln. Nova Hedwigia **9**, 503–582.
Knowles, M. C. (1913). The marine and maritime lichens of Howth. Scient. Proc. R. Dubl. Soc. **14**, 79–143.
Laundon, J. R. (1956). The lichen ecology of Northamptonshire. In "The First Fifty Years. A History of Kettering & District Naturalists Society & Field Club", pp. 89–96. Kettering and District Naturalists Society and Field Club, Kettering.
Laundon, J. R. (1958). The lichen vegetation of Bookham Common. Lond. Nat. **37**, 66–79.
Laundon, J. R. (1967). A study of the lichen flora of London. Lichenologist **3**, 277–327.
Laundon, J. R. (1970). London's lichens. Lond. Nat. **49**, 20–69.
LeBlanc, F. (1963). Quelques sociétés ou unions d'épiphytes du sud du Québec. Can. J. Bot. **41**, 591–638.
Mattick, F. (1951). Wuchs- und Lebensformen, Bestand- und Gesellschaftsbildung der Flechten. Bot. Jb. **75**, 378–424.
McLean, R. C. (1915). The ecology of the maritime lichens at Blakeney Point, Norfolk. J. Ecol. **3**, 129–148.

McVean, D. N. and Ratcliffe, D. A. (1962). "Plant Communities of the Scottish Highlands". H.M.S.O., Edinburgh and London.
Meijer-Drees, E. (1953). A tentative design for rules of phytosociological nomenclature. *Vegetatio* **3**, 205–214.
Motyka, J. (1926). Die Pflanzenassociationen des Tatragebirges. VI. Teil. Studien über epilithischen Flechtengesellschaften. *Bull. int. Acad. pol. Sci. Lett., sér.* **B, 1926,** 189–227.
Ochsner, F. (1928). Studien über die Epiphyten-Vegetation der Schweiz. *Jb. St. Gall. naturw. Ges.* **63** (2), 1–108.
Prince, C. R. (1974). A study of a lichen synusium on the Sands of Forvie, Scotland. *Nova Hedwigia* **25**, 719–736.
Ratcliffe, D. A. (1968). An ecological account of atlantic bryophytes in Britain. *New Phytol.* **67**, 365–439.
Rayner, R. W., ed. (1976) ["1975"]. "The Natural History of Pagham Harbour. Part II". Bognor Regis Natural History Society, Bognor Regis.
Ried, A. (1960a). Stoffwechsel und Verbreitungsgrenzen von Flechten. I. Flechtenzonierung an Bachufern und ihre Beziehungen zur jährlichen Überflutungsdauer und zum Mikroklima. *Flora, Jena* **148**, 612–638.
Ried, A. (1960b). Stoffwechsel und Verbreitungsgrenzen von Flechten. II. Wasser- und Assimilationshaushalt, Entquellungs- und Submersion-resistenz von Krustenflechten. *Flora, Jena* **149**, 345–385.
Rose, F. and James, P. W. (1974). Regional studies on the British lichen flora. I. The corticolous and lignicolous species of the New Forest, Hampshire. *Lichenologist* **6**, 1–72.
Rose, F. and Wallace, E. C. (1974). Changes in the bryophyte flora of Britain. *In* "The Changing Flora and Fauna of Britain" (D. L. Hawksworth, ed.), pp. 27–46. Academic Press, London and New York.
Santesson, R. (1939). Über die Zonationsverhältnisse der lakustrinen Flechten einiger Seen im Angebodagebiet. *Medd. Lunds Univ. Limnol. Inst.* **1939** (1), 1–70.
Seaward, M. R. D. (1975). Lichen flora of the West Yorkshire conurbation. *Proc. Leeds phil. lit. Soc. sci. sect.* **10**, 141–208.
Sheard, J. W. (1968). The zonation of lichens on three rocky shores at Inishowen. *Proc. R. Ir. Acad.,* **B, 66,** 101–112.
Sheard, J. W. and Ferry, B. W. (1967). The lichen flora of the Isle of May. *Trans. Proc. bot. Soc. Edinb.* **40**, 268–282.
Shimwell, D. W. (1971). "The Description and Classification of Vegetation". Sidgwick and Jackson, London.
Stafleu, F. A. *et al.*, eds. (1972). International Code of Botanical Nomenclature adopted by the eleventh International Botanical Congress, Seattle, August 1969. *Regnum Vegetabile* **82**, 1–426.
Swinscow, T. D. V. (1968). Pyrenocarpous lichens: 13. Freshwater species of *Verrucaria* in the British Isles. *Lichenologist* **4**, 34–54.
Swinscow, T. D. V. (1971). Pyrenocarpous lichens: 15. Key to *Polyblastia* Massal. in the British Isles. *Lichenologist* **5**, 92–113.

Syers, J. K. (1964). A study of soil formation on Carboniferous limestone with particular reference to lichens as pedogenic agents. Ph.D. thesis, University of Durham.

Tansley, A. G. (1939). "The British Isles and their Vegetation". Cambridge University Press, Cambridge.

Watson, W. (1918a). Cryptogamic vegetation of the sand-dunes of the west coast of England. *J. Ecol.* **6**, 126–143.

Watson, W. (1918b). The bryophytes and lichens of calcareous soil. *J. Ecol.* **6**, 189–198.

Watson, W. (1919). The bryophytes and lichens of freshwater. *J. Ecol.* **7**, 71–83.

Watson, W. (1925). The bryophytes and lichens of arctic alpine vegetation. *J. Ecol.* **13**, 1–26.

Watson, W. (1932). The bryophytes and lichens of moorland. *J. Ecol.* **20**, 284–313.

Watson, W. (1936a). The bryophytes and lichens of British woods. Part I. Beech woods. *J. Ecol.* **24**, 139–161.

Watson, W. (1936b). The bryophytes and lichens of British woods. Part II. Other woodland types. *J. Ecol.* **24**, 446–478.

Wilmanns, O. (1966). Die Flechten- und Moosvegetation des Spitzbergs. *Spitzb. Tübingen* **3**, 244–277.

Wirth, V. (1968). Soziologie, Standortsökologie und Areal des *Lobarion pulmonariae* im Südschwarzwald. *Bot. Jb.* **88**, 317–365.

Wirth, V. (1972). Die Silikatflechten-Gemeinschaften im außeralpinen Zentraleuropa. *Diss. Bot., Lehre* **17**, 1–306, *1–9*.

Yarranton, G. A. (1967). A quantitative study of the bryophyte and macrolichen vegetation of the Dartmoor granite. *Lichenologist* **3**, 392–408.

11 | Lichen Conservation in Britain

OLIVER L. GILBERT

I. Introduction	415
II. The Current Situation	416
III. Habitat Destruction and Disturbance	417
IV. Environmental Pollution	419
V. Conservation Priorities	421
VI. Conservation Research	424
A. The Dynamic Nature of Lichen Populations	424
B. Manipulation of Environmental Factors	424
C. Manipulation of Species	428
VII. An Outline Strategy	432
References	434

I. Introduction

In a world where much of mankind is living under conditions of deprivation, it is not easy to justify the allocation of money and resources for the protection of native plants and animals. This is particularly the case when the object of conservation is a relatively inconspicuous and little studied group of organisms of negligible economic importance. Lichen conservationists have therefore to rely strongly on general arguments equally appropriate for the rest of nature. Unfortunately, experience has shown that none of these arguments carry much weight when they conflict with developments addressed to urgent problems of food supply, housing, defence, employment or recreation.

Ecologists have recently appreciated that it may be possible to make out a case for the conservation of all groups on the diversity/stability hypothesis—namely, the more species there are in an ecosystem the more stable and predictable it will be. Though the evidence for this is incomplete, there are strong theoretical reasons for arguing that there are no useless species and it is in our long-term interests to work towards a wise coexistence with all living things. If this is correct the present *ad hoc* conservation measures should be seen as an important "holding and learning operation". In addition to arguments based on economics, there are deep-seated cultural reasons for conserving wildlife and its habitat. These arguments are not fully worked out but are connected with expanding mankind's perception and comprehension of the universe, and giving a sense of purpose and achievement to life (Ratcliffe, 1976).

In an unpublished submission to the Nature Conservancy Council in 1972, the British Lichen Society suggested a number of reasons why lichen conservation is of importance. Though such reasons are on their own a rather narrow base for projecting conservation arguments, the following points from it are worth emphasizing:

1. Lichens cannot at present be permanently cultured, grown in botanic gardens, stored in seed banks or be artificially maintained for long periods. The only method of guaranteeing the perpetuation of communities or individual species is to safeguard their existence under natural conditions in the field.

2. For its size Britain has a very varied lichen flora, including a strong "Atlantic element", of which more fragments survive than in many adjacent parts of industrialized north-west Europe, so lichen conservation work in Britain is of international importance.

3. Lichen-rich sites frequently show a high degree of correlation with areas of general ecological interest, though not necessarily with a rich phanerogamic flora. This makes their acquisition particularly attractive.

4. It can also be pointed out that no organism lives alone: there is a tangled web of interrelationships and a considerable number of fungal and animal species depend on lichens for their survival.

II. The Current Situation

The lichen flora of Britain is now well known and stands at just under 1400 species. Of these, 40 (*c.* 3% of the total flora) have not been seen this century and are presumed to have become extinct (Hawksworth *et al.*, 1974). This aggregation hides, however, a level of extinction running as high as 25% or more in a number of recently well worked counties (see Hawksworth, 1975), and considerably higher over smaller areas. Species density varies widely but away from urban or intensively farmed areas, 150–250 species are usually present in any 10×10 km^2 (38 sq. miles) unit and in the best areas numbers are even higher. Devonshire (2600 sq. miles) with 606 spp. (Hawksworth, 1975) and the Isle of Mull with an exceptional *c*. 680 spp. in 367 sq. miles (P. James, in press) are two of the richest districts discovered, but as they are also two of the best-worked areas, it is likely that a number of equally rich places remain to be found.

Against this background of intensive distributional studies, little work specifically directed at lichen conservation has been attempted in Britain, though, due to a frequent coincidence with sites of high ecological interest, many important lichen communities are present in the 150 national nature reserves and the 900 county naturalist trust reserves which together cover nearly 100,000 acres (45,000 ha) of the country. In 1968 the British Lichen Society set up a Conservation Committee which keeps an up-to-date graded list of sites worthy of conservation on their lichen content alone and the Nature Conservancy Council is kept informed of those in the top categories. In some cases, local conservation bodies are also notified of important lichen sites in their area so they can act in a watchdog capacity, as accidental destruction is possibly the major threat to localities where the scientific interest is represented by small plants and animals.

The present list of designated sites is based mainly on the enthusiasm and interests of a small band of workers and within the next few years, work done in connection with the Society's Mapping Scheme promises to provide the raw material from which a more objective conservation strategy can be drawn up. The problem of conserving Britain's lichens, however, involves considerably more than identifying a series of potential nature reserves. While nature reserves are invaluable for protecting rare assemblages of lichens from habitat destruction, as a primary land use they will never cover more than a small proportion of the country and they produce their own problems. Once a reserve is declared, management and particularly increased educational, scientific and public pressures may have to be faced, so lichen conservation research must keep pace with reserve acquisition. The greatest problem facing conservation is that even when sited in a nature reserve, lichen vegetation is not safe from the far-reaching effects of aerial pollution. Thus the conservation of major sites should ideally be carried out on a country-wide or international basis.

III. Habitat Destruction and Disturbance

Lichens have a strictly limited ability to respond to abrupt change; consequently in any particular habitat they reach their greatest abundance and diversity where conditions have been stable over a long period. The current pressure on land which comes particularly from an expanding and increasingly mechanical farming and forestry industry, together with urbanization and mineral extraction, has militated against this necessary habitat continuity in many parts of the country.

Lowland terricolous habitats appear to be at greatest risk. The progressive fragmentation and restriction of the Breckland heaths (54,000 acres in 1880, to 19,000 acres in 1968), the Dorset heaths (two-thirds reduction in the period 1811–1960), the Surrey heaths, and chalk grassland (now only 3·3% of the outcrop under grassland) provide just a few examples, and the more demanding lichen species of these habitats are slow to invade more recently established communities. It is probable that in lowland Britain these ancient lichen-rich vegetation types will become scarce outside nature reserves.

Mature deciduous woodland has also been under constant and increasingly severe pressure. This culminated in the present century when one-third of our standing timber was felled during each of the two world wars. Since then, much of what was left has been converted to conifers. Short-rotation forestry, whether of coniferous or deciduous trees, provides woodland particularly unfavourable for lichens, and Rose (1974) has pointed out that even mature oak forest known to have been clear felled and replanted with oak as long ago as the late eighteenth century, still has a greatly reduced lichen flora compared with old uneven-aged oak forest (Table I). A number of the rarer species appear to be intolerant of even short periods of desiccation which makes them particularly sensitive to destruction of the canopy.

Other habitats e.g. upland, maritime and freshwater, which at present seem secure could be threatened in the future, especially by new demands for power and industrial development in depressed areas. Few habitats

TABLE I.

Lichen content of oak woodland related to history of management (after Rose, 1974).

	No. in sample	Average no. of taxa in 1 km² or less
Ancient mixed oak forests	36	118
Mature oak clear felled and replanted during the eighteenth and nineteenth centuries	5	50
Old coppice woodland with oak standards	11	42

seem entirely safe from environmental disturbance in a small, heavily industrialized country.

Occasionally the destruction of natural habitats is compensated for by the large-scale construction of artificial ones which allow a number of species to increase and give rise to new communities. Quarrying for building-stone produces a wide range of secondary habitats which increase the lichen richness of many districts far beyond any destruction incurred at the quarry sites. These sites may in time themselves become refugia for uncommon lichens, e.g. *Solorina saccata* at Barnack Hills and Holes, Northamptonshire (Laundon, 1964). Hawksworth *et al.* (1974) list over 80 saxicolous lichens which occur on buildings, bridges, roofs etc. in south-east England where natural saxicolous habitats are very rare or absent. This effect is especially well seen in churchyards, the conservation value of which has recently received publicity (Barker, 1972), on walls marking property boundaries and on ornamental stonework in the grounds of stately homes. Because there have been gains as well as losses, the effects of habitat destruction are not always easy to evaluate. The majority of lichen vegetation in Britain is affected by man to some extent which makes it all the more important to protect the remaining examples of near natural communities. A great deal of interest is, however, also present in the man-made or cultural landscape where the lichen vegetation reflects our history and technology to a remarkable degree. Each age produces its own landscape capital which favours different lichen assemblages, and though the older artefacts tend to be the richest, even today new lichen patterns are being formed which will be valued by future generations. The neolithic heathlands, Bronze Age megaliths, Gothic churches, ornamental parks, lime-mortared stone walls (now being replaced by cement mortar or wire and posts) and today's asbestos-cement roofs all support lichen communities worthy of conservation in their own right. Such secondary habitats, many of which are already an accepted part of the humanized landscape, will for this reason often be easier to preserve than primary ones.

IV. Environmental Pollution

Habitat destruction is not the major threat to our lichen flora—overlying it are the devastating effects of pollution. Lichens possess an inherent ability to concentrate traces of pollutants until they become toxic, so a necessary adaptation to life in inhospitable habitats places the group in great jeopardy. Air pollution, in particular sulphur dioxide, fluorides, wind-blown fertilizers and certain agricultural sprays have severely impoverished the lichen flora of large areas of lowland Britain. The effects of these aerial pollutants have recently been reviewed in Ferry *et al.* (1973) where details of their toxicity, distance over which they can be transported from source and precise ecological consequences can be found. At fairly low levels of pollution the ability of many species to colonize new sites becomes impaired, so apparently intact communities have only a relict status, although encouragingly Skye and Hallberg (1969) have reported that if sulphur dioxide levels are drastically reduced the lichen flora shows signs of immediate recovery. Not all lichen communities are equally sensitive to environmental pollution, which has led to the situation where no epiphytic lichen communities of note survive in Derbyshire or in the Craven District of Yorkshire, but the limestone cliffs in those regions still carry assemblages of national importance.

Mapping studies (Hawksworth *et al.*, 1973; Seaward, 1973) have amply shown the scale of the decline caused by environmental pollution and recently it has been quantified for sites in Surrey (Laundon, 1973a), Huntingdonshire (Laundon, 1973b) and Norfolk (Coppins and Lambley, 1974). This is an important extension of previous work and the alarming rates of decline reported give added urgency for action on the regularly postponed plans to curb pollution. Much of the most important epiphytic lichen vegetation in lowland Britain is probably already of relict status. A few studies have shown that there has been little change this century in certain remote areas, but it is difficult to find sufficiently detailed early surveys to pursue this line of enquiry.

Despite the much publicized reduction in urban levels, sulphur dioxide —the major hazard to lichens—is continuing to spread out over new ground as a result of increasing urbanization and the erection of new power stations with very tall chimneys. Lichenologists view with great concern plans to site further power stations burning sulphur-rich fossil fuel and other polluting industries such as oil refineries and petrochemical complexes in those few parts of the country which have so far escaped contamination.

Occasionally intense local pollution produces new and interesting lichen assemblages worthy of conservation because they are species-rich or include rare lichens. Poelt and Huneck (1968) have reported that wooden vine supports in the Uberetsch region of Austria, if regularly drenched with copper-containing fungicidal sprays, develop a lichen assemblage containing *Acarospora anomala*, *Lecanora vinetorum*, *Sarcogyne simplex* and *Stereocaulon nanodes*. Further examples of interesting lichen communities developing on sites artificially contaminated by heavy metals can be found associated with lead/copper/zinc/barium mine waste heaps over a hundred years old. In the Mendip Hills (Brown, 1973) and the Pennines these have become well colonized by lichens which are very rare or absent in the surrounding countryside, including such species as *Biatorella campestris*, *Cetraria islandica*, *Lecidea atrata*, *L. geophana*, *Pannaria pezizoides*, *Peltigera venosa*, *Stereocaulon dactylophyllum*, *S. nanodes*, *S. pileatum* and fertile *Baeomyces roseus*. Lichen communities on metal-rich substrates in Britain are discussed further in Chapter 10.

There is a little-studied pollution effect associated with alkaline-dust emission from limestone quarries and cement works which can raise the pH of tree bark for several miles around the source. In moderately polluted parts of the country, this enables many sulphur dioxide-sensitive epiphytes to survive in generally hostile areas so trees in the neighbourhood of alkaline-dust sources often carry the richest corticolous communities in a region (Gilbert, 1976). This and the other examples quoted above, illustrating how environmental pollution occasionally acts to produce anomalous lichen-rich vegetation of some conservation value (i.e. mine waste heaps which are under constant threat of reworking), must not divert attention from the fact that its normal effect is to cause severe impoverishment.

V. Conservation Priorities

In the last decade, a great deal of survey work has been accomplished and it is now possible to identify a number of lichen-rich sites which Britain has an especial responsibility to safeguard. These localities hold examples of ecological systems which in the author's opinion should be regarded as of probable international importance since they appear better represented in this country than elsewhere in Europe or else their centre of distribution is here.

An example of the first type are remnants of the deciduous summer forests which formerly covered much of the Atlantic seaboard of western Europe. These climax forests have now largely been converted to prosperous agricultural land but many experts are of the opinion that in the New Forest, Hampshire, we have one of the best remaining examples of ancient forest of its type, i.e. oceanic, on the lowland plains of Europe. Its extent (36 km^2), uneven-age structure, relatively undisturbed state, well documented history and immensely rich epiphytic lichen flora (259 spp.) mark it as of prime conservation importance (Tubbs, 1968; Rose and James, 1974).

The landscaped parklands which surround many English stately homes provide another example of a habitat better represented in this country than elsewhere. Their landscapes, dotted with trees, some planted, some primeval, some acidic, others eutrophicated, can support a higher density of epiphytic lichens per unit area than is found even in old woodland. The lavish use of stone for enclosure and ornament provides an ample habitat for saxicolous lichens, and old lawns, banks and rockeries hold a variety of terricolous species. The survival and stability of these parklands and formal gardens are chiefly the result of their being in one ownership for long periods (and hence under one system of management), the family regarding them as a setting for the house. This often resulted in a low intensity of land use being practised, while the popularity of sheltered sites has meant that many are protected from the worst effects of air pollution. Examples which contain remnants of wooded deer parks enclosed in medieval times—now areas of old pollards and maiden trees in open-closed canopy, usually over bracken or rough grassland, are particularly rich (Rose, 1974). This type of contrived landscape did not become popular on the Continent until the late eighteenth century where it was known as *Le Jardin Anglais* or *Englischer Garten*.

On a Eurasian scale, the British Isles is outstanding for its strong representation of plants which depend on an oceanic environment. For example, the Atlantic zone of Europe is universally regarded as the type locality for dwarf shrub heaths which until recently constituted a major element of the vegetation in lowland parts of the region. Over much of this area, notably Sweden, Denmark, the Netherlands, Belgium and Germany, their extent is being greatly reduced by conversion to farmland or forestry. Though originally man-made—the result of progressive podzolization following forest destruction—they have had time to develop a rich and distinctive lichen flora, of which more may survive in Britain than elsewhere; *Baeomyces, Catillaria, Cetraria, Cladonia, Cornicularia, Lecidea* and *Micarea* species are characteristic of this habitat in Britain. The European endemic *Cladonia impexa* (Fig. 1), for example, has a major part of its population in these heathlands, the lichen ecology of which is little studied, although Coppins and Shimwell (1971) have pointed out how important management can be in controlling their composition.

There is evidence that other Atlantic lichen communities have a major part of their distribution in Britain which, due to its off-shore position, represents an end-point to a climatic gradient of continental scale, i.e. strongly oceanic temperate. Examples of these are discussed in detail in Chapter 10.

National sites should contain the best examples of particular types of lichen vegetation to be found in the British Isles, and ideally the whole field of variation in lichen vegetation found in this country should be protected. On a European scale, the terricolous communities of Breckland, boreal coniferous forest assemblages in central Scotland and much of our

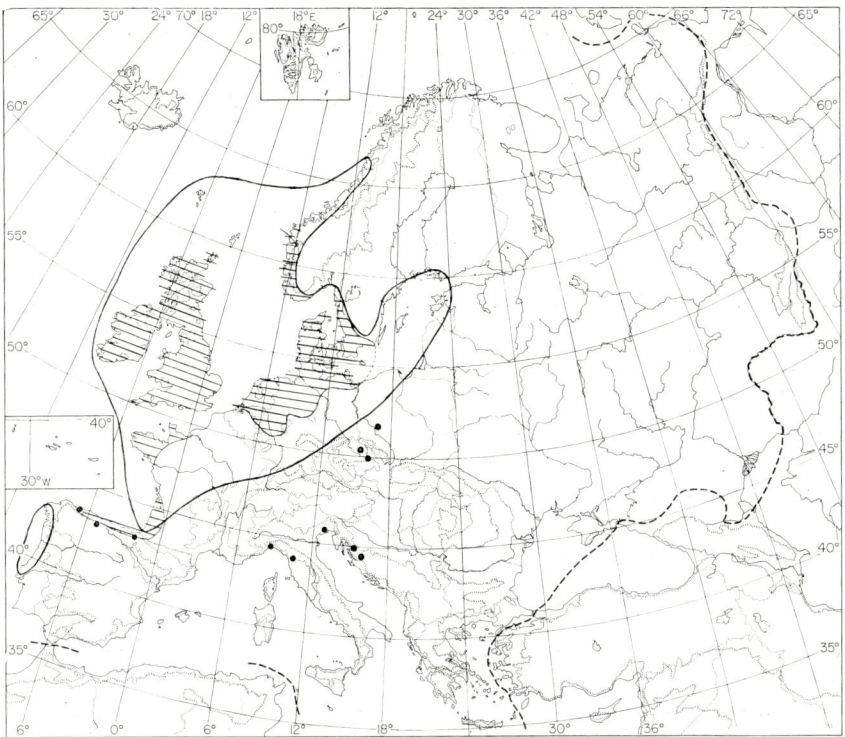

FIG. 1. World distribution of *Cladonia impexa* enclosed by continuous thick line with dots for outlying localities (after Ahti, 1961); and the main area in which lowland heaths occur (hatched). After Gimingham (1972).

montane lichen vegetation appear impoverished, but on a national scale they are exciting places where assemblages of rarities might be seen. An example of this at the species level is provided by *Cladonia stellaris*, which in Britain is a most sought after, possibly extinct, alpine rarity with an almost completely circumpolar distribution (see p. 154). The Nature Conservancy Council is directed by the *National Parks and Access to the Countryside Act 1949* and the *Nature Conservancy Act 1973* with protecting and managing such sites.

Regional, county and local sites are also the concern of the Nature Conservancy Council, County Naturalists' Trusts, local authorities and land owners. They could be designated to protect a single, very rare, or distinctive species but a far greater contribution can be made if habitats containing a diverse, or even aesthetically pleasing (e.g. churchyards and old mine waste heaps), lichen vegetation are protected.

VI. Conservation Research

Survey work on the British lichen flora is well in hand and because of this, conservation priorities are becoming clearer. As reserve acquisition proceeds, however, problems of management will appear and work on this topic has hardly started. Studies on community structure, composition and dynamics, and on the autecology of our rarer species are urgently needed. For example, in some woods the lichen interest is centred on a few large, senile, decaying trees. Is this their natural pattern of distribution—rare and capricious—or are they the last remnants of a primeval woodland lichen flora and now of relict status? If the latter, what habitat factors need manipulating to increase their vigour? With the present state of knowledge it would be premature to write on conservation practice but some research relevant to management will be reviewed.

A. *The Dynamic Nature of Lichen Populations*

In his classic work on pattern and process in plant communities, Watt (1947) studied seven terricolous habitats, five of which contained lichens; he found that the lichens had a dynamic relationship with the dominant plant which played a critical role in determining their abundance. This relationship is illustrated for dry calcareous grassland in Fig. 2, where it is clear that the lichens are particularly associated with degenerate phases of the dominant *Festuca* plants, but are rare or absent when the grass tussocks are in the building phase. To ensure the survival of the full complement of terricolous lichens in these habitats, it is important that the dominant should be present as an uneven-aged stand and any factor which interferes with (i.e. shortens) the downgrade series, such as burning, could be a serious threat to their continued existence. Similar but much longer cyclical rhythms related to the life-span of trees are known to be important in controlling the distribution and abundance of epiphytes in woodland but require more detailed investigation.

B. *Manipulation of Environmental Factors*

The environment can be manipulated locally to encourage or discourage lichens though it will be some time before this is practised in other than a crude manner. The examples given below outline a few possibilities once the relevant research has been undertaken.

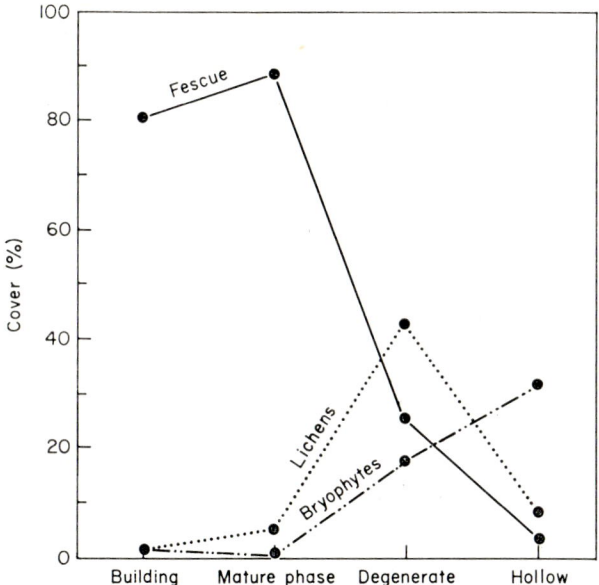

FIG. 2. Graphical presentation of data from dry lichen-rich *Festuca* grassland in Breckland (after Watt, 1947). The phases represent a cyclical time sequence related to the initiation, growth and decay of *Festuca ovina* tussocks.

1. Soil Nutrients

Experiments showing that low soil-nutrient levels are important in the maintenance of "lichen-rich pasture" have been performed on Braunton Burrows dune system in Devonshire. Over a 2-year period Willis (1963) made substantial applications of major nutrients (N, P and K) to the dwarf (1·5 cm tall) uniform vegetation dominated by *Cladonia furcata*, *C. rangiformis*, *Diploschistes scruposus*, *Fulgensia*, *Toninia*, *Collema* and *Leptogium* species and bryophytes. At the end of the 2-year experimental period, a lush growth of *Festuca rubra* and *Poa pratensis* had taken over the plots and many lichens were extinct (Table II). In the "dry dune pasture" at Braunton, nutrient application caused the cover of *Peltigera canina* to drop from 10% to just a trace in 2 years. The mechanism for lichen exclusion appeared to be changes in light and humidity at ground level. This research indicated how relatively more important low nutrient availability was than water stress or grazing in maintaining the lichen interest of this important site. The same conclusion has been reached concerning lichen-rich areas of chalk grassland on Salisbury Plain (Wells et al., 1976). A similar phenomenon has been observed, but at the lower

TABLE II.

The effect of the addition of mineral nutrients on the vegetation of lichen-rich pasture, Braunton Burrows (after Willis 1963).

	June 1958	Oct. 1958	June 1959	June 1960
Bare ground (%)	5	T	T	T
Higher plants (%)	30	70	75	90
Bryophytes (%)	10	T	T	T
Lichens (%)	50	25	20	5
Average no. spp./unit area	23	–	19	17
Average height of vegetation (cm)	1·5	8	15	10

Average scores of the main constituents of the vegetation of the treated areas given as subjective estimates of relative bulk expressed as a percentage. T = trace. Start of the experiment 19 June, 1958, final recording 20 June, 1960.

end of the pH scale, in Sheffield where certain lawns contain luxuriant mats of *Peltigera* and spreads of *Cladonia*. Investigation revealed that the lichen-rich swards were restricted to lawns on acid, nutrient-deficient soils (O. L. Gilbert, unpublished). The implications of these observations are that fertilizer or lime should be used only sparingly on sites important for terricolous lichens or on certain ornamental lawns which owe their character to a high biomass of lichens.

2. Agricultural Sprays

The effect of agricultural sprays has not received serious study. Many lichenologists have observed moribund lichens adjacent to sprayed farmland but cause and effect has not been proved. Evidence that some widely used herbicides may not be strongly toxic to terricolous lichens has been seen at two sites in Northumberland where little-used garden paths annually treated with Simazine (2-chloro-4,6-bis(ethylamino)-1,3,5-triazine) are thickly covered with *Cladonia furcata* and *Baeomyces rufus*. At another site luxuriant *Peltigera canina*, *P. polydactyla* and *P. spuria* occur on garden paths regularly treated with Prefix (2,6-dichlorothiobenzamide, chlorothiamide).

Following these observations, some experiments (O. L. Gilbert, unpublished data) were set up in Derbyshire to test the effect of the three commonly used herbicides, paraquat/diquat (general), MCPA (scrub) and 2,4-D (selective), on a range of lichens. The results (Table III)

TABLE III.

The effect of a single application of three herbicides on a range of lichens under field conditions.
Dosages: $5 \times$ recommended rate; recommended rate; $\frac{1}{4} \times$ recommended rate.
Application August, 1973; final recording August, 1974.

Dominant Lichens	MCPA			Paraquat/Diquat			2,4-D		
	$5\times$	$1\times$	$\frac{1}{4}\times$	$5\times$	$1\times$	$\frac{1}{4}\times$	$5\times$	$1\times$	$\frac{1}{4}\times$
Sandstone rocks and walls *Lecanora conizaeoides, Ochrolechia androgyna, Parmelia saxatilis, P. omphalodes, Rhizocarpon geographicum*	Severe damage[a]	No damage		No damage			No damage		
Limestone walls *Caloplaca aurantia, C. heppiana, Aspicilia calcarea, Lecanora dispersa, Verrucaria sphinctrina*	No damage			No damage			No damage		
Acid heathland *Cladonia chlorophaea, C. coccifera, C. floerkeana, Lecidea granulosa, L. uliginosa*	?	Slight damage[b]	No damage	?	Slight damage[b]	No damage	?	Slight damage[b]	No damage
Oak trees *Hypogymnia physodes, Lecanora conizaeoides, Parmelia saxatilis, P. sulcata, Pseudevernia furfuracea*	Severe damage[c]	No damage		No damage			No damage		
Concrete posts *Lecanorion dispersae Xanthoria parietina, Physcia adscendens*	No damage			No damage			No damage		

[a] After 4 weeks the lichens had become discoloured and appeared moribund. Later many became detached and flaked off but 20% still alive a year later.

[b] No clear effect but about 10% in each plot were moribund at the final recording, compared to 5% in the controls. This may have been associated with the death of the surrounding *Calluna*.

[c] After 4 weeks the lichens had become discoloured (bleached, brown or reddish) and over the next few months became detached.

show that only MCPA applied at many times the recommended rate had any adverse effects on the lichens tested. It must be emphasized that in these experiments the effect of repeated doses was not investigated, but that the single applications included doses many times heavier than those normally experienced from spray drift. The conclusion is that the lichens tested are resistant to single applications of the three herbicides, though massive contamination with MCPA can be lethal. This raises the possibility that certain herbicides could be potentially useful in lichen conservation work. They have already been used in a conservation role to prevent scrub regrowth in plagioclimax grassland (Moore, 1968).

3. Oil Spills

A growing mass of evidence suggests that the detergents hosed in great quantities on to rocks and beaches during clean-up operations following oil spills do considerably more damage to shore lichens than crude oil (Brown, 1972; Ranwell, 1968). Brown (1974) records that 7 years after the *Torrey Canyon* disaster, lichen communities on the lower shore at Caerthillian Cove, Cornwall, had shown considerable recovery while at higher levels recolonization had hardly started. As oil spills affect large areas indiscriminately (140 miles of the Cornish coast severely polluted with oil in 1967), work of this type is clearly as urgent as acquiring coastal nature reserves.

C. Manipulation of Species

Conservation work in most fields is concerned primarily with manipulating the environment; the direct manipulation of species is disapproved of especially by botanists. With our increasing awareness of how much man has been involved in shaping our present vegetation and with increasing threats even to remote areas, it is gradually becoming accepted that conservation should include direct action to help particularly rare or attractive species to survive. The artificial pollination and spreading of seed which enabled the Kentish colony of the monkey orchid (*Orchis simia*) to increase from one plant in 1955 to over 250 ten years later shows the potential of this approach. Safeguards against the abuse of re-introduction have been discussed by the Society for the Promotion of Nature Reserves (1970) and the whole matter is sympathetically reviewed by Duffey (1974).

Recently I have started research on selected lichens to learn something of their population dynamics and to discover how easy it would be to transplant them if the need arose.

1. *Bryoria fuscescens*

This species has become exceedingly rare in the polluted English Midlands but is abundant on a single sandstone boulder at Birchover, Derbyshire. In June, 1971, two small specimens from this population, each 3 cm long,

FIG. 3. Expanding lichen populations. A. A single specimen of *Bryoria fuscescens* transplanted using "Isopon" (black) on to a sandstone boulder gave rise to ten new colonies in the first 3 years (scale × ½). B. *Usnea subfloridana*. Slow natural spread on an ash bole from a single plant believed to have established in the 1960s. The smaller thalli usually survive only a year or two (scale × ½).

were cemented on to an adjacent boulder using Isopon glass fibre of the type used for repairing car bodies. Three years later the transplants were still thriving and had given rise to a number of new colonies. The pattern of spread (Fig. 3 A) has been downwards producing a streak of young plants the density of which falls off sharply with distance. By 1976, the streak of young plants had coalesced into a continuous strand 25 cm long, and six further new colonies, up to 8 cm long, had appeared either side and above the original.

2. *Usnea subfloridana*

The dynamics of the only known Derbyshire population has been studied for 4 years. Spread from the two original thalli (believed to have established in the late 1960s) has been similar to that recorded for *Bryoria* (Fig. 3 B).

3. *Parmelia perlata, Ramalina fastigiata, R. fraxinea*

These three species have been transplanted from *Fraxinus* on to *Fraxinus* in localities where they were rare. All have survived in a healthy state and shown considerable positive growth but no new colonies have been observed in the first 5 years.

4. *Lobaria pulmonaria*

The distribution of this spectacular lichen has decreased sharply in Britain, and transplant experiments with it have been undertaken by D. L. Hawksworth and F. Rose (Hawksworth, 1971). For 8 years a transplanted specimen has grown vigorously and multiplied on a Hampshire garden tree. Work on the dynamics of relict *Lobaria* populations at three sites in the north of England is still at an early stage (O. L. Gilbert, unpublished) but it has been discovered that the rate of turnover in a colony can be high. On isolated trees the alternate, abrupt wetting and drying regime makes the lobes curl and uncurl vigorously so they become loosened and may hang down or drop off the bark/moss substratum. Twenty-five transplants of *Lobaria* have shown only 30% survival after 3 years, mainly due to poor technique: many dropped off or were killed by the adhesive. Two or three, however, have produced single new colonies a few inches down the trunk.

5. Saxicolous Habitats

Transplanting in this habitat appears relatively simple, at least for foliose species. Richardson (1967) seems to have been the first to attempt it when for taxonomic reasons he stuck 25 colonies of *Xanthoria* on to a scraped asbestos-cement roof using the resin glue "Araldite". After 18 months, 24 were still thriving. Recently *Hypogymnia physodes*, *Parmelia glabratula*, *P. saxatilis*, *Platismatia glauca* and *Pseudevernia furfuracea* have been transplanted on to sandstone wall tops using a variety of glues. All are still healthy and firmly attached after 5 years. Opportunities may sometimes be presented for transplanting intact saxicolous communities, a technique which is described by Seaward (1976).

6. Terricolous Habitats

Attempts to transplant turfs containing *Cetraria islandica* from an opencast fluorspar mine in Derbyshire into reinstated grassland behind the drag line failed due to rabbits selecting the turfs for scraping and defecation.

The glues investigated during my transplant work were "clear Bostik", "black Bostik", "Evostik", "Araldite" and "Isopon fibreglass". All caused a slow necrosis in species at the point of contact with the glue; the chance of a successful transplant is thus highest if the lichen is transplanted with its parent substratum attached so the glue never makes contact with the living lichen. Alternatively the glue can be applied sparingly to only part of the underside so the lichen has a good chance of becoming attached by rhizina growth. Necrosis was only slight with Evostik and Araldite.

A promising technique is the application of lichen fragments in suspension. Dry *Parmelia saxatilis* crushed up by hand and mixed with water was "painted" on to sandstone walls. This resulted in the establishment of a number of new colonies. Though the great majority of the lichen fragments are blown away or washed into dark crevices where they decay, a small proportion get held in rugose surfaces or attached to cobwebs and after only 4 weeks some become anchored by rhizinal growth. Suspensions of *Hypogymnia physodes* fragments "painted" on to wooden palings also resulted in the establishment of young thalli. A limitation to this method has recently come to light: young newly established fragments have all been destroyed at many sites due to invertebrate grazing, especially by psocids.

It seems that transplanting even in an empirical way can achieve a certain amount of success, so rather than see a rare lichen disappear from a particular area because of a land use change or an unusual combination of natural conditions such as wind-throw or the appearance of a tree disease, it should be possible to establish it elsewhere in the landscape. As autecological information is most unlikely to be available to act as a guide during such operations, it will, for the time being, be important to record and report on all attempts.

VII. An Outline Strategy

As ecosystems are complex entities composed of many interesting groups of plants and animals, it is clearly unrealistic to consider the conservation of lichens on their own. It is disquieting, however, that in the past, most decisions made in connection with wildlife conservation have ignored them completely. This unsatisfactory state of affairs has been due mainly to a lack of readily available information, so it is essential for lichenologists to ensure that future decision makers are kept informed of the now considerable body of lichen site data. Raw data, however, are of severely limited value—they require analysis before recommendations can be made. A start has been made on evaluation by the Conservation Committee of the British Lichen Society whose on-going survey of important lichen sites is updated regularly and re-analysed every 5 years.

The normal procedure for a "Conservation Review" which I see as the single most important activity for those interested in lichen conservation today, is to concentrate initially on data collection, this stage being followed by comparative site assessment and finally by the determination of adequate representation, i.e. how many of each type of habitat should be preserved. At some stage a decision has to be made on the most useful way of disseminating the results.

Data collection, the task of surveying the country for key lichen sites, has been seriously under way for some 6 years. We are still in a "grand period of exploration". The formal recording of these primary scientific data is less well advanced, the information being scattered in reports, on mapping cards, in published papers or even still in notebooks or people's heads. The preferred system involves the completion of special proformas, but often due to a feeling that the field work is not complete or identifications are awaited, there is a reluctance to fill these in. If a number of main habitat groups are recognized, namely coastland, woodland, lowland grassland and heath, lowland saxicolous, peatland, upland, aquatic and man-made (e.g. churchyards, enclosure landscapes), it can be appreciated that our knowledge of many is still inadequate.

The second (analysis) stage, when the quality of sites is compared with others, becomes increasingly difficult as the number to be considered grows. This is a stage where value judgements figure prominently and a good deal of subjective and personal opinion is involved. How, for instance, should anomalous sites with a large number of rare species be assessed against an extensive area of more typical habitat? Applying Ratcliffe's (1971) criteria—of extent, diversity, naturalness, rarity, fragility, representativeness or typicalness within the field of ecological variation, research and education value, recorded history and potential—is a great help, though for lichen-rich areas a weighting for fragility (sensitivity to adversity) may be necessary. As long as the raw data are retained, however, evaluation can be repeated when superior methodologies emerge or the aims of conservation change.

The determination of adequate representation is also difficult, but decisions in this area depend to some extent on how much money and other resources are devoted to nature conservation. Some sites, highly valued locally, would not qualify for protection at all if located in a region with a more extensive representation of the particular ecosystem. Despite these difficulties some preliminary selection has started (e.g. Richardson, 1975).

After the selection and safeguarding of a site, on-going management should be practised. Scientific land management for lichens is still in its infancy though some general guide-lines can be found (Gilbert, 1975), and possibilities for manipulating lichen communities are sometimes a by-product of general ecological research. By continuing with traditional low-intensity systems of land management, much can be done to maintain the *status quo* in nature reserves without understanding all the interactions. I do not believe that in Britain, legislation such as *The Conservation of Wild Creatures and Wild Plants Act 1975*, which included lichens, or strictures against overcollecting, can play a very large part in the protection of lichens. It is interesting to note that in Poland there is legislation protecting corticolous lichens, but for medicinal reasons not conservationist ones (Gawtowska and Gawtowska, 1973).

A policy of conservation, through the creation of a network of nature reserves, is considered by some as an admission of defeat; they feel that the whole landscape should be managed by methods sympathetic to wildlife. This attractive, but rather idealistic, view would require a much greater public awareness of conservation. This could be achieved by education on a broad front and parallel with a conservation review: it will be important to direct energies into writing popular books, articles, nature-trail notes etc., to help raise the general level of interest in lichens. This approach has done much, for instance, to increase appreciation of hedgerows. High priority should therefore be given to fostering a popular and informed interest in the group, being mindful that there are important aesthetic, as well as scientific, benefits to be gained from lichen conservation.

References

Ahti, T. (1961). Taxonomic studies on reindeer lichens (*Cladonia*, subgenus *Cladina*). *Annls bot. Soc. zool.-bot. fenn. Vanamo* **32** (1), i–iv, 1–160.

Barker, G. M. A. (1972). "Wildlife Conservation in the Care of Churches and Churchyards". Church Information Office, London.

Brown, D. H. (1972). The effect of Kuwait crude oil and a solvent emulsifier on the metabolism of the marine lichen *Lichina pygmaea*. *Mar. Biol.* **12**, 309–315.

Brown, D. H. (1973). The lichen flora of the lead mines at Charterhouse, Mendip Hills. *Proc. Bristol Nat. Soc.* **32**, 267–274.

Brown, D. H. (1974). Field and laboratory studies on detergent damage to lichens at the Lizard, Cornwall. *Cornish Studies* **2**, 33–40.

Coppins, B. J. and Lambley, P. W. (1974). Changes in the lichen flora of the parish of Mendlesham, Suffolk, during the last fifty years. *Suffolk nat. Hist.* **16**, 319–335.

Coppins, B. J. and Shimwell, D. W. (1971). Cryptogamic complement and biomass in dry *Calluna* heath of different ages. *Oikos* **22**, 204–209.

Duffey, E. (1974). "Nature Reserves and Wildlife". Heinemann Educational Books.

Ferry, B. W., Baddeley, M. S. and Hawksworth, D. L., eds. (1973). "Air Pollution and Lichens". Athlone Press, London.

Gawtowska, J. and Gawtowska, R. M. (1973). Protection of the resources of industrial and medicinal plants. *In* "Protection of Man's Natural Environment" (W. Szafer and W. Michajtow, eds), pp. 418–432. Warsaw Polish Scientific Publishers.

Gilbert, O. L. (1975). "Wildlife Conservation and Lichens". Devon Trust for Nature Conservation, Exeter.

Gilbert, O. L. (1976). An alkaline dust effect on epiphytic lichens. *Lichenologist* **8**, 173–178.

Gimingham, C. H. (1972) "Ecology of Heathlands". Chapman and Hall, London.

Hawksworth, D. L. (1971). *Lobaria pulmonaria* (L.) Hoffm. transplanted into Dovedale, Derbyshire. *Naturalist, Hull* **1971**, 127–128.

Hawksworth, D. L. (1975). The changing lichen flora of Leicestershire. *Trans. Leicester lit. phil. Soc.* **68**, 32–56.

Hawksworth, D. L., Rose, F. and Coppins, B. J. (1973). Changes in the lichen flora of England and Wales attributable to pollution of the air by sulphur dioxide. *In* "Air Pollution and Lichens" (B. W. Ferry, M. S. Baddeley and D. L. Hawksworth, eds), pp. 330–367. Athlone Press, London.

Hawksworth, D. L., Coppins, B. J. and Rose, F. (1974). Changes in the British Lichen Flora. *In* "The Changing Fauna and Flora of Britain" (D. L. Hawksworth, ed.), pp. 47–78, Academic Press, London and New York.

Laundon, J. R. (1964). Semi-natural vegetation in Northamptonshire. *J. Northampt. nat. Hist. Soc.* **34**, 268–281.

Laundon, J. R. (1973a). Changes in the lichen flora of Bookham Commons with increased air pollution and other factors. *Lond. Nat.* **52**, 82–92.

Laundon, J. R. (1973b). Lichens. *In* "Monkswood. A Nature Reserve Record" (R. C. Steele and R. C. Welch, eds), pp. 95–100. Nature Conservancy, London.

Moore, N. W. (1968). The value of pesticides for conservation and ecology. *In* "Some Safety Aspects of Pesticides in the Countryside" (N. W. Moore and P. W. Evans, eds), pp. 104–108. Joint ABMAC/Wild Life Education and Communications Committee, London.

Poelt, J. and Huneck, S. (1968). *Lecanora vinetorum* nova spec., ihre Vergesellschaftung, ihre Ökologie und ihre Chemie. *Öst. bot. Z.* **115**, 411–422.

Ranwell, D. S. (1968). Lichen mortality due to "Torrey Canyon" oil and decontamination measures. *Lichenologist* **4**, 55–56.

Ratcliffe, D. A. (1971). Criteria for the selection of nature reserves. *Advmt Sci., Lond.* **27**, 294–298.

Ratcliffe, D. A. (1976). Thoughts towards a philosophy of nature conservation. *Biol. Conserv.* **9**, 45–53.

Richardson, D. H. S. (1967). The transplantation of lichen thalli to solve some taxonomic problems in *Xanthoria parietina* (L.) Th.Fr. *Lichenologist* **3**, 386–391.

Richardson, D. H. S. (1975). "The Vanishing Lichens". David and Charles, Newton Abbot.

Rose, F. (1974). The epiphytes of oak. *In* "The British Oak: its History and Natural History" (M. G. Morris and F. H. Perring, eds), pp. 250–273. Classey, Faringdon.

Rose, F. and James, P. W. (1974). Regional studies on the British Lichen Flora. 1. The corticolous and lignicolous species of the New Forest, Hampshire. *Lichenologist* **6**, 1–72.

Seaward, M. R. D. (1973). Distribution maps of lichens in Britain. *Lichenologist* **5**, 464–480.

Seaward, M. R. D. (1976). Performance of *Lecanora muralis* in an urban environment. *In* "Lichenology, Problems and Progress" (D. H. Brown, D. L. Hawksworth and R. H. Bailey, eds), pp. 323–357. Academic Press, London and New York.

Skye, E. and Hallberg, I. (1969). Changes in the lichen flora following air pollution. *Oikos* **20**, 547–552.

Society for the Promotion of Nature Reserves (1970). "A Policy on Introductions to Nature Reserves". Conservation Liaison Committee, Technical Publication No. 2.

Tubbs, C. R. (1968). "The New Forest, An Ecological History". David and Charles, Newton Abbot.

Watt, A. S. (1947). Pattern and process in the plant community. *J. Ecol.* **35**, 1–22.

Wells, T. C. E., Sheail, J., Ball, D. F. and Ward, L. K. (1976). Ecological studies on the Porton Ranges: relationships between vegetation, soils and land-use history. *J. Ecol.* **64**, 589–626.

Willis, A. J. (1963). Braunton Burrows: The effects on the vegetation of the addition of mineral nutrients to the dune soils. *J. Ecol.* **51**, 353–374.

Appendix A: A Bibliographic Guide to the Lichen Floras of the World

DAVID L. HAWKSWORTH

I. Introduction	438
II. Lichen Floras of the World	439
A. Africa	439
B. Atlantic Islands	444
C. Asia	445
D. Australasia and Pacific	452
E. U.S.S.R.	455
F. Europe	461
G. North America	476
H. Central America and West Indies	485
I. South America	487
J. Antarctica	491
III. World Monographs	493
IV. Sources of Further Information	496
A. Literature Compilations	496
B. Lichenological Journals	497
C. Exsiccatae	498
D. Herbaria	499
Acknowledgements	500
References	500

I. Introduction

In the early days of lichenology it was possible to include details of all known taxa and their distributions in a single volume (e.g. Acharius, 1810, 1814), but as interest in the group increased and the large number of species came to be appreciated, the literature has swelled to almost unmanageable proportions. Fortunately, a number of lichenologists have been at pains to compile detailed lists of publications (see pp. 496–497), but even a study of these is now an extremely laborious task.

Compared with the information now available for vascular plants on a world-wide scale (e.g. Good, 1974), our knowledge of lichens is, nevertheless, meagre, except for a few groups which have been intensively studied in recent years (see pp. 493–496). Workers in different countries, and those wishing to determine lichens from them, need to know at least the major floristic studies concerned with their regions. For the vascular plants a list of the major national floras was prepared by Blake and Atwood (1942, 1961) but, apart from the information included in Ainsworth (1971), no compilation of the major lichen floras has been attempted in recent years.

The information currently available on the lichens of different countries varies very widely. No check-lists are extant for many countries and floras aimed to be used by non-specialists are even scarcer. Much of the published information is rather old and the identifications frequently have to be treated with some caution. Europe is by far the best known continent lichenologically but even here some well studied countries (e.g. Sweden) still lack modern comprehensive check-lists and floras.

As will be apparent from the lists of publications given below, an enormous amount of basic taxonomic and floristic research still has to be carried out in the lichens. An introduction to the procedures to be adopted in the preparation and publication of taxonomic and floristic studies in the lichens is provided by Hawksworth (1974).

In selecting titles for this Appendix, I have endeavoured to include, where available, the most recent check-lists, comprehensive floras, introductory floras and bibliographies. Detailed surveys of the lichen vegetation are also included in a few cases where these will prove useful. For countries which lack comprehensive publications, available lists are cited, but in cases where there are very many of these, preference is given to recent papers with extensive bibliographies citing other papers on the area. Some regional monographs on particular families and genera have also been included and a few world monographs requiring study for material from all countries are listed on pp. 493–496. Because of limitations of space, monographs have been restricted primarily to those dealing with very many species, major genera of macrolichens and large geographical regions. Further references to monographs can be traced through Ainsworth (1971). Keys to the world's genera of lichens are included in Smith (1921) and Zahlbruckner (1926) but these works are now somewhat dated.

The lists presented here are selective and not to be treated as definitive. All publications listed have been checked by me in the original, unless otherwise indicated, except in the case of the U.S.S.R. where many references were supplied by Dr T. Ahti. Additional sources of references to further papers are discussed on pp. 496–500.

For countries and continents where many papers are cited these are listed in the following categories: (a) check-lists, comprehensive floras, introductory floras, regional bibliographies etc.; (b) regional monographs; and (c) major distributional, ecological, habitat and vegetational surveys. Larger islands and island-groups are listed independently and not under countries of which they are dependents.

II. Lichen Floras of the World

A. Africa

Dodge, C. W. (1953–1971). Some lichens of tropical Africa I–V. *Ann. Mo. bot. Gdn* **40**, 271–401 (1953); **43**, 381–396, **44**, 1–76 (1956–1957) [II. *Usnea*]; **46**, 39–193 (1959) [III. Parmeliaceae]; *Beih. Nova Hedwigia* **12**, 1–282 (1964) [IV. Dermatocarpaceae to Pertusariaceae]; **38**, 1–225 (1971) [V. Lecanoraceae to Physciaceae]. [Keys; descriptions]

Stizenberger, E. (1890–1895). Lichenaea africana. *Ber. Tät. St. Gall. naturw. Ges.* **1888–1889**, 105–249 (1890); **1889–1890**, 133–268 (1891); **1891–1892**, 86–96 (1893); **1893–1894**, 215–264 (1895). [Check-list with some 1650 spp.]

Algeria

Durrieu de Maisonneuve, M. C. and Montagne, J. P. F. C. (1846–1869). "Exploration scientifique de l'Algérie, Botanique, I. Cryptogamie". 631 pp. Imprimerie Impériale, Paris. [Lichens pp. 198–295 (1846–1849); descriptions]

Faurel, L., Ozenda, P. and Schotter, G. (1951–1954). Matériaux pour la flore lichénologique d'Algérie et de Tunisie, I–III. *Bull. Soc. Hist. nat. Afr. N.* **42**, 62–112 (1951) [I. (Caliciaceae, Cypheliaceae, Peltigeraceae, Pertusariaceae)]; **44**, 12–50 (1953) [II. (Graphidaceae)]; **45**, 275–298 [III. (Arthoniaceae, Dirinaceae, Roccellaceae)].

Faurel, L., Ozenda, P. and Schotter, G. (1953). Matériaux pour la flore lichénologique d'Algérie et de Tunisie. In "Desert Research", pp. 310–317. Israel Research Council, Jerusalem. [Not seen]

Flagey, C. (1896). Catalogue des lichens de l'Algérie. In "Flore de l'Algérie" (J. A. Battandier and L. Trabut, eds), Vol. 2 (1), pp. vii–xii, 1–140. Jourdan, Algiers. [Detailed flora]

Angola

Salisbury, G. (1971). The *Thelotremata* of Angola and Moçambique. *Revta Biol.* **7**, 271–280.

Tavares, C. N. (1953). Lichens from Angola and Moçambique—I. *Port. Acta biol.*, **B, 4**, 154–161. [Including bibliography]

Central and East Africa

Ahti, T. (1977). The *Cladonia gorgonena* group and *Cladonia gigantea* in East Africa. *Lichenologist* **9**, 1–15.

Cengia Sambo, M. C. (1938). Licheni del Kenia e del Tanganica raccolti dai Rev. Padri della consolata. *Nuovo G. bot. ital.* **45**, 364–387.

Duvigneaud, P. (1952). Les Usnées barbues et le *Crossopterygo-Usneetum* des savannes du Bas Congo. *Bull. Soc. r. bot. Belg.* **85**, 99–114.

Duvigneaud, P. (1956). Les *Stereocaulon* des hautes montagnes du Kivu. *Lejeunia, Mém.* **14**, 1–114.

Frey, E. (1967). Die lichenologischen Ergebnisse der Forschungreisen des Dr. Hans Ullrich Stauffer in Zentralafrika (Virunga-Vulkane 1954/1955) und Sudafrica-Australien-Ozeanien-USA. *Bot. Jb.* **86**, 209–255.

Klement, O. (1962). Zur Flechten-Vegetation von Tanganjika. *Stuttg. Beitr. Naturk.* **85**, 1–8.

Krog, H. and Swinscow, T. D. V. (1975). Parmeliaceae, with the exclusion of *Parmelia* and *Usnea*, in East Africa. *Norw. J. Bot.* **22**, 115–123.

Krog, H. and Swinscow, T. D. V. (1976). The genus *Ramalina* in East Africa. *Norw. J. Bot.* **23**, 153–175.

Motyka, J. (1961). *Usneae* a R. A. Maas Geesteranus in Africa orientali et australi anno 1949 collectae. *Persoonia* **1**, 415–431.

Swinscow, T. D. V. and Krog, H. (1974). *Usnea* subgenus *Eumitria* in East Africa. *Norw. J. Bot.* **21**, 165–185.

Swinscow, T. D. V. and Krog, H. (1975a). The genus *Pyxine* in East Africa. *Norw. J. Bot.* **22**, 43–68.

Swinscow, T. D. V. and Krog, H. (1975b). The *Usnea undulata* aggregate in East Africa. *Lichenologist* **7**, 121–138.

Swinscow, T. D. V. and Krog, H. (1976a). The *Usnea bornmuelleri* aggregate in East Africa. *Norw. J. Bot.* **23**, 23–31.

Swinscow, T. D. V. and Krog, H. (1976b). The genera *Anaptychia* and *Heterodermia* in East Africa. *Lichenologist* **8**, 101–136.
Swinscow, T. D. V. and Krog, H. (1976c). The genus *Coccocarpia* in East Africa. *Norw. J. Bot.* **23**, 251–259.
Zahlbruckner, A. and Hauman, L. (1936). Les lichens des hautes altitudes au Ruwenzori. *Mém. Inst. r. colon. belge, sect. sci. tech.*, 8° *coll.* **5**, 1–31.

Chad

Werner, R. G. (1950). Lichenes. *Mém. Inst. fr. Afr. noire* **8**, 18–21.

Egypt

Sickenberger, E. (1901). Contributions à la flore d'Égypte. *Mém. Inst. égypt.* **4**, 167–335. [Lichens pp. 319–331]
Werner, R. G. (1966). Notes de lichénologie libano-syrienne, VIII et égyptienne. *Bull. Soc. bot. Fr.* **113**, 74–83. [Including bibliography]

Ethiopia

See also under Central and East Africa.

Cengia Sambo, M. (1937a, b). Lichenes africae orientalis italicae. I–II. *Nuovo G. bot. ital.* **44**, 456–470 (1937a) [Parte I. I licheni del' Abissinia meridionale e della Somalia]; **44**, 471–483 (1937b) [Parte II. I licheni dell' Eritrea e dell' Abissinia settentrionale].
Motyka, J. and Pichi-Sermolli, R. (1952). *Usneae* in missione ad Lacum Tana et Semièna R. Pichi-Sermolli anno 1937 lectae. *Webbia* **8**, 383–404.
Winnem, B. (1975). *Parmelia* subgenus *Amphigymnia* in Ethiopia. *Norw. J. Bot.* **22**, 139–166.

Guinea

Des Abbayes, H. (1958). Lichens récoltés en Guinée Française et en Côte d'Ivoire. IX. *Bull. Inst. fr. Afr. noire*, **A, 20** (1), 1–27. [Including bibliography]
Vězda, A. (1973–1975). Foliicole Flechten aus der Republik Guinea (W-Africa). I–III. *Čas. slezsk. Mus. Opave*, **A, 22**, 67–90 (1973); **23**, 173–190 (1974); **24**, 117–126 (1975).

Ivory Coast

See Des Abbayes (1958) under Guinea.

Kenya

See Central and East Africa.

Kerguelen Islands

Crombie, J. M. (1877). Revision of the Kerguelen lichens collected by Dr. Hooker. *J. Bot., Lond.* **15,** 101–107.

Crombie, J. M. (1879). Transit of Venus Expedition 1874/1875. Botany of Kerguelen Island: Lichens. *Phil. Trans. r. Soc.* **168,** 46–52.

Libya

Trotter, A. (1950). La mico-lichenologia e la fitopatologia nel quadro del popolamento biogeographico della Libya. *Delpinoa* **3,** 155–174. [Including bibliography]

Madagascar

See Malagasy Republic.

Malagasy Republic

Des Abbayes, H. (1961). Lichens récoltés à Madagascar et à la Réunion (Mission H. des Abbayes, 1956) I. Introduction. II. Parméliacées. *Mém. Inst. scient. Madagascar,* **B, 10,** 81–121. [Including bibliography]

Marion Island

Huntley, B. J. (1971). Vegetation. *In* "Marion and Prince Edward Islands" (R. A. Dyer, J. F. Winterbottom and E. M. van Zinderen Baker, eds), pp. 91–140. Balkema, Cape Town. [Not seen; reference supplied by Dr D. C. Lindsay]

Mauritius

Crombie, J. M. (1877). Lichenes insulae Rodriguesii. *J. Linn. Soc., Bot.* **15,** 431–445.

Daruty, A. (1873). Lichens de Maurice récoltés par M. A. Daruty et déterminés par M. H. A. Weddell, D.M.P. *Trans. R. Soc. Arts Sci. Maurit.,* n.s. **7,** 163–166.

Lindau, G. (1908). Lichenes von Madagascar, Mauritius und den Comoren. *In* "Reise in Ostafrica in dem Jahren 1903–1905" (A. Voeltzkow, ed.), Vol. 3, pp. 1–14. Schweizerbartsche, Stuttgart.

Moçambique

See under Angola.

Morocco

Asta, J., Clauzade, G. and Ozenda, P. (1972). Lichens du Sud-Ouest marocain. *Revue bryol. lichén.* **38,** 299–303. [42 spp.]

Gattéfossé, J. and Werner, R. G. (1931). Catalogus lichenum marocanorum adhuc cognitorum. *Bull. Soc. Sci. nat. Maroc* **11,** 187–257. [Check-list]

Werner, R. G. (1955). Contribution à la flore cryptogamique du Maroc.—XIX. *Bull. Soc. Sci. nat. phys. Maroc* **35,** 19–67. [Keys; many spp.]

Werner, R. G. (1972). Lichens et champignons de la plaine Marocaine. *Bull. Acad. Soc. Lorr. sci.* **11,** 83–97. [Including bibliography]

Werner, R. G. (1976). Amendement ou maintien de certaines déterminations lichéniques marocaines. *Bull. Soc. bot. Fr.* **123,** 433–440.

Nigeria

Thorold, C. A. (1952). The epiphytes of *Theobroma cacao* in Nigeria in relation to the incidence of Black-pod disease (*Phytophthora palmivora*). *J. Ecol.* **40,** 125–142. [Including foliicolous lichens]

Réunion

See under Malagasy Republic.

Rhodesia

van der Byl, P. A. (1931). 'n Lys van korsmosse (Lichenes) versamel in die Uni van Suid-Africa en in Rhodesië gedurende die tydperk 1917–1929. *Annale Univ. Stellenbosch,* **A, 9** (3), 1–17.

St Paul and Amsterdam Islands

Nylander, W. (1875). Liste des lichens recueillis par M. G. de l'Isle, aux îles Saint-Paul et d'Amsterdam et description des espèces nouvelles. *C. r. hebd. Séanc. Acad. Sci., Paris,* **D, 81,** 725–726.

Nylander, W. (1886). Lichenes Insulae Sancti Pauli. *Flora, Jena* **69,** 318–322.

Somalia

See Cengia Sambo (1937a) under Ethiopia.

South Africa

Almborn, O. (1966). Revision of some lichen genera in southern Africa I. *Bot. Notiser* **119,** 70–112. [Including bibliography and details of South African material in herbaria]

Doidge, E. M. (1950). The South African fungi and lichens to the end of 1945. *Bothalia* **5,** 1–1094. [Lichens pp. 225–376; check-list; bibliography]

Hale, M. E. (1971). Studies on *Parmelia* subgenus *Xanthoparmelia* (Lichens) in South Africa. *Bot. Notiser* **124,** 343–354.

Mattick, F. (1970). Flechtenbestände der Nebelwüste und Wanderflechten der Namib. *Namib und Meer* **1,** 35–44.

Tanzania

See also Central and East Africa.

Müller [Argoviensis], J. (1894). Lichenes Usambarenses. *Bot. Jb.* **20**, 238–298.

Vězda, A. (1975). Foliikole Flechten aus Tanzania (Öst-Afrika). *Folia geobot. phytotax., Praha* **10**, 383–432.

Tunisia

See also Faurel *et al.* (1951–1954) under Algeria.

Hue, A. l'Abbé (1897). Lichenes. *In* "Exploration Scientifique de la Tunisie, Botanique [2], Catalogue raisonné des plantes cellulaires de la Tunisie" (N. Patouillard, ed.), pp. 136–151. Imprimerie Nationale, Paris.

Werner, R. G. (1951). Les origines de la flore lichénique de la Tunisie d'après nos connaissances actuelles. *Revue bryol. lichén.* **20**, 200–207.

Uganda

See Central and East Africa.

Zaire

See Central and East Africa.

B. Atlantic Islands

Açores

Degelius, G. (1941). Lichens from the Azores, mainly collected by Dr. H. Persson. *Göteborgs K. Vetensk.- o. VitterSamh. Handl.*, **6B, 1** (7), 1–46.

Des Abbayes, H. (1947). Lichens des Iles Açores récoltés par V. et P. Allorge. *Revue bryol. lichén.* **16**, 105–112. [Including bibliography]

Navás, L. (1909). Líquenes de las islas Azores. *Brotéria* **8**, 46–52.

Canary Islands

Champion, C. L. (1976). Algunos líquenes nuevos para las Islas Canarias. *Vieraea* **6**, 25–32.

Follmann, G. (1976). Lichen flora and lichen vegetation of the Canary Islands. *In* "Biogeography and Ecology in the Canary Islands" (G. Kunkel, ed.), pp. 267–286. W. Junk, The Hague. [Including extensive bibliography]

Klement, O. (1965). Zur Kenntnis der Flechtenvegetation der Kanarischen Inseln. *Nova Hedwigia* **9**, 503–582.

Østhagen, H. and Krog, H. (1976). Contribution to the lichen flora of the Canary Islands. *Norw. J. Bot.* **23**, 221–242.

Pitard, C.-J. and Harmand, J. l'Abbé (1911). Contribution à l'étude des lichens des Îles Canaries. *Bull. Soc. bot. Fr.* **58**, Mém. **22**, 1–72.

Vainio, E. A. (1924). Lichenes Teneriffenses anno 1921 a F. Boergesen collecti. *K. dansk. Vidensk. Selsk. Skr.* **6**, 392–398.

Madeira

Navás, P. L. (1913). Synopsis de los líquenes de las islas de Madera. *Brotéria, sér. bot.* **9**, 69–82; **10**, 51–100; **11**, 5–32, 121–134, 202–210.

Tavares, C. N. (1952). Contributions to the lichen flora of Macronesia. I. Lichens from Madeira. *Port. Acta biol.*, **B, 3**, 308–391.

Tavares, C. N. (1965). Ihla da Madeira o meio e a flora. *Revta Fac. Ciênc. Univ. Lisb.*, **C, 13**, 51–174. [Bibliography]

São Tomé and Príncipe

Nylander, W. (1889). "Lichenes insularum Guineensium (San Thomé, do Principe, das Cabras)", 54 pp. Schmidt, Paris.

Tristan da Cunha

Jørgensen, P. M. (1977). Foliose and fruticose lichens from Tristan da Cunha. *Skr. norske Vidensk-Akad. mat.-nat. Kl.*, n.s. **36**, 1–40. [Including bibliography].

Wace, N. M. and Dickson, J. H. (1965). The terrestrial botany of the Tristan da Cunha Island. *Phil. Trans. R. Soc.*, **B, 249**, 273–360.

C. Asia

Hertel, H. (1977). Gesteinbewohnende Arten der Sammelgattung *Lecidea* (Lichenes) aus Zentral-, Ost-, und Südasien. *Ergebn. ForschUnternehmens Nepal Himalaya* **6**, 145–378.

Yoshimura, I. (1971). The genus *Lobaria* of Eastern Asia. *J. Hattori Bot. Lab.* **34**, 231–364.

Afghanistan

Jacquemin-Roussard, M. and Kilbertus, G. (1971). Quelques lichens d'Afghanistan: ébauche écologique. *Bull. Acad. Soc. Lorr. sci.* **10**, 59–65.

Poelt, J. and Wirth, V. (1968). Flechten aus dem nordöstlichen Afghanistan gesammelt von H. Roemer im Rahmen der Deutsch Wakhan-Expedition 1964. *Mitt. bot. StSamml., Münch.* **7**, 219–261. [82 spp.]

Bahrein

Lamb, I. M. (1936). Lichens from Bahrein Island. *J. Bot., Lond.* **74**, 346–351.

Bangladesh

See Awasthi (1965) under India.

Borneo

See also under Singapore.

Krempelhuber, A. von (1875). Lichenes quos legit O. Beccari in insulis Borneo et Singapore annis 1866 et 1867. *Nuovo G. bot. ital.* **7**, 5–67.

China

Chao, C.-D. (1964). A preliminary study on Chinese *Parmelia. Acta phytotax. sin.* **9**, 139–166.

Chao, C.-D., Hsü, L.-W. and Sun, Z. M. (1974). Species novae *Usneae* sinicae. *Acta phytotax. sin.* **13**, 90–107. [Keys 60 spp.]

Lu, D.-A. (1958–1959). Notes on Chinese lichens 1–2. *Acta phytotax. sin.* **7**, 263–269 (1958) [1.—*Peltigera*]; **8**, 178–179 (1959) [2.—Umbilicariaceae].

Magnusson, A. H. (1940, 1944). Lichens from Central Asia. 1–2. *Rept sci. Exp. N.W. Prov. China*, XI (*Bot.*) **1**, 1–167 (1940); **2**, 1–68 (1944).

Moreau, C. and Moreau, M. (1951). Lichens de Chine. *Revue bryol. lichén.* **20**, 183–199.

Nylander, W. and Crombie, J. M. (1884). On a collection of lichens made in Eastern Asia by the late Dr. A. C. Maingay. *J. Linn. Soc., Bot.* **20**, 49–69.

Paulson, R. (1928). Lichens from Yunnan. *J. Bot., Lond.* **66**, 313–319.

Sato, M. (1952). Lichenes Khinganenses: or a list of the lichens collected by Prof. T. Kira in the Great Khingan Range, Manchuria. *Bot. Mag., Tokyo* **65**, 769–770.

Vej Tzjan-Czunj and Chen, J. R. (1974). Materials for a lichen flora of the Mount Jolmo Lungma region in southern Tibet, China. *In* "Report of the Chinese Mt. Jolmo Lungma expedition 1966–1968", pp. 173–182. Academia Sinica, Peking. [79 spp. from Mt Everest; not seen; reference supplied by Dr T. Ahti]

Zahlbruckner, A. (1930). Lichenes. *Symbolae sinicae* **3**, 1–254.

Zahlbruckner, A. (1934). Nachträge zur Flechtenflora Chinae. *Hedwigia* **74**, 195–213.

Himalaya

See under China, India and Nepal.

India

Awasthi, D. D. (1965). Catalogue of the lichens from India, Nepal, Pakistan, and Ceylon. *Beih. Nova Hedwigia* **17**, 1–137. [Check-list 1310 spp.; bibliography]

Awasthi, D. D. and Singh, K. P. (1972–1975). Additions to the lichen flora of India [I.], III. *Geophytology* **1**, 97–102 (1972); **5**, 110–112 (1975).

Chopra, G. L. (1934). Lichens of the Himalayas, Part I. Lichens of Darjeeling and the Sikkim Himalayas. *Publs Dep. Bot. Univ. Punjab* **4**, i–viii, 1–105. [Including illustrations]

Roychowdhury, K. N. (1977) ["1976"] Unrecorded species of lichens from India. *Bull. bot. Surv. India* **15**, 132–136.

Schubert, R. and Klement, O. (1966). Beitrag zur Flechtenflora von Nord- und Mittelindien. *Nova Hedwigia* **11**, 1–73.

Singh, K. P. (1973). Additions to the lichen flora of India—II. *Kavaka* **1**, 43–46.

Indonesia

See under Borneo, Java, Malaysia, New Guinea and Sumatra.

Iran

Rechinger, K. H., Baumgartner, J., Petrak, F. and Szatala, Ö. (1940). Ergebnisse einer botanische Reise nach dem Iran, 1937. *Annls naturh. Mus. Wien* **50** (1939), 410–536. [Lichens pp. 521–533]

Steiner, J. (1910). Lichenes Persici coll. a cl. Consule Th. Strauss. *Annls mycol.* **8**, 212–245.

Szatala, Ö. (1957). Prodromus einer Flechtenflora des Irans. *Annls hist.-nat. Mus. natn. hung.*, n. ser. **8**, 101–154. [248 spp.; keys larger genera]

Weber, W. A. (1965). Iranian plants collected by Per Wendelbo in 1959. VIII. Lichenes. *Årbok Univ. Bergen., mat.-nat. ser.* **1964** (14), 1–8.

Iraq

Schubert, R. (1973). Notizen zur Flechtenflora des nördlichen Mesopotamien (Irak). *Feddes Repert.* **83**, 585–589.

Steiner, J. (1921). Lichenes aus Mesopotamien und Kurdistan sowie Syrien und Prinkipo. *Annln naturh. Mus. Wien* **34**, 1–68.

Israel

See also Santesson (1942) under Syria.

Alon, G. and Galun, M. (1971). The genus *Caloplaca* in Israel. *Israel J. Bot.* **20**, 273–292.

Galun, M. (1970). "The Lichens of Israel", 116 pp. Israel Academy of Sciences and Humanities, Jerusalem. [Keys; descriptions; bibliography; illustrations, some in colour]

Japan

(a)

Sato, M. (1959–1965). Catalogus lichenum japonicorum (ed. 2). *Miscnea bryol. lichen., Nichinan* **1** (19), 11–12, (20), 9–12, (21), 7–10; **2**, 27–28, 89–92, 107–108, 121–124, 166–168, 182–184; **3**, 9–12, 27–30, 43–44, 77–80, 95–96, 125–128, 159–160, 173–176, 189–191. [Check-list 1308 spp.; also issued separately in 1965]

Shibata, S. (1971). Commemorating the 90th birthday of Professor Y. Asahina. *Chem. Pharm. Bull., Tokyo* **19**, v–vii. [Including bibliography of Asahina's papers from 1941]

Yoshimura, I. (1974) "Lichen Flora of Japan in Colour" [Hoikusha's Illustrations for Naturalists no. 52], 365 pp. Hoikusha Publishing, Osaka. [Keys; descriptions; all macrolichens and spp. genera microlichens; bibliography; illustrations, 528 spp. in colour; Japanese with Latin names]

(b)

Asahina, Y. (1950). "Lichens of Japan. Vol. I. Genus *Cladonia*", 255 pp. Hirokawa Publishing, Tokyo.

Asahina, Y. (1952). "Lichens of Japan. Vol. II. Genus *Parmelia*", 162 pp. Research Institute for Natural Resources, Tokyo.

Asahina, Y. (1956). "Lichens of Japan. Vol. III. Genus *Usnea*", 129 pp. Research Institute for Natural Resources, Tokyo.

Asahina, Y. (1971). "Atlas of Japanese *Cladoniae*", 14 pp. Research Institute for Natural Resources, Tokyo. [Monochrome photographs; 27 plates]

Kashiwadani, H. (1975). The genera *Physcia, Physconia* and *Dirinaria* (Lichens) of Japan. *Ginkgoana* **3**, 1–77.

Kurokawa, S. (1959–1961). *Anaptychiae* (lichens) and their allies in Japan (1)–(6). *J. Jap. Bot.* **34**, 117–124, 174–184; **35**, 91–96, 240–243, 353–358; **36**, 51–56.

Nakanishi, M. (1966). Taxonomical studies on the family Graphidiaceae of Japan, *J. Sci. Hiroshima Univ.*, **B**(2), **11**, 51–126.

Oshio, M. (1968). Taxonomical studies on the family Pertusariaceae of Japan. *J. Sci. Hiroshima Univ.*, **B**(2), **12**, 81–163.

Java

See also Montagne and van den Bosch (1857) under Sumatra.

Boedijn, K. B. (1940). The mycetozoa, fungi and lichens of the Krakatau group. *Bull. bot. Gdns Buitenz.*, sér. 3, **16**, 358–429. [13 spp.]

Groenhart, P. (1936). Beiträge zur Kenntnis der Javischen Flechten I–III. *Ned. kruidk. Archf* **46**, 690–784.

Zahlbruckner, A. (1943). Flechtenflora von Java. *Beih. Feddes Repert.* **127**, 1–80.

Zahlbruckner, A. and Mattick, F. (1956). Flechtenflora von Java, 2. Teil. *Willdenowia* **1**, 433–528.

Lebanon

Werner, R. G. (1955–1966). Notes de lichénologie libano-syrienne I–VIII. *Bull. Soc. bot. Fr.* **101,** 355–360 (1955); **102,** 350–356 (1956); **103,** 461–467 (1956); **104,** 321–326 (1957); **105,** 238–243 (1958); **106,** 332–337 (1959); **110,** 311–315 (1963); **113,** 74–83 (1966).

Malaysia

Groenhart, P. (1941–1954). Malaysian lichens. I–IV. *Bull. bot. Gdns Buitenz.*, sér. 3, **17,** 198–203 (1941); *Reinwardtia* **1,** 33–39 (1950), 197–198 (1951); **2,** 385–402 (1954).

Sammy, N. (1975). An annotated list of lichens from Penang. *Malay. Nat. J.* **28,** 214–216.

Mongolian People's Republic

Ahti, T. (1976). The lichen genus *Cladonia* in Mongolia. *J. Jap. Bot.* **51,** 365–373.

Golubkova, N. S. (1971). K flore lishaĭnikov Mongol'skoĭ Narodnoĭ Respubliki. *Bot. Zh. SSSR* **56,** 777–786.

Golubkova, N. S. and Tsogt, U. (1974a). O lishaĭnikakh yuzhnykh pustyn' Mongol'skoĭ Narodnoĭ Respubliki. *Bot. Zh. SSSR* **59,** 43–52.

Golubkova, N. S. and Tsogt, U. (1974b). O lishaĭnikakh Doliny Ozer Mongol'skoĭ Narodnoĭ Respubliki. *Nov. Sist. Nizsh. Rast.* **11,** 181–294.

Schubert, R. and Klement, O. (1971). Beitrag zur Flechtenflora der Mongolischen Volksrepublik. *Feddes Repert.* **82,** 187–262. [333 spp.; bibliography]

Nepal

See also Awasthi (1965) under India*.

Des Abbayes, H. (1974). *Cladonia* du Nepal. *Ergebn. ForschUnternehmens Nepal Himalaya* **6,** 111–115.

Asahina, Y. (1966). Lichenes. *In* "The Flora of Eastern Himalaya" (H. Hara, ed.), pp. 593–610. University of Tokyo Press, Tokyo.

Bystrek, J. (1969). Die Gattung *Alectoria*. *Ergebn. ForschUnternehmens Nepal Himalaya* **6,** 17–24.

Jahns, H. M. and Seelen, E. J. R. (1974). *Baeomyces*-Funde aus dem Himalaya. *Ergebn. ForschUnternehmens Nepal Himalaya* **6,** 101–108.

Kurokawa, S. (1974). *Anaptychia* of the Nepal-Himalaya. *Ergebn. Forsch-Unternehmens Nepal Himalaya* **6,** 109–110.

Lamb, I. M. (1966). Die Gattung *Stereocaulon*. *Ergebn. ForschUnternehmens Nepal Himalaya* **1,** 349–352.

Mitchell, M. [E.] (1974). Die Gattung *Leptogium* sect. *Mallotium* im Himalaya. *Ergebn. ForschUnternehmens Nepal Himalaya* **6,** 121–126.

Poelt, J. (1966a). Die Lobaten Arten der Sammelgattung *Lecanora*. *Ergebn. ForschUnternehmens Nepal Himalaya* **1** (3), 187–202.

* And also Poelt [*Ergebn. ForschUnternehmens Nepal Himalaya* **6,** 447–458, 1977] for a full list of papers based on his collections.

Poelt, J. (1966b). Die Gattung *Ochrolechia*. *Ergebn. ForschUnternehmens Nepal Himalaya* **1** (4), 251–261.

Poelt, J. (1974). Die Gattungen *Physcia, Physciopsis* und *Physconia*. *Ergebn. ForschUnternehmens Nepal Himalaya* **6**, 57–144.

Poelt, J. and Reddi, B. V. (1969). *Candelaria* und *Candelariella*. *Ergebn. Forsch-Unternehmens Nepal Himalaya* **6**, 1–16.

Schmidt, A. (1974). *Chaenotheca* und *Coniocybe*. *Ergebn. ForschUnternehmens Nepal Himalaya* **6**, 133–134.

Vězda, A. and Poelt, J. (1974). Die Gattungen *Dimerella* und *Pachyphiale*. *Ergebn. ForschUnternehmens Nepal Himalaya* **6**, 127–132.

Pakistan

See also Awasthi (1965) under India.

Ahmad, S. (1965). A preliminary contribution to the lichen-flora of West Pakistan. *Biologia, Lahore* **11**, 21–47.

Hawksworth, D. L. and Mahmood, T. (1971). Some lichens from coniferous forests in West Pakistan. *Pakist. J. scient. ind. Res.* **14**, 113–115.

Philippines

Herre, A. W. C. T. (1958). New records of Philippine and other tropical Pacific lichens with descriptions of five new species. *Philipp. J. Sci.* **86** (1957), 13–35. [108 spp.]

Herre, A. W. C. T. (1963). The lichen genus *Usnea* and its species at present known from the Philippines. *Philipp. J. Sci.* **92**, 41–76.

Sbarbaro, C. (1938). Catalogus quorundam lichenum in Insulis Philippinensibus ab M. Ramos, G. Edano etc. annis 1918–1926 lectorum. *Archo bot.-Sist. Fito-geogr. Genet.* **14**, 45–51.

Vainio, E. A. (1909–1923). Lichenes insularum Philippinarum I–IV. *Philipp. J. Sci.*, **C**, **4**, 651–662 (1909); **8**, 99–137 (1913); *Annls Acad. sci. fenn.* **A**, **15** (6), 1–368 (1921); **19** (5), 1–84 (1923). [Descriptions]

Saudi Arabia

See also under Yemen.

Abu-Zinada, A. H. and Hawksworth, D. L. (1975) ["1974"]. A contribution to the lichen flora of Saudi Arabia. *Bull. Fac. Sci. Riyadh Univ.* **6**, 224–233.

Singapore

See also under Borneo.

Nylander, W. (1891). "Sertum lichenaeae e Labaun et Singapore", 48 pp. Schmidt, Paris.

Sri Lanka (Ceylon)

See Awasthi (1965) under India.

Sumatra

Montagne, C. and Van Den Bosch, R. B. (1857). Lichenes. *In* "Plantae Junghuhniae" (F. A. W. Miquel, ed.), pp. 427–494. Leiden. [This volume was issued in five parts between 1851 and 1858]

Syria

See also Steiner (1921) under Iraq, and Werner (1955–1966) under Lebanon.

Santesson, R. (1942). Some lichens from Palestine and Syria. *Ark. Bot.* **30B** (5), 1–5.

Taiwan

Wang-Yang, J.- R. and Lai, M.- J. (1973). A checklist of the lichens of Taiwan. *Taiwania* **18,** 83–104. [Check-list 396 spp.; bibliography]
Wang-Yang, J.-R. and Lai, M.-J. (1976). Additions and corrections to the lichen flora of Taiwan. *Taiwania* **21,** 226–228.
Zahlbruckner, A. (1933). Flechten der Insel Formosa. *Feddes Repert.* **31,** 194–224; **33,** 22–68. [Descriptions of some spp. included]

Thailand

Vainio, E. A. (1909). Lichenes. *Bot. Tidsskr.* **29,** 104–152. [93 spp.; descriptions]
Vainio, E. A. (1921). Lichenes in summo monte Doi Sutep (circ. 1675 m.s.m.) in Siam boreali anno 1904 a D:re C. C. Hosseo collecti. *Annls bot. Soc. zool.-bot. fenn. Vanamo* **1** (3), 33–55.

Turkey

See also Steiner (1921) under Iraq.

Pišút, I. (1970). Interessante Flechtenfunde aus der Turkei. *Preslia* **42,** 379–383.
Szatala, Ö. (1927). Lichenes Turciae asiaticae a Patre Prof. Stefano Selinka in insula Burges Adassi (Antigoni) lecti. *Magy. bot. Lapok* **26,** 20–22.
Szatala, Ö. (1960). Lichenes Turciae asiaticae ab Victor Pietschmann collecti. *Sydowia* **14,** 312–325.

Vietnam

Des Abayes, H. (1964). Lichens nouveaux ou intéressants du Vietnam. *Revue bryol. lichén.* **32,** 216–222.

Müller [Argoviensis], J. (1891). Lichenes Tonkinenses a cl. B. Balansa lecti. *Hedwigia* **30**, 181–189.
Schmid, M. (1974). "Végétation du Viet-nam" [Mémoires Orstom no. 75], 243 pp. Orstom, Paris. [Lichens pp. 136–137]

Yemen

Steiner, J. (1903). Bearbeitung der von O. Simony 1898 und 1899 in Südarabien, auf Sokotra und den benachbarten Inseln gesammelten Flechten. *Denskr. kaiser. Akad. wiss. Wien, mat.-nat. Kl.* **71** (1902), 93–102.

D. Australasia and Pacific

Australia

Cheel, E. (1903–1907). Bibliography of Australian lichens. *J. Proc. R. Soc. N.S.W.* **37**, 172–182 (1903); **40**, 141–154 (1907). [Including New Zealand and South Sea Islands]
Filson, R. B. (1969). A review of the genera *Teloschistes* and *Xanthoria* in the lichen family Teloschistaceae in Australia. *Muelleria* **2**, 65–115.
Filson, R. B. (1976). Australian lichenology: a brief history. *Muelleria* **3**, 183–190.
Weber, W. A. and Wetmore, C. M. (1972). Catalogue of the lichens of Australia exclusive of Tasmania. *Beih. Nova Hedwigia* **41**, i–vi, 1–137. [Check-list; bibliography]

Queensland

Shirley, J. (1888–1890). The lichen flora of Queensland I–IV. *Proc. R. Soc. Qd* **5**, 80–110 (1888); **6**, 5–55, 129–145, 165–218 (1889); IV [Separately printed only] (1890).

South Australia

Rogers, R. W. (1974). Lichens from the T. G. B. Osborn vegetation reserve at Kronamore in arid South Australia. *Trans. R. Soc. S. Austral.* **98**, 113–123. [Including key]
Rogers, R. W. and Lange, R. T. (1972). Soil surface lichens in arid and subarid south-eastern Australia I. Introduction and floristics. *Aust. J. Bot.* **20**, 197–213. [Including key]

Tasmania

Bratt, G. C. and Cashin, J. A. (1975). Additions to the lichen flora of Tasmania I. *Pap. Proc. R. Soc. Tasm.* **109**, 17–20.
Bratt, G. C. and Cashin, J. A. (1976). Additions to the lichen flora of Tasmania II. *Pap. Proc. R. Soc. Tasm.* **110**, 139–148.

Martin, W. (1965). Comparison of lichen flora of New Zealand and Tasmania. *Trans. R. Soc. N.Z., Bot.* **3**, 1–6.

Wetmore, C. M. (1964). Catalogue of the lichens of Tasmania. *Revue bryol. lichén.* **32** (1963), 223-264. [Check-list; bibliography]

Victoria

Wilson, F. R. M. (1891). On lichens collected in the colony of Victoria, Australia. *J. Linn. Soc., Bot.* **28**, 353–374.

Western Australia

Bibby, P. and Smith, G. G. (1955). A list of lichens of Western Australia. *J. Proc. R. Soc. West. Aust.* **39**, 28–29.

Sammy, N. C. and Smith, G. G. (1974). An annotated list of lichens from the coastal limestone near Perth, Western Australia. *J. R. Soc. West. Aust.* **57**, 38–42.

Cook Islands

Sbarbaro, C. (1939). Aliquot lichenes oceanici in Cook insulis (Tonga, Rarotonga, Tongatabu, Eua) collecti. *Archo bot. Sist. Fito-geogr. Genet.* **15**, 100–104.

Easter Island

Follmann, G. (1962). Die Flechtengesellschaften der Osterinsel. *Ber. dt. bot. Ges.* **75**, 245–260. [Including bibliography]

Skottsberg, C. (1956). Derivation of the flora and fauna of Juan Fernandez and Easter Island. Part 2. Easter Island. *In* "The Natural History of Juan Fernandez and Easter Island" (C. Skottsberg, ed.), Vol. 1, pp. 406–439. Almqvist and Wiksells, Uppsala. [Lichens pp. 415–416; check-list 23 spp.]

Galápagos Islands

See under South America.

Gilbert Islands

Moul, E. T. (1959). The bryophytes and lichens of Onotoa, Gilbert Islands. *Bryologist* **61**, 370–373.

Hawaii

Klement, O. (1966). Zur Kenntnis der Flechtenflora und -vegetation des Hawaii Archipels 1: Lanai. *Nova Hedwigia* **11**, 245–283. [Including bibliography]

Magnusson, A. H. (1955). A catalogue of the Hawaiian lichens. *Ark. Bot., ser. 2*, **3**, 223–402.

Juan Fernández Islands

Skottsberg, C. (1956). Derivation of the flora and fauna of Juan Fernandez and Easter Island. Part 1. The Juan Fernandez Islands. *In* "The Natural History of Juan Fernandez and Easter Island" (C. Skottsberg, ed.) Vol. 1, pp. 193–405. Almqvist and Wiksells, Uppsala. [Lichens pp. 243–254, 289–292; check-list 194 spp.]

New Caledonia

Harmand, J. l'Abbé (1911–1912). Lichens receuillis dans la Nouvelle Calédonie ou en Australie par le R.P. Pionnier. *Bull. Séanc. Soc. Sci. Nancy*, sér. 3, **12**, 124–144 (1911); **13**, 37–64 (1912).
Smith, A. L. (1922). [Plants from New Caledonia] Lichens. *J. Linn. Soc., Bot.* **46**, 71–87. [Including bibliography]

New Guinea

Kashiwadani, H. (1975). Enumeration of *Anaptychiae* and *Parmeliae* of Papua New Guinea. *In* "The Botanical Expedition to Papua New Guinea", pp. 75–83. National Science Museum, Tokyo.
Lindau, G. (1923). Lichenes novo-guineenses. *Bot. Jb.* **58**, 250–254.
Szatala, Ö. (1956). Prodrome de la flore lichénologique de la Nouvelle Guinée. *Annls hist.-nat. Mus. natn. hung.*, ser. n. **7** (1955), 15–50. [224 spp.]

New Zealand

(a)

Galloway, D. J. (1974). A bibliography of New Zealand lichenology. *N.Z. J. Bot.* **12**, 397–422. [437 refs]
Martin, W. (1966). Census catalogue of the lichen flora of New Zealand. *Trans. R. Soc. N.Z., Bot.* **3**, 139–159. [Check-list; bibliography]
Martin, W. (1968). Supplement to census catalogue of New Zealand lichens. *Trans. R. Soc. N.Z., Bot.* **3**, 203–208.
Martin, W. and Child, J. (1972). "Lichens of New Zealand", 193 pp. Reed and Reed, Wellington. [Keys; notes; illustrations, some in colour]
Murray, J. (1962–1963). Keys to New Zealand lichens. Parts 1–3. *Tuatara* **10**, 120–128 (1962); **11**, 46–56, 98–109 (1963). [Keys to genera]
Sato, M. (1966). Revision of the New Zealand lichens (1). *Miscnea bryol. lichen., Nichinan* **4**, 45–48. [Including bibliography]

(c)

Fineran, B. A. and Dodge, C. W. (1973). Lichens from the Southern Alps, New Zealand II. Records from the Mt Cook District. *Pacific Sci.* **27**, 274–280. [Including bibliography of recent papers]

Mark, A. F. and Bliss, L. C. (1970). The high-alpine vegetation of Central Otago, New Zealand. *N.Z. J. Bot.* **8**, 381–451.

Martin, W. (1970). The lichen flora of the Dunedin botanical subdistrict. *Trans. R. Soc. N.Z., Biol.* **11**, 243–255.

Murray, J. (1960). Studies of New Zealand lichens 1–3. *Trans. R. Soc. N.Z., Bot.* **88**, 177–195 [Coniocarpineae], 197–210 [Teloschistaceac], 381–399 [Peltigeraceae].

Auckland Islands

Fineran, B. A. (1971). A catalogue of bryophytes, lichens, and fungi collected on the Auckland Islands. *J. R. Soc. N.Z.* **1**, 215–229. [Including bibliography]

Papua

See under New Guinea.

Samoa

Zahlbruckner, A. (1908). Die Flechten der Samoa-Inseln. *Denkschr. Akad. Wiss. Wien, math.-nat. Kl.* **81**, 222–287.

Tahiti

Vainio, E. A. (1924). Lichenes A. W. Setchell et H. E. Parks in Insula Tahiti a 1922 collecti. *Univ. Calif. Publs Bot.* **12**, 3–15.

E. U.S.S.R.*

Abramov, I. I., ed. (1971–1975). "Opredelitel' lishaĭnikov SSSR". 3 vols. Nauka, Leningrad. [Handbook of the lichens of the U.S.S.R.; keys; descriptions; illustrations; numerous contributors; not yet complete, Vol. 1 deals with Pertusariaceae, Lecanoraceae and Parmeliaceae, Vol. 2 is an introduction to lichenology, and Vol. 3 deals with Caliciales, Thelotremataceae, Pannariaceae, Stictaceae, Lichinaceae etc.]

Elenkin, A. (1901, 1904). Lichenes florae Rossiae et regionum confinium orientalium. I–IV. *Trudȳ imp. S.-Peterb. bot. Sada* **19**, 1–52; **24**, 1–118. [Labels of an exsiccata provided with extensive notes on distribution etc.]

Komarnitzkiĭ, N. A., Tomin, M. P. and Krasil'nikov, N. A. (1960). "Opredelitel' nizshikh rastenii." Vol. 5. Moscow. [Keys; not seen]

Oxner, A. N. (1946). Nemoral'noi element v likhenoflore sovetskoi Arktiki. *Mater. Istor. Florȳ Rastit. SSSR* **2**, 475–490. [Phytogeographic elements]

* The assistance of Dr T. Ahti in compiling most references cited in this section is gratefully acknowledged.

Rassadina, K. A. (1950). Tzetrariya (*Cetraria*) SSSR. *Trudȳ bot. Inst. Akad. Nauk SSSR*, ser. *II*, **5**, 171–304.
Savicz, V. P. (1950a). Podvodnẏe lishaĭniki. *Trudȳ bot. Inst. Akad. Nauk SSSR*, ser. *II*, **5**, 148–170. [Keys to numerous aquatic lichens]
Savicz, V. P. (1950b). Konspekt k flore lishaĭnikov sem. Umbilicariaceae v SSSR. *Notul. syst. Inst. cryptog. Horti bot. petropol.* **6**, 97–108.
Smirnova, Z. N. (1962). "Kormovẏe lishaĭniki Kraĭnego Severa SSSR (kratkiĭ opredelitel')", 72 pp. Sel'khozizdat, Leningrad. [Common macrolichens of tundra important as reindeer forage; keys; illustrations]
Tomin, M. P. (1930). Opredelitel' lishaĭnikov solontzevatȳkh pochv v polupust. obl. Yugo-Vostoka. *In* "Yubileĭnyĭ sbornik prof. B. A. Kellera", pp. 3–10. Voronezh. [Keys to semi-desert lichens in south-east European U.S.S.R. and Kazakhstan]
Tomin, M. P. (1937). "Opredelitel' kustistȳkh i listovatȳkh lishaĭnikov SSSR", 312 pp. Izdatel'stvo Akademii Nauk BSSR, Moscow. [The only macrolichen flora covering all the U.S.S.R.]
Trass, H. (1972). A list of Cladoniaceae in the lichen flora of the U.S.S.R. *Folia cryptog. Eston.* **1**, 111–118. [Check-list 97 spp.]

Armenian S.S.R.

Nikogosyan, V. G. (1966). Novẏe dannẏe po likhenoflore Armenii. *Biol. Zh. Armenii* **19**, 106–113. [Including references to earlier papers]

Azerbaijan S.S.R.

Barchalov, S. O. (1969). "Listovatẏe i kustistẏe lishaĭniki Azerbaĭdzhana", 287 pp. Akademiya Nauk Azerbaĭdzhanskoĭ SSR, Baku. [Macrolichens; keys; descriptions; illustrations]
Barchalov, S. O. (1975). "Likhenoflora Talȳsha (Obschchaya Chast')", 151 pp. Akademiya Nauk Azerbaĭdzhanskoĭ SSR, Baku.
Novruzov, V. S. (1972). Novẏe vidȳ lishaĭnikov dlya Azerbaĭdzhanskoĭ SSR iz Kubinskogo i Kusarskogo raionov. *Nov. Sist. Nizsh. Rast.* **9**, 301–303. [56 taxa new to the republic]

Belorussian S.S.R. (*White Russia*)

Gorbach, N. V. (1965). "Opredelitel' listovatȳkh i kustistȳkh lishaĭnikov BSSR", 180 pp. Akademiya Nauk BSSR, Minsk. [Macrolichens; keys; descriptions; illustrations]
Gorbach, N. V. (1973). "Lishaĭniki Belorussii. Opredelitel'", 368 pp. Nauka i tekhnika, Minsk. [Keys; descriptions; illustrations]

Estonian S.S.R.

Räsänen, V. (1931). Die Flechten Estlands. I. *Ann. Acad. Sci. fenn.*, A, **34** (4), 1–162.

Sõmermaa, A. (1972). "Ecology of Epiphytic Lichens in Main Estonian Forest Types" [Scripta mycologica, Tartu No. 4], 117 pp. Academy of Sciences of the Estonian S.S.R., Tartu. [In English]

Trass, H. (1958). Eesti NSV kladooniate (põdrasamblike) määraja. *Loodusuur. Seltsi NSV Tead. Akad., Abiks Loodus.* **39**, 1–115. [*Cladonia* spp.; keys; descriptions; illustrations]

Trass, H. (1970). The elements and development of the lichen-flora of Estonia. *Trans. Tartu St. Univ.* **268**, *Pap. Bot.* **9**, 5–233. [Including check-list; maps; illustrations etc.; extensive English summary]

European U.S.S.R.

Dombrovskaya, A. V. (1970a). "Konspekt florȳ lishaĭnikov Murmanskoĭ oblastii severo-vostochnoĭ Finlyandii", 118 pp. Nauka, Leningrad. [Annotated check-list including north-west Finland]

Dombrovskaya, A. V. (1970b). "Lishaĭniki Khibin", 184 pp. Nauka, Leningrad. [Tabular keys; illustrations]

Dombrovskaya, A. A. and Shlyakov, R. N. (1967). "Lishaĭniki i mikhi severa evropeĭskoĭ chasti SSSR Kratkii opredelitel'", 182 pp. Nauka, Leningrad. [Northern European U.S.S.R.; keys; illustrations]

Elenkin, A. A. (1906–1911). Flora lishaĭnikov Srednei Rossii 1–4. *Izv. Estestvennoistor. Muz. Sheremetevoi* **3**, 1–184 (1906); **4** (1907); **8** (1911). [Important detailed account of lichens of central Russia]

Golubkova, N. S. (1966). "Opredelitel' lishaĭnikov srednei polosȳ evropeĭskoĭ chasti SSSR", 256 pp. Nauka, Moscow and Leningrad. [Keys; descriptions; illustrations]

Kuvaev, V. B. (1970). Lishaĭniki i mikhi Pripolyarnogo Urala i prilegayushchikh ravnin. *Trudȳ Inst. ekol. Rast. Zhiv. Ural'sk* **70**, 93–133. [Includes summary of literature records of Ural lichens]

Nikol'skii, P. N. (1929). Obzor literaturȳ o lishaĭnikakh Vyatskogo kraya. *Izv. glav. bot. Sada SSSR* **28**, 609–623. [Kirov region]

Mereschkowsky, K. S. (1919). Contribution à la flore lichénologique des environs de Kazan. *Hedwigia* **61**, 183–241.

Mereschkowsky, K. S. (1920). K poznaniyu lishaĭnikov okrestnostei Kazani. *Trudȳ bot. Muz. Ross. Akad. Nauk* **18**, 93–142.

Norrlin, J. P. (1876). Flora Kareliae Onegensis II. *Meddn Soc. Fauna Flora fenn.* **1**, 1–46. [Karelian A.S.S.R.]

Räsänen, V. (1939). Die Flechtenflora der nördlichen Küstengegend am Laatokka-See. *Suomal. eläin-ja kasvit. Seur. van.-kasvit. Julk.* **12** (1), 1–240. [South-west Karelian A.S.S.R.]

Rassadina, K. A. (1930). O lishainikakh b. Petergofskogo uezda Leningradskoĭ gubernii. *Trudȳ bot. Muz. Akad. Nauk SSSR* **22**, 223–271. [Leningrad region]

Tomin, M. P. (1956). "Opredelitel' korkovȳkh lishaĭnikov evropeĭskoĭ chasti SSSR", 532 pp. Izdatel'stvo Akademii Nauk BSSR, Minsk. [Crustose spp.; keys; descriptions; illustrations]

Franz Josef Land (Zemlya Frantsa Iosifa)

Lynge, B. (1931). Lichens collected on the Norwegian Scientific Expedition to Franz Jozef Land 1930. *Skr. Svalbard Ishavet* **38**, 1–31.

Georgian S.S.R.

Inaschvilli, T. (1969). Species nonnullae novae raraeque lichenum Caucasicarum. *Not. syst. geogr. Inst. bot. Tbiliss* **27**, 10–13.
Inaschvilli, T. (1970). Lichenes novae pro Georgia. *Not. syst. geogr. Inst. bot. Tbiliss* **28**, 10–13.
Inaschvilli, T. (1971). Novi et curiosi lichenes pro Georgia. *Not. syst. geogr. Inst. bot. Tbiliss* **29**, 24–25.
Pakhunova, V. G. (1933). Beiträge zur Studium der Flechten Georgiens. *Trudy tbiliss. bot. Inst.* **1**, 301–348. [Georgian with German summary]
Pakhunova, V. G. (1956). Likhenoflora Racha-Lechumi. *Trudy tbiliss. bot. Inst.* **18**, 139–180.
Steiner, J. (1919). Flechten am Transkaukasien. *Annls mycol.* **17**, 1–32.
Szatala, Ö. (1942). Lichenes in Peninsula Taurica et in Caucaso ab F. Kamienski, D. Sosnowsky et E. Koenig collecti. *Borbásia* **4**, 70–96.
Tomin, M. P. (1933). Materialy k poznaniyu lishaĭnikov gornykh lesov Zakavkaz'ya. *Trudy tbiliss. bot. Inst.* **1**, 353–373. [Keys to epiphytic spp. in Transcaucasia]
Vainio, E. A. (1889). Lichenes e Caucaso et in peninsula Taurica annis 1884–1885 ab H. Lojka et M. a Déchy collecti. *Természetr. Füz.* **22**, 269–343.
Voronov, Y. (1916). Materialy k lishaĭnikovoĭ flore Kavkaza. *Izv. kavkaz. Muz.* **9**, 203–224.

Kamchatka and Kuril Islands

Asahina, Y. (1934). Lichens of the Northern Kurile Islands. *Bull. biogeogr. Soc. Japan* **4**, 339–342.
Du Rietz, G. E. (1929). The lichens of the Swedish Kamtchatka-expedition. *Ark. Bot.* **22** (13), 1–25.
Sato, M. (1936). Notes on the lichen flora of Tisima or the Kuriles. *Bot. Mag., Tokyo* **50**, 610–617.
Savicz, V. P. (1924). Die Cladonien Kamtschatkas. *Feddes Repert.* **19**, 337–372.
Trass, H. H. (1963). K flore lishaĭnikov Kamchatki I. *In* "Issledovanie prirody Dal'nego Vostoka" (E. Parmasto, ed.), pp. 170–220. Akademiya Nauk Estonskoĭ SSR, Tallinn.

Kazakh S.S.R.

Andreeva, E. I. (1961). Materialy k flore lishaĭnikov Akmolinskoĭ oblasti. *Trudy Inst. bot. Akad. Nauk Kazak.* **9**, 221–236.
Andreeva, E. I. (1963). Lishaĭniki melkosopochnoĭ chasti pustyni Betpak-Dala. *Trudy Inst. bot. Akad. Nauk Kazak.* **15**, 178–203.

Kirghiz S.S.R.

Andreeva, E. I. (1959). Lishaĭniki Issȳk-Kul'skoĭ kotlovinȳ i ikh geograficheskoe raspredelenie. *Trudȳ Inst. Geogr. Leningr.* **75,** 144–155.

Brotherus, V. F. (1898). Contributions à la flore lichénologique de l'Asie Centrale. *Öfvers. finska VetenskSoc. Förh.* **40,** 1–13.

Mereschkowsky, K. S. (1911). Likhenologicheskaya poezdka v Kirgizskie stepȳ (gora Bogdo). *Trudȳ Obshch. Estest. imp. kazan. Univ.* **43** (5), 1–41.

Latvian S.S.R.

Bruttan, A. (1870). Lichenen Est-, Liv- und Kurlands. *Arch. Naturk. Liv.-Est.- u. Kurlands, ser.* 2, **7,** 163–326.

Piterans, A. V. (1970). Flora lishaĭnikov dolina reki Daugava. *Uchen. Zap. latv. gos. Univ.* **127,** 139–164.

Piterans, A. V. and Vimba, E. (1970). "Zemāko augu pētīšanas vēsture Latvijas PSR", 80 pp. Latviĭskiĭ Gosudarstvennyĭ Universitet, Riga. [Including bibliography of Latvian lichenology]

Lithuanian S.S.R.

Minkevičius, A. (1963). Medžiaga Lietuvos TSR kerpiu florai. *Liet. TSR Mokslų Akad. Darb., ser. biol.,* **3,** 79–95.

Räsänen, V. (1946). Lichenes a Professor Dr Kaarlo Linkola anno 1931 in Lituania [*sic!*] collecti. *Kuopion Luon. Ystäv. Yhdist. Julk., ser.* **B, 2** (3), 1–16.

Murmansk Region

See European U.S.S.R.

Novaya Zemlya

Lynge, B. (1928). Lichens from Novaya Zemlya. *Rep. scient. Results Norw. Exped.* **43,** 1–299.

Saghalien (Sakhalin)

Sato, M. (1936). Notes on the lichen flora of Minami-Karahuto, or the Japanese Saghalien. *Bull. biogeogr. Soc. Japan* **6,** 97–121.

Siberia and the Soviet Far East

See also Kamchatka and Kuril Islands and Saghalien.

Kochmaryova, A. P., Gorovoi, P. G. and Samoilenko, I. N. (1973). Lichens. *In* "Flora, Vegetation and Plant Resources of the Far East. Bibliography (1928–1969)", pp. 123–127, 374–377. Far East Science Center, Vladivostok. [Bibliography; in Russian and English]

Lokinskaya, M. A. (1970). Naibolee rasprostranennȳe vidȳ lishaĭnikov na severo-vostoke SSSR. *Vod. grib. Sibiri Dal'nego Vostoka* **1** (3), 233–246. [Magadan Region and Yakutia]

Makarova, I. I. (1973). Lishaĭniki Iul'tinskogo raiona Chukotskogo natzional'-nogo okruga. *Nov. Sist. Nizsh. Rast.* **10**, 249–258. [Western Chukot Peninsula]

Malme, G. O. A. (1932). Lichenes orae Sibiriae borealis inde ab insula Minin usque ad promentorium Ryrkajpia in expeditione Vegae lecti. *Ark. Bot.* **25A** (2), 1–42.

Oxner, A. N. (1940). Lishaĭniki baseĭnŷ rik Leni, Yani, Indigirki ta Pidvennogo Pribaikallya. I–II. *Zh. Inst. Bot. Kŷyiv* **31**, 117–139; *Bot. Zh.* **1**, 77–100, 313–324; **1** (3–4), 31–57. [Yakutia and Irkutsk Region]

Oxner, A. N. and Blum, O. B. (1971). K flore lishainikov sovetskogo Dal'nego Vostoka. I. Sem. Peltigeraceae. *Nov. Sist. Nizsh. Rast.* **8**, 249–263. [Including keys]

Piin, T. H. and Trass, H. H. (1971). Napochvennŷe lishaĭniki okrestnosteĭ Tarei (zapadnŷĭ Taĭmŷr). *In* "Biogeotzenozŷ taĭmŷrskoĭ tundrŷ i ikh produktivnost", pp. 151–160. Nauka, Leningrad. [Terricolous spp. of Taimyr Region]

Popova, T. G., Skabichevskii, A. P., Vasil'eva, L. N., Nozdrenko, M. V., Savicz, V. P. and Bardunov, L. V. (1967). Polveka izucheniya nizshikh rastenii Sibiri i Dal'nego Vostoka. *Izv. sib. Otdel. Akad. Nauk SSSR, ser. biol.-meditz. nauk*, **5** (1), 12–24. [Bibliography of research in last 50 years]

Rassadina, K. A. (1936). Likhenologicheskiĭ ocherk Baikal'skikh beregov. *Trudŷ bot. Inst. Akad. Nauk SSSR, ser. II*, **3**, 625–662. [Lake Baykal]

Rassadina, K. A. (1940) ["1938"]. Materialŷ k flore lishaĭnikov Altaya. *Trudŷ bot. Inst. Akad. Nauk SSSR, ser. II*, **4**, 295–321. [Altay Mountains]

Savicz, V. P. and Elenkin, A. A. (1950). Vvedenie k flore lishaĭnikov aziatskoĭ chasti SSSR. *Trudŷ bot. Inst. Akad. Nauk SSSR* **2** (6), 181–343. [Asiatic U.S.S.R.; bibliography of 401 lichen papers to 1926 with abstracts and species lists]

Vainio, E. A. (1896). Lichenes in Sibiria meridionali collecti. *Acta Soc. Fauna Fl. fenn.* **13**, 1–20. [North Sayan Mountains]

Vainio, E. A. (1909). Lichenes in viciniis hibernae expeditionis Vegae prope pagum Pitlekai in Sibiria septentrionali a D:re E. Almquist collecti. *Ark. Bot.* **8** (4), 1–175.

Vainio, E. A. (1928). Enumeratio lichenum in viciniis fluminis Konda (circ. 60° lat. bor.) in Siberia occidentali crescentium. *Ann. Acad. Sci. fenn., ser. A*, **27** (6), 65–122.

Vodop'yanova, N. V. (1973). Osnovnŷe lishaĭnikovŷe gruppirovki Gornoĭ Shorii. *In* "Vodorosli, gribŷ i lishaĭniki lesostepnoĭ i lesnoĭ zon Sibiri" (T. G. Popov, ed.), pp. 119–127. Nauka, Novosibirsk. [Lichen communities in south-central Siberia]

Tajik S.S.R.

Akramova, R. H. (1971). Lishaĭniki. [*In* "Flora i rastitel'nost' yshchel'ya reki Varzob" (P. N. Ovczinnikov, ed.)]. *Trudŷ Inst. bot. Akad. Nauk Tajik. SSR* **22**, 231–232.

Golubkova, N. S. (1973). Pervȳĭ sistematicheskiĭ spisok lishaĭnikov Vostochnogo Pamira. *Nov. Sist. Nizsh. Rast.* **10,** 206–223. [115 spp. from Pamir Mountains]

Turkmen S.S.R.

Dzhuraeva, Z. (1971). O lishaĭnikakh tzentral'nogo Kopet-Daga. *Mater. VI Simp. mikol. likhen. Pribalt. resp.*, *Riga* **1,** 10–14.
Dzhuraeva, Z. (1974). Priurochennost' lishaĭnikov k razlichnȳm tipam pochv. *Scripta mycol.*, *Tartu* **5,** 206–209. [41 terricolous spp. from Turkmenian deserts]
Vainio, E. A. (1887). Plantae Turcomanicae a G. Radde et A. Walter collectae. Lichenes. *Acta Horti petropol.* **10,** 551–562.

Ukrainian S.S.R.

See also under Georgian S.S.R.

Makarevicz, M. F. (1963). "Analiz likhenoflori Ukrayin'skikh Karpat", 263 pp. Akademiï Nauk Ukraïnskoï RSR, Kiev. [Phytogeography; many maps]
Mereschkowsky, K. S. (1920a). Spisok lishaĭnikov Krȳma. *Trudȳ bot. Muz. Ross. Akad. Nauk* **18,** 143–176. [Crimea]
Mereschkowsky, K. S. (1920b). Enumeratio lichenum in peninsula Taurica hucusque cognitorum. *Bull. Soc. bot. Fr.* **67,** 186–197, 284–295. [Crimea; including bibliography]
Oxner, A. N. (1937). "Viznachnik lishaĭnikīv URSR", 342 pp. Akademiï Nauk Ukraïnskoï RSR, Kiev. [Keys; descriptions; illustrations]
Oxner, A. N. (1956). "Flora lishaĭnikīv Ukraïni, **1**", 495 pp. Akademiï Nauk Ukraïnskoï, Kiev. [Keys; descriptions; illustrations]
Oxner, A. N. (1968). "Flora lishaĭnikīv Ukraïni, **2** (1)", 500 pp. Naukova Dumka, Kiev. [Keys; descriptions; illustrations]

Uzbek S.S.R.

Shafeev, N. G. (1953). K poznaniyu lishaĭnikov Ferganskoĭ dolinȳ. *Notul. syst. Inst. cryptog. Horti bot. petropol.* **9,** 17–26.
Vainio, E. A. (1904). Lichenes ab Ove Paulsen praecipue in provincia Ferghana (Asia media) et a Boris Fedtschenko in Tjanschan a. 1898 et 1899 collecti. *Bot. Tidsskr.* **26,** 241–250.

F. Europe*

(a)

Anders, J. (1928). "Die Strauch- und Laubflechten Mitteleuropas", 217 pp. Gustav Fischer, Jena. [Keys; descriptions; many photographs; macrolichens only; reprint 1975, Asher, Amsterdam]

*Because of restrictions of space, titles included in this section have been strictly limited; details of most European monographs are included in Poelt (1969) listed under Section (a).

Frey, E. (1969). "Flechten" [Hallwag-Taschenbücher No. 89], 64 pp. Hallwag, Bern and Stuttgart. [Notes; colour photographs; common spp., mainly macrolichens]
Gams, H. (1967). "Kleine Kryptogamenflora III. Flechten (Lichenes)", 244 pp. Gustav Fischer, Stuttgart. [Keys; illustrations]
Olivier, H. l'Abbé (1907–1909). Lichens d'Europe 1–2. *Mém. Soc. natn. Sci. nat. math. Cherbourg* **36**, 77–274 (1907); **37**, 29–200 (1909). [Keys; descriptions]
Poelt, J. (1969). "Bestimmungschlüssel europäischer Flechten", 757 pp. J. Cramer, Lehre. [Keys; valuable bibliography]

(b)

Degelius, G. (1954). The lichen genus *Collema* in Europe. *Symb. bot. upsal.* **13** (2), 1–499.
Hertel, H. (1967). Revision einiger calciphiler Formenkreis der Flechtengattung *Lecidea. Beih. Nova Hedwigia* **24**, 1–155.
Krog, H. and James, P. W. (1977). The genus *Ramalina* in Fennoscandia and the British Isles. *Norw. J. Bot.* **24**, 15–43.
Motyka, J. (1958). Lichenum generis *Alectoria* Ach., subgenus *Bryopogon* (L.) Th. Fr. in Europa media (descriptiones specierum). *Fragm. flor. geobot.* **3**, 205–231.
Motyka, J. (1960). Przeglad gatunków rodzaju *Ramalina* Ach. środkowej i zachodniej Europy.—Conspectus. *Fragm. flor. geobot.* **6**, 645–682.
Nádvorník, J. (1942). Systematische Übersicht der mitteleuropäischen Arten der Flechtenfamilie Caliciaceae. *Studia bot. čsl.* **5**, 6–40.
Poelt, J. (1953–1975). Mitteleuropäische Flechten I–X. *Mitt. bot. StSamml., Münch.* **1**, 230–238, 323–332; **2**, 46–56, 273–283, 386–399; **3**, 568–584; **4**, 171–197; **8**, 191–210; **12**, 1–32. [Critical notes on many taxa; some keys; index to the series in the last part]
"Rabenhorst's Kryptogamen-Flora von Deutschland, Österreich und der Schweiz", Vol. 9. Akademie Verlag, Leipzig.
[Vol. **9** includes:
   Erichsen, C. F. E. (1935–1936). Pertusariaceae. *Ibid.* **5** (1), 319–728.
   Frey, E. (1933). Cladoniaceae (unter Ausschluss der Gattung *Cladonia*), Umbilicariaceae. *Ibid.* **4** (1), 1–426.
   Gyelnik, V. K. (1940a). Lichinaceae. *Ibid.* **2** (2), 1–110.
   Gyelnik, V. K. (1940b). Heppiaceae. *Ibid.* **2** (2), 111–134.
   Gyelnik, V. K. (1940c). Pannariaceae. *Ibid.* **2** (2), 135–272.
   Hillmann, J. (1935). Teloschistaceae. *Ibid.* **6** (1), 1–36.
   Hillmann, J. (1936). Parmeliaceae. *Ibid.* **5** (3), 1–309, *1–10*.
   Keissler, K. von (1933). Moriolaceae. *Ibid.* **1** (1), 1–43.
   Keissler, K. von (1936–1938). Pyrenulaceae bis Mycoporaceae, Coniocarpineae. *Ibid.* **1** (2), 1–846.
   Keissler, K. von (1958–1960). Usneaceae. *Ibid.* **5** (4), 1–755.
   Lynge, B. (1935). Physciaceae. *Ibid.* **6** (1), 37–188.

Magnusson, A. H. (1935). Acarosporaceae und Thelocarpaceae. *Ibid.* **5** (1), 1–318.
Redinger, K. (1936–1938). Arthoniaceae, Graphidaceae, Chiodectonaceae, Dirinaceae, Roccellaceae, Lecanactidaceae, Thelotremaceae, Diploschistaceae und Coenogoniaceae. *Ibid.* **2** (1), 1–404.
Sandstede, H. (1931). Die Gattung *Cladonia*. *Ibid.* **4** (2), 1–531.
Zschacke, H. (1933–1934). Epigloeaceae, Verrucariaceae und Dermatocarpaceae. *Ibid.* **1** (1), 44–695.]
[Keys; descriptions; illustrations; all families not covered]
Runemark, M. (1956a, b). Studies in *Rhizocarpon* I–II. *Op. bot. Soc. bot. Lund.* **2** (1), 1–152 [I. Taxonomy of the yellow species in Europe]; **2** (2), 1–150 [II. Distribution and ecology of the yellow species in Europe].
Schmidt, A. (1970). Anatomisch-taxonomische Untersuchungen an europäischen Arten der Flechtenfamilie Caliciaceae. *Mitt. Staatsinst. Allg. Bot. Hamburg* **13**, 111–166.
Tibell, L. (1971). The genus *Cyphelium* in Europe. *Svensk bot. Tidskr.* **65**, 138–164.
Wunder, H. (1974). Schwarzfrüchtige, saxicole Sippen der Gattung *Caloplaca* (Lichenes, Teloschistaceae) in Mitteleuropa, dem Mittelmeergebiet und Vorasien. *Biblthca lich., Lehre* **3**, 1–186.

(c)

Barkman, J. J. (1958). "Phytosociology and Ecology of Cryptogamic Epiphytes", 628 pp., Van Gorcum, Assen.
Klement, O. (1955). Prodromus der mitteleuropaischen Flechtengesellschaften. *Beih. Feddes Repert.* **135**, 5–194.
Schauer, T. (1965). Ozeanische Flechten in Nordalperaum. *Port. Acta Biol.*, **B, 8** (1), 17–229.
Wirth, V. (1972). Die Silikatflechten-Gemeinschaften in außeralpine Zentraleuropa. *Diss. Bot., Lehre* **17**, 1–306, *1–9*.

Albania

Szatala, Ö. and Timkó, G. (1926). Additamenta ad floram Albaniae. *Magyar Tud. Akad. Balkán-kutat.* **3**, 151–179.

Austria

Dalla Torre, K. W. von and Sarnthein, L. von (1902). "Die Flechten (Lichenes) von Tirol, Vorarlberg und Liechtenstein" [Flora der gefürsteten Graftschaft Tirol, des Landes Vorarlberg und des Fürstentumes Liechtenstein IV], 936 pp. Wagnersche Universitätsbuchhandlung, Innsbruck. [Keys; descriptions]

Bear Island (*Medvezh'ii ostrov*)

Lynge, B. (1926). Lichens from Bear Island (Bjørnøya). *Resultater Norske Spitsbergeneksped.* **1** (9), 1–78.

Belgium

Barkman, J. J. (1963). De epifyten-flora en -vegetatie van Midden-Limburg (België). *Ver. K. ned. Akad. Wet.*, II, **54** (4), 1–46.

De Sloover, J. and Lambinon, J. (1965). Contribution à l'étude des lichens corticoles du bassin de la Dendre. *Bull. Soc. r. Bot. Belg.* **98**, 229–273.

Duvigneaud, P. and Giltay, L. (1938). Catalogue des lichens de Belgique. *Bull. Soc. r. Bot. Belg.* **70**, *Suppl.*, 1–52. [Check-list 480 spp.; bibliography]

Lambinon, J. (1969). "Les Lichens" 2nd ed., 196 pp. Les Naturalistes Belges, Bruxelles. [Reprinted from *Naturalistes Belg.* **49**, 205–280, 449–558 (1968).] [Keys; illustrations]

British Isles

See United Kingdom.

Bulgaria

Motyka, J. and Železova, B. (1963). Monographische Untersuchungen der Gattung *Usnea* in Bulgarien. *Izv. bot. Inst. Sof.* **10**, 67–120.

Pišút, I. (1967–1969). Príspevok k poznaniu lišajníkov Bulharska I–II. *Rer. nat. Mus. natn. Slov.*, *Bratislava* **13** (2), 3–10 (1967); **15** (1), 27–37 (1969).

Popnikolov, A. and Železova, B. (1964). "Flora na B"lgariya-Lishei", 517 pp. Narodna Prosveta, Sofia. [Keys; descriptions; illustrations, some in colour]

Corsica

Kalb, K. (1976). Flechtenfunde aus Korsika. *Herzogia* **4**, 55–63.

Maheu, J. and Gillet, A. (1926). "Lichens de l'est de la Corse", 114 pp. Berthier, Dijon. [300 spp.]

Werner, R. G. (1973). Étude phyto- et paléogéographique de la flore lichénique d'une île, la Corse. *Revue bryol. lichén.* **39**, 293–343. [Including check-list 776 spp.; bibliography]

Werner, R. G. and Deschatres, R. (1974). Contribution à l'étude des lichens de la Corse. III. *Bull. Soc. bot. Fr.* **121**, 299–318.

Crete

Kleinig, H. (1966). Beitrag zur Kenntnis der Flechtenflora von Kreta. *Nova Hedwigia* **11**, 513–526. [Check-list 240 spp.; bibliography]

Rondon, Y. (1969). Contribution à l'étude des lichens de l'île de Crète. *Revta Fac. Ciênc. Univ. Lisb.*, **2C, 16**, 105–117.

Czechoslovakia

Černohorský, Z., Nádvorník, J. and Servít, M. (1956). "Klíc k Určování Lišejníků ČSR I" 156 pp. Nakladatelství Československí Akademie Věd., Prague. [Keys; descriptions; macrolichens only]

Pišút, I. (1961). Bemerkungen über einige Arten der Flechtengattung *Cladonia* in der Slowakei. *Acta Fac. Rerum nat. Univ. comen., Bratisl.*, **B, 6**, 513–531.
Pišút, I. (1968). Die Arten der Flechtengattung *Collema* G. H. Web. in der Slowakei. *Acta Fac. Rerum nat. Mus. nat. Slov., Bratisl.* **14** (2), 5–71.
Pišút, I. (1970). Doplnky k poznaniu lišajníkov Slovenska 6. *Acta Fac. Rerum nat. Mus. nat. Slov., Bratisl.* **16**, 31–40.
Pišút, I. (1971). Verbreitung der Arten der Flechtengattung *Lobaria* (Schreb.) Hue in der Slowakei. *Acta Fac. Rerum nat. Mus. nat. Slov., Bratisl.* **17**, 105–130.
Servít, M. (1954). "Československé lišejníky Čeledi Verrucariaceae", 249 pp. Nakladatelství Československé Akademie, Prague. [Including keys; descriptions]
Suza, J. (1925). Nástin zeměpisného rozšíření lišejníků na Moravě vzhledem k poměrum Evropským. *Přírodov. fak. Masarykova univ. Brnŏ* **55**, 1–152. [Including check-list for Moravia]
Vězda, A. (1970). Neue oder wenig bekannte Flechten in der Tschechoslowakei. I. *Folia geobot. phytotax., Praha* **5**, 307–337.

Denmark

See also Scandinavia.

Branth, J. S. D. and Rostrup, E. (1869). Lichenes Daniae eller Danmarks laver. *Bot. Tidsskr.* **3**, 127–298. [Including descriptions]
Degelius, G. (1965). Lavfloran i Hald Egeskov (Jylland). Ett bidrag till de danska ekskogresternas naturhistoria. *Bot. Tidsskr.* **61**, 1–21.
Christiansen, M. S. (1947). Bidrag til Danmarks lavflora. I. *Bot. Tidsskr.* **48**, 172–191.
Erichsen, C. F. E. (1942). Neue dänische Flechten. *Annls mycol.* **40**, 140–149. [Including bibliography]
Galløe, O. (1927–1954, 1972). "Natural History of the Danish Lichens", 10 vols. Aschehoug etc., Copenhagen. [Descriptions; illustrations, many in colour, 1397 plates]

Eire

See Ireland.

Faeroe Islands

Degelius, G. (1966). Notes on the lichen flora of the Faroe Islands. *Acta Horti gothoburg.* **28** (1), 1–13.
Hansen, K. (1968). Lichens in the Faeroes. *Bot. Tidsskr.* **63**, 305–318. [Including bibliography]

Finland

See also under U.S.S.R. (Murmansk) and Scandinavia.

Collander, R., Erkamo, V. and Lehtonen, P., eds (1973). Bibliographia botanica fenniae 1901–1950. *Acta Soc. Fauna Flora fenn.* **81,** 1–647. [Bibliography]

Hakulinen, R. (1963). "Jäkäläkasvio", 253 pp. Werner Sönderström Osakeyhtiö, Helsinki and Porvoo. [Macrolichens; keys; descriptions; illustrations]

Koskinen, A. (1955). "Über die Kryptogamen der Bäume, besonders die Flechten, im Gewässergebiet das Päijänne sowie an den Flüssen Kalajoki, Lestijoki und Pyhäjoki. Floristische, soziologische und ökologische Studie I", 176 pp. Privately printed, Helsinki. [Epiphytic spp.; not seen, reference supplied by Dr T. Ahti]

Räsänen, V. (1927). Über Flechtenstandorte und Flechtenvegetation im westlichen Nordfinnland. *Suomal. eläin- ja kasvit. Seur. van. Julk.* **7,** 1–202.

Räsänen, V. (1951). "Suomen Jäkäläkasvio"[*Kuopion Luon. Ystäv. Yhdist.*, **A,** No. 5], 158 pp. Kuopion Kansallinen, Kirjapaino and Kuopiosaa. [Keys; illustrations]

Vainio, E. A. (1921–1934). Lichenographia fennia: I–IV. *Acta Soc. Fauna Flora fenn.* **49** (2), 1–274 (1921); **53** (1), 1–340 (1922); **57** (1), 1–238 (1927); **57** (2), 1–531 (1934). [Keys; descriptions]

France

(a)

Guillaumot, M. (1951). Flore des lichens de France et de Grande-Bretagne. *Encycl. Biol.* **42,** 1–604. [Keys; brief descriptions]

Harmand, J. l'Abbé (1905–1913). "Lichens de France". 5 vols, 1185 pp. Klincksieck, Paris. [Keys; descriptions]

Olivier, H. l'Abbé (1897–1900). "Exposé Systématique et descriptions des lichens de l'ouest et du Nord-Ouest de la France". 2 vols. and suppl. 352, 426, 32 pp. Klincksieck, Paris. [Keys; descriptions]

Ozenda, P. and Clauzade, G. (1970). "Les Lichens, Étude biologique et flore illustrée", 802 pp. Masson et Cie, Paris. [Keys 2200 spp.; bibliography; illustrations]

(c)

Des Abbayes, H. (1934). La végétation lichénique du Massif Armoricain, Étude chorologique et écologique. *Bull. Soc. Sci. nat. Ouest Fr.*, sér. 5, **3,** 1–267.

Asta, J., Clauzade, G. and Roux, C. (1972). Premier aperçu de la végétation lichénique du Parc National de la Vanoise. *Trav. Sci. Parc Natl. Vanoise* **2,** 73–105. [392 spp.]

Bouly de Lesdain, M. (1910). "Recherches sur les Lichens des environs de Dunkerque". [Thèse Université de Paris, sér. A no. 625], 301 pp. Michel, Dunkirk.

Coppins, B. J. (1971). Field meeting in Brittany. *Lichenologist* **5,** 146–169. [498 taxa]

Massé, L. J. C. (1964). Recherches phytosociologiques et écologiques sur les lichens des schistes rouges cambriens des environs de Rennes (I.-et-V.). *Vegetatio* **12**, 103–122.

Franz Josef Land

See Zemlya Frantsa Iosifa (U.S.S.R.).

Germany (*German Federal Republic and German Democratic Republic*)

Bertsch, K. (1964). "Flechtenflora von Südwestdeutschland", 251 pp. Eugen Ulmer, Stuttgart. [Including keys; illustrations]
Erichsen, C. F. E. (1957). "Flechtenflora von Nordwestdeutschland" 411 pp. Gustav Fischer, Stuttgart. [Keys; descriptions]
Grummann, V. (1963). "Catalogus lichenum germaniae", 208 pp. Gustav Fischer, Stuttgart. [Check-list 2169 spp.; distribution; bibliography]
Hillmann, J. and Grummann, V. (1957). Flechten. *In* "Kryptogamenflora der Mark Brandenburg VIII", 898 pp. Borntraeger, Berlin and Nikolassee. [Keys; descriptions; illustrations]
Wirth, V. (1975). Neue und bemerkenswerte Flechtenfunde in Deutschland. *Ber. bayer. bot. Ges.* **46**, 111–123.
Wirth, V. (1976). Veränderungen der Flechtenflora und Vegetation in der Bundesrepublik Deutschland. *SchrReihe Vegetationskde* **10**, 177–202.

Greece

See also Crete.

Harmand, J. l'Abbé and Maire, R. (1909). Contribution à l'étude des lichens de la Grèce. *Bull. Séanc. Soc. Sci. Nancy* **1909** (6), 1–36.
Degelius, G. (1956). Studies in the lichen family Collemataceae. II. On the *Collema* flora of the mainland of Greece. *Svensk bot. Tidskr.* **50**, 496–512.
Krause, W. and Klement, O. (1962). Zur Kenntnis der Flora und Vegetation auf Serpentinstandorten des Balkans 5. Flechten und Flechtengesellschaften auf Nord-Euböa (Griechland). *Nova Hedwigia* **4**, 182–262.
Rondon, Y. (1970). Contribution à l'étude des lichens du Péloponèse. *Port. Acta Biol.*, **B**, **11**, 38–50.
Steiner, J. (1919). Beiträge zur Kenntnis der Flora Griechlands. C. Lichenes. *Verh. zool.-bot. Ges. Wien* **69**, 52–101.
Szatala, Ö. (1941, 1959). Contributions à la connaissance des lichens de la Grèce. I–II. *Borbásia* **3**, 113–136 (1941) [I. La presqu'île Athos (Hagion Oros)]; *Annls hist.-nat. Mus. natn. hung.* **51**, 121–144 (1959) [II. Mont Olympe].
Wilmanns, O. and Phitos, D. (1960). Zur Epiphytenflora des Parnes. *Dassika Chronika* **18/19**, 1–8. [Greek with German summary]

Greenland

Dahl, E. (1950). Studies in the macrolichen flora of south-west Greenland. *Meddr Grønland* **150** (2), 1–176.
Dahl, E., Lynge, B. and Scholander, P. F. (1937). Lichens from southeast Greenland. *Skr. Svalbard Ishavet* **70,** 1–76.
Hansen, K. (1962). Macrolichens from central west Greenland collected on the botanical expedition in 1958. *Meddr Grønland* **163** (6), 1–64. [Including valuable bibliography]
Hansen, K. (1971). Lichens in south Greenland, distribution and ecology. *Meddr Grønland* **178** (6), 1–84.
Lynge, B. (1937). Lichens from west Greenland collected chiefly by Th. M. Fries. *Meddr Grønland* **118** (8), 1–225.
Lynge, B. (1940). Lichens from north east Greenland. II. Microlichens. *Skr. Svalbard Ishavet* **81,** 1–143.
Lynge, B. and Scholander, P. F. (1932). Lichens from north east Greenland. *Skr. Svalbard Ishavet* **41,** 1–116.

Hungary

Fóriss, F. (1957). Új zuzmófajok és fajváltozatok Magyarország flórájában. *Bot. Közl.* **47,** 67–76. [112 spp. new to Hungary]
Gallé, L. (1968). Deutung und richtige Bezeichung der aus Ungarn Beschriebenen Flechtenzönosen. *Acta Bot. Acad. scient. Hung.* **14,** 29–40. [Phytosociology]
Hazslinszký, F. A. (1884). "A Magyar birodalom Zuzmó-Flórája", 304 pp. Ungar. Natur. Ges., Budapest.
Szatala, O. (1927–1942). Lichenes Hungariae. I–III. *Folia crypt.* **1,** 337–434 (1927), 833–928 (1930); **2,** 267–460 (1942).
Verseghy, K. (1958). Die endemische Flechten der Karpaten und des Karpatenbeckens. *Annls hist.-nat. Mus. natn. hung.* **50,** 65–73. [103 "endemic" spp.]
Verseghy, K. (1965–1966). *Squamaria-* und *Squamarina-*Arten in Ungarn. [I-] II. *Bot. Közl.* **52,** 124–129 (1965); **53,** 11–23 (1966).
Verseghy, K. (1970–1972). *Gasparrinia-*Arten in Ungarn. I–III. *Bot. Közl.* **57,** 23–29 (1970); **58,** 21–28 (1971); **59,** 13–18 (1972).
Verseghy, K. (1973). *Caloplaca-*Arten in Ungarn. *Studia Bot. Hung.* **8,** 33–64.

Iceland

Degelius, G. (1957). The epiphytic lichen flora of the birch stands in Iceland. *Acta Horti gothoburg.* **22,** 1–51.
Galløe, O. (1932). The lichen flora and lichen vegetation of Iceland. *Botany of Iceland* **2,** 103–247.
Kristinsson, H. (1972). Additions to the lichen flora of Iceland I. *Acta Bot. Isl.* **1,** 43–50.
Kristinsson, H. (1973–1974). Recent literature on the botany of Iceland I–II. *Acta Bot. Isl.* **2,** 67–76 (1973); **3,** 102–104 (1974).

Lynge, B. (1940). Lichens from Iceland. I. Macrolichens. *Skr. Norske Vid.-Akad. Oslo, mat.-nat. kl.* **1940** (7), 1–56.

Ireland

See also under United Kingdom.

Fenton, A. F.-G. (1969). The lichens of Northern Ireland. *Ir. Nat. J.* **16**, 110–127. [Including bibliography]

Knowles, M. C. (1929). The lichens of Ireland. *Proc. R. Ir. Acad.*, **B, 38**, 179–434. [Check-list; bibliography; distribution]

Mitchell, M. E. (1961). L'élément eu-océanique dans la flore lichénique du sud-ouest d'Irlande. *Revta Biol.* **2**, 177–256.

Mitchell, M. E. (1971). "A Bibliography of Books, Pamphlets and Articles relating to Irish Lichenology, 1727–1970", 76 pp. Privately printed, Galway. [Bibliography 422 publications indexed by subject and vice-county]

Porter, L. (1948). The lichens of Ireland (Supplement). *Proc. R. Ir. Acad.*, **B, 51**, 347–386.

Seaward, M. R. D. (1975–1977). Contributions to the lichen flora of south-east Ireland—I–II. *Proc. R. Ir. Acad.*, **B, 75**, 185–205; **77**, 119–134.

Italy

Jatta, A. (1900). "Sylloge lichenum italicorum", 662 pp. Vecchi, Trani.

Jatta, A. (1909–1911). Lichenes. *In* "Flora italica cryptogama", Vol. 3, pp. i–xxii, 1–958. Società Botanica Italiana, Casiano. [Keys; descriptions]

Kalb, K. (1970). Flechtengesellschaften der vorderen Ötztaler Alpen. *Diss. Bot., Lehre* **9**, 1–118.

Sbarbaro, C. (1955). Novae lichenum species in Italiae (praesertim in Liguria) inventae annis 1922–1955. *Annali Mus. civ. Stor. nat. Giacomo Doria* **68**, 114–126.

Jan Mayen

Sheard, J. W. (1962). A contribution to the lichen flora of Jan Mayen. *Lichenologist* **2**, 76–85. [Including bibliography]

Jugoslavia

See Yugoslavia.

Luxembourg

Feltgen, E. (1901–1902). Merch sowie die nächste und weitre Umgebung zum Gebrauch für Naturfreunde. *Soc. nat. Luxemb.* **11**, 394. [Not seen; 165 spp.]

Koltz, J.-P.-J. (1897). "Prodrome de la Flore du Grand-Duché de Luxembourg, Lichenées" [*Mém. Soc. bot. Luxemb.* **13**, 91–349]. 2 vols. [517 spp.]

Malta

Sommiers, S. and Gatto, A. C. (1915). Flora Melitensis nova. *Boll. R. Orto bot. Palermo, n.s.* **1** (2), 343–364.

Netherlands

Barkman, J. J. (1958). "Phytosociology and Ecology of Cryptogamic Epiphytes", 628 pp. Van Gorcum, Assen.

Hennipman, E. (1968). De Nederlandse *Cladonia*'s (Lichenes). *Wet. Meded. K. ned. natuurh. Veren.* **79**, 1–53.

Maas Geesteranus, R. A. (1948). Revision of the lichens of the Netherlands I. Parmeliaceae. *Blumea* **6** (1947), i–viii, 1–199. [Including bibliography of Dutch lichenology]

Maas Geesteranus, R. A. (1952). Revision of the lichens of the Netherlands II. Physciaceae. *Blumea* **7**, 206–287.

Maas Geesteranus, R. A. (1954–1958). Notes on Dutch lichens I–II. *Blumea* **7**, 570–592 (1954); *Suppl.* **4**, 178–187 (1958).

Norway

See also Scandinavia.

Degelius, G. (1955). The lichen flora on calcareous substrata in southern and central Nordland (Norway). *Acta Horti gothoburg.* **20**, 35–56.

Havaas, J. (1910). Beiträge zur Kenntnis der westnorwegischen Flechtenflora, I. *Bergens Mus. Årb.* **1909** (1), 1–36.

Jørgensen, P. M. and Ryvarden, L. (1970). Contribution to the lichen flora of Norway. *Årbok Univ. Bergen, mat.-nat. ser.* **1969** (10), 1–24.

Lynge, B. (1921). Studies on the lichen flora of Norway. *Skr. Norske Vid.-Akad. Oslo, mat.-nat. kl.* **1921** (7), 1–252. [Macrolichens]

Østhagen, H. (1976). Nye utbredelsesdata for norske makrolav. *Blyttia* **34**, 189–203.

Novaya Zemlya

See under U.S.S.R.

Poland

(a)

Motyka, J. (1960–1964). Porosty (Lichenes). *In* "Flora Polska" **3** (2), 1–500 (1964) [Cladoniaceae]; **4** (2), 1–414 (1964) [Acarosporaceae, Umbilicariaceae]; **5** (1), 1–274 (1960) [Parmeliaceae]; **5** (2), 1–353 (1962) [Usneaceae]. Polska Akademia Nauk Instytut Botanica, Warsaw. [Keys; descriptions; illustrations]

Nowak, J. and Tobolewski, Z. (1975). "Porosty Polskie", 1177 pp. Polska Akademia Nauk Instytut Botaniki, Warsaw and Krakow. [Keys; descriptions; illustrations; a comprehensive flora]

Tobolewski, Z. R. (1965). Wykas porostów dotychczas stwierdzonych w Polsce (wraz z bibliografia lichenologiczna). *Pozn. Tow. Przyj. Nauk, Prace Kom. Biol.* **24** (3), 1–62. [Check-list 1198 spp.; bibliography 272 refs.]

(b)

Tobolewski, Z. R. (1966). Rodzina Caliciaceae w Polsce. *Pozn. Tow. Przyj. Nauk, Prace Kom. Biol.* **24** (5), 1–105.

(c)

Fabiszewski, J. (1968). Porosty Śnieżnika Kłodzkiego i Gór Bialskich. *Monogr. Bot., Warsaw* **26**, 1–115.

Motyka, J. (1925). Die Pflanzenassoziationen des Tatra-Gebirges. II. Teil: Die epilithischen Assoziationen der nitrophilen Flechten im Polnischen Teile der Westtatra. *Bull. int. Acad. Cracovie*, **B**, **1924**, 835–850.

Nowak, J. (1961). Porosty Wyżny (Jury) Krakowsko-Częstochowskiej. *Monogr. Bot., Warsaw* **11** (2), 1–128.

Olech, M. (1973). Porosty Beskidu Sądeckiego. *Zesz. nauk. Uniw. jagielloński.* **116**, *Bot.* **1**, 87–192.

Tobolewski, Z., ed. (1971–1977). Lichens (Lichenes). *In* "Atlas of Geographical Distribution of Spore-plants in Poland" (J. Szweykowski and T. Wojterski, eds), **3** (1), 1–31 (1971); **3** (2), 1–23 (1974); **3** (3), 1–25 (1976); **3** (4), 1–42 (1977). Polska Akademia Nauk, Komitet Botaniczny i Instytut Botaniki, Poznań. [Distribution maps, 3 (1) includes list of 1203 spp. to be mapped]

Portugal

Sampaio, G. (1921). Novas contribuições para o estudo dos líquenes Portugueses. *Brotéria, sér. bot.* **19**, 12–35. [Including bibliography]

[Sampaio, G.] (1970). Miscelanea dos trabalhos sobre líquenes. *Publ. Inst. bot. Gonçalo Sampaio, Porto, sér.* 3A, **20**, i–vii, 1–228. [Reprint of collected papers by Sampaio mainly on Portuguese lichens]

Tavares, C. N. (1945a). Líquenes da Serra da Estrêla. *Brotéria, sér. ciên. nat.* **14**, 14–24.

Tavares, C. N. (1945b). Contribuição para o estudo das Parmeliáceas Portuguesas. *Port. Acta Biol.*, **B**, **1**, 1–210.

Tavares, C. N. (1950). Líquenes da Serra do Gerês. *Port. Acta Biol.*, **B**, **3**, 1–189. [295 spp.]

Tavares, C. N. (1956). Notes lichénologiques—IX. *Revta Fac. Ciênc. Univ. Lisb.*, **C**, **5**, 123–134. [See also others in this series in various journals.]

Tavares, C. N. (1965). The genus *Pannaria* in Portugal. *Port. Acta Biol.*, **B**, **8**, 1–16.

Rumania (Romania)

Moruzi, C., Petria, E. and Mantu, E. (1967). Catalogul lichenilor din România. *Lucr. Grăd. bot. Buc.* **1967**, 1–389. [Check-list 1393 spp.; 201 refs., bibliography]

Moruzi, C. and Toma, N. (1971). "Licheni.—Determinator de Plante Inferioare", 221 pp. Editura Didactica si Pedagogica, Bucurest. [Keys; descriptions; illustrations]

Scandinavia

See also under Denmark, Finland, Norway and Sweden.

(a)

Christiansen, M. S., Krusenstjerna, E. von and Waern, M. (1976). "Vår Flora i Färg. Kryptogamer," 325 pp. Almqvist and Wiksell, Stockholm. [Semi-popular guide; coloured illustrations]

Dahl, E. and Krog, H. (1973). "Macrolichens of Denmark, Finland, Norway and Sweden", 185 pp. Universitetsforlaget, Oslo, Bergen and Tromsø. [Keys; illustrations; bibliography]

Fries, Th. M. (1871–1874). "Lichenographia scandinavica". 2 vols., 639 pp. Berling, Uppsala. [Descriptions; the most recent flora for all Scandinavia including crustose spp.]

Magnusson, A. H. (1929). "Flora över Skandinaviens busk- och bladlavar", 127 pp. Norstedt, Svenska Bokförlaget, Stockholm. [Macrolichens; keys; descriptions; some illustrations]

Magnusson, A. H. (1936). Lavar. *In* "Förteckning över Skandinaviens Växter" (Lunds Botaniska Förening, ed.) **4**, 1–93. Gleerups, Lund. [Check-list for Denmark, Norway and Sweden]

Magnusson, A. H. (1950). "Tillägg och ändringar (jan. 1950) till Skandinaviens växter. 4. Lavar", 28 pp. Uppsala. [Mimeographed supplement to the 1936 check-list]

(b)

Gilenstam, C. (1969). Studies in the lichen genus *Conotrema*. *Ark. Bot.*, ser 2, **7**, 149–179.

Hakulinen, R. (1954). Die Flechtengattung *Candelariella* Müller Argoviensis mit besonderer Berücksichtigung ihres Auftretens und ihrer Verbreitung in Fennoskandien. *Annls bot. Soc. zool.-bot. fenn. Vanamo* **27** (3), 1–127.

Magnusson, A. H. (1952a, b). Key to the species of *Lecidea* in Scandinavia and Finland. I–II. *Svensk bot. Tidskr.* **46**, 178–198 (1952a) [I. Saxicolous species]; **46**, 313–323 (1952b). [II. Non-saxicolous species].

Moberg, R. (1977). The lichen genus *Physcia* and allied genera in Fennoscandia. *Symb. bot. upsal.* **22** (1), 1–108.

Nordin, I. (1972). "*Caloplaca*, sect. *Gasparrinia* i Nordeuropa, Taxonomiska och ekologiska studier", 184 pp. Skriv Service AB, Uppsala.

Santesson, R. (1939). Amphibious pyrenolichens 1. *Ark. Bot.* **29A** (10), 1–67.

Tibell, L. (1969). The genus *Cyphelium* in northern Europe. *Svensk bot. Tidskr.* **63**, 465–485.

(c)

Ahlner, S. (1948). Utbredningstyper bland nordiska barrträdslavar. *Acta phytogeogr. suec.* **22**, i–x, 1–257.

Almborn, O. (1948). Distribution and ecology of some south Scandinavian lichens. *Bot. Notiser, Suppl.* **1** (2), 1–252.

Degelius, G. (1935). Das ozeanische Element der Strauch- und Laubflechtenflora von Skandinavien. *Acta phytogeogr. suec.* **7,** i–xii, 1–411.

Du Rietz, G. E. (1945). Lichenes. *In* "Vilda växter in Norden. Mossor, lavar, svampar, alger" (J. A. Nannfeldt and G. E. Du Rietz, eds), 67–185. Bokförlaget Natur och Kultur, Stockholm. [Detailed accounts of selected spp.]

Hakulinen, R. (1962). Die Flechtengattung *Umbilicaria* in Ostfennoskandien und angrenzenden Teilen Norwegens. *Annls bot. Soc. zool.-bot. fenn. Vanamo* **32** (6), 1–87.

Hakulinen, R. (1964). Die Flechtengattung *Lobaria* Schreb. in Ostfennoskandien. *Annls bot. fenn.* **1,** 202–213.

Hakulinen, R. (1965). Über die Verbreitung und das Vorkommen einiger nördlichen Erd- und Steinflechten in Ostfennoskandien. *Aquilo, ser. Bot.* **3,** 22–66.

Spain

Des Abbayes, H. (1946). Lichens d'Espagne récoltés de 1926 à 1935 par M. et Mme P. Allorge. *Revue bryol. lichén.* **15,** 79–86.

Bailey, R. H. (1970). Some lichens from northern Spain. *Revue bryol. lichén.* **74,** 983–986.

Degelius, G. (1966). Lichens from the summit of the Picacho de Veleta (Sierra Nevada, Spain). *Svensk bot. Tidskr.* **60,** 338–340.

Klement, O. (1965). Flechtenflora und Flechtenvegetation der Pityusen. *Nova Hedwigia* **9,** 434–501. [Including bibliography for Balearic Islands]

Lázaro é Ibiza, B. (1906). "Compendio de la Flora Espanõla". 2nd ed., Vol. 1. Libería de los Sucessores de Hernando, Madrid. [Lichens pp. 427–487; descriptions 456 spp.]

Tavares, C. N. (1959). Lichens from Spain I. *Revta Fac. Ciênc. Univ. Lisb.,* sér. 2a, **C, 7,** 53–74. [Including bibliography]

Werner, R. G. (1975). Étude écologique et phytogéographique sur les lichens d'Espagne méridionale. *Revue bryol. lichén.* **41,** 55–82. [Including annotated bibliography]

Spitzbergen

Hertel, H. and Ullrich, H. (1976). Flechten von Amsterdamöya (Svalbard). *Mitt. bot. StSamml., Münch.* **12,** 417–512. [Including bibliography]

Lynge, B. (1938). Lichens from the west and north coasts of Spitsbergen and the North-East Land. *Skr. Norske Vid.-Akad. Oslo, mat.-nat. kl.* **1938** (6), 1–136. [Including bibliography]

Lynge, B. (1940). Et bidrag til Spitzbergens lavflora. *Skr. Svalbard Ishavet* **79,** 1–22.

Mattick, F. (1950). Die Flechten Spitsbergens. *Polarforschung* **2,** 261–273.

Sweden

See also Scandinavia.

Almborn, O. (1952). Key to the sterile corticolous crustaceous lichens occurring in south Sweden. *Bot. Notiser* **1952**, 239–263.

Degelius, G. (1939). Die Flechten von Norra Skaftön. *Uppsala Univ. Årsskr.* **1939** (11), 1–206.

Hasselrot, T. E. (1953). Nordliga lavar i Syd- och Mellansverige. *Acta phytogeogr. suec.* **33**, i–vii, 1–200.

Krok, Th. O. B. N. and Almqvist, S., eds. (1969). "Svenska flora för skolor. II. Kryptogamer utom ormbunkväxter" 8th ed., (E. Almqvist, ed.), 390 pp. Svenska Bokförlaget, Stockholm. [Lichens pp. 124–177 ed. O. Almborn; keys; descriptions; lst ed. published 1917]

Magnusson, A. H. (1923–1955). New or interesting Swedish lichens I–XV. *Bot. Notiser* **1923**, 401–416; **1924**, 377–391; **1926**, 237; **1927**, 115–127; **1929**, 110–122; **1930**, 459–476; **1932**, 417–444; **1934**, 457–479; **1937**, 124–140; **1939**, 302–314; **1942**, 1–18; **1945**, 304–314; **1948**, 401–412; **1951**, 64–82; **1955**, 292–306.

Magnusson, A. H. (1952). Lichens from Torne Lappmark. *Ark. Bot., ser.* 2, **2** (2), 45–249. [Keys; 790 spp.]

Santesson, R. (1939). Über die Zonationverhältnisse der lakustrinen Flechten einiger Seen in Anebodagebiet. *Medd. Lunds Univ. Limnol. Inst.* **1**, 1–70.

Ursing, B. (1971). "Svenska växter i text och bild. Kryptogamer", 530 pp. Nordisk rotogravyr, Stockholm. [Lichens pp. 131–205, ed. R. Santesson; keys; descriptions; coloured illustrations; 1st ed. published 1949]

Switzerland

Frey, E. (1952–1959). Die Flechtenflora und -Vegetation des Nationalparks im Unterengadin I–II. *Ergebn. wiss. Unters. schweiz. NatnParks* **3**, 361–503 (1952) [I. Die diskokarpen Blatt- und Strauchflechten]; **6**, 241–319 (1959) [II. Die Entwicklung der Flechtenvegetation auf photogrammetrisch Kontrollierten.].

Frey, E. (1959–1963). Beiträge zu einer Lichenenflora der Schweiz I–III. *Ber. schweiz. bot. Ges.* **69**, 156–345 (1959) [I. Cladoniaceae, Parmeliaceae]; **73**, 389–503 (1963) [II. III. Die Familie Physciaceae].

Ochsner, F. (1928). Studien über die Epiphyten-Vegetation der Schweiz. *Ber. Tät. St. Gall. naturw. Ges.* **63**, 1–108.

Stizenberger, E. (1882–1883). Lichenes helvetici eorumque stationes et distributio. *Ber. Tät. St. Gall. naturw. Ges.* **22**, 255–522 (1882); **23**, 201–327 (1883).

United Kingdom

See also Ireland.

(a)

Duncan, U. K. (1970). "Introduction to British Lichens", 366 pp. Buncle, Arbroath. [Keys; descriptions; bibliography; illustrations]

Hawksworth, D. L. (1970). Guide to the literature for the identification of British lichens. *Bull. Br. mycol. Soc.* **4**, 73–95. [Papers listed under generic names; also issued separately interleaved]
Hawksworth, D. L. and Seaward, M. R. D. (1977). "Lichenology in the British Isles 1568–1975", 244 pp. Richmond Publishing, Richmond. [Bibliography]
James, P. W. (1965). A new check-list of British lichens. *Lichenologist* **3**, 95–153. [Also issued separately with interleaves]
James, P. W. (1966). A new check-list of British lichens—Additions and corrections I. *Lichenologist* **3**, 242–247.
James, P. W. (1971). New or interesting British lichens: 1. *Lichenologist* **5**, 114–148.
Smith, A. L. (1918–1926). "A Monograph of the British Lichens". 2nd ed., 2 vols (1918, 1926), 519, 447 pp. British Museum (Natural History), London. [Descriptions; illustrations; the most recent "definitive" flora]
Watson, W. (1953). "Census Catalogue of British Lichens", 110 pp. British Mycological Society, London. [Distribution by vice-county; bibliography]

(b)

Hawksworth, D. L. (1972). Regional studies in *Alectoria* (Lichenes) II. The British species. *Lichenologist* **5**, 181–261.
Sheard, J. W. (1964). The genus *Buellia* de Notaris in the British Isles [excluding section *Diploica* (Massal.) Stiz.]. *Lichenologist* **2**, 225–262.
Sheard, J. W. (1967). A revision of the lichen genus *Rinodina* (Ach.) Gray in the British Isles. *Lichenologist* **3**, 328–367.
Swinscow, T. D. V. (1960–1971). Pyrenocarpous lichens: 1–15. *Lichenologist* **1**, 169–178 (1960), 242–250 (1961); **2**, 6–56 (1962) [3. The genus *Porina* in the British Isles], 156–166 [4. Guide to the British species of *Staurothele*], 167–171 (1963), 276–283 (1964); **3**, 42–54 (1965) [7. On the genus *Microglaena* Körb.; with G. M.-Jones], 55–64 (1965) [8. The marine species of *Arthopyrenia* in the British Isles], 72–83 (1965) [9. Notes on various species], 233–235 (1966), 415–417 (1967) [11. A new species of *Arthopyrenia*], 418–422 (1967) [12. The genus *Geisleria* Nitschke]; **4**, 34–54 (1968) [13. Freshwater species of *Verrucaria* in the British Isles], 218–233 (1970) [14. *Arthopyrenia* Massal. sect. *Acrocordia* (Massal.) Müll. Arg. in the British Isles]; **5**, 92–112 (1971) [15. Key to *Polyblastia* Massal. in the British Isles]
Wade, A. E. (1965). The genus *Caloplaca* Th. Fr. in the British Isles. *Lichenologist* **3**, 1–28.

(c)

See also Chapter 10 in this volume.
British Lichen Society (1973–1977). Distribution maps of lichens in Britain. *Lichenologist* **5**, 464–480 (1973); **6**, 169–199 (1974); **7**, 180–192 (1975); **9**, 175–187 (1977). [Maps numbered serially; various authors]
Coppins, B. J. (1976). Distribution patterns shown by epiphytic lichens in the British Isles. *In* "Lichenology: Progress and Problems" (D. H. Brown,

D. L. Hawksworth and R. H. Bailey, eds), pp. 249–278. Academic Press, London and New York.

Fletcher, A. (1975). Keys to marine and maritime lichens in the British Isles I–II. *Lichenologist* **7**, 1–52 [I. Siliceous rocky shore species], 73–115 [II. Calcareous and terricolous species]

Hawksworth, D. L., Coppins, B. J. and Rose, F. (1974). Changes in the British lichen flora. *In* "The Changing Flora and Fauna of Britain" (D. L. Hawksworth, ed.), pp. 47–78. Academic Press, London and New York.

James, P. W. (1970). The lichen flora of shaded acid rock crevices and overhangs in Britain. *Lichenologist* **4**, 309–322.

Laundon, J. R. (1962–1963). The taxonomy of sterile crustaceous lichens in the British Isles. 1–2. *Lichenologist* **2**, 57–67 (1962) [1. Terricolous species]; **2**, 101–151 (1963) [2. Corticolous and lignicolous species].

Rose, F. and James, P. W. (1974). Regional studies on the British lichen flora I. The corticolous and lignicolous species of the New Forest, Hampshire. *Lichenologist* **6**, 1–72.

Yugoslavia

Hofmann, F., Nowak, R. and Winkler, S. (1974). Substrate dependence of calcareous and silicate rock inhabiting lichens of the island Ciovo, Yugoslavia. *J. Hattori bot. Lab.* **38**, 313–325.

Kušan, F. (1953). "Prodromus Flore Lišaja Jugoslavije", 595 pp. Jugoslovenska Akademija Znanosti i Umjetnosti, Zagreb. [Check-list 1159 spp.; bibliography]

Pišút, I. (1967). Notizen zur Flechtenflora Mazedoniens. *Fragm. Balcanica, Skopje* **6**, 53–56. [Including bibliography]

G. North America

(a)

Anderson, R. A. (1974). Additions to the lichen flora of North America—III. *Bryologist* **77**, 41–47.

Hale, M. E. (1969). "How to Know the Lichens", 226 pp. Brown, Dubuque, Iowa. [Keys; illustrations; maps; macrolichens only]

Hale, M. E. and Culberson, W. L. (1970). A fourth check-list of the lichens of the continental United States and Canada. *Bryologist* **73**, 499–543. [2735 spp.]

(b)

Ahti, T. (1969). Notes on brown species of *Parmelia* in North America. *Bryologist* **72**, 233–239.

Brodo, I. M. and Hawksworth, D. L. (1977). *Alectoria* and allied genera in North America. *Opera bot. Soc. bot. Lund.* **42**, 1–164.

Culberson, W. L. (1964) ["1963"]. A summary of the lichen genus *Haematomma* in North America. *Bryologist* **66**, 224–237.

Culberson, W. L. and Culberson, C. F. (1967). A new taxonomy for the *Cetraria ciliaris* group. *Bryologist* **70**, 158–166.

Esslinger, T. L. (1973). Chemical and taxonomic studies on some corticolous members of the lichen genus *Cetraria* in western North America. *Mycologia* **65**, 602–613.

Harris, R. C. (1973). The corticolous pyrenolichens of the Great Lakes Region. *Michigan Bot.* **12**, 3–68.

Henssen, A. (1963). The North American species of *Placynthium*. *Can. J. Bot.* **41**, 1687–1724.

Howard, G. E. (1970). The lichen genus *Ochrolechia* in North America north of Mexico. *Bryologist* **73**, 93–130.

Imshaug, H. A. (1957). The lichen genus *Pyxine* in North and Middle America. *Trans. Am. microsc. Soc.* **76**, 246–269.

Imshaug, H. A. and Brodo, I. M. (1966). Biosystematic studies on *Lecanora pallida* and some related lichens in the Americas. *Nova Hedwigia* **12**, 1–59.

Jordan, W. P. (1973). The genus *Lobaria* in North America north of Mexico. *Bryologist* **76**, 225–251.

Magnusson, A. H. (1944). Some species of *Caloplaca* from North America. *Bot. Notiser* **1944**, 63–79.

Sheard, J. W. (1974). The genus *Dimelaena* in North America north of Mexico. *Bryologist* **77**, 128–141.

Sierk, H. A. (1964). The genus *Leptogium* in North America north of Mexico. *Bryologist* **67**, 245–317.

Thomson, J. W. (1950). The species of *Peltigera* of North America north of Mexico. *Am. Midl. Nat.* **44**, 1–68.

Thomson, J. W. (1963). The lichen genus *Physcia* in North America. *Beih. Nova Hedwigia* **7**, 1–172.

Thomson, J. W. (1967). The lichen genus *Baeomyces* in North America north of Mexico. *Bryologist* **70**, 285–298.

Thomson, J. W. (1968) ["1967"]. "The Lichen Genus *Cladonia* in North America", 172 pp. University of Toronto Press, Toronto.

Tibell, L. (1975). The Caliciales of boreal North America. *Symb. bot. upsal.* **21** (2), 1–128.

Wetmore, C. M. (1960). The lichen genus *Nephroma* in North and Middle America. *Publs Mich. St. Univ. Mus., ser. biol.* **1** (11), 369–452.

Wetmore, C. M. (1970). The lichen family Heppiaceae in North America. *Ann. Mo. bot. Gdn* **57**, 158–209.

(c)

Imshaug, H. A. (1957). Alpine lichens of western United States and adjacent Canada. I. The macrolichens. *Bryologist* **60**, 177–272.

Phillips, W. L. and Stuckey, R. L. (1976). "Index to Plant Distribution Maps in North American Periodicals through 1972", xxvii + 686 pp. G. K. Hall, Boston.

Thomson, J. W. (1972). Distribution patterns of American arctic lichens. *Can. J. Bot.* **50**, 1135–1156.

Canada

Alberta

Bird, C. D. (1972). "A Catalogue of the Lichens reported from Alberta, Saskatchewan and Manitoba". Revised edition, 49 pp. University of Calgary Department of Biology, Calgary. [Check-list 542 spp.; bibliography; mimeographed publication]

Bird, C. D. (1973). Species collected in Alberta on the first 1971 foray of the American Bryological and Lichenological Society. Part I. Introduction and lichens. *Bryologist* **76**, 388–402.

Bird, C. D. and Marsh, A. H. (1972). Phytogeography and ecology of the lichen family Cladoniaceae in southwestern Alberta. *Can. J. Bot.* **50**, 915–933.

Bird, C. D. and Marsh, A. H. (1973a). Phytogeography and ecology of the lichen family Umbilicariaceae in southwestern Alberta. *Can. J. Bot.* **51**, 2169–2175.

Bird, C. D. and Marsh, A. H. (1973b). Phytogeography and ecology of the lichen family Parmeliaceae in southwestern Alberta. *Can. J. Bot.* **51**, 261–288.

British Columbia

Bird, C. D. and Bird, R. D. (1973). Lichens of Saltspring Island, British Columbia. *Syesis* **6**, 57–80.

Otto, G. F. and Ahti, T. (1967). "Lichens of British Columbia, Preliminary Checklist", 40 pp. University of British Columbia, Vancouver. [Check-list 569 spp.; bibliography; mimeographed publication]

Manitoba

See also Bird (1972) under Alberta.

Stringer, P. W. and Stringer, N. H. L. (1974). Seventeen lichens new to Manitoba. *Bryologist* **77**, 243–245.

Newfoundland

Ahti, T. (1974). Notes on the lichens of Newfoundland. 3. Lichenological exploration. *Ann. bot. fenn.* **11**, 89–93. [Bibliography]

Northwest Territories

Ahti, T., Scotter, G. W. and Vänskä, H. (1973). Lichens of the Reindeer Preserve, Northwest Territories, Canada. *Bryologist* **76**, 48–76.

Barrett, P. E. and Thomson, J. W. (1975). Lichens from a high arctic coastal lowland, Devon Island, N.W.T. *Bryologist* **78**, 160–167.

Hale, M. E. (1954). Lichens from Baffin Island. *Am. Midl. Nat.* **51**, 232–264.

Lynge, B. (1947). Lichenes. *Bull. natn. Mus. Can.* **97**, 298–369. [Canadian eastern Arctic]

Thomson, J. W. (1970). Lichens from the vicinity of Coppermine, Northwest Territories. *Can. Fld Nat.* **84,** 155–164.

Thomson, J. W., Scotter, G. W. and Ahti, T. (1969). Lichens of the Great Slave Lake Region, Northwest Territories, Canada. *Bryologist* **72,** 137–177.

Nova Scotia

Lamb, I. M. (1954). Lichens of Cape Breton Island, Nova Scotia. *A. Rept. natn. Mus. Can. Bull.* **132,** 239–313.

McDonald, J. (1973). A catalogue of the A. H. MacKay lichen collection with a short biography of A. H. MacKay. *Curatorial Rept. Nova Scotia Mus.* **16,** 1–13. [Mimeographed; reference supplied by Dr T. Ahti]

Ontario

Ahti, T. (1964). Macrolichens and their zonal distribution in boreal and arctic Ontario. *Ann. bot. fenn.* **1,** 1–35.

Brodo, I. M. (1967–1972). Lichens of the Ottawa area. I–III. *Trail & Landscape* **1,** 40–45, 114–115, 118–121; **6,** 15–26. [Keys, including crustose spp.; line drawings]

Québec

Lepage, E. (1972). Nouveau catalogue des lichens du Québec. *Naturaliste can.* **99,** 533–550.

Saskatchewan

See Bird (1972) under Alberta.

Yukon

Bird, C. D. (1967). "A Catalogue of the Lichens reported from the Yukon", 16 pp. University of Calgary Department of Biology, Calgary. [Check-list 145 spp.; bibliography; mimeographed]

Hoefs, M. and Thomson, J. W. (1972). Lichens from the Kluane Game Sanctuary, S.W. Yukon Territory. *Can. Fld Nat.* **86,** 249–252.

St Pierre and Miquelon

See also Lepage (1972) under Québec.

Le Gallo, P. C. (1952). Lichens des îles Saint-Pierre et Miquelon (Première série). *Revue bryol. lichén.* **21,** 144–172.

U.S.A.

Culberson, W. L. (1955). A guide to the literature on the lichen flora and vegetation of the United States. *U.S. Dept. Agric., Agric. Res. Serv., Pl. Dis. Sect., Sp. Publ.* **7,** 1–54. [Comprehensive bibliography with titles listed under names of states]

Fink, B. (1935). "The Lichen Flora of the United States", 426 pp. University of Michigan Press, Ann Arbor. [Keys; descriptions; illustrations; the most recent "comprehensive" flora]

Hale, M. E. (1961). "Lichen Handbook, A Guide to the Lichens of Eastern North America", 178 pp. Smithsonian Institution, Washington, DC. [Including keys spp. macrolichens, gen. crustose lichens; illustrations]

Nearing, G. G. (1947). "The Lichen Book, Handbook of the Lichens of Northeastern United States", 648 pp. Privately printed, Ridgewood, N.J. [Chartkeys; descriptions; illustrations]

Alabama

Mohr, C. (1901). Plant life of Alabama. *Contr. U.S. natn. Herb.* **6,** 1–912. [Lichens pp. 263–284]

Alaska

Cummings, C. E. (1904). The lichens of Alaska. *In* "Harriman Alaska Expedition" (J. Cardot *et al.*, eds), Vol. 5, pp. 67–152. Doubleday, Page and Co., New York. [Keys]

Degelius, G. (1937). Lichens from southern Alaska and the Aleutian Islands, collected by Dr. E. Hultén. *Medd. Göteborgs Bot. Trädg.* **12,** 105–144.

Krog, H. (1968). The macrolichens of Alaska. *Skr. norsk. Polarinst.* **144,** 1–180.

Murray, B. M. (1974). "Catalog of Bryophytes and Lichens of central Brooks Range, Alaska. A literature review", 46 pp. University of Alaska Museum, Anchorage. [Mimeographed; not seen, reference supplied by Dr T. Ahti]

Arizona

Nash, T. H. III (1975). Lichens of Maricopa County, Arizona. *J. Arizona Acad. Sci.* **10,** 119–125. [Including keys]

Nash, T. H. III and Johnsen, A. B. (1975). Catalog of the lichens of Arizona. *Bryologist* **78,** 7–24.

Arkansas

Hale, M. E. (1957). Corticolous lichen flora of the Ozark Mountains. *Trans. Kansas Acad. Sci.* **60,** 155–160.

California

Hasse, H. E. (1913). The lichen flora of southern California. *Contr. U.S. natn. Herb.* **17**, i–xii, 1–132.

Herre, A. W. C. T. (1910). The lichen flora of the Santa Cruz Peninsula, California. *Proc. Wash. Acad. Sci.* **12**, 27 269.

Sigal, L. L. and Toren, D. R. (1974). New distribution records of lichens in California. *Bryologist* **77**, 469–470.

Tucker, S. C. and Kowalski, D. T. (1975). New state records of lichens from northern California. *Bryologist* **78**, 366–368.

Colorado

Anderson, R. A. and Carmer, M.-B. (1974). Additions to the lichen flora of Colorado. *Bryologist* **77**, 216–222.

Carmer, M.-E. (1975). Corticolous lichens of riparian deciduous trees in the Central Front Range of Colorado. *Bryologist* **78**, 44–56.

Shushan, S. and Anderson, R. A. (1970). Catalog of the lichens of Colorado. *Bryologist* **72** (1969), 451–483.

Connecticut

Evans, A. W. (1950). Notes on the *Cladoniae* of Connecticut IV. *Rhodora* **52**, 77–123.

Evans, A. W. and Meyrowitz, R. (1926). Catalogue of the lichens of Connecticut. *Bull. Conn. St. geol. nat. Hist. Surv.* **37**, 1–56.

Florida

Eckfeldt, J. W. and Calkins, W. W. (1887). The lichen-flora of Florida. *J. mycol.* **3**, 121–126.

Moore, B. (1968). The macrolichen flora of Florida. *Bryologist* **71**, 161–266.

Hawaii

See under Australasia and Pacific.

Idaho

Schroeder, N. E. and Schroeder, G. J. (1975). Catalog of the lichens of Idaho. *Bryologist* **78**, 32–43.

Illinois

Skorepa, A. C. (1970). Lichenological records from central and northern Illinois. *Trans. Ill. St. Acad. Sci.* **63**, 78–82. [Including bibliography]

Indiana

Herre, A. W. C. T. (1944). Lichens known from Indiana. *Proc. Indiana Acad. Sci.* **53** (1943), 81–95.

Iowa

Greene, W. (1907). "Plants of Iowa", 264 pp. Iowa State Horticultural Society, Des Moines, Iowa. [Lichens pp. 95–110]

Maine

Degelius, G. (1940). Contributions to the lichen flora of North America I. Lichens from Maine. *Ark. Bot.* **30A**(1), 1–62.

Massachusetts

Weaver, R. E. (1975). Lichens: mysterious and diverse. *Arnoldia* **35**, 133–159. [Keys and photographs of macrolichens]

Michigan

See also Harris (1973) under North America (a).

Hedrick, J. (1940). Lichens of northern Michigan. *Pap. Mich. Acad. Sci.* **25** (1939), 47–65.
Hedrick, J. and Lowe, J. L. (1936). Lichens from Isle Royale, Lake Superior. *Bryologist* **39**, 73–91.
Thomson, J. W. (1951). Some lichens from the Keweenaw Peninsula, Michigan. *Bryologist* **54**, 17–53.

Minnesota

Fink, B. (1910). The lichens of Minnesota. *Contr. U.S. natn. Herb.* **14**, i–xvii, 1–269. [Keys; descriptions; illustrations]

Mississippi

Rogers, K. E. and Skorepa, A. C. (1974). Notes on the lichens of Mississippi. *J. Mississippi Acad. Sci.* **19**, 148–153.

Missouri

Gier, L. J. and Kendrick, J. (1973). Missouri lichens. *Trans. Kansas Acad. Sci.* **75**, 207–217.

Nebraska

Webber, H. J. (1890). Catalogue of the flora of Nebraska. *Rept. Nebraska St. Bd. Agr.* **1889**, 37–162. [Not seen; lichens pp. 54–59]

New Jersey

Britton, N. L. (1899). Catalogue of plants found in New Jersey. *Final Rep. geol. Surv. New Jers.* **2** (1), 27–642. [Lichens pp. 357–384]

New Mexico

Egan, R. S. (1972). Catalog of the lichens of New Mexico. *Bryologist* **75**, 17–35.

New York

Brodo, I. M. (1968). The lichens of Long Island, New York: A vegetational and floristic survey. *Bull. N.Y. St. Mus. Sci. Serv.* **410**, i–x, 1–330. [Including keys]

North Carolina

Degelius, G. (1941). Contributions to the lichen flora of North America. II. The lichen flora of the Great Smoky Mountains. *Ark. Bot.* **30A** (3), 1–80.

Ohio

Showman, R. E. (1973). The foliose and fruticose lichen flora of the Ohio River Valley between Gallipolis, Ohio, and Parkerburg, West Virginia. *Ohio J.Sci.* **73**, 357–363.
Showman, R. E. (1977). Additions to the lichen flora of Ohio. *Ohio J. Sci.* **77**, 26–27.
Taylor, J. C. (1967). Lichens of Ohio. Part 1. Foliose lichens. *Biol. Notes Ohio Biol. Surv.* **3**, i–iv, 1–151. [Keys; descriptions; illustrations; maps]
Taylor, J. C. (1968). Lichens of Ohio. Part 2. Fruticose and cladoniiform lichens. *Biol. Notes Ohio Biol. Surv.* **4**, 153–227, 1–22. [Keys; descriptions; illustrations; maps]

Oklahoma

Thomson, J. W. (1961). Lichens collected in Oklahoma at the time of the American Bryological Society meetings. *Bryologist* **64**, 255–262.

Oregon

Magnusson, A. H. (1939). Western American lichens, mainly from Oregon. *Medd. Göteborgs Bot. Trädg.* **13**, 237–253.

South Dakota

Wetmore, C. M. (1968). Lichens of the Black Hills of South Dakota and Wyoming. *Publs Mich. St. Univ. Mus., biol. ser.* **3** (4), 209–464. [Including keys]

Tennessee

See also Degelius (1941) under North Carolina.

Phillips, H. C. (1963). Foliose and fruticose lichens from Tennessee. *Bryologist* **66**, 77–79.

Phillips, H. C. (1970). An annotated list of foliose and fruticose lichens in Land Between the Lakes. *J. Tennessee Acad. Sci.* **45**, 97–109. [Including brief descriptions of 82 spp.]

Phillips, H. C. (1974). "Lichens and Ferns of Land Between the Lakes", 60 pp. Tennessee Valley Authority, Clarksville. [Including coloured plates of 47 macrolichens]

Texas

Egan, R. S. (1977). New and additional lichen records from Texas II. *Bryologist* **80**, 136–142.

Wetmore, C. M. (1976). Macrolichens of Big Bend National Park, Texas. *Bryologist* **79**, 296–313.

Utah

Flowers, E. (1954). Some lichens of Utah. *Proc. Utah Acad. Sci.* **31**, 101–105.

Washington

Howard, G. E. (1950). "Lichens of the State of Washington", 191 pp. University of Washington Press, Seattle. [Keys; descriptions]

Thomson, J. W. (1969). "A Catalogue of Lichens of the State of Washington", 59 pp. Wisconsin. [Check-list; bibliography]

West Virginia

See Showman (1973) under Ohio.

Wisconsin

Brodo, I. M. (1967). Lichens collected in Wisconsin on the 1965 foray of the American Bryological Society. *Bryologist* **70**, 208–227. [Including bibliography]

Wyoming

See under South Dakota.

H. Central America and West Indies

Imshaug, H. A. (1959). Catalogue of Central American lichens. *Bryologist* **59**, 69–114. [Mexico 997 spp.; Central America 635 spp.; West Indies 1752 spp.]

Bahamas

See also West Indies.

Riddle, L. W. (1920). Lichens. *In* "The Bahama Flora" (N. L. Britton and C. F. Millspaugh, eds), pp. 522–553. Privately printed, New York. [Keys]

Bermuda

See West Indies.

British Honduras

Hedrick, J. (1939). Lichens from British Honduras collected by E. B. Mains. *Pap. Mich. Acad. Sci.* **24**, 9–15. [Including bibliography]

Costa Rica

Dodge, C. W. (1933). The foliose and fruticose lichens of Costa Rica. I. *Ann. Mo. bot. Gdn* **20**, 373–467.

Cuba

See also West Indies.

Vězda, A. and Samek, V. (1972). Cladonien der Kiefernwälder im NO der Provinz Oriente, Cuba. *Acta mus. siles.*, **A, 21,** 1–18.

Dominica

See also West Indies.

Hale, M. E. (1971). Morden-Smithsonian Expedition to Dominica: the lichens (Parmeliaceae). *Smithson. Contr. bot.* **4,** 1–25.
Hale, M. E. (1974). Morden-Smithsonian Expedition to Dominica: the lichens (Thelotremataceae). *Smithson. Contr. bot.* **16,** i–iii, 1–46.

El Salvador

Nowak, R. and Winkler, S. (1972). Foliicole Flechten von El Salvador, C.A. *Revue bryol. lichén.* **38,** 269–279.

Guadeloupe

See West Indies.

Jamaica

See also West Indies.

Dix, W. L. (1957). Jamaican lichens. Some unreported collections. *Bryologist* **60,** 154–165.

Mexico

Brizuela, F. and Guzmán, G. (1971). Estudios sobre los líquenes de México, II. *Boln Soc. Mexic. micol.* **5,** 79–103.

González de la Rosa, M. E. and Guzmán, G. (1976). Estudios sobre los liquenes de México, III. Observaciones sobre especies no consideradas anteriormente. *Boln Soc. Mexic. micol.* **10,** 27–64.

Guzmán, L. D. de, Brizuela, F. and Guzmán, Y. G. (1972). Estudios sobre los líquenes de México, I. Notas sobre algunas especies. *An. Esc. nac. Cienc. biol., Méx.* **19,** 9–30. [Including bibliography]

Imshaug, H. A. (1956). Catalogue of Mexican lichens. *Revue bryol. lichén.* **25,** 321–385. [Check-list; bibliography]

Wirth, M. and Hale, M. E. (1963). The lichen family Graphideaceae in Mexico. *Contr. U.S. natn. Herb.* **36,** 63–119.

Puerto Rico

See West Indies.

Trinidad

See West Indies.

West Indies

Imshaug, H. A. (1955). The lichen genus *Buellia* in the West Indies. *Farlowia* **4,** 473–512.

Imshaug, H. A. (1957). Catalogue of West Indian lichens. *Bull. Inst. Jamaica, sci. ser.* **6,** 1–153. [Check-list 1751 spp.; bibliography; history of studies in each island]

Taylor, R. M. (1972). The lichen genus *Baeomyces* in the West Indies. *J. Hattori bot. Lab.* **35,** 303–311.

Welch, W. H. and Crum, H. A. (1971). Recent cryptogamic collections in the West Indies. *Revue bryol. lichén.* **37** (1970), 223–235.

I. South America

The following series of papers is based essentially on material collected by the Regnellian Expedition of 1892–1894 which visited most of the main South American countries. Many contributions include keys and detailed descriptions.

Lynge, B. (1914). Die Flechten der ersten Regnellschen Expedition. Die Gattungen *Pseudoparmelia* gen. nov. und *Parmelia* Ach. *Ark. Bot.* **13** (13), 1–172.
Lynge, B. (1924a). On South American *Anaptychiae* and *Physciae*. *Skr. Norske Vid.-Akad. Oslo, mat.-nat. kl.* **1924** (16), 1–47.
Lynge, B. (1924b). On some South American lichens of the genera *Parmelia, Candelaria, Theloschistes* and *Pyxine. Nyt Mag. Naturv.* **62**, 83–97.
Malme, G. O. A. (1897). Die Flechten der ersten Regnellschen Expedition. I. Die Gattung *Pyxine* (Fr.) Nyl. *Bih. K. Svenska Vet. Akad. Handl.*, III, **23** (13), 1–52.
Malme, G. O. A. (1902). Die Flechten der ersten Regnellschen Expedition. II. Die Gattung *Rinodina* (Ach.) Stizenb. *Bih. K. Svenska Vet. Akad. Handl.*, III, **28** (1), 1–53.
Malme, G. O. A. (1924). Die Collematazeen des Regnellschen Herbars. *Ark. Bot.* **19B** (8), 1–29.
Malme, G. O. A. (1925). Die Pannariazeen des Regnellschen Herbars. *Ark. Bot.* **20A** (3), 1–23.
Malme, G. O. A. (1926). Lichenes blastenospori Herbarii Regnelliani. *Ark. Bot.* **20A** (9), 1–51.
Malme, G. O. A. (1928). Lichenes pyrenocarpi aliquot in Herbario Regnelliano asservati. *Ark. Bot.* **22A** (6), 1–11.
Malme, G. O. A. (1927). *Buelliae* itineris Regnelliani primi. *Ark. Bot.* **21A** (14), 1–42.
Malme, G. O. A. (1929a). *Pyrenulae* et *Anthracothecia* Herbarii Regnelliani. *Ark. Bot.* **22A** (11), 1–40.
Malme, G. O. A. (1929b). *Porinae* et *Phylloporinae* in Itinere Regnelliano primo collectae. *Ark. Bot.* **23A** (1), 1–37.
Malme, G. O. A. (1934a). Die Ramalinen der ersten Regnellschen Expedition. *Ark. Bot.* **26A** (12), 1–9.
Malme, G. O. A. (1934b). Die Gyalectazeen der ersten Regnellschen Expedition. *Ark. Bot.* **26A** (13), 1–10.
Malme, G. O. A. (1934c). Die Stictazeen der ersten Regnellschen Expedition. *Ark. Bot.* **26A** (14), 1–18.
Malme, G. O. A. (1935). *Bacidiae* itineris Regnelliani primi. *Ark. Bot.* **27A** (5), 1–40.
Malme, G. O. A. (1936a). *Lecideae* Expeditionis Regnellianae primae. *Ark. Bot.* **28A** (7), 1–53.
Malme, G. O. A. (1936b). *Pertusariae* Expeditionis Regnellianae primae. *Ark. Bot.* **28A** (9), 1–27.

Malme, G. O. A. (1940). Lichenes nonnulli in Expeditione Regnelliana prima collecti. *Ark. Bot.* **29A** (6), 1–35.

Redinger, K. (1933–1940). Die Graphidineen Flechten der ersten Regnell'schen Expedition nach Brasilien 1892–94. I–IV. *Ark. Bot.* **25A** (13), 1–20 (1933) [I. *Glyphis, Medusulina* und *Sarcographa*]; **26A** (1), 1–105 (1934) [II. *Graphina* und *Phaeographina*]; **27A** (3), 1–103 (1935) [III. *Graphis* und *Phaeographis* nebst einem Nachrage zu *Graphina*]; **29A** (19), 1–52 (1940) [IV. *Opegrapha*].

Redinger, K. (1936). Thelotremaceae brasilienses. *Ark. Bot.* **28A** (8), 1–122.

Santesson, R. (1942a). The South American *Cladinae*. *Ark. Bot.* **30A** (10), 1–27.

Santesson, R. (1942b). The South American *Menegazziae*. *Ark. Bot.* **30A** (11), 1–8.

Santesson, R. (1943). South American *Calicia* collected during the first Regnellian Expedition in 1892–94. *Ark. Bot.* **30A** (14), 1–12.

Santesson, R. (1944). Contribution to the lichen flora of South America. *Ark. Bot.* **31A** (7), 1–28.

Argentina

Grassi, M. M. (1952) ["1950"]. Contribución al catálogo de líquenes Argentinos, I. *Lilloa* **24**, 5–294. [Check-list 942 spp.; bibliography]

Lamb, I. M. (1955). New lichens from Northern Patagonia, with notes on some related species. *Farlowia* **4**, 423–471.

Lamb, I. M. (1958). La vegetación liquénica de los Parques Nacionales Patagónicos. *An. Parq. nac., B. Aires* **7**, 3–188. [Including keys; descriptions]

Osorio, H. S. (1968–1977). Contributions to the lichen flora of Argentina—I–IX. *Bryologist* **71**, 285–286 (1968) [I. Some lichens from the province of Buenos Aires]; *Comun. bot. Mus. Hist. nat. Montev.* **4** (48), 1–5 (1969a) [II. Lichens from the Province of Misiones]; *Bryologist* **72**, 409–410 (1969b) [III. Additions]; **73**, 392–394 (1970) [IV. New or additional records]; *Comun. bot. Mus. Hist. nat. Montev.* **4**, (54), 1–2 (1970) [V. Some new records]; *Revue bryol. lichén.* **41**, 83–85 (1975) [VI. Lichens from Concordia, Entre Rios Province]; *Comun. bot. Mus. Hist. nat. Montev.* **4** (57), 1–4 (1975) [and L. I. Ferraro, VII. New and noteworthy records from the Province of Corrientes]; *Bryologist* **79**, 358–360 (1976) [VIII. Lichens from Punta Lara, Buenos Aires Province]; *Mycotaxon* **4**, 331–334 (1977) [and L. I. Ferraro, IX. Some lichens from the Provinces of Santa Fé and Santiago de Estero]

Bolivia

Nylander, W. (1859). Lichenes in regionibus exoticis quibusdam vigentes exponit synopticis enumerationibus. *Annls Sci. nat., Bot., sér.* 4, **11**, 205–264.

Nylander, E. (1861). Additamentum ad Lichenographiam Andium Boliviensium. *Annls Sci. nat., Bot., sér.* 4, **15**, 365–382.

Brazil

Cengia Sambo, M. (1940). Licheni del Brasile. *Annali Bot.* **22**, 19–41.

Mattick, F. (1956). Auf den Spuren des Lichenologen Wainio in Brasilien: Das Carassa Gebirge. *Willdenowia* **1**, 404–432.
Osorio, H. S. (1974). Contribution to the lichen flora of Brazil. 1. New or additional records. *Revta Fac. Sci. Univ. Lisboa*, C, **17**, 447–450.
Redinger, K. (1933). Neue und wenig bekannte Flechten aus Brasilien. *Hedwigia* **73**, 54–67.
Vainio, E. A. (1890). Étude sur la classification naturelle et la morphologie des lichens du Brésil. *Acta Soc. Fauna Flora fenn.* **7** (1), i–xxix, 1–247 (1890a) [Pars prima]; **7** (2), 1–256 (1890b) [Pars secunda]. [Including descriptions]
Zahlbruckner, A. (1902). Studien über brasilianische Flechten. *Sber. Akad. Wiss. Wien, mat.-nat. Kl.*, I, **111**, 357–432.

Chile

Follmann, G. (1961–1966). Catálogo de los líquenes de Chile. Parte I–IV. *Revta Universitaria [Universidad de Chile]* **46**, 173–203 (1961) [I]; **47**, 63–97 (1962) [II]; **49**, 17–65 (1964) [III]; **50/51**, 33–74 (1966) [IV]. [Check-list; bibliography]
Follmann, G. (1967). Die Flechtenflora der nordchilenischen Nebeloase Cerro Moreno. *Nova Hedwigia* **14**, 215–281.
Follmann, G. (1968). Felsbewohnende Arthoniaceen der chilenischen Pazifikküste. *Willdenowia* **4**, 365–382.
Follmann, G. and Redón, J. (1972). Ergänzungen zur Flechtenflora der nordchilenischen Nebeloasen Fray Jorge und Talinay. *Willdenowia* **6**, 431–460.
Gay, C. (1852). "Historia Física y Política de Chile, Botánica", Vol. 8. Museo de Historia Natural de Santiago, Paris and Santiago. [Lichens pp. 53–228 by C. Montagne; descriptions]
Redón, J. (1972). Beobachtungen zur Geographie und Ökologie der chilenischen Flechtenflora. *J. Hattori bot. Lab.* **37**, 153–167.
Redón, J. (1974). Observaciones sistemáticas y ecológicas en líquenes del Parque Nacional "Vicente Perez Rosales". *An. Mus. Hist. nat. Valparaiso* **7**, 169–225.
Redón, J. and Follmann, G. (1972). Beobachtungen zur Verbreitung chilenischer Flechten VI. Revision einiger Arten der Krustenflechtenfamilie Lecanactidaceae. *Philippia* **1**, 186–193.

Colombia

Lindau, G. (1912). Beitrag zur Kenntnis der Flechten von Kolumbien. *Mém. Soc. neuchât. Sci. nat.* **5**, 57–66.
Nowak, R. and Winkler, S. (1975). Foliicolous lichens of Chocó, Colombia, and their substrate abundance. *Lichenologist* **7**, 53–58.
Nylander, W. (1863). Lichenographiae novo-granatensis prodromus. *Acta Soc. Sci. fenn.* **7**, 415–504.
Nylander, W. (1867). Lichenes additamentum. *In* Prodromus florae Novo-Granatensis ou énumeration des plantes de la Nouvelle-Grenade avec description des espèces nouvelles (J. Triana and J. E. Planchon, eds). *Annls Sci. nat., Bot., sér.* 5, **7**, 301–354. [467 spp.]

Ecuador

Müller [Argoviensis], J. (1879). Les lichens Neo-Grenadins et Ecuadoriens, récoltés par M. Ed. André. *Revue mycol., Toulouse* **1**, 160–171.

Zahlbruckner, A. (1905). Flechten im Hochlande Ecuadors gesammelt von Prof. D. Hans Meyer im Jahr 1903. *Beih. bot. Zbl.* **19**, 75–84.

Falkland Islands

See also Grassi (1952) under Argentina.

Imshaug, H. A. (1968). Expedition to Falkland Islands, 1968. *Antarctic Jl U.S.* **4**, 247–248.

Galápagos Islands

Weber, W. A. (1966). Lichenology and bryology in the Galápagos Islands, with check list of the lichens and bryophytes thus far reported. *In* "The Galápagos" (R. I. Bowman, ed.), pp. 190–200. University of California Press, Berkeley and Los Angeles.

Paraguay

Osorio, H. S. (1970). Lichens from Cantera, south Paraguay. *Comun. bot. Mus. Hist. nat. Montev.* **4** (50), 1–3. [Including bibliography]

Patagonia

See Argentina.

Peru

De Castañeda, R. R. (1969). Lista de líquenes de la Provincia de Trujillo (Departamento La Libertad, Perú). *Bol. Soc. bot. Libertad* **1** (2), 45–54.

Herre, A. W. C. T. (1944). The South American lichens collected by the second University of California Botanical Garden Expedition to the Andes. *Revta Universtaria, Cuzco* **33** (87), 547–564.

Soukup, J. (1965). Lista de líquenes de Peru. *Biota, Lima* **6**, 28–45.

Tierra del Fuego

See also Grassi (1952) under Argentina.

Hawksworth, D. L. and Moore, D. M. (1969). Some lichens from Tierra del Fuego with notes on their chemical constituents. *Bryologist* **72**, 247–251. [Including bibliography of recent papers]

Uruguay

Osorio, H. S. (1972). Contribution to the lichen flora of Uruguay. VII. A preliminary catalogue. *Comun. bot. Mus. Hist. nat. Montev.* **4** (56), 1–46. [Check-list 367 spp.; bibliography]

Osorio, H. S. (1975). Contribution to the lichen flora of Uruguay. VIII. Additions and corrections. *Comun. bot. Mus. Hist. nat. Montev.* **4** (59), 1–12.

Venezuela

Dennis, R. W. G. (1965). Fungi Venezuelani: VII. *Kew Bull.* **19**, 231–273.

Dennis, R. W. G. (1970). Fungus flora of Venezuela and adjacent countries. *Kew Bull., add. ser.* **3**, i–xxxiv, 1–531. [Ecological notes]

Hertel, H. (1974). Krustenflechten aus Venezuela. *Mitt. bot. StSamml., Münch.* **11**, 405-430.

Reyes, C. R. (1974). Nota adicional acerca del Catálogo de los líquenes de Venezuela. *Bryologist* **77**, 248–249. [Foliicolous spp.]

Reyes, C. R. and Skorepa, A. (1974). Contribución a la flora liquenológica del Macarao, Venezuela I. *Bryologist* **77**, 257.

Vareschi, V. (1973). Resultados liquenológicos de excursiones efectuadas en Venezuela No. 3. Catálogo de los líquenes de Venezuela. *Acta bot. venez.* **8**, 177–245. [Check-list; bibliography]

J. Antarctica

Darbishire, O. V. (1923). Lichens. *Nat. Hist. Rep. Br. antarct. Terra Nova Exped., Bot.* **3**, 29–76. [Check-list 208 spp.]

Dodge, C. W. (1948). Lichens and lichen parasites. *Rep. B.A.N.Z. antarct. Exped.,* **B, 7,** 1–276. [Keys; descriptions]

Dodge, C. W. (1973). "Lichen Flora of the Antarctic Continent and Adjacent Islands", 399 pp. Phoenix Publishing, Canaan, New Hamps. [Keys; descriptions]

Dodge, C. W. and Baker, G. E. (1938). The second Byrd Antarctic Expedition—Islands", 399 pp. Phoenix Publishing, Canaan, New Hamps. [Keys; descrip-[Descriptions; illustrations]

Filson, R. (1966). The lichens and mosses of MacRobertson Land. *A.N.A.R.E. Sci. Rep.,* **B,** *bot.* **82,** 1–169. [Descriptions; illustrations]

Filson, R. (1974). Studies in Antarctic lichens II. Lichens from the Windmill Islands. *Muelleria* **3**, 9–36.

Golubkova, N. S. and Savicz, V. P. (1966). Lichenes familiae Umbilicariaceae e parte orientali Antarcticae. *Nov. Sist. Nizsh. Rast.* **3**, 257–263.

Golubkova, N. S. and Schapiro, I. A. (1970). De chemotaxonomia specierum antarcticum nonnullarum generis *Neuropogon* Nees et Fw. *Nov. Sist. Nizsh. Rast.* **7**, 277–282.

Golubkova, N. S. and Simonov, I. M. (1972). Lishaĭniki Oazisa Schirmakhera. *Trudy sov. Antarkt. Eksped.* **60**, 317–327.

Golubkova, N. S., Savicz V. P. and Simonov, I. M. (1968) Lishaĭniki zapadnoĭ chasti Zemli Enderbi. *Trudy sov. Antarkt. Eksped.* **38**, 247–253.

Kashiwadani, H. (1970). Lichens from the Prince Olav Coast, Antarctica. *J.A.R.E. Sci. Rep.*, **E, 30**, 1–21. [Descriptions; illustrations]

Lindsay, D. C. (1971, 1975). Notes on Antarctic lichens: IV, VIII. *Bull. Br. Antarct. Surv.* **25**, 99–100 (1971); **39**, 115–118 (1975).

Lindsay, D. C. (1972). Lichens from Vestfjella, Dronning Maud Land. *Meddr norsk Polarinst.* **101**, 1–21.

Murray, J. (1963). Lichens from Cape Hallett area, Antarctica. *Trans. R. Soc. N.Z., Bot.* **2**, 59–72.

Auckland Islands

See under New Zealand.

Bouvetøya

Holdgate, M. W., Tillbrook, P. J. and Vaughan, R. W. (1968). The biology of Bouvetøya. *Bull. Br. Antarct. Surv.* **15**, 1–7.

Falkland Islands

See under South America.

Kerguelen Islands

See under Africa.

Marion Island

See under Africa.

South Georgia

Darbishire, O. V. (1912). The lichens of the Swedish Antarctic Expedition. *Wiss. Ergebn. schwed. Südpolarexped.* **4** (11), 1–74.

Lindsay, D. C. (1974) ["1973"]. South Georgian microlichens: I. The genera *Buellia* and *Rinodina*. *Bull. Br. Antarct. Surv.* **37**, 81–89.

Lindsay, D. C. (1975a) ["1974"]. The macrolichens of South Georgia. *Br. Antarct. Surv. Sci. Rep.* **89**, 1–92. [Keys; descriptions; communities; illustrations; including some microlichens]

Lindsay, D. C. (1975b) ["1974"]. New taxa and new records of lichens from South Georgia. *Bull. Br. Antarct. Surv.* **39**, 13–20.

South Orkney Islands, South Sandwich Islands, South Shetland Islands and the Antarctic Peninsula

These areas are treated together as they appear to form a distinct phytogeographical element separate from continental Antarctica and the subantarctic islands (e.g. South Georgia).

Allison, J. S. and Smith, R. I. L. (1973). The vegetation of Elephant Island, South Shetland Islands. *Bull. Br. Antarct. Surv.* **33/34,** 185–212.
Follmann, G. (1965). Una asociación nitrófila de líquenes epipetricos de la Antártica Occidental con *Ramalina terebrata* Tayl. et Hook. como espécie caracterizante. *Publnes Inst. antart. chil.* **4,** 1–18.
Lamb, I. M. (1948). Antarctic pyrenocarp lichens. *'Discovery' Rep.* **25,** 1–30. [Keys; descriptions; illustrations]
Lamb, I. M. (1964). Antarctic lichens I. The genera *Usnea, Ramalina, Himantormia, Alectoria, Cornicularia. Br. Antarct. Surv. Sci. Rep.* **38,** 1–34. [Keys; descriptions; illustrations]
Lamb, I. M. (1968). Antarctic lichens II. The genera *Buellia* and *Rinodina. Br. Antarct. Surv. Sci. Rep.* **61,** 1–129. [Keys; descriptions; illustrations]
Lindsay, D. C. (1969). New records for Antarctic Umbilicariaceae. *Bull. Br. Antarct. Surv.* **21,** 61–69. [Key; descriptions]
Lindsay, D. C. (1971–1974) ["1971–1973"]. Notes on Antarctic lichens: I–III, V–VII. *Bull. Br. Antarct. Surv.* **24,** 11–19, 115–118, 119–120; **26,** 77–80; **28,** 43–48; **36,** 105–114.
Redón, J. (1969). Nueva asociación de líquenes muscicolas de la Antártica Occidental, con *Sphaerophorus tener* Laur., como espécie caracterizante. *Boln Inst. antart. chil.* **4,** 5–11. [Many spp.]
Smith, R. I. L. (1973) ["1972"]. Vegetation of the South Orkney Islands with special reference to Signy Island. *Br. Antarct. Surv. Sci. Rep.* **68,** 1–124.
Smith, R. I. L. and Corner, R. W. M. (1973). Vegetation of the Arthur Harbour —Argentine Islands region of the Antarctic Peninsula. *Bull. Br. Antarct. Surv.* **33/34,** 89–122.

III. World Monographs

There is no comprehensive list of all important papers on each genus available at the present time, although the most useful can be traced through the entries under generic names in Ainsworth (1971). A guide to those most valuable to British students (including pertinent non-British literature) is provided by Hawksworth (1970). Various regional monographs of particular groups have been cited in the preceding section of this Appendix (pp. 439–493) and many of these include detailed accounts of taxa also present in other areas. This section includes citations of some of the available world monographs or treatments covering other very large areas (i.e. not confined to a particular continent). As in the selection of regional works, preference is given here to publications dealing with macrolichens, although detailed accounts of some genera of microlichens are also included. The taxonomic and geographical areas dealt with in many of these will be clear from their titles.

Ahti, T. (1961). Taxonomic studies on reindeer lichens (*Cladonia*, subgenus *Cladina*). *Annls bot. Soc. zool.-bot. fenn. Vanamo* **32** (1), 1–160.

Ahti, T. (1966). *Parmelia olivacea* and the allied non-isidiate and non-sorediate corticolous lichens in the Northern Hemisphere. *Acta Bot. fenn.* **70,** 1–68.

Awasthi, D. D. (1975). A monograph of the lichen genus *Dirinaria*. *Biblthca lich.*, Lehre **2,** 1–120.

Bitter, G. (1901). Zur Morphologie und Systematik von *Parmelia*, Untergattung *Hypogymnia*. *Hedwigia* **40,** 171–274.

Culberson, W. L. and Culberson, C. F. (1968). The lichen genera *Cetrelia* and *Platismatia* (Parmeliaceae). *Contr. U.S. natn. Herb.* **34,** 449–558.

Darbishire, O. V. (1898). Monographia *Roccelleorum*. *Biblthca bot.* **45,** 1–103.

Degelius, G. (1974). The lichen genus *Collema* with special reference to the extra-European species. *Symb. bot. upsal.* **20** (2), 1–215.

Eigler, G. (1969). Studien zur Gliederung der Flechtengattung *Lecanora*. *Diss. bot.*, Lehre **4,** i–iii, 1–195.

Forssell, K. B. T. (1885). Beiträge zur Kenntnis der Anatomie und Systematik der Gloeolichenen. *Nova Acta reg. soc. sci. upsal.*, III, **13** (6), 1–118.

Hale, M. E. (1965). A monograph of *Parmelia* subgenus *Amphigymnia*. *Contr. U.S. natn. Herb.* **36,** 193–358.

Hale, M. E. (1968). A synopsis of the lichen genus *Pseudevernia*. *Bryologist* **71,** 1–11.

Hale, M. E. (1975a). A revision of the lichen genus *Hypotrachyna* (Parmeliaceae) in tropical America. *Smithson. contr. Bot.* **25,** i–iii, 1–73.

Hale, M. E. (1975b). A monograph of the lichen genus *Relicina* (Parmeliaceae). *Smithson. contr. Bot.* **26,** i–iii, 1–32.

Hale, M. E. (1976a). A monograph of the lichen genus *Pseudoparmelia* Lynge (Parmeliaceae). *Smithson. contr. Bot.* **31,** i–iii, 1–62.

Hale, M. E. (1976b). A monograph of the lichen genus *Bulbothrix* Hale (Parmeliaceae). *Smithson. contr. Bot.* **32,** i–iii, 1–29.

Hale, M. E. (1976c). A monograph of the lichen genus *Parmelina* Hale (Parmeliaceae). *Smithson. contr. Bot.* **33,** 1–60.

Hale, M. E. and Kurokawa, S. (1964). Studies on *Parmelia* subgenus *Parmelia*. *Contr. U.S. natn. Herb.* **36,** 121–191.

Hedlund, T. (1892). Kritische Bemerkungen über einige Arten der Flechtengattungen *Lecanora*, *Lecidea* und *Micarea*. *Bih. K. svenska VetenskAkad. Handl.*, III, **18** (3), 1–104.

Henssen, A. (1963). Eine Revision der Flechtenfamilien Lichinaceae und Ephebaceae. *Symb. bot. upsal.* **18** (1), 1–123.

Hertel, H. (1968–1975). Beiträge zur Kenntnis der Flechtenfamilie Lecideaceae I–VI. *Herzogia* **1,** 25–39 (1968), 321–329 (1969); **2,** 37–62 (1970), 231–261 (1971), 479–515 (1973); **3,** 365–406 (1975).

Hertel, H. (1975). Ein vorläufiger Bestimmungsschlüssel für die kryptothallinen, schwarzfrüchtigen, saxicolen Arten der Sammelgattung *Lecidea* (Lichenes) in der Holarktis. *Decheniana* **127,** 37–78.
Hue, l'Abbé (1924). Monographia Crocyniarum. *Bull. Soc. bot. Fr.* **71,** 311–402.
Jahns, H. M. (1970). Remarks on the taxonomy of the European and North American species of *Pilophorus* Th. Fr. *Lichenologist* **4,** 199–213.
Jørgensen, P. M. (1972). Further studies in *Alectoria* sect. *Divaricatae* DR. *Svensk bot. Tidskr.* **66,** 191–201.
Jørgensen, P. M. (1975). Contributions to a monograph of the *Mallotium*-hairy *Leptogium* species. *Herzogia* **3,** 433–460.
Kurokawa, S. (1962). A monograph of the genus *Anaptychia*. *Beih. Nova Hedwigia* **6,** 1–115.
Kurokawa, S. (1973). Supplementary notes on the genus *Anaptychia*. *J. Hattori bot. Lab.* **37,** 563–607.
Lamb, I. M. (1947). A monograph of the lichen genus *Placopsis* Nyl. *Lilloa* **13,** 151–288.
Lamb, I. M. (1951). On the morphology, phylogeny, and taxonomy of the lichen genus *Stereocaulon*. *Can. J. Bot.* **29,** 522–584.
Lamb, I. M. and Ward, A. (1974). A preliminary conspectus of the species attributed to the imperfect lichen genus *Leprocaulon* Nyl. *J. Hattori bot. Lab.* **38,** 499–553.
Lettau, G. (1932–1937). Monographische Bearbeitung einiger Flechtenfamilien. *Beih. Feddes Repert.* **69** (1–2), 1–250. [Diploschistaceae, Gyalectaceae etc.]
Llano, G. A. (1950). "A Monograph of the lichen family Umbilicariaceae in the Western Hemisphere" [Navexos P-831], 281 pp. Office of Naval Research, Washington, D.C.
Magnusson, A. H. (1929). A monograph of the genus *Acarospora*. *K. svenska VetenskAkad. Handl.*, III, **7** (4), 1–400.
Magnusson, A. H. (1933). A monograph of the genus *Ionapsis*. *Acta Horti gothoburg.* **8,** 1–47.
Magnusson, A. H. (1939). Studies in species of *Lecanora*, mainly the *Aspicilia gibbosa* group. *K. svenska VetenskAkad. Handl.*, III, **17** (5), 1–182.
Magnusson, A. H. (1940). Studies in species of *Pseudocyphellaria*—the *crocata* group. *Acta Horti gothoburg.* **14,** 1–36.
Magnusson, A. H. (1944). Studies in the *ferruginea*-group of the genus *Caloplaca*. *Göteborgs K. Vetensk.- o. VitterhSamh. Handl.*, **B, 3** (1), 1–17.
Magnusson, A. H. (1947). Studies in non-saxicolous species of *Rinodina*, mainly from Europe and Siberia. *Acta Horti gothoburg.* **17,** 191–338.
Magnusson, A. H. (1956). A second supplement to the monograph of *Acarospora* with keys. *Göteborgs K. Vetensk.- o. VitterhSamh. Handl.*, **6B, 6** (17), 1–34.
Motyka, J. (1936–1938). "Lichenum generis *Usnea* studium monographicum, Pars generalis". 2 vols, 651 pp. Leopoli.
Motyka, J. (1947). Lichenum generis *Usnea* studium monographicum, Pars generalis. *Annls Univ. Mariae Curie-Skłodowska*, **C, 1,** 277–476.
Martin, W. (1965). The lichen genus *Cladia*. *Trans. R. Soc. N.Z., Bot.* **3** (2), 7–12.

Oberwinkler, F. (1970). Die Gattungen der Basidiolichenen. *Vortr. bot. Ges.* [*Dtsch. bot. Ges.*], N.F., **4**, 139–169.
Poelt, J. (1966). Zur Kenntnis der Flechtengattung *Physconia* in der Alten Welt und ihrer Beziehungen zur Gattung *Anaptychia*. *Nova Hedwigia* **12**, 107–135.
Salisbury, G. (1972). *Thelotrema* Ach. sect. *Thelotrema* 1. The *T. lepadinum* group. *Lichenologist* **5**, 262–274.
Santesson, R. (1952). Foliicolous lichens I. A revision of the taxonomy of the obligately foliicolous, lichenized fungi. *Symb. bot. upsal.* **12** (1), 1–590.
Thomson, J. W. (1967). Notes on *Rhizocarpon* in the Arctic. *Nova Hedwigia* **14**, 421–481.
Vainio, E. A. (1887–1897). Monographia Cladoniarum universalis. I–III. *Acta Soc. Fauna Flora fenn.* **4**, 1–510 (1887); **10**, 1–499 (1894); **14**, 1–268 (1897).
Verseghy, K. (1962). Die Gattung *Ochrolechia*. *Beih. Nova Hedwigia* **1**, 1–146.
Vězda, A. (1965–1973). Flechtensystematische Studien I–IX. *Preslia* **37**, 127–143 (1965a) [I. Die Gattung *Petractis* Fr.], 237–245 (1965b) [II. *Absconditella*, eine neue Flechtengattung]; *Folia geobot. phytotax.*, Praha **1**, 154–174 (1966a) [III. Die Gattung *Ramonia* Stiz. und *Gloeolecta* Lett.], 311–340 (1966b) [IV. Die Gattung *Gyalidea* Lett.]; **2**, 311–317 (1967a) [V. Die Gattung *Ramonia* Stiz. Zusätze], 383–396 (1967b) [VI. Die Gattung *Sagiolechia* Massal.]; **7**, 203–215 (1972) [VII. *Gyalideopsis*, eine neue Flechtengattung]; **8**, 311–316 (1973a) [VIII. Drei neue Arten der Gyalectaceae sensu amplo aus Neu-Guinea], 417–424 (1973b) [IX. Die Gattung *Ramonia* Stiz. Zusätze 2].
Vězda, A. and Wirth, V. (1976). Zur Taxonomie der Flechtengattung *Micarea* Fr. em. Hedl. *Folia geobot. phytotax.*, Praha **11**, 93–102.

IV. Sources of Further Information

A. Literature Compilations

A comprehensive account of the earlier lichenological literature is provided by Krempelhuber (1867–1872) in which details of papers on particular geographical regions are listed. The compendia of Ciferri (1957–1960) and Lindau and Sydow (1908–1918) between them list papers under authors' names published to the end of 1930. In these two works, however, papers dealing with fungi are also treated and so a detailed search of them is a formidable task. Abstracts of lichenological papers were included in Part I of each issue of *Just's Botanischer Jahresbericht* (59 vols, 1874–1939) covering the literature appearing in the years 1873–1931, inclusive. The abstracts provided by Zahlbruckner for this publication are particularly valuable and cover the years 1884 to 1931 (Vols 13–59).

Citations of literature references to all taxa are included in Zahlbruckner (1921–1940). Lamb (1963), Hawksworth (1972) and the Commonwealth

Mycological Institute (1971 →) indicate the country of origin of all newly described lichen taxa listed.

Grummann (1974) lists the main papers of lichenologists published into the early 1960s, arranging them under authors' names.

The most important continuing bibliography of lichenological literature is that of Culberson (1951 →). Shorter lists of recent publications are included in issues of the *Bulletin of the British Lichen Society* (1958 →) and the *Revue bryologique et lichénologique* (1928 →). Detailed abstracts of many lichenological papers compiled by G. Follmann appear at intervals in *Excerpta Botanica*, **C** (1959 →).

Lichenological papers are also sometimes cited in various more general abstracting journals (for details of these see Hawksworth, 1974) but as these usually feature in Culberson's series an examination of them from the lichenological standpoint is of only limited value.

Smith (1921) provides a detailed account of lichenology up to that time and the bibliography in that book is of continuing value; supplements to this were provided in a series of articles which include data on major floras published since 1921 (Smith, 1923, 1925, 1927, 1931, 1933). Attention is also drawn to the literature reviews of Bioret (1921) covering the period 1910–1919, and des Abbayes (1953) covering that from 1939 to 1952. The series of review articles by Poelt (1955–1974), now being continued by Hertel (1976), includes synopses of major floristic studies arranged under geographical heads.

In addition to the above sources, those concerned with the preparation of lichen floras for particular areas should consult journals of Museums and Natural History Societies covering them and older floras concerned primarily with vascular plants (which sometimes have lichen lists appended). Further information on the preparation of lichen floras is included in Hawksworth (1974).

B. Lichenological Journals

Lichenological papers appear in a great many scientific journals published in many different countries. A few journals are, however, devoted entirely or partly to lichens and these are extremely valuable sources of information. Details of those currently appearing are given here.

Bibliotheca lichenologica **1** → (1974 →). J. Cramer, Lehre. [Monographs]

The Bryologist **1** → (1898 →). The American Bryological and Lichenological Society [Formerly The Sullivant Moss Society (1899–1949) and The American Bryological Society (1949–1969)], Brooklyn. [Index to vols 1–60 (A. C. Crum, ed.) published separately, 1959, 217 pp.; to vols 61–75 (N. G. Miller, ed.) published separately, 1976, 193 pp.] Quarterly.

Bulletin of the British Lichen Society **1** → (1958 →). British Lichen Society, London. Mimeographed newsletter, each number with separate pagination. Twice-yearly.

Folia cryptogamica Estonica **1** → (1972 →). Societas Investigatorum Rerum Naturae Academiae Scientarum R.P.S.S. Estoniae, Tartu.

Herzogia **1** → (1968 →). J. Cramer, Lehre. Journal of the Bryologisch-Lichenologischen Arbeitsgemeinschaft für Mitteleuropa. Twice-yearly.

International Lichenological Newsletter **1** (1967 →). International Association for Lichenology, Worcester, Massachusetts. Twice-yearly.

Journal of the Hattori Botanical Laboratory **1** → (1967 →). Hattori Botanical Laboratory, Nichinan. Annual.

Lichen [*News Bulletin of the Lichenological Society of Japan*] **1** → (1972 →). Lichenological Society of Japan. In Japanese. Twice-yearly.

The Lichenologist **1** → (1958 →). Academic Press, London and New York. Journal of the British Lichen Society. Twice-yearly.

Miscellanea bryologica et lichenologica **1** → (1955 →). Hattori Botanical Laboratory, Nichinan. Thrice-yearly.

Mycotaxon **1** → (1974 →). Mycotaxon, Ithaca, New York. Quarterly.

Nova Hedwigia **1** → (1959 →). J. Cramer, Lehre. [See also the *Beihefte* series **1** → (1962 →).]

Novosti Sistematiki Nizshikh Rastenič **1** → (1964 →) [*Novitates systematicae plantarum non vascularum*]. Institutum Botanicum nomine V. L. Komarovii, Academia Scientarum U.R.S.S. [Formerly *Notulae systematicae ex Instituto Cryptogamico Horti Botanici Petropolitani*, 1922–1963], Leningrad. Annual.

Revue bryologique et lichénologique **1** → (1928 →). Muséum National d'Histoire Naturelle, Paris.

C. Exsiccatae

Exsiccatae, sets of dried specimens distributed to major herbaria and individuals as gifts or in exchange, are a very valuable source of information on the lichen flora of regions they cover. Most lichen exsiccatae have been concerned with particular geographical areas but a few, including some still continuing (e.g. Vězda's *Lichenes selecti exsiccatae*, 1960→; that of Follmann cited below) are world-wide. In the case of modern exsiccatae, printed matter bearing the information on the labels of the packets distributed is usually either issued separately or published independently in a journal.

Lichenologists are fortunate in having Lynge's very detailed accounts of the world's *Lichenes exsiccati* at their disposal (Lynge, 1913, 1915–1922, 1939). This monumental work not only lists the names on each collection distributed in each exsiccata, but provides a comprehensive index to them.

Sayre (1969) provides a list of all general cryptogamic and lichen exsiccatae which includes detailed information on the dates of publication of each part (decade or fascicle), references to further information on them, indications of the data on the labels and other pertinent comments. Both Lynge and Sayre indicate where the sets they studied were. Culberson (1955) provides information on lichen exsiccatae retained in North American herbaria.

The following exsiccatae have started to appear or appeared since the work of Sayre (1969):

Anderson, R. A. (1976 →). "Lichens of Western North America". Distributed by University of Denver, Colorado.

Brodo, I. M. (1970 →). "Lichenes Canadenses exsiccati". Distributed by the National Herbarium of Canada, Ottawa. [See Brodo, *Bryologist* **74**, 151–153 (1971); **79**, 385–405 (1976).]

Filson, R. (1975 →). "Lichenes Antarctici exsiccati". Distributed by the National Herbarium of Victoria. [See Filson, *Muelleria* **3**, 146–158 (1975).]

Follmann, G. (1968–1970). "Lichenes exsiccati a Museo Botanico Berolinensi editi". Distributed by the Kryptogamenabteilung, Botanisches Museum, Berlin. [See Follmann, *Willdenowia* **4**, 383–390, 391–397 (1968); **5**, 15–21 (1968); **6**, 17–24 (1970).]

Follmann, G. (1973 →). "Lichenes exsiccati selecti a Museo Historiae Naturalis Casselensi editi". Distributed by the Naturkundemuseum im Ottoneum, Kassal. [A continuation of Follmann (1968–1970) cited above (fasc. IV, No. 81 →); see Follmann, *Philippia* **2**, 13–21 (1973), 139–146 (1974), 221–235 (1975).]

Henssen, A. (1969 →). "Lichenes cyanophili exsiccati". Distributed by the Botanisches Institut der Universität Marburg.

"Lichenes Groenlandici Exsiccati" (1972 →). Distributed by the Botanical Museum, University of Copenhagen.

Nowak, J. (1971? →). "Lichenes Poloniae meridionalis exsiccati". Distributed by the Botanical Institute of the Polish Academy of Sciences, Kraków. [See Nowak, *Fragm. flor. geobot.* **21**, Suppl., 569–596 (1976).]

Poelt, J. (1975 →). "Plantae Graecenses". Distributed by the Institut für Systematische Botanik der Universität Graz.

Verseghy, K. (1969 →). "Lichenes exsiccati". Distributed by the Hungarian Natural History Museum, Budapest. [See Verseghy, *Fragm. bot.* **7**, 67–76 (1969).]

Verseghy, K. (1969 →). "Lichenotheca parva". Distributed by the Hungarian Natural History Museum. [A continuation of Gyelnik's exsiccata of the same title (1937).]

Yoshimura, I. (1972 →) "Lichenes Japonici exsiccati". Distributed by the Hattori Botanical Laboratory, Nichinan.

D. Herbaria

Those involved in the compilation of regional floras and check-lists will need not only to locate published data but also the specimens providing the basis of the published accounts and other unpublished collections from the area. Unfortunately there is no single entirely comprehensive list of lichen herbaria. The *Collector's Index* being produced by The International Association for Plant Taxonomy and Nomenclature will eventually provide details of collectors whose material is represented in most of the larger herbaria but is not yet complete (Lanjouw and Stafleu, 1954, 1957; Chaudri *et al.*, 1972; Vegter, 1976). Information on the location of lichen herbaria is, however, included in the works of Stafleu (1967), Grummann (1974), Hawksworth (1974) and Holmgren and Keuken (1974–1976). Hawksworth and Seaward (1977) deal with the British lichen herbaria; the work of Kent (1957) is also valuable for that country although it is primarily concerned with vascular plants.

In modern floristic accounts authors usually indicate where their collections are preserved and it is to be hoped that this will become a standard practice. It has been remarked that "Lists of records that cannot be verified are mere waste paper" (Dennis, 1968).

Acknowledgements

I am very grateful to Dr M. E. Hale for reading the manuscript; to Dr D. C. Lindsay and Dr P. K. Verseghy for their assistance with the entries for Antarctica and Hungary, respectively; and particularly to Dr T. Ahti for his painstaking comments on the whole, the U.S.S.R. section and information on some exsiccatae.

References

Des Abbayes, H. (1953). Travaux sur les lichens parus de 1939 à 1952. *Bull. Soc. bot. Fr.* **100**, 83–123.
Acharius, E. (1810). "Lichenographia universalis". Göttingen. [Reprinted 1976 Richmond Publishing, Richmond.]
Acharius, E. (1814). "Synopsis methodica Lichenum". Lund.
Ainsworth, G. C. (1971). "Ainsworth & Bisby's Dictionary of the Fungi". 6th ed. Commonwealth Mycological Institute, Kew.
Bioret, G.-F.-M. (1921). Revue des travaux sur les lichens de 1910 à 1919. *Rev. gén. bot.* **33**, 63–76, 146–160, 214–220, 264–272, 328–336, 372–396.
Blake, S. F. and Atwood, A. C. (1942, 1961). "A Geographical Guide to Floras of the World". 2 vols. U.S. Department of Agriculture [Publs Nos 401, 797], Washington, D.C.

Chaudri, M. N., Vegter, I. H. and De Wal, C. M. (1972). Index Herbariorum. Part II (3). Collectors I-L. *Regnum Vegetabile* **86**, i–xxii, 297–473.

Ciferri, R. (1957–1960). "Thesaurus literaturae mycologicae et lichenologicae, Supplementum 1911–1930". 4 vols. Papia, Cortina.

Commonwealth Mycological Institute (1971 →). "Index of Fungi" **4** →. Commonwealth Mycological Institute, Kew.

Culberson, W. L. (1951 →). Recent literature on lichens—1→. *Bryologist* **54** →.

Culberson, W. L. (1955). "Lichenes exsiccati" in herbariis Americae septentrionalis asservati. *Bryologist* **62**, 45–62.

Dennis, R. W. G. (1968). "British Ascomycetes". J. Cramer, Lehre.

Good, R. (1974). "The Geography of the Flowering Plants". 4th ed. Longmans, London.

Grummann, V. (1974). "Biographisch-bibliographisches Handbuch der Lichenologie". J. Cramer, Lehre.

Hawksworth, D. L. (1970). Guide to the literature for the identification of British lichens. *Bull. Br. mycol. Soc.* **4**, 73–95.

Hawksworth, D. L. (1972). "Index of Fungi Supplement. Lichens 1961–1969". Commonwealth Mycological Institute, Kew.

Hawksworth, D. L. (1974). "Mycologist's Handbook". Commonwealth Mycological Institute, Kew.

Hawksworth, D. L. and Seaward, M. R. D. (1977). "Lichenology in the British Isles 1568–1975". Richmond Publishing, Richmond.

Hertel, H. (1976). Systematik der Flechten. *Fortschr. Bot., Berl.* **38**, 280–297.

Holmgren, P. K. and Keuken, W. (1974). Index Herbariorum. Part I. The herbaria of the world. Sixth edition. *Regnum Vegetabile* **92**, i–vii, 1–397.

Holmgren, P. K. and Keuken, W. (1975). Index Herbariorum. Ed. 6. Geographical arrangements of the herbaria listed in Index herbariorum, Edition 6. *Taxon* **24**, 543–551.

Holmgren, P. K. and Keuken, W. (1976). Index Herbariorum. Additions to "The herbaria of the world" ed. 6. *Taxon* **25**, 517–524.

Kent, D. H. (1957). "British Herbaria". Botanical Society of the British Isles, London.

Krempelhuber, A. von (1867–1872). "Geschichte und Litteratur der Lichenologie". 3 vols. Wolf, Munich.

Lamb, I. M. (1963). "Index nominum Lichenum inter annos 1932 et 1960 divulgatorum". Ronald Press, New York.

Lanjouw, J. and Stafleu, F. A. (1954). Index Herbariorum. Part II. Collectors A–D. *Regnum Vegetabile* **2**, 1–174.

Lanjouw, J. and Stafleu, F. A. (1956). Index Herbariorum. Part II (2). Collectors E–H. *Regnum Vegetabile* **9**, 175–295.

Lindau, G. and Sydow, P. (1908–1918). "Thesaurus litteraturae mycologicae et lichenologicae". 5 vols. Borntraeger, Berlin.

Lynge, B. (1913). On the world's "Lichenes exsiccati". *Nyt Mag. Naturvid.* **51**, 95–122.

Lynge, B. (1915–1922). Index specierum et varietatum "Lichenum exsiccatorum". *Nyt Mag. Naturvid.* **53,** *1–112* (1916), **54,** *113–304* (1916) [Pars I, 1]; **55,** *305–384* (1917), **56,** *385–464* (1918), **57,** *465–559* (1919) [Pars 1, 2]; **58,** *1–96* (1920), **59,** *97–192* (1921), **60,** *193–316* (1922) [Pars II]. [Issued in parts in the main journal but with an independent pagination; also issued bound as three volumes with the pagination as above.]

Lynge, B. (1939). Index collectionum "Lichenes exsiccati". Supplementum I. *Nyt Mag. Naturvid.* **79,** 233–323.

Poelt, J. (1955–1974). Systematik der Flechten. *Fortschr. Bot., Berl.* **17,** 220–238 (1955); **18,** 75–82 (1956); **20,** 56–62 (1958); **23,** 49–56 (1961); **25,** 60–70 (1963); **27,** 328–340 (1965); **30,** 275–290 (1968); **32,** 256–270 (1970); **34,** 361–380 (1972); **36,** 263–276 (1974).

Sayre, G. (1969). Cryptogamae exsiccatae—An annotated bibliography of published exsiccatae of Algae, Lichenes, Hepaticae and Musci. *Mem. N.Y. Bot. Gdn* **19,** 1–174.

Smith, A. L. (1921). "Lichens". Cambridge University Press, Cambridge. [Reprinted in 1975 by Richmond Publishing, Richmond.]

Smith, A. L. (1923). Recent work on lichens. *Trans. Br. mycol. Soc.* **8,** 193–206.

Smith, A. L. (1925). Recent work on lichens. *Trans. Br. mycol. Soc.* **10,** 133–152.

Smith, A. L. (1927). Recent lichen literature. *Trans. Br. mycol. Soc.* **12,** 231–275.

Smith, A. L. (1931). Recent lichen literature. *Trans. Br. mycol. Soc.* **15,** 193–235.

Smith, A. L. (1933). Recent lichen literature. *Trans. Br. mycol. Soc.* **18,** 93–126.

Stafleu, F. A. (1967). Taxonomic literature. A selective guide to botanical publications with dates, commentaries and types. *Regnum Vegetabile* **52,** i–xx, 1–556.

Vegter, I. H. (1976). Index Herbariorum. Part II (4). Collectors M. *Regnum Vegetabile* **93,** 475–576.

Zahlbruckner, A. (1921–1940). "Catalogus lichenum universalis". 10 vols. Borntraeger, Berlin.

Zahlbruckner, A. (1926). Lichenes (Flechten), B. Spezieller Teil. *In* "Die Natürlichen Pflanzenfamilien" 2nd ed. (A. Engler and K. Prantl, eds), Vol. 8, pp. 61–270. Engelmann, Leipzig.

Appendix B: Selected Glossary

MARK R. D. SEAWARD
and DAVID L. HAWKSWORTH

As well as providing definitions of the terms employed in this book, we have defined others used in lichen ecology (excluding eco-physiology) and biogeography (excluding pedology, geology and climatology), as no comprehensive glossary of these appears to exist. Terms used in the description of lichens themselves and in their taxonomy/nomenclature have been compiled by Ainsworth (1971) and Hawksworth (1974) respectively, and further terms relating to the description and classification of communities are discussed in Barkman (1958, 1973) and Shimwell (1971). In naming lichen phytogeographic elements, the terminology currently used by phytogeographers concerned with vascular plants should be adopted wherever possible (e.g. Good, 1974). Ahti *et al.* (1968) provide a valuable classification of vegetation types in north-west Europe and also discuss the application of further terms used (or potentially useful) in accounts of lichen distribution. Phytogeographic terminology relating to the British Isles, and definitions of terms relating to maritime situations, are discussed in Ratcliffe (1968) and Fletcher (1973) respectively. The following works are recommended as additional reference sources: Kenneth (1963), Stearn (1973) and the Special Committee for the International Biological Programme (1974).

Words differing from one another in only a stem, prefix or suffix are usually represented by one entry in the glossary; for example, only *poleophilous* has been defined—*poleophobous* and *poleotolerant* then being self-explanatory.

Our thanks are due to Dr M. J. Crawley, Dr A. Fletcher, Dr D. C. Lindsay and Dr P. B. Topham for their help in the preparation of this glossary.

References

Ahti, T., Hämet-Ahti, L. and Jalas, J. (1968). Vegetation zones and their sections in northwestern Europe. *Annls bot. fenn.* **5**, 169–211.
Ainsworth, G. C. (1971). "Ainsworth & Bisby's Dictionary of the Fungi", 6th ed. Commonwealth Mycological Institute, Kew.
Barkman, J. J. (1958). "Phytosociology and Ecology of Cryptogamic Epiphytes". Van Gorcum, Assen.
Barkman, J. J. (1973). Synusial approaches to classification. *In* "Ordination and Classification of Communities" (R. H. Whittaker, ed.), pp. 435–491. W. Junk, The Hague.
Fletcher, A. (1973). The ecology of maritime (supralittoral) lichens on some rocky shores of Anglesey. *Lichenologist* **5**, 401–422.
Good, R. (1974). "The Geography of the Flowering Plants", 4th ed. Longmans, London.
Hawksworth, D. L. (1974). "Mycologist's Handbook". Commonwealth Mycological Institute, Kew.
Kenneth, J. H. (1963). "A Dictionary of Biological Terms", 8th ed. Oliver and Boyd, Edinburgh.
Ratcliffe, D. A. (1968). An ecological account of Atlantic bryophytes in the British Isles. *New Phytol.* **67**, 365–439.
Shimwell, D. W. (1971). "The Description and Classification of Vegetation". Sidgwick and Jackson, London.
Special Committee for the International Biological Programme (1974). "Quantities, Units and Symbols. Recommendations for Use in IBP Synthesis". International Council of Scientific Unions, London.
Stearn, W. T. (1973). "Botanical Latin", 2nd ed. David and Charles, Newton Abbot.

Glossary

Abundance: cover, usually expressed in phytosociological records on a numerical scale; see Braun-Blanquet Scale and Domin Scale.
Accessibilité (Fr.): the totality of conditions prevailing at a particular site that may influence the possibility of diaspores (q.v.) reaching the site (phytogeographic concept).
Accidental: used of species occurring fortuitously on a substratum not usual for them (e.g. a saxicolous species spreading on to adjacent soil).
Acidophilous: preferring acidic substrata or habitats.
Adaptation: the process by which a species becomes adjusted, genotypically or phenotypically, to a particular set of environmental factors.
Aerohygrophilous: requiring high atmospheric humidity.

Affinity: similarity; used in phytosociology and phytogeography in the comparison of communities or floras, often by numerical methods.

Affinity Index (A): used in assessment of the similarity between communities or floras, e.g.

$$A = \frac{c}{a \times b}$$

where a = number of species in the first, b = number of species in the second community, and c = number of species common to both.

Alliance: phytosociological taxon comprising one or more associations; the first name of an alliance ends in the suffix *-ion* (e.g. *Xanthorion parietinae*).

Allopatric: occupying different, non-overlapping geographical areas; cf. *Sympatric*.

Amphi-Atlantic: occurring on both sides of the Atlantic Ocean, usually only in the coastal areas.

Angara: ancient continental land-mass resulting from the break-up of Pangaea (q.v.).

Anheliophilous: preferring diffuse light.

Anombrophyte: species avoiding direct rain.

Anthropochorous: associated with human agencies.

Anthropogenic: of factors made or influenced by man; sometimes applied to species favouring habitats created or influenced by man, but synanthropic (q.v.) is to be preferred.

Arboreal: of trees; see also *Epiphytic*.

Artenpaare (Germ.): species pair (q.v.).

Association (*ass.*): communities characterized by floristic composition, constant and differential species; the basic unit of phytosociological classification equivalent to that of the "species" in idiotaxonomy (q.v.); the first name of an association ends in the suffix *-etum* (e.g. *Parmelietum furfuraceae*).

Association element: the constant species of an association.

Aufnahme (pl.–n) (Germ.): record (phytosociological).

Autecology: study of the ecological relationships of individual organisms.

Autogenic succession: succession brought about by the vegetation itself.

Azonal: of vegetation or soil patterns related to more local geographical features (e.g. mountain range); cf. *Zonal*.

Baldwin effect: process by which advantageous phenotypic changes are succeeded by similar genotypically determined changes.

Basiphilous: preferring base-rich substrata.

Biomass: the weight of living material in a particular habitat.

Biotype: a particular genotype expressed in one or more individuals.

Bipolar: occurring in both Arctic and Antarctic regions but disjunct to varying degrees between the polar regions.

Bodenvagen (Germ.): used of species with no preference for a particular substratum; see also *Omnicolous*.

Bole: tree trunk below the main branches, or more restrictedly the enlarged lower part of the tree trunk; ambiguous and best avoided.

Boreal: pertaining to the circumpolar bioclimatic zone, also called the northern coniferous zone or taiga.

Braun-Blanquet Scale: six-category scale for estimating cover and abundance of a species in a record (q.v.): + (under 1%), 1 (1–5%), 2 (6–25%), 3 (26–50%), 4 (51–75%), 5 (76–100%).

Buffer capacity: the degree to which a substratum, thallus or solution is able to resist changes in its pH caused by acidic or alkaline additives.

Calcareous: having a high calcium content; improperly used as a term contradistinctive to siliceous (q.v.).

Calcicole: an inhabitant of a calcium-rich substratum.

Calciphilous: preferring substrata rich in calcium compounds.

Characteristic species, espèce caractéristique (Fr.), *Charakterart* (Germ.), *ledart* (Swed.): species used in the delimitation of a community (phytosociological); see also *Exclusive, Preferential* and *Selective species*.

Chelation: sequestration; formation of complex organic compounds with loosely-bound cations (e.g. by lichen acids).

Chorology: the study of geographical distributions of plants and animals.

Circumboreal: used of species occurring in boreal regions in a broad latitudinal zone around the North Pole.

Circumneutrophilous: preferring more or less neutral substrata.

Circumpolar: used of species occurring in a broad latitudinal zone in Arctic and subarctic, or Antarctic and subantarctic, regions.

Class: phytosociological taxon comprising one or more orders; usually the highest rank adopted in phytosociological systems; names of classes end in the suffix *-eae* (e.g. *Epipetreae*).

Climatoid: of epiphytes with a wide edaphic and a narrow climatic amplitude.

Climax community: the final stage of an undeflected succession; in a particular climatic region each habitat tends to support a particular climax community.

Cline: a continuous, usually linear, intergradation in some character(s) shown by the individuals of a species or community in relation to environmental or geographical factor(s).

Coefficient Générique (Fr.): see *Richness Index*.

Coenotype: a community with a constant composition but lacking faithful or differential species.

Cold desert: a floristic region distinguished by physiological drought induced by low temperature and low precipitation, the flora exhibiting a high ratio of cryptogams to phanerogams, with timber-producing plants totally absent.

Community: (1) two or more species growing together in a particular habitat; (2) a neutral phytosociological term not denoting a particular rank.

Competition: the struggle of two or more species to exploit the same, limiting resource(s).

Coniophilous: preferring substrata enriched by dust containing excreta.

Consociation: subdivision of an association (q.v.) greater in rank than society (q.v.) and characterized by one physically dominant species.

Constancy: the frequency of a species within records of a particular community type, usually expressed on a six-category scale: r (under 1%), I (1–20%), II (21–40%), III (41–60%), IV (61–80%), V (81–100%).

Constant species: a species invariably present in a particular community.

Continental: occurring in climatic conditions (severe winters and hot summers) prevailing in the interior regions of a continent.

Continuum: an intergradation between two or more extremes, whether linear or not.

Cortège Moyen Spécifique (Fr.) (*CMS*): the mean number of lichen and bryophyte species growing with a particular species in the survey area; used in calculation of Index of Atmospheric Purity (q.v.).

Corticolous: growing on or inhabiting bark.

Cosmopolitan: widely distributed on a world scale; occurring at least in the temperate zone of both hemispheres.

Cover: the extent to which a species occupies the available surface area; often expressed as a percentage or by a six (e.g. Braun-Blanquet Scale, q.v.) or eleven (e.g. Domin Scale, q.v.) category scale.

Crown: that part of the tree above the trunk (cf. bole), or the uppermost branches of a tree.

Cyclic succession: a serial development of communities in which the original community is eventually restored.

Déclinant (Fr.): an ecological subdivision of a subassociation.

Decorticate: with the bark no longer present.

Deflected succession: a succession which, instead of progressing to the climax community (q.v.), is affected by some factor causing it to proceed to a different community.

Delimitation: phytosociologically the process of determining the extent (boundaries, limits, scope) of particular syntaxa (q.v.).

Diaspore: any propagule, sexual or asexual.

Differential species: non-characteristic species used in distinguishing closely related communities.

Disjunct: of a population of a species widely separated geographically or otherwise from other populations of the same species.

Domin Scale: an eleven-category scale for estimating cover and abundance of species in a record (q.v.): + (single individual), 1 (a very few individuals), 2 (sparsely distributed), 3 (frequent but cover less than 4%), 4 (cover 4–10%), 5 (cover 11–25%), 6 (cover 26–33%), 7 (cover 34–50%), 8 (cover 51–75%), 9 (cover 76–90%), 10 (cover 91–100%).

Dominance, Dominanz (Germ.): the extent to which a species predominates in a community.

Ecad: phenotypically distinct population (q.v.) adapted to, or found only in, particular ecological situations; an ecad is not accorded formal taxonomic recognition; see also *Morphotype*.

Ecocline: a gradient of variation in anatomical, morphological, physiological or other characters correlated with a gradient in one or more ecological factors; sometimes restricted to genetically based variations.

Ecologically aggressive: applied to lichens which show a low degree of specificity to substrata and a high capacity for rapid dispersal and re-establishment; such species occur in a wide range of communities and can compete successfully with seemingly more specialized plants.

Ecosystem: the system consisting of all the living organisms of an area, their non-living environment, and the interactions between them.

Ecotone: boundary line or transitional area between two communities.

Ecotype: genetic races adapted to particular ecological requirements; ecotypes are often accorded formal names in accepted taxonomic ranks (e.g. species, subspecies, varieties).

Edaphic: concerned with the soil.

Emersed: emerged; used by some authors in contradistinction to submersed/submerged.

Endemic: occurring only in a single, usually small, geographic area.

Endolithic: of saxicolous lichens with at least their phycobiont living within the substratum.

Endophloeodic, endophloeodal, endophloeoic: of lichens with the part of the thallus containing the phycobiont living inside the periderm.

Endophyllous: subcutaneous (used of foliicolous lichens where the mycobiont is partly parasitic on the leaves of its host).

Endoxylic: growing largely within wood.

Epidendric: of epiphytes on shrubs and trees.

Epigeic: growing on soil; on the ground (terrestrial).

Epilichenes (Germ.): collective term for epilithic (q.v.), epixylic (q.v.), epiphloeodic (q.v.) and epigeic (q.v.) lichens; see also *Kryptolichenes*.

Epilithic: of saxicolous lichens growing on the surface of substrata.
Epiphloeodic, epiphloeodal, epiphloeoic: of lichens with the phycobiont living above the outermost layers of bark.
Epiphyllous: growing on the surfaces (usually the upper) of leaves, the mycobiont not penetrating the leaf epidermis.
Epiphytic: growing non-parasitically on plants (usually woody); see also *Epidendric, Arboreal.*
Epixylic: growing on wood surfaces.
Epizoic: growing on animals.
Erratic: of usually unattached organisms moved around by wind or other physical agencies.
-etalia: suffix denoting a syntaxon in the rank of order.
-etum: suffix denoting a syntaxon in the rank of association or union.
Euhydrophilous: requiring submergence in freshwater for most of the year.
Euramerica: ancient continental land-mass resulting from the break-up of Pangaea (q.v.).
Euryionic: with a large pH amplitude; cf. *Stenoionic.*
Eurysubstratic: of species occurring on an extremely wide range of substrata.
Eutrophiated: see *Eutrophicated.*
Eutrophicated: nutrient-enriched (correctly used of water); cf. *Hypertrophicated.*
Exclusive species: characteristic species (q.v.) completely or almost confined to a particular community.
Extant: of species or communities still present in an area.
Extinct: of species or communities no longer present in an area.
Facies: (1) communities with a dominant, but lacking other constant species (q.v.); (2) part of a community (phytosociology) or species (idiotaxonomy) with particular distinctive features (characters or composition), of little or no syn- or idio- taxonomic significance—used in Chapter 10 in this sense; (3) a phytosociological unit below the rank of nodum (q.v.).
Facultative: used of a species able to adopt a particular mode of nutrition in a particular situation, e.g. a parasite becoming a saprophyte after the death of its host; cf. *Obligate.*
Faithful species: species only occurring in a particular community.
Federation: phytosociological taxon (Uppsala School) including one or more unions (q.v.); usually taken as equivalent in rank to alliance (q.v.); ending in the suffix *-ion* but not always based on a generic name, e.g. *Physodion.*

Fensterflechten (Germ.): "window lichens", so named because their upper cortex is flush with the soil surface and acts like a window admitting light to the buried algal layers.

Fidelity: the degree of restriction of a species to a particular community type; usually expressed on a I (accidental or rare) – V ($\pm$ entirely restricted to) point scale.

Flora: (1) the plant kingdom; (2) the plants in a particular geographic region or site; (3) a list, usually annotated, of the plants in a particular geographic region or site.

Floristic: based on the species present regardless of their frequency.

Foliicolous: growing on leaves.

Formation: phytosociological taxon (Uppsala School) usually taken as equivalent to that of class (q.v.).

Foveolate: pitted (e.g. used to describe endolithic lichens whose ascocarps form pits in the substratum).

Frequency: the number of occurrences of a species in a record (q.v.) or community.

Genotype: the genetic (hereditary) constitution of an individual.

Geoplese: tree-base epiphytes.

-Gesellschaft (Germ.): applied in phytosociology to a community of a rank equivalent to that of subassociation (q.v.); e.g. *Lepraria candelaris*—Gesellschaft.

Gibber plain (Australian): rock-strewn area overlying soils in arid lands.

Gondwanaland: ancient southern continental land-mass formed from the break-up of Pangaea (q.v.); existed in Permo-Carboniferous period (270–340 million years ago), and later fragmented to form Australasia, Antarctica, most of Africa, South America and India.

Growth curve: a graph of a measurement of growth against time or size; used, for example, in lichenometry (q.v.) in determinations of the age of lichen thalli.

Growth form: life-form or habit (q.v.).

Habit: the morphological form assumed by an individual or species.

Habitat: a particular site with a specific set of conditions; natural place of occurrence of an organism; a loosely applied term.

Hamada (North African): see *Gibber plain*

Heliophilous: preferring direct sunlight.

Hemerophilous: of species encouraged by human activity (e.g. building of walls, planting of trees).

Hemiboreal: of the southern taiga (q.v.).

Heterogeneity Coefficient: N/n, where $N =$ total number of species in a table of records, and $n =$ average number of species per record (q.v.).

Holarctic: of the entire north temperate and Arctic zones.

Host specificity: extent to which a species is confined to a particular host (implying it has a nutritional or other relationship to it).

Hygrochasic: changing form on absorption of water.

Hygrophilous: moisture-loving; growing in habitats with high humidity.

Hyperepiphytes: epiphytes attached directly to other epiphytes.

Hypertrophiated: see *Hypertrophicated.*

Hypertrophicated: over-enriched with nutrients; cf. *Eutrophicated.*

Hypolithic: see *Endolithic.*

Hypophloeodic, hypophloeodal, hypophloeoic: see *Endophloeodic.*

Hypophyllous: growing on the lower surfaces of leaves.

Idiotaxonomy: classification of organisms; cf. *Syntaxonomy.*

Index of Atmospheric Purity (I.A.P.):

$$I.A.P. = \frac{n}{100}\left(\sum_{n}^{1} Q \times f\right), \text{ or } \sum_{n}^{1} (Q \times f)\bigg/10,$$

where f = frequency of each species at the site, n = number of species at the site, and Q = Index of Toxiphoby (q.v.) or, now usually, Cortège Moyen Spécifique (q.v.).

Index of Ecological Continuity (I.E.C.): the percentage of 20 selected old-forest indicator species (q.v.) occurring in a particular site; *Revised Index of Ecological Continuity (R.I.E.C.):* the percentage occurrence of up to a maximum of 20 out of a total list of 30 selected old-forest indicator species; *I.E.C.* and *R.I.E.C.* values have proved valuable in assessing the relative importance of sites for lichens characteristic of little disturbed woodlands in the British Isles.

Index of Lichen Abundance (I.L.A.):

$$I.L.A. = \sum_{n}^{1}\left(\frac{Qa}{Qs} \times C\right) \times 10,$$

where C = cover on a scale chosen by the investigator, n = number of species at the site, Qa = the average number of species growing with a species regardless of substratum and Qs = the average number of species growing with a particular species on the substratum under consideration.

Index of Toxiphoby, Index of Poleophoby, Index of Poleotolerance: qualitative scales ranking species according to their tolerance of polluted areas; formerly used in the calculation of the Index of Atmospheric Purity (q.v.).

Index species: see *Characteristic species.*

Indicator species: a species whose presence or absence shows whether or not particular conditions prevail or have prevailed at a site.

-ion: suffix denoting a syntaxon in the rank of alliance (q.v.) or federation (q.v.).

Isopleth: contour-like line joining points of an equivalent concentration or level of a particular factor (e.g. sulphur dioxide, rainfall).

Isotherm: contour-like line joining points with the same average temperature.

Kryptolichenes (Germ.): collective term for endolithic (q.v.), endoxylic (q.v.) and endophoeoic (q.v.) lichens; cf. *Epilichenes*.

Lacustrine: pertaining to lakes.

Late Cenozoic era: geological period of time covering the Quaternary period and the later part of the Tertiary; when applied to Antarctica it covers the last 10 million years or so (Pliocene–Pleistocene–Recent epochs).

Lichen desert: area in a town or around a pollution source from which all lichens, or at least all macrolichens on trees, are absent.

Lichenicolous: growing on lichens.

Lichenometry: measurement of lichen growth; specifically used as a technique for dating rock surfaces by the size of lichen thalli.

Life-form: the physical relationship between a morphological form and its substratum, e.g. endophloeodal, hypophloeodal, epilithic, endolithic, crustose, foliose, fruticose.

Lignicolous: growing on decorticate wood (lignum).

Littoral: of the zone on rocky shores which is emersed (q.v.) and submersed by the tide on sheltered shores, but may be continuously emersed on very exposed shores.

Lusitanian: formerly used of species mainly confined to northern Spain and Portugal extending northwards to include the extreme south and west of Britain.

Marine: see *Littoral*.

Maritime: see *Supralittoral*.

Mesic-supralittoral: wettest part of the supralittoral (q.v.), which is occasionally washed by splashing waves.

Mesotrophic: of a water-body which is intermediate between nutrient-poor (oligotrophic) and nutrient-rich (eutrophic).

Metallophytes: species confined to substrata with very high heavy metal contents; see also *Pseudometallophytes*.

Microhabitat: a particular ecological niche within a more broadly defined habitat, e.g. rock underhang, leaf surface, bark crevice.

Minimal area: the least area which includes all the characteristic species (q.v.) of a community; see also *Species-area curve*.

Morphotype: a morphologically differentiated group of specimens of a taxon, of undetermined or no taxonomic importance.

Muscicolous: growing on or over mosses.

Neutrophilous: preferring basic or neutral substrata.

Nitrophilous: preferring substrata rich in nitrogenous compounds.

Nodum (pl. *noda*): (1) peak (e.g. concentration of records) in a continuum; (2) a neutral phytosociological term for a community of any rank; (3) a syntaxonomic rank below that of association (q.v.).

Obligate: restricted to a particular host, substratum or method of nutrition; cf. *Facultative.*

Old-forest indicator species: species normally found only in mature or extremely old stands of forest or ancient parklands; used in calculation of Index of Ecological Continuity (q.v.).

Oligotrophic: of a water-body which is poor in nutrients; see also *Mesotrophic.*

Ombrophilous: rain-loving.

Omnicolous: not confined to specific substrates.

Order: a syntaxon below the rank of class comprising one or more alliances (phytosociological); names of orders end in the suffix *-etalia* (e.g. *Leprarietalia*).

Ordination: method of numerical analysis, summarizing similarities between communities, and species-distributions.

Ornithocoprophilous: preferring substrata (or substrates) rich in bird excrement.

Oroarctic: pertaining to a treeless vegetational zone north of the orohemiarctic zone (q.v.).

Oroboreal: pertaining to an altitudinal zone in mountains which closely corresponds to a latitudinal boreal zone.

Orohemiarctic: pertaining to a vegetational zone having few trees and north of the boreal (coniferous forest) zone; = subarctic of many authors.

-osum: suffix denoting an infra-associational syntaxon (e.g. the rank of subassociation).

Oxydated, oxidated: of thalli characterized by reddish-brown deposits of iron oxides often derived from the substratum (see Chapter 2).

Palaearctic: pertaining to a region comprising most of the Old World north of the tropics (i.e. Europe, Asia north of the Himalayas, northern Arabia and Africa north of the Sahara).

Pangaea: the original solitary continental land-mass that existed until the Carboniferous period (340 million years ago), and later fragmented through continental rafting to form Gondwanaland, Angara and Euramerica.

Pantropical: occurring in tropical regions throughout the world.

Parasite: an organism living in or on another living organism (host) and obtaining its nutrients from it to the detriment of the host; see also *Pathogen.*

Parasymbiont: an organism symbiotic with a pre-existing symbiosis (e.g. a lichenicolous fungus not damaging its host).

Pathogen: a parasite able to cause disease, and usually death, in its host.

Pedogenesis: the process of soil formation.

Pedogenic: soil-forming; sometimes incorrectly termed as pedogenetic.

Permanent quadrats: quadrats marked out and studied at intervals over a period of time (usually several years).

Pesticides: used as a general term to include fungicides, herbicides, insecticides and similar chemicals employed in the control of pests or diseases of crops; an alternative preferred usage is plant protective chemicals.

Phenotype: the outward appearance of an individual (i.e. the response of the genotype to environmental factors).

Phorophyte: the host tree of an epiphyte.

Photophilous: light-loving.

Phycotype: either one of two morphologically different structures formed by the interaction of a single mycobiont with different phycobionts.

Phyllosphere: the microhabitat of a leaf and its component organisms.

Phytocoenosis: phytosociological taxon comprising one or more synusiae (q.v.).

Phytosociology: the study of plant communities; usually applied only to syntaxonomic investigations.

Pioneer: the first species or communities establishing themselves on a particular substratum.

+ *(plus):* (1) species occurring in the community outside the quadrat limits of a record (q.v.); (2) an indication of cover value (see *Domin Scale*).

±: more or less.

Pluviophilous: rain-loving.

Poikilohydric: conforming to external changes in water availability.

Poikilothermic: conforming to external changes in temperature.

Poleophilous: favouring urban habitats.

Poleotolerance of lichen synusiae (P):

$$P = \sum_{i=1}^{n} \frac{a_i \times c_i}{C_i},$$

where a_i = the degree of poleotolerance of each species (assessed on a ten-category scale), C_i = the total lichen cover, c_i = the percentage cover of each species, and n = the number of species.

Polymorphic, polymorphous: occurring in a variety of morphological forms.
Population: individuals of a species growing together in close proximity in a particular habitat.
Preferential species: characteristic species (q.v.) occurring $\pm$ abundantly in several communities, but reaching their optimal development in only one of them.
Primary species: see *Species pair.*
Pseudometallophytes: Facultative metallophytes; species tolerant of, and thriving on, substrata with very high heavy metal contents but not confined to them; see also *Metallophytes.*
Psychrophilous: preferring cold habitats or climates.
Pure stand: a community in which a single species predominates (usually with less than six other species present, these having a low cover value).
Quadrat: a circumscribed area for sampling or study; a rectangle or square marked out in a community within which species and such attributes as frequency-abundance are determined.
Recent Epoch: geological period of time covering the last 5000 years approximately.
Record: (1) a list of species in a community with degrees of presence (phytosociological); (2) a report of a species from a particular locality (biogeographical).
Relevé (Fr.): record (phytosociological).
Relic (noun): a species or community remaining after the loss, from whatever cause, of its associates; cf. relict (adjective).
Rhytidome: dead outer bark tissues of a host tree.
Richness Index (R.I.): $R.I. = \dfrac{\text{number of genera} \times 10}{\text{number of species}}$
Saprobe: see *Saprophyte.*
Saprophyte: An organism deriving its nutrients from dead or diseased organisms or other decaying organic material; often instrumental in the process of decay.
Saxicolous: growing on or inhabiting rocks.
Secondary species: see *Species pair.*
Selective species: characteristic species (q.v.) found most frequently in one community, but occurring sparsely in others.
Sere: one plant community in a succession of communities.
Siliceous: used of rocks with a high silicate content; often misused to imply non-basic or acidic.
Silicicolous: growing on siliceous rocks.
Skiophilous (*sciophilous*): shade-loving.

Sociability: the pattern of occurrence of a species within a community indicated in phytosociological records on a five-point scale; placed after a full-stop following a cover value by the Braun-Blanquet School (e.g. 2.3): 1 (growing singly), 2 (grouped or tufted), 3 (small patches), 4 (extensive patches), 5 (pure populations); cf. *Abundance.*

Sociation (soc.): a community lacking faithful (q.v.) or differential (q.v.) species but having a single dominant and some constant species (Uppsala School); usually taken as a rank less than that of subassociation (q.v.) or union (q.v.).

Society: (1) subdivision of a consociation (q.v.) characterized by subordinate species; (2) a community in which a single species predominates permanently or seasonally; (3) sociation (q.v.).

Sociology: the study of communities.

Sørensen Coefficient (K):

$$K = \frac{200 \times c}{a + b},$$

where a = number of species in one region, b = number of species in other region, and c = number of species common to both regions; used numerically to compare similarities of floras or other species lists.

Species-area curve: a graph of the number of species found against the size of the area surveyed; used to determine minimal area (q.v.).

Species pair: concept that existing species reproducing mainly, or only, by vegetative means (secondary species) have been derived from extinct or extant species with similar chemistries and reproducing mainly, or only, by sexual means (primary species).

Stand: a community growing in a particular locality; sometimes used in the sense of record (phytosociological).

Stem-flow: rainwater flowing down the trunk of a tree.

Stenoionic: with a small pH amplitude; cf. *Euryionic.*

Subarctic: see *Orohemiarctic.*

Subassociation (subass.): a non-geographical subdivision of an association; equivalent in rank to that of variant (q.v.).

Subcuticular: penetrating beyond the cuticle of the host leaf.

Subhydrophilous: requiring periodic submergence in freshwater; amphibious.

Submesic-supralittoral: seldom splashed zone above mesic-supralittoral (q.v.), dominated by *Xanthoria parietina.*

Substitution: the phytogeographical replacement of one species by another, often closely allied, one (i.e. vicariad, q.v.); *regional substitution:* geographical replacement in a single large area (e.g. northern Europe v. Mediterranean); *intraregional (local) substitution:* habitat replacement, both species occupying the same general area.

Substrate: the material or medium in or on which an organism grows (and from which it usually derives nutrients); used more widely as synonymous (perhaps incorrectly) with substratum (pl. substrata); specifically used as a chemical term defining a medium or matrix.

Substrate switch: the transition of a species from growing on one substrate to another, usually in response to a change in environmental factor(s), e.g. microclimatic, pollution.

Substratohygrophilous: species requiring substrata with a high moisture content.

Substratoid: epiphytes with a narrow edaphic and wide climatic amplitude.

Subvariant: geographical subdivision (phytosociological) of a variant (q.v.).

Succession: the replacement of one community by another in the same site.

Supracuticular: not penetrating the cuticle of the host leaf.

Supralittoral: zone on rocky shores never submersed by the tide but containing non-terrestrial lichens; it has three subzones: submesic-, mesic- and xeric-supralittoral.

Sympatric: occupying the same geographical area; cf. *Allopatric.*

Synanthropic: of species most commonly occurring near human dwellings.

Synecology: study of the ecological relationships of communities or ecosystems.

Synergistic: of factors acting together to produce a different, often more marked, effect than they would independently.

Synsystematics: the study of the classification (taxonomy *and* nomenclature) of communities; see also *Syntaxonomy.*

Syntaxon (pl. *syntaxa*): a plant community of any rank; a neutral term not denoting any particular rank; equivalent to taxon in idiotaxonomy.

Syntaxonomy: classification and naming of communities, contrasting with idiotaxonomy, the classification of organisms.

Synusia (pl. *synusiae*): a community composed of species with similar habits and ecological requirements.

Systematics: the study of the taxonomy *and* nomenclature of species or communities.

Taiga: the boreal zone (q.v.) or that dominated by coniferous forest.

Taxon (pl. *taxa*): a taxonomic (idiotaxonomic) group of any rank; a neutral term; cf. *Syntaxon.*

Taxonomic spectrum (T): expression of the number of genera, and species within them, in a particular community or regional flora; see also *Richness Index.*

Terrestrial: on or of the land or earth (used in a more limited sense to describe a zone of rocky sea-shores where species predominate which are also widespread in inland sites).

Terricolous: growing on the ground.

Throughfall: rainwater falling directly to the ground; cf. *Stem-flow*.

Toxiphobous: intolerant of pollution.

Transect: a line or narrow belt, usually along an ecological or topographical gradient, along which sampling is carried out.

Transplant: material moved in a living state, with or without its immediate substrate, to a new site.

Type record: record (q.v.) serving as the nomenclatural type of a syntaxon in the rank of association or below.

Ubiquitous: used of a species or community occurring $\pm$ everywhere on a world-wide or more local scale.

Union: a community consisting of one or more species of a similar habit (Uppsala School); phytosociological taxa given this rank are usually taken as equivalent to that of association (q.v.).

Vagant: used of unattached lichens (see Chapter 2).

Vagrant: see *Vagant*.

Variant (*var.*): a geographical subdivision of an association (phytosociological), equivalent in rank to that of subassociation (q.v.); this use of the abbreviation *var.* is not to be confused with the idiotaxonomic rank of variety.

Vicariad: a taxon derived from another closely related to it but confined to a different ecological or geographical area.

Vitricolous: growing on glass.

Wanderflechten (Germ.): "wandering lichens"; see *Erratic* and *Vagant*.

Xeric: pertaining to a limited water supply or drought conditions.

Xeric-supralittoral: zone of rocky shore receiving sea-water only as wind-borne spray; drought-prone part of the shore.

Xerophyte: plant capable of living in dry places, with morphological adaptations to reduce water loss (i.e. xeromorphism).

Zonal: of major latitudinal vegetation and soil patterns related to the climate; cf. *Azonal*.

Taxonomic Index

For names of syntaxa see Subject Index.

A

Abies, 168
  pectinata, 324
Abrothallus parmeliarum, 16
Acacia tortilis, 240
Acari, 79–83, 88, 96–99
Acarospora, 11, 14–15, 26–27, 54, 218–219
  anomala, 420
  atrata, 386
  bella, 221
  cervina, 221
  fuscata, 262, 277, 283, 371, 378–379, 381–382
  glaucocarpa, 350, 363
  gwynnii, 193, 197
  heppii, 275, 350
  macrospora, 350
  schleicheri, 19
  sinopica, 33, 286, 383
  smaragdula, 24, 283, 383
    var. *lesdainii*, 383
  strigata, 221
Acer, 48, 147, 168, 308, 331
  campestre, 347
  pseudoplatanus, 348
Acrocordia, 305
  conoidea, 350, 355, 357–358
  gemmata, 325, 346
  monensis, 397
  salweyi, 350, 387, 401
Adineta barbata, 73

Aedes, 95
Aegithalos caudatus, 127
Agama atricolis, 124
Agrestria, 22
Agyrium rufum, 170
Alaskozetes antarcticus, 97
Alces alces, 134
Alectoria, 15, 49, 129, 261, 305, 306, see also *Bryoria* and *Pseudephebe*
  "*jubata*", 169
  nigricans, 190–191, 201, 377, 406
  ochroleuca, 134, 406
  sarmentosa, 151, 340
    ssp. *vexillifera*, 405–407
Alnus, 147, 168, 315, 324, 327, 330, 334
Amazilia yucatanensis, 126
Ameronothrus maculatus, 85, 92
Amphibia, 122–123
Amphigerontia bifasciata, 75, 77
Amphipoda, 98
Anaptychia, 103, 305, see also *Heterodermia ciliaris*, 160, 334, 342, 345–346, 348
  var. *melanosticta*, 160
  fusca, 262, 268, 348, 378, 386
  mamillata, 160, 386
Andreaea, 49
  alpina, 373
  nivalis, 373
  petrophila, 376
  rothii, 376
  rupestris, 376–377

Aneides
  aeneus, 122, 124
  lugubris, 122
Anolis ortonii, 125
Anopheles, 95
Anthrocothecium pyrenuloides, 315
Antitrichia curtipendula, 325
Anurophorus laricis, 75
Aphanostigma pyri, 102
Aphelenchoides parietinus, 72
Apterygota, 75
Arachnida, 82–83
Araeolaimidae, 72
Araneae, 82–83
Archella, 71
Archilochus
  alexandri, 126
  colubris, 126
Archisotoma brucei, 97
Arctomia delicatula, 327
Arctostaphylos uva-ursi, 402
Arion, 84
Arthonia, 241
  arthonioides, 370
  aspersella, 315
  atlantica, 388
  caesia, 74
  clemens, 10
  didyma, 325
  impolita, 261, 305, 307, 309–310, 346
  leucopellaea, 309
  lobata, 387–388
  lurida, 316–317
  phaeobaea, 385
  punctiformis, 315–316
  radiata, 315, 317–318
  spadicea, 317
  stellaris, 316
  tumidula, 315, 317
Arthoniales, 304
Arthopyrenia, 83, 389, see also *Acrocordia*
  fallax, 315
  halodytes, 15, 103, 349, 389
  punctiformis, 315

Arthopyrenia—continued
  saxicola, 350
  subarenisida, 402
Arthothelium, 315
  ilicinum, 316
  ruanum, 315
Arthropoda, 97, 109
Asahinea scholanderi, 152
Ascaris suum, 95
Aspicilia, 11, 14–15, 22–24, 26, 379, 394–395
  affinis, 230
  alboradiata, 23
  caesiocinerea, 378–379
  calcarea, 33, 90, 221, 255–256, 283, 350, 353, 358, 395, 427
  cinerea, 171
  contorta, 256, 350, 395
  desertorum, 136
  disserpens, 23
  esculenta, 19, 22, 136, 221, 230–231
  gibbosa, 267, 283, 392
  jussufii, 136
  lacustris, 391–392
  laevata, 391
  leprosescens, 378, 385
  prevostii, 350
  verrucosa, 400
  virginea, 23
Astigmata, 79, 82
Astragalus danicus, 399
Athelia arachnoidea, 322
Atriplex
  stipitata, 242
  vesicaria, 242
Attagenus piceus, 94
Aves, 125–128
Azilia, 82

B

Bacidia, 218–219, 262, see also *Micarea*
  affinis, 325
  biatorina, 325
  chlorococca, 162, 261, 305, 316, 320
  circumspecta, 261

Taxonomic Index

Bacidia—continued
  cuprea, 350
  herbarum, 398
  incompta, 345
  inundata, 256–258, 265–267, 274, 284, 391–393
  muscorum, 256, 396–398
  phacodes, 343
  rubella, 342–343
  sabuletorum, 261, 350, 362–363, 396–397, 400, 402
  scopulicola, 387
  stipata, 194
  umbrina, 54, 171, 256, 258, 260, 265–266, 282, 284, 362–363, 379, 383
Baeomyces, 422
  roseus, 191, 403, 406, 421
  rufus, 32, 275, 369–370, 402–403, 406, 426
Balanus balanoides, 83
Barbilophozia attenuata, 329
  floerkii, 329
Bats, 134–136
Bazzania trilobata, 329
Belgica antarctica, 97
Belonia russula, 362
Betula, 147, 168, 264, 315, 327, 336
  ermanii, 147, 169
  pubescens, 328
    ssp. *tortuosa*, 147
Biatorella
  campestris, 421
  fossarum, 362, 400
  moriformis, 261
  ochrophora, 325
Biston betularia, 106–107
  f. *carbonaria*, 106–107
  f. *insularia*, 106–107
  f. *typica*, 106–107
Bivalvia, 83–84, 98
Blattella germanica, 94
Botrychium lunaria, 399
Bruceia pulverina, 78
Bryobia, 89, 102
  praetiosa, 89

Bryophyta, 348, 358, 396
Bryoria, 49, 58, 129, 162, 305, 306, 430
  bicolor, 374
  capillaris, 327, 340
  chalybeiformis, 200
  fremontii, 136, 154
  furcellata, 155, 159
  fuscescens, 260, 264, 335–336, 342, 374, 429
  lanestris, 261
  smithii, 330
Bryum
  alpinum, 392
  capillare, 310, 398
Buellia, 24, 54, 188, 218–219, 228, 305, 342
  aethalea, 363, 371, 387, 394–395
  alboatra, 256, 261, 278–279, 362, see also *B. epipolea*
  aspersa, 394–395
  asterella, 398
  canescens, 162, 228, 255–256, 258, 274, 277–281, 287, 307, 309–310, 343, 346, 348, 350, 354, 360, 378, 386
  coniops, 201
  disciformis, 150
  epigaea, 154–155, see also *B. asterella*
  epipolea, 350
  frigida, 186, 197–198
  griseovirens, 260, 318
  leptoclinoides, 388
  pulverea, 320
  punctata, 55, 255–256, 258, 260, 265, 307, 343, 346, 378, 394
  russa, 205
  schaereri, 308
  stellulata, 279, 379, 386, 395
  subalbula, 221
    var. *fuscocapitellata*, 240
  subdisciformis, 387
  verruculosa, 387, 394–395
Buteo
  lineatus, 126
  lineatus elegans, 126
  swainsonii, 126

C

Caliciales, 304–305, 307
Calicium
  abietinum, 261, 307
  glaucellum, 261, 264, 307
  hyperellum, 307, see also *C. viride*
  quercinum, 261
  salicinum, 261, 308
  trabinellum, 261
  viride, 308–309, 311, 336, 346
Callidina, 73
Calluna, 50, 53, 57, 320–321, 402–404, 406, 427
  vulgaris, 402
Caloplaca, 33, 80, 97, 218–219, 305, 342, 349–350, 353, 355–356, 358, 378, 384
  arnoldii, 387
  atrosanguinea, 240
  aurantia, 234, 350, 353–354, 427
  aurantiaca, 40, 256, 261, 279, 362–363, see also *C. ruderum*
  cerina, 221, 256, 261, 277, 283, 319
  chalybaea, 274, 350
  cirrochroa, 350, 355, 358
  cirrochrooides, 205
  citrina, 255–256, 258, 260, 262, 265, 268, 273, 279, 282–284, 287, 350, 361, 363, 395–397
  decipiens, 262, 273, 283, 350, 355
  ehrenbergii, 234
  ferruginea, 92, 386
  flavovirescens, 362
  furfuracea, 261
  heppiana, 33, 35, 268, 272–273, 277–279, 284, 350, 353, 358, 427
  holocarpa, 255–256, 260, 279, 284, 350, 395
    var. *pyrithroma*, 279
  lactea, 350
  lithophila, 265
  littorea, 387
  luteoalba, 240, 256, 283, 345
  marina, 384–385
  microthallina, 384

Caloplaca—continued
  ochracea, 356, see also *C. tetrasticha*
  regalis, 194
  ruderum, 273, 350
  saxicola, 161, 221, 256, 262, 278–279, 283, 350, 354, 358
  scopularis, 378, 385
  stillicidiorum, 256, 355, 362, 400
  teicholyta, 280, 350, 353
  tetrasticha, 350, 356
  thallincola, 384–385
  variabilis, 256, 350, 353
  velana, 350
  verruculifera, 262, 378, 385
  vitellinula, 256, 279
  vitricolor, 280
Calytpe anna, 126
Camelus, 134
Camisia segnis, 80
Campecopea hirsuta, 83, 88
Camptothecium
  lutescens, 398
  sericeum, 325, 347
Candelaria concolor, 74, 256, 261, 342
Candelariella, 287
  arctica, 378, 385
  aurella, 256, 262, 268, 273, 279, 282, 284, 350, 355
  coralliza, 378
  medians, 92, 273–275, 350, 361
  reflexa, 343
  vitellina, 171, 255–256, 260, 270, 273, 275, 282–284, 332, 343, 346, 362–363, 371, 378–379, 381–383
Capra ibex, 134
Capreolus capreolus, 134
Carabodes labyrinthicus, 109
Carex
  arenaria, 399
  ericetorum, 399
  paniculata, 403
  rupestris, 400
Carlina vulgaris, 399
Carpinus, 317
Catillaria, 169, 422
  atropurpurea, 325

Catillaria—continued
  chalybeia, 262, 279, 366, 379, 385, 390–392, 394–395
  contristans, 406
  corymbosa, 193
  erysiboides, 279
  graniformis, 261–262, 264
  griffithii, 261, 264, 307–308, 310, 332–333, 346
  lenticularis, 350, 358, 363
  lightfootii, 261, 318
  littorella, 387
  melanobola, 265–266
  sphaeroides, 283, 325
Cavernularia
  hultenii, 151, 158–159
  lophyrea, 153, 163
Cebysa leucoteles, 103
Cecidomyiidae, 79
Centaurium erythraea, 399
Centropyxis, 71
Ceratozetes cisalpinus, 89
Ceratrocha cornigera, 73
Cerobasis guestfalicus, 93
Cervus
  elaphus, 134
  elaphus roosevelti, 133
Cetraria, 15, 49, 96–97, 165, 259, 305, 306, 422
  annae, 152
  aurescens, 153
  canadensis, 153, 164
  chlorophylla, 55, 154, 162, 335–336, 338, 374
  ciliaris, 155, 159, 162
  commixta, 374
  cucullata, 129, 134
  delisei, 167, 373, 406
  ericetorum, 165, 406
  halei, 159
  hepatizon, 374
  islandica, 129, 132, 134, 136, 165, 167, 234, 403, 406–407, 421, 431
  juniperina, 93, 164, 337
  kamczatica, 152
  komarovii, 152

Cetraria—continued
  laevigata, 152, 165
  merrillii, 153
  nivalis, 95, 129, 134, 149–150, 234, 373, 405–407
  pinastri, 168, 261, 264, 337
  pseudocomplicata, 152
  richardsonii, 20, 26, 152
  sepincola, 261, 337–338
  viridis, 164
Cetrelia, 305
  olivetorum, 326, 329–330
Chaenotheca
  brunneola, 307, 310
  carthusiae, 308
  chrysocephala, 261
  ferruginea, 261, 308–309
  hispidula, 308
  laevigata, 308
  phaeocephala, 261, 264
  trichialis, 307–308
Chaenothecopsis debilis, 261
Chamaea fasciata, 126
Chelicerca galapagensis, 76
Chelonethi, 83
Chiodecton
  myrticola, 388
  petraeum, 388
Chondrina avenacea, 108
Chondropsis, 238
  semiviridis, 19, 23, 26, 221, 227–231, 233–237
Chrysopa, 103
Chthamalus stellatus, 83
Cirripedia, 83, 98
Cladina, see *Cladonia* subgen. *Cladina*
Cladium mariscus, 265
Cladonia, 15, 36, 47, 50–51, 54, 57–58, 71, 73–75, 79, 128, 134–135, 149, 165–167, 169, 171, 203, 259, 264–265, 275, 305, 313, 330, 338, 370, 374, 394, 401, 403, 405–406, 422, 426
  subgen. *Cladina*, 36, 165, 170, 259, 407
  aberrans, 154

Cladonia—continued
  alaskana, 152, 158, 165
  alpestris, see *C. stellaris*
  amaurocrea, 129, 165
  arbuscula, 16, 36, 129, 159, 165, 167, 403, 405
    ssp. *beringiana*, 164
  bacillaris, 169, 261, 265, 403
  bacilliformis, 158, 169
  bellidiflora, 151, 405–407
  boryi, 153
  botrytes, 169, 313, 403
  caespiticia, 314
  carneola, 169, 313, 403
  cenotea, 169, 313, 403
  cervicornis, 374, 401, 403
  chlorophaea, 54, 256, 265, 275, 313, 329, 374, 401, 403–404, 427
  coccifera, 338, 371, 374, 381–382, 403, 405–407, 427
  coniocraea, 150, 168, 260, 310, 312–314, 322, 329, 369, 403
  conista, 256
  conistea, 401
  cornuta, 149, 165
  crispata, 149, 165, 403, 407
    var. *cetrariaeformis*, 374, 403
  cristatella, 153, 157, 165
  cyanipes, 57
  deformis, 313, 406
  digitata, 313
  elongata, 57, 234
  fimbriata, 54, 91–92, 256, 261, 274, 313, 322
  floerkeana, 151, 403, 427
  foliacea, 72, 152, 234, 398, 401–402
  fragillissima, 405
  furcata, 57, 167, 256, 261, 313, 374, 398, 401, 403–404, 425–426
  glauca, 403
  gonecha, 165, 313, 403, 405–406
  gracilis, 167, 313, 374, 403, 405–406
    var. *dilatata*, 165
    var. *nigripes*, 164
  impexa, 36, 164, 261, 403–407, 422–423

Cladonia—continued
  kanewskii, 152
  luteoalba, 406
  macilenta, 258, 260, 265, 313–314, 369
  mitis, 36, 129, 149, 165
  multiformis, 153, 167
  nemoxyna, 313
  nylanderi, 401
  ochrochlora, 91, 329
  pacifica, 164
  parasitica, 310, 314
  phyllophora, 149
  pityrea, 261, 265, 382, 401, 403
  pocillum, 159, 360, 397–398, 400, 402
  polydactyla, 403, 405
  pseudomacilenta, 151
  pseudorangiformis, 155, 165
  pseudostellata, 152, 164
  pyxidata, 54, 56, 73, 256, 261, 313, 338, 401
  rangiferina, 36, 129, 132, 149–150, 156, 163, 165, 167, 194, 201, 234
  rangiformis, 152, 397–398, 400–402, 425
  scabriuscula, 158, 167, 313
  squamosa, 167, 313, 338, 369, 371, 374, 382, 403, 405
    var. *allosquamosa*, 403
  stellaris, 36, 57, 71, 129, 133, 135–137, 154, 156, 165–166, 423
    var. *aberrans*, 155, 164, see also *C. aberrans*
  strepsilis, 151, 403
  subcervicornis, 374
  subfurcata, 167
  subulata, 275, 403
  "*sylvatica*", 234
  tenuis, 151, 403, 405–406
  terrae-novae, 164
  uncialis, 57, 129, 149, 164–165, 406
    ssp. *dicraea*, 164, 374, 403, 407
  verticillata, 401, 403, see also *C. cervicornis*
Cladophora, 71

Clathroporina calcarea, 350, 356, 358, 363
Clausilia, 88
Clemensia albata, 78
Coccinella, 102
Coleoptera, 78–79
Collema, 34, 228, 244, 327, 350, 359, 425
   auriculatum, 350, 359–360, 396
   ceraniscum, 400
   coccophorum, 240, 244
   crispum, 279, 350, 359–360, 397
   cristatum, 350, 359, 363
   fasciculare, 327
   flaccidum, 256
   fluviatile, 393
   furfuraceum, 327
   multipartitum, 351, 355, 359
   nigricans, 327
   occultatum, 327
      var. *populinum,* 164
   polycarpon, 351, 359, 363
   pulchellum, 155
   subflaccidum, 327
   tenax, 351, 396–398, 400
   tuniforme, 351, 359
   undulatum, 351, 359
Collembola, 75, 94, 96–99, 101, 105, 109
Cololejeunea minutissimus, 332
Colpoda, 71
Comacla simplex, 78
Coniocybe
   furfuracea, 53, 308, 367–368
   sulphurea, 308
Conocephalum conicum, 369
Contopus
   pertinax, 126
   virens, 126
Cora pavonia, 234
Coriscium viride, 405
Cornicularia, 58, 422
   aculeata, 191, 194, 201, 261, 264, 329, 377, 398, 403, 406
   divergens, 373, 406
   muricata, 403, 406

Cornicularia—continued
   normoerica, 373–374, 376–377
   odontella, 155
Corylus, 315, 323, 326, 396
Crambidia
   allegheniensis, 78
   casta, 78
   pallida, 78
Cryptognathus lagena, 80
Cryptopygus antarcticus, 97
Cryptostigmata, 79–81, 85–86, 88–89, 92, 98, 109
Curculionidae, 103
Cyanophyceae, 392, 397
Cymatomera, 104
Cynanthus latirostris, 126
Cynodontium polycarpon, 376
Cyphelium, 264, 304
   inquinans, 260, 264, 305, 320–321
   notarisii, 259, 261–262, 321
   ocellatum, 261
   sessile, 16
   tigillare, 170, 261
Cyrtolaelaps racovitzai, 82, 97
Cystocoleus niger, 194, 201, 367–370

D

Dactylina, 203, 205
Daidalotarsonemus, 80
Dendriscocaulon, 14
Dendroica
   cerulea, 126
   dominica, 126
   fusca, 126
   occidentalis, 126
   striata, 127
   townsendi, 126
Dermaptera, 76
Dermatina
   quercus, 53, 315
   swinscowii, 315
Dermatocarpon, 14, 218–219
   cinereum, 400
   fluviatile, 391–392
   hepaticum, 221, 396–398, 400, 402

Dermatocarpon—continued
  lachneum, 244, see also *D. hepaticum*
  meiophyllum, 391
  miniatum, 221, 351, 356, 359
  rivulorum, 392
Desmazieria, see also *Ramalina*
  pulchribarbara, 231
Desmococcus, 75–77, 281, 320
  vulgaris, 322
Diaspididae, 102
Dicranella heteromalla, 368
Dicranoweissia cirrata, 334
Dicranum
  fuscescens, 329
  scoparium, 329–330
Dimelaena
  oreina, 38
  radiata, 24
Dimerella lutea, 325
Diplophyllum albicans, 329, 368–369
Diploschistes, 219, 224
  caesioplumbeus, 279, 380, 387
  gypsaceus, 351, 355
  ocellatus, 223, 225, 240–241
  scruposus, 26, 221, 223, 225, 234, 262, 369–371, 379, 425
    var. *bryophilus*, 398, 402
  steppicus, 234
Diptera, 79, 98
Dirina
  repanda, 351, 356
  stenhammeri, 280, 351, 356, 358, 387
Dirinaria picta, 125
Ditrichum flexicaule, 399
Ditylenchus, 72
  intermedius, 72
Dometorina, 94
  plantivaga, 80, 85–86, 92, 109
Drepanocladus, 96
Dryas octopetala, 400

E

Elipsocus
  hyalinus, 76
  mclachlani, 76–77, 91, 93

Elipsocus—continued
  westwoodii, 76–77
Embioptera, 76
Embolidium italicum, 170
Empetrum, 164
  nigrum, 402, 406–407
Empididae, 79
Encalypta streptocarpa, 399
Encephalographa cerebrina, 351
Enchytraeidae, 73
Endocarpon, 219, 244
  pusillum, 22
Endopterygota, 75
Enterographa
  crassa, 310, 317
  hutchinsiae, 368, 387, 390
Ephebe lanata, 392
Epicriidae, 82
Epidiaspis leperii, 102
Eremastrella tobleri, 228
Erica
  cinerea, 402
  tetralix, 402
Erigeron acer, 399
Erioderma pedicellatum, 151
Eriophorum vaginatum, 131
Eriophyoidea, 82
Eucleidae, 103
Eudorylaimus carteri, 72
Eugenes fulgens, 126
Euglypha, 71
Euopsis granatina, 362
Euphagus carolinus, 127
Euphorbia, 71
Euphrasia officinalis, 399
Evernia
  esorediosa, 163
  mesomorpha, 129, 155, 159, 169
  prunastri, 16, 90, 93, 127, 137, 154, 168, 256, 260, 325, 329, 332–333, 335–336, 339, 341, 346, 348
Exopterygota, 75

F

Fagus, 309, 313, 315, 317, 331, 334, 340, 396–397

Fagus—continued
   sylvatica, 337
Festuca, 424
   ovina, 399, 425
   rubra, 425
Ficus repens, 280
Fontinalis antipyretica, 393
Forficula auricularia, 90
Fraxinus, 308, 317, 323, 327, 331, 333, 337, 347, 429–430
   excelsior, 337
Frullania
   dilatata, 315, 332, 369
   fragillifolia, 325
   tamarisci, 325, 329
Fulgensia, 398, 400, 402, 425
   bracteata, 221
   fulgens, 221, 398
Fuscidea
   cyathoides, 56, 369–371, 374, 376–377, 381, 387
   kochiana, 372–373
   taeniarum, 377
   tenebrica, 374, 376, 387

G

Galium verum, 399
Gastropoda, 83, 88, 98
Geisleria jamesii, 273
Gentianella amarella, 399
Geochelone elephantopus, 125
Gloeocapsa, 33
Graphidaceae, 15
Graphina anguina, 316
   ruiziana, 315
Graphis, 39, 54, 313
   elegans, 316–317
   scripta, 38, 40, 316–318
Grimmia, 56
   campestris, 80
   doniana, 373, 377
   trichophylla, 377
Gyalecta
   cupularis, 355, see also *G. jenensis*
   flotowii, 346

Gyalecta—continued
   geoica, 400
   jenensis, 351, 355, 358, 363, 390
Gyalectina, 305
   carneolutea, 343
Gyalidea
   fritzei, 362, 392
   hyalinescens, 365
Gymnomitrion, 377
Gymnomitrium cocinnatum, 305
Gymnopholus, 103
   subgen. *Symbiopholus*, 103
   lichenifer, 103
Gyrophora cylindrica, 376, see also *Umbilicaria cylindrica*

H

Haematomis mexicana, 78
Haematomma, 55
   elatinum, 317–318, 325
   ochroleucum, 366–367, 370
      var. *porphyrium*, 310, 346, 366
   ventosum, 56, 373–374
Halozetes belgicae, 97
Hedera, 312
   helix, 280
Hedwigia ciliata, 171, 377
Helix hortensis, 90
Hemerobius, 103
Hemimetabola, 75
Hemiptera, 76
Heppia, 218–219, 228
   lutosa, 244
Heterodermia, 241
   leucomelos, 340
   obscurata, 326
Hildenbrandtia, 393
Himantormia lugubris, 194, 204
Holometabola, 75
Homoptera, 102
Huilia
   albocaerulescens, 363, 366, 369, 371, 379, 387, 391
   crustulata, 262, 370
   hydrophila, 391

Huilia—continued
  macrocarpa, 262, 336, 370–371, 383, 391
  "f. *oxydata*", 383
Hyadesia halophila, 82
Hyadesidae, 82
Hyale nilssoni, 83
Hyla
  avivoca, 122
  versicolor, 122–123
Hylocharis leucotis, 126
Hylocichla
  minima, 126
  ustulata, 126
Hymenoptera, 79
Hypnum cupressiforme, 168, 310, 329, 332, 334, 347, 369, 376
  var. *filiforme*, 330
Hypogastrura packardi, 75, 91, 101
  tullbergi, 75
Hypogymnia, 305, 306
  bitteri, 155, 158
  duplicata, 153
  enteromorpha, 163
  fragillima, 152
  intestiniformis, 234, 374, 376
  mundata, 152
  physodes, 38, 91, 93–94, 129, 150, 156, 159, 162–163, 167, 171, 256, 258, 260, 262, 274, 284, 322, 329, 332–333, 335–336, 341, 369, 374, 404, 427, 431
  tubulosa, 151, 163, 260, 329, 335–336, 374
Hypoprepia
  fucosa, 78, 136
  miniata, 78, 136
Hypsibius
  oberhaeuseri, 74
  tuberculatus, 74, 109
Hysteriales, 304
Hysterium angustatum, 310, 332

I

Icmadophila ericetorum, 191, 403, 405

Ilex, 315–316
Illice
  nexa, 78
  subjecta, 78
  unifascia, 78
Insecta, 75–79, 97–98
Ionaspis
  epulotica, 351, 363
  suaveolens, 392
Isopoda, 83, 88, 98
Isoptera, 76
Isothecium myosuroides, 310, 325, 329–330, 374
Ixoreus naevius, 126

J

Juncus trifidus, 50

K

Kochia sedifolia, 240–241
Koeleria gracilis, 399

L

Lama huanacos, 134
Lambdina fiscellaria fiscellaria, 102
  fiscellaria lugubrosa, 102
Larix, 147, 331
  decidua, 75
  europaea, 336
  gmelinii, 164
  laricina, 164
  sibirica, 164
Lasaea, 83–84
  rubra, 84, 99, 102
Lasallia
  papulosa, 153
  pennsylvanica, 155
  pustulata, 32, 151, 234, 372, 379, 381–382
Lasiurus cinereus, 135–136
Lecanactidaceae, 304–305

Lecanactis
  abietina, 308–310
  amylacea, 309
  corticola, 309
  dilleniana, 366, 387
  dryophila, 309
  grumulosa, 356
  hemispherica, 273
  homalotropum, 315, 325
  monstrosa, 387–388
  premnea, 309–311
Lecania
  aipospila, 385
  brialmontii, 194
  cyrtella, 256, 265, 343
  erysibe, 256, 262, 277, 279, 283–284, 351, 385
  olivacella, 279
  rupicola, 387
Lecanora, 20, 24, 26, 71, 79, 86, 203, 218–219, 262, 318, 350, see also *Aspicilia*
  actophila, 385
  alphoplacum, 56
  andrewii, 378
  atra, 279, 366, 372, 379, 386–388
  badia, 262, 267, 371, 379
  baranowii, 20–21
  campestris, 255–256, 262, 279–280, 283–284
  carpinea, 256, 315, 318
  chlarona, 256, 260, 264, 318
  chlarotera, 256, 260, 305, 315, 318–319, 332–333, 343, 346, 348
  chrysoleuca, 20, see also *L. rubina*
  coilocarpa, 158
  confusa, 260, 305, 315, 318–319
  conizaeoides, 55, 76–77, 82, 87, 91, 93, 107, 162, 257–258, 260, 265, 268, 274–275, 280, 284, 287–288, 305, 316, 319–322, 332, 336, 346, 383, 427
  crenulata, 277, 279, 351, 363

Lecanora—continued
  dispersa, 35, 54, 161, 255–258, 260, 265, 268, 273, 277–279, 282, 284, 289, 346, 351, 354–355, 358, 387, 394–395, 398, 427
    f. *dissipata*, 268
  epanora, 262, 286, 383
  epibryon, 49, 406
  expallens, 162, 260, 264, 284, 307–308, 310, 318, 332–333, 336, 346, 348, 363
  farinaria, 259, 261
  farinosa, 234
  fugiens, 378, 387
  fuscescens, 150, 158
  gangaleoides, 366, 379, 387–388
  grumosa, 366, 379
  hagenii, 161, see also *L. dispersa*
  haydenii, 20
  helicopis, 385
  intricata, 163, 171, 262, 264, 284, 371–374, 379
    var. *soralifera*, 262, 277, 336, 371–372, 383
  intumescens, 318
  jamesii, 261, 318, 325
  laevis, 319
  leptacina, 373
  mauroides, 377
  melanophthalma, 20–21
  muralis, 35, 38, 45, 89–90, 221, 255–256, 258, 260, 262, 268–273, 275, 283, 345, 351, 361–363, 378
  orosthea, 366, see also *Lecidea orosthea*
  pallida, 93, 305, 318, 335
  piniperda, 260, 264
  pityrea, 319, see also *L. conizaeoides*
  poliophaea, 378, 385
  polytropa, 205, 262, 264, 275, 277, 284, 370–374, 379, 381–383, 394–395
  populicola, 160
  praepostera, 387–388
  quercicola, 325
  rubina, 94–95, see also *L. chrysoleuca*

Lecanora—continued
  rupicola, 10, 86, 262, 366, 379, 381, 387
  saligna, 256, 261
  sambuci, 343
  sarcopisioides, 261
  saxicola, see *L. muralis*
  sordida, 379, see also *L. rupicola*
  straminea, 378, 385
  subaurea, 286, 383
  subcarnea, 366
  subfusca, 92, 318, see also *L. chlarotera*
    var. *allophana*, 88
  subfuscata, 261
  subintricata, 261
  symmicta var. *sorediosa*, 205
  tenera, 387–388
  varia, 260, 305, 319–321
  vinetorum, 420
Lecanorales, 304
Lecidea, 14, 24, 169–170, 203, 218, 262, 422, see also *Fuscidea, Huilia, Lecidella, Micarea* and *Trapelia*
  assimilata, 406
  atrata, 33, 262, 286, 373, 383, 421
  auriculata, 32
  caesioatra, 406
  cinnabarina, 317, 325
  confluens, 373
  decipiens, 220–221, 223–225, 244, 398, 400, 402
  diducens, 32, 387
  elabens, 158, 170
  enteroleuca, 265
  erratica, 262, 267, 370, 394–395
  euphorea, 319
  friesii, 261, 322
  fuscoatra, 171, 371
  geophana, 421
  glaucolepidea, 405
  goniophila, 283
  granulosa, 49, 54, 56, 149, 165, 256, 260, 266, 310, 312, 314, 322, 370, 402–403, 406, 427

Lecidea—continued
  illita, 362–363
  insularis, 10, 379
  jurana, 351, 355
  lapicida, 170, 373, 377, 383
    "f. *oxydata*", 383
  leucophaea, 262, 371–372, 377
  lithophila, 170, 363, 370–372, 377
  lucida, 262, 268, 274, 277, 284, 366–367
  niveoatra, 364
  ochrococca, 261
  oligotropha, 165, 403
  orosthea, 262, 366, 379, 388
  ostreata, 322, see also *L. scalaris*
  pantherina, 371–374, 377, 383
  pelobotryon, 363, 371
  phaeops, 365, 391
  plana, 370
  pulcherrima, 152
  pycnocarpa, 373
  quernea, 260, 310, 332–333, 343, 346, 348
  scalaris, 162, 260, 305, 308, 322, 336
  silacea, 33, 383
  speirea, 362–363, 390
  stenotera, 406
  sulphurea, 10, 274, 374, 379, 381, 387
  symmicta, 256, 260, 305, 315, 318–319, see also *Lecanora symmicta* var. *sorediosa*
  taylori, 365, 391
  templetonii, 355, 400
  tenebricosa, 318
  tenebrosa, 372–374, 377, 379, 387
  tessellata, 25–26
  tumida, 262, 277, 369–371, 373, 379, 387, 391, 394–395
  turgidula, 260
  uliginosa, 49, 54, 165, 256, 260, 265–266, 310, 312, 314, 370, 402–403, 427
  umbonata, 362–363
  umbrina, 274
  wallrothii, 401
  watsonii, 395–396

Lecidella
  elaeochroma, 256, 260, 305, 315, 318–319, 332, 343, 346
  pulveracea, 261
  scabra, 256, 262, 264, 363
  stigmatea, 351, 362–363, 390, 395–396
  subincongrua, 262, 380, 387
Leciographa parellaria, 16
Ledermuelleria, 80
Lehmannia marginata, 91, 93
Lejeunea ulicina, 315
Lemanea, 393
Lempholemma botryosum, 352, 360
Leontodon taraxacoides, 399
Lepidoptera, 76, 78, 102, 105–107
Lepidozia reptans, 329
Lepraria, 304–305, 351, 356–358, 401
  candelaris, 279, 308, 310, 332, 346, 348
  chlorina, 366
  crassissima, 351, 357–358
  incana, 87, 162, 274, 284, 308–312, 321, 332, 336, 346, 351, 357–358, 366–369, 383
  membranacea, 367–369, 383
  neglecta, 170, 374
Leproplaca, 357
  chrysodeta, 351, 357–358
  xantholyta, 351, 356–358
Leptogium, 326–327, 401, 425
  azureum, 327
  brebissonii, 326
  burgessii, 326–327
  cyanescens, 151, 327
  hibernicum, 315, 327
  lichenoides, 325, 351, 359, 397, 402
  plicatile, 351
  saturninum, 327
  schraderi, 396
  sinuatum, 397, 402
  subtile, 396
  teretiusculum, 325, 359
Lepus timidus, 257
Lerina incarnata, 78
Letharia vulpina, 94, 154, 234

Leucodon sciuroides, 347
Lichenobius littoralis, 78
Lichenoconium, 16
Lichina, 83–84, 98, 349
  confinis, 385, 388
  pygmaea, 73, 75, 83–84, 88–89, 98–99, 109, 389
Linum catharticum, 399
Lipura, 75
Lithographa
  dendrographa, 325, 327, 343
  tesserata, 362–3
Lithosiidae, 76, 78
Littorina, 84
Lobaria, 304, 315, 326, 430
  amplissima, 323, 325
  japonica, 152
  laetevirens, 325–326, 387
  oregana, 100–101, 134, 153
  pulmonaria, 74, 91, 93–94, 323, 325–326, 430
  quercizans, 155
  sachalinensis, 152
  scrobiculata, 265, 325–326
Lopadium
  fecundum, 400
  pezizoides, 326
Lophocolea heterophylla, 310
Lophozia ventricosa, 329
Loricula elegantula, 76, 100
Lotia leucoptera, 127
Lotus corniculatus, 399
Lumbricillus, 73

M

Machilidae, 75
Macrobiotus hufelandii, 74
Macrotrachela ehrenbergii, 73
Magellozetes antarcticus, 97
Marmota caligata, 135
Massalongia carnosa, 194, 392
Mastodia tesselata, 201
Maudheimia petronia, 91–92
Medicago lupulina, 399
Meessiinae, 78, 110

Megatherium, 257
Megophrys, 122
Menegazzia, 305
  terebrata, 326, 329–330
Mergus merganser, 126
Mesodorylaimus bastiani, 72
Mesopsocus
  immunis, 75, 93
  unipunctatus, 75, 106–107
Mesostigmata, 79, 82, 97
Metzgeria
  fruticulosa, 332
  furcata, 310, 315, 332
Micarea, 169, 329, 422
  clavulifera, 367
  denigrata, 256, 260
  globularis, 261
  lignaria, 260, 262
  melaena, 261, 406
  nitschkeana, 256, 258, 261, 266, 320
  polioides, 367
  prasina, 261, 264, 314
  semipallens, 367
  subviridescens, 401
  sylvicola, 367
  turfosa, 406
  violacea, 261
Microcaeculus hispanicus, 80
Microglaena
  larbalestieri, 391
  muscorum, 397
  sphinctrinoides, 400
Microphysidae, 100
Microthelia micula, 317
Microtus pennsylvanicus terraenovae, 135
Milnesium tardigradum, 74
Mniobia, 73
Mnium hornum, 310
Moelleropsis nebulosa, 401
Molinia coerulea, 403
Mollusca, 74, 83–85, 88, 93, 109
Moschus moschiferus, 134
Mycobates parmeliae, 80, 86
Mycoblastus, 305
  sanguinarius, 261, 329–330, 336

Mycoporellum sparsellum, 316
Myriapoda, 83
Mytilus edulis, 84

N

Neckera
  complanata, 325
  pumila, 325
Nematoda, 72–73, 110
Neomachilis halophila, 75
Nephroma, 167, 304
  arcticum, 167, 406
  bellum, 158
  helveticum var. *sipeanum*, 152
  laevigatum, 325, 387
Neuroptera, 79
Normandina pulchella, 310, 325–326
Nostoc, 100, 360
Notida pavida, 103
Nuttallornis borealis, 126

O

Ocellularia subtile, see *Thelotrema subtile*
Ochrolechia, 55, 58, 305
  androgyna, 329–330, 335–336, 369, 374, 406, 427
  antarctica, 188
  frigida, 167, 188, 201, 205, 406–407
  geminipara, 406
  parella, 16, 261, 274, 279, 348, 379, 386, 397
  tartarea, 329–330, 374, 406
  turneri, 162, 260, 336, 346
  yasudae, 261, 332–333, 344, 346, 348
Odocoileus
  hemionus, 133
  hemionus columbianus, 133
Oligochaeta, 73, 98
Ommatocepheus ocellatus, 80–81
Omphalodina, 20, 26
Omphalodium
  arizonicum, 21–22
Oncopeltus fasciatus, 94

Oniscus murarius, 90
Ononis repens, 399
Opegrapha, 39
  atra, 261, 315–318, 348
  calcarea, 351, 356
  cesareensis, 387
  chevallieri, 351
  confluens, 387
  gyrocarpa, 367–368, 387
  lichenoides, 313
  lithyrga, 368, 387, 390
  lyncea, 309–310
  mougeotii, 351, 356
  niveoatra, 313
  ochrocheila, 312–313
  persoonii, 351
  prosodea, 309–310, 343
  rufescens, 317
  saxatilis, 279, 351, 355, 366
  saxicola, 279, 351, 355–356, 366
  saxigena, 368
  sorediifera, 325
  subelevata, 356
  varia, 346
  vermicellifera, 312–313
  viridis, 317
  vulgata, 312, 316–317
  zonata, 367–368, 387, 390
Ophniospora atrata, 373
Oppia loxolineata, 97
Orchis simia, 428
Oreamnos americanus, 134
Oreodoxa, 100
Oribatei, 80
Oribatula parisi, 80
Ornithopus perpusillus, 399
Orthodontium lineare, 310, 314
Orthoptera, 76, 104
Orthotrichum, 315
  affine, 315
  anomalum, 376
  diaphanum, 347
  lyellii, 325, 334
  pulchellum, 315
  striatum, 315
  tenellum, 315

Oryctolagus cuniculus, 257
Ovibos moschatus, 134
Oxystyla undata, 85, 93–94

P

Pachyphiale cornea, 304, 325
Panagrolaimus rigidus, 72
Pannaria, 304, 326
  hookeri, 194, 201, 362
  ignobilis, 327
  mediterranea, 325
  microphylla, 387
  pezizoides, 421
  pityrea, 325
  rubiginosa, 163, 325–326
Pannariaceae, 315
Paralorryia mali, 80
Parisotoma octooculata, 97
Parmelia, 38–41, 71, 73–74, 76, 79–80, 82, 100, 103, 218, 220, 230–231, 282, 375, 381
  acetabulum, 152, 160, 305, 334, 342, 344–346
  albertana, 153, 155
  alpicola, 41, 373–374
  amphixantha, 221
  australiensis, 19, 230
  baltimorensis, 37–38, 75, 91, 94, 101
  borreri, 333, 343
  britannica, 378, 382
  camtschadalis, 230
  caperata, 37–38, 46, 90, 237, 260, 265, 305, 310, 331–333, 344, 346, 348
  carporrhizans, 343, see also *P. quercina*
  centrifuga, 41, 79, 83, 97–98, 108–109
  chlorochroa, 20, 26, 220, 230
  conspersa, 33, 37, 40, 51, 86, 171, 221, 262, 275, 379–382
  convoluta, 19, 221, 230–231
  corei, 234
  crinita, 103, 325, 329
  delisei, 387
  disjuncta, 382

Parmelia—continued
 elegantula, 306, 344, 346
 endochlora, 326, 330
 exasperata, 261, 284, 318
 exasperatula, 261, 264, 345
 furfuracea, 335, see also *Pseudevernia furfuracea*
 gerlachei, 194
 glabratula, 260, 310, 325, 329, 332–333, 336, 346, 431
  ssp. *fuliginosa*, 366, 371, 374, 379–381, 387
 glomellifera, 380
 incurva, 372–373
 isidiotyla, see *P. verruculifera*
 laciniatula, 261, 305, 342, 344, 346
 laevigata, 305, 326, 328–330
 loxodes, 381–382, 387
 magna, 234
 molliuscula, 230
 mougeotii, 262, 275, 381–382
 multispora, 153
 olivacea, 150, 159, 162–163
  var. *albopunctata*, 164
 omphalodes, 262, 373–375, 381, 427
  var. *discordans*, 164
 pachyderma, 234
 pastillifera, 343
 perforata, 94
 perlata, 127, 256, 265, 305, 331–333, 341, 346, 348, 430
 physodes, 334, see also *Hypogymnia physodes*
 prolixa, see *P. pulla*
 pseudosaxatilis, 152
 pulla, 387
 quercina, 305, 342, 343
 reddenda, 325
 reptans, 221
 reticulata, 103, 305, 332–334, 386
 revoluta, 20, 265, 305, 325, 332, 334
 rudecta, 153
 ryssolea, 152, 230

Parmelia—continued
 saxatilis, 38, 92, 128, 150, 162, 170–171, 204, 260, 268–269, 276, 306, 322, 325, 329, 332–333, 335–336, 346, 369, 371, 374–375, 381–382, 386, 427, 431
 scabrosa, 281–282
 septentrionalis, 150, 338
 sibirica, 153
 sinuosa, 326
 soredians, 260, 305, 332–334, 344
 squarrosa, 155
 stenophylla, 80
 stygia, 374
 subaurifera, 256, 260, 264, 284, 318, 333, 346, 348
 subolivacea, 153
 subrudecta, 332–333, 346, 348
 sulcata, 38, 150, 162, 168, 256, 260, 274, 310, 332–333, 335–336, 341, 346, 348, 382, 427
 taractica, 171
 taylorensis, 305, 326, 329–330
 tiliacea, 152, 261, 345–346
 tinctorum, 276
 trabeculata, 153
 ulophyllodes, 155
 ushuaiensis, 194
 vagans, 220, 230–231
 verruculifera, 86, 262, 380, 382, 387
 wyomingiaca, 153
Parmeliella, 304, 326
 atlantica, 325–326
 corallinoides, 325
 plumbea, 325–326, 387
Parmeliopsis, 305, 306, 337
 aleurites, 260, 264, 320, 337
 ambigua, 55, 162, 168, 260, 320, 322, 335–337
 hyperopta, 168, 261, 336–337
 placorodia, 153
Parula americana, 126
Parus hudsonicus, 126
Patella, 84, 109
Peltigera, 16, 34, 47, 49, 54, 58, 135, 165, 167, 401, 426

Peltigera—continued
  aphthosa, 167
  canina, 54, 56, 94, 167, 256, 314, 398, 425–426
  collina, 151, 325
  evansiana, 153
  horizontalis, 151, 325
  malacea, 167
  polydactyla, 54, 167, 314, 426
  praetextata, 314, 325
  rufescens, 56, 398, 402
  scabrosa, 167
  spuria, 54, 275, 426
  venosa, 421
Penicillium, 257
Periplaneta americana, 94
Pertusaria, 55, 91, 218–219, 316–317
  albescens, 94, 332–333, 346
    var. *corallina*, 332, 346
  amara, 75, 85, 87, 91, 93, 260, 262, 279, 283, 316, 329, 332–333, 335, 346, 363
  ceuthocarpoides, 380, 387
  chiodectonoides, 387
  coccodes, 16, 40, 260, 333, 346
  corallina, 262, 369, 371–372, 374, 377
  dactylina, 406
  dealbata, 371
  flavicans, 279, 380, 388
  flavida, 333
  glomerata, 362
  graphica, 78
  hemisphaerica, 310, 317, 325, 333
  hymenea, 92, 310, 317, 325, 346
  lactea, 371, 374
  leioplaca, 312, 317–318
  leucosora, 279–280
  monogona, 374, 387
  oculata, 406
  ophthalmiza, 330
  pertusa, 75, 91, 94, 261, 264, 317, 325, 332–333, 337, 346
  pseudocorallina, 262, 363, 371–372, 374, 380, 387
  rupicola, 92
  velata, 325

Petractis clausa, 351, 355, 396–397
Petrobia latens, 82
Petrobius, 75
Peucedramus taeniatus, 126
Phaeographis
  dendritica, 316–317
  lyellii, 316
Phainopepla nitens, 126
Phalangida, 83
Phalangium opilio, 83
Phasmidae, 76
Phauloppia lucorum, 80, 91–92, 98
Phenacomys
  intermedius, 134
  longicaudatus, 134
Philodina, 73, 87
Philotarsus picicornis, 75
Phlyctis
  agelaea, 318
  argena, 310, 318, 332–333, 336, 348
Phragmites australis, 265
Phylloporina elaeospila, 278
Phylloxera, 102
Physcia, 38, 41, 55, 74, 78, 86, 97, 103, 241, 258, 274, 305, 342, 344, 361
  adscendens, 255–256, 258, 260, 265–266, 274, 284, 287, 344, 347–348, 351, 354, 396–397, 427
  aipolia, 344, 347–348
  caesia, 16, 92, 255–256, 261, 265, 273, 283, 345, 351, 361, 378, 397
  ciliata, 163
  clementei, 348
  dubia, 262, 279, 351, 361, 378
  leptalea, 256, 283, 347–349
  lithotea, 279
  millegrana, 74, 153
  nigricans, 351, 361
    var. *tremulicola*, 161
  orbicularis, 92, 161, 256, 258, 260, 262, 265, 273–274, 279, 283–284, 287, 344–345, 347, 351, 354, 361
  setosa, 128
  subobscura, 378, 387
  tenella, 255–257, 260, 283, 339, 344, 347–348, 351, 397

Physcia—continued
  tribacia, 344–345, 348, 360, 378
  tribacioides, 348–349
  wainioi, 351, 361, 378
Physciopsis, 305
  adglutinata, 256, 279, 345, 347–348
Physconia, 305, 342, 344, 361
  enteroxantha, 261, 344, 347, 351
  farrea, 256, 344
  grisea, 344, 347, 351, 361
  pulverulenta, 256, 334, 347–348, 351
Phytomonas, 71
Phytoseiidae, 82
Picea, 147, 169, 331
  abies, 151
    ssp. *obovata*, 164
  glauca, 164, 168
  mariana, 164, 168–169
  maritima, 56
Pilophorus strumaticus, 371
Pilosella officinarum, 399, 401
Pinus, 147, 169, 309, 320
  banksiana, 164, 169
  contorta ssp. *latifolia*, 164
  sylvestris, 164, 169, 264, 331, 399
    ssp. *scotica*, 336
Pirnodus detectidens, 80, 85–86, 92
Placidiopsis custanii, 397
Placopsis, 201–202
  contortuplicata, 188
  gelida, 34, 191, 275, 371
Placynthium
  dolichoterum, 362
  flabellosum, 392
  lismorense, 359, 363
  nigrum, 159, 279, 351, 357, 359, 390, 402
  pannariellum, 362, 392
  subradiatum, 351
  tantaleum, 390
  tremniacum, 351
Plagiochila
  punctata, 329–330
  spinulosa, 330
Plantago coronopus, 401

Platismatia, 305
  glauca, 55, 151, 162, 167, 191, 234, 260, 262, 329, 335–336, 374, 431
  herrei, 153
  norvegica, 151, 158, 374, 406
  tuckermanii, 153
Plecoptera, 76
Plectus, 72
  cirratus, 72–73
Pleureta alpina, 73
Pleurococcus, 320, see also *Desmococcus*
Poa pratensis, 425
Pohlia nutans, 56
Polioptila
  caerulea, 126
  melanura, 126
Polyblastia, 353, 389
  agraria, 397, 402
  albida, 351, 353
  cruenta, 392
  cupularis, 351, 353, 362
  dermatodes, 353, 397
  diminuta, 353
  gelatinosa, 397, 402
  inumbrata, 363
  quartzina, 392
  schraderi, 351
  scotinospora, 362–363, 390
  sendtneri, 400
  theleodes, 362–363, 390
  tristicula, 360, 397, 402
  vouauxii, 256
  wheldonii, 397, 402
Polychidium muscicola, 392
Polycoccum
  galligenum, 16
  trypethelioides, 16
Polytrichum formosum, 329
Polyxenus, 83
Populus, 147, 160, 168
  balsamifera, 147
  tremuloides, 147, 169
Porella platyphylla, 347
Porina
  ahlesiana, 390

Porina—continued
 chlorotica, 279, 367–368, 387, 390
  var. *carpinea*, 316–317
  var. *persicina*, 351, 356, 358
 coralloidea, 325
 curnowii, 387
 guentheri var. *grandis*, 391
  var. *lucens*, 391
 hibernica, 325–326
 interjungens, 391
 lectissima, 367–368, 390
 leptalea, 317, 325
 mammillosa, 406
Porocyphus kenmorensis, 391
Prasiola quadrata, 385
Prosopis juliflora var. *velutina*, 240
Prostigmata, 79–80, 82, 88, 97–98
Protoblastenia
 immersa, 351, 395–396
 incrustans, 351, 353, 355
 metzleri, 351, 356, 395–396
 monticola, 351, 362–363, 395–396
 rupestris, 108, 262, 358–359, 362–363, 390, 395–396, 398
  var. *calva*, 362
 siebenhaarina, 362
Protozoa, 71–72, 110
Protura, 75
Provertex delamarei, 86
Prunella vulgaris, 399
Prunus
 avium, 39
 spinosa, 319, 394
Psaltriparus
 melanotis, 126
 minimus, 126
Pseudechiniscus suillus, 74
Pseudephebe
 minuscula, 38–39, 41, 195–197, 201, 206
 pubescens, 194, 201, 206, 373–374, 377
Pseudevernia, 305
 consocians, 153
 furfuracea, 102, 137, 152, 162, 167, 260, 264, 335–336, 427, 431

Pseudocyphellaria, 201, 315, 326
 anomala, 153
 anthrapsis, 153
 aurata, 340
 crocata, 158, 326
 intricata, 326
 lacerata, 326
 thouarsii, see *P. intricata*
Psocoptera, 75–77, 91–93, 100, 109–110
Psora ostreata, see *Lecidea scalaris*
Psoroma hypnorum, 200
Psorotichia, 228
 pictava, 282
 schaereri, 351, 359
Pterogonium gracile, 325
Pterygota, 75
Ptilidium pulcherrimum, 168
Ptychoglene
 phrada, 78
 tenuimargo, 78
Ptychomitriun polyphyllum, 377
Pycnothelia papillaria, 151, 403
Pygoctenucha
 funerea, 78
 terminalis, 78
Pyramidula rupestris, 108
Pyrenopeziza lettaui, 16
Pyrenula
 dermatodes, 315
 laevigata, 315
 nitida, 258, 312, 317
 nitidella, 317
Pyrocephalus rubinus, 126

Q

Quercus, 48, 147, 168, 264, 309, 311, 315, 317, 323, 326–328, 331, 333, 337, 340
 petraea, 328
 robur, 333

R

Rachophorus, 122
Racodium rupestre, 367–369

Radula complanata, 315
Ramalina, 49, 74, 80, 97, 133, 136, 241, 260, 263, 305, 339, 342, 345
  baltica, 319, 339, 347–348
  calicaris, 261, 339, 342, 348
  cuspidata, 387
  duriaei, 339
  farinacea, 127, 163, 234, 260, 264, 310–311, 325, 332–333, 339, 347–348
  fastigiata, 86, 152, 160, 261, 321, 332, 339, 347–348, 430
  fraxinea, 135, 152, 157, 160, 234, 261, 339, 430
  kullensis, 83
  maciformis, 233–238
  menziesii, 133, 135–136
  minuscula, 155
  pollinaria, 339
  polymorpha, 279, 378
  pulchribarbara, 20
  roesleri, 159
  scopularis, 386, see also *R. siliquosa*
  siliquosa, 262, 267–268, 386–388
  sinensis, 155
  subfarinacea, 163, 256, 378, 387
  terebrata, 194–195, 204
  thrausta, 159
Rangifer, 132
  tarandus, 72, 101, 129, 133, 166
Regulus satrapa, 126
Reptilia, 124–125
Reuterella helvimacula, 76–77, 91, 93, 100
Rhabdolaimus terrestris, 72
Rhacomitrium, 49
  fasciculare, 376–377
  heterostichum, 329, 369, 376–377
    var. *gracilescens*, 373
  lanuginosum, 373, 405–406
  protensum, 392
Rhizocarpon, 54, 56, 88, 170, 189, 391
  alpicola, 372–373
  badioatrum, 373
  constrictum, 380, 386–387, 394

Rhizocarpon—continued
  geminatum, 392
  geographicum, 32, 37, 39, 41, 56, 170, 262, 283, 366, 370–374, 377, 380–381, 387, 427
  grande, 170
  hochstetteri, 374
  laevatum, 391–392
  lecanorinum, 171, 371, 374, 376
  obscuratum, 150, 170, 262, 267, 363, 370, 374, 380, 383, 387, 391–392, 394
    var. *reductum*, 395
  oederi, 33, 286, 383
  petraeum, 362–363, 390
  polycarpon, 262, 374, 376
  riparium, 377
  rittokense, 56
  umbilicatum, 33, 351, 353
  viridiatrum, 380
Rhytidium rugosum, 399
Rinodina, 218–219, see also *Dimelaena*
  archaeoides, 201
  atrocinerea, 380, 387
  bischoffii, 256, 351
  confragosa, 387
  exigua, 261
  isidioides, 326
  luridescens, 387
  nimbosa, 201
  oxydata, 368
  petermannii, 194
  roboris, 310, 325, 333, 347
  sophodes, 315, 318
  subexigua, 255–256, 262, 265, 273, 277–280, 282, 284, 378, 385
  subglaucescens, 387
Roccella, 94
  cervicornis, 231
  fuciformis, 388
  montagnei, 137
  phycopsis, 388
  tinctoria, 137
Rodentia, 134–135
Rotatoria, 73
Rotifera, 73

Taxonomic Index

Rumex acetosella, 401
Rupricapra rupricapra, 134

S

Sagiolechia rhexoblephara, 400
Salix, 168, 308, 315, 341
   atrocinerea, 326
   herbacea, 405-406
   repens, 319
Sambucus, 345
Saproglyphidae, 82
Sarcogyne
   privigena, 32
   regularis, 255, 351, 395-396, 398
   simplex, 32, 420
Sarothamnus, 394
Scapania gracilis, 329-330
   umbrosa, 369
Schaereria cinereorufa, 371
Schismatomma
   decolorans, 307-312, 347
   niveum, 309
   virgineum, 308-309
Scirtothrips dorsalis, 102
Sciurus carolinensis, 257
Sclerophyton circumscriptum, 387-388
Scytonema, 392
Sedum acre, 399
Selasphorus
   platycerus, 126
   rufus, 126
   sasin, 126
Seligeria, 397
Senecio jacobaea, 399
Setophaga ruticilla, 127
Solenopsora
   candicans, 351, 353
   holophaea, 387
   vulturiensis, 387, 401
Solorina, 34, 167
   crocea, 73, 157
   saccata, 159, 355, 400, 419
   spongiosa, 400
Sorbus, 168, 315
Sphaerophorus, 305, 330

Sphaerophorus—continued
   fragilis, 376
   globosus, 191, 194, 201, 314, 328-329, 369, 374
Sphenobilus minutus, 368
Sphinctrina microcephala, 261-262
Sporidesmium, 77
Sporopodium fuscoluteum, 362
Squamarina
   crassa, 234, 355, 360, 398, 400, 402
     f. *pseudocrassa*, 402
   lentigera, 221, 398
Staurothele, 11, 15, 18
   caesia, 351
   clopima, 18
   fissa, 391-393
   hymenogonia, 351, 395-396
   rupifraga, 351
   succedens, 351, 390
Stenocybe
   pullatula, 315
   septata, 316
Stereocaulon, 16, 34, 49, 167, 371, 383
   alpinum, 201, 234
   dactylophyllum, 151, 370-371, 383, 421
   delisei, 371, 383
   evolutum, 370, 383
   glabrum, 194
   intermedium, 152
   nanodes, 383, 420-421
   paschale, 129, 134, 149, 156, 165-166
   pileatum, 274-275, 285-286, 371, 383, 392, 421
   sasakii, 152
   saxatile, 155, 406
   tomentosum, 149
   vesuvianum, 262, 275, 383
   vulcani, 34, 50
Stereotydeus villosus, 97
Stichococcus, 18
Sticta, 304, 315, 326
   canariensis, 88, 326, 387
   dufourii, 88, 326, see also *S. canariensis*

Sticta—continued
 filix, 14
 fuliginosa, 326
 limbata, 325
 sylvatica, 325
Stigonema, 392
Synalissa symphorea, 351, 359

T

Taraxacum laevigatum, 399
Tardigrada, 73–74, 110
Taxus, 309
Teloschistes, 305, 342, 348–349
 brevior, 231
 chrysophthalmus, 261, 319, 348–349
 flavicans, 319, 340, 348
 lacunosus, 233
 peruensis, 20
 villosus, 348
Tetranychidae, 82, 102
Tetranychus lapidus, 80, 82
Tettigoniidae, 76, 104
Thamnolia vermicularis, 128, 191
 var. *subuliformis*, 406
Thecamoebae, 71
Thelidium, 353
 aeneovinosa, 392
 decipiens, 351, 353, 359, 363, 390, 395–396
 fumidum, 392
 incavatum, 351, 353, 395
 macrocarpum, 397
 microcarpum, 396
 papulare, 351, 360
 pyrenophorum, 351, 360, 362–363, 391
Thelocarpon
 epibolum, 405
 laureri, 256
 magnussonii, 256
 olivaceum, 256
Thelopsis
 melathelia, 362, 400
 rubella, 325

Thelotrema
 lepadinum, 304, 316–317, 325
 monosporum, 315
 subtile, 315, 326–327
Tholurna dissimilis, 154
Thymus, 399
Thyrea radiata, 362
Thysanoptera, 76
Thysanura, 73
Tineidae, 78, 95
Tineola bisselliella, 94
Tomasellia
 gelatinosa, 315
 ischnobela, 315
Toninia, 219, 425
 aromatica, 351, 360, 387, 402
 candida, 360
 caradocensis, 322
 coeruleonigricans, 221, 398, 402
 havaasii, 406
 leucophaeopsis, 383
 lobulata, 398, 402
 mesoidea, 387
 squalescens, 406
 squalida, 406
 tristis, 406
Tornabenia
 atlantica, 261
 ephebaea, 231
Tortula laevipila, 347
Trachycephalus lichenatus, 122
Trapelia
 coarctata, 256–257, 268, 275, 363, 369–371, 374, 381
 involuta, 392
 moorei, 371, 392
 ornata, 382
Trentepohlia, 368
 iolithus, 171
Tricherememaeus serratus, 87
Trichomonas, 95
Trinema, 71
Tydeus tilbrooki, 80, 97
Tylenchus
 davainei, 73
 filiformis, 72–73

U

Ulex, 320
Ulmus, 74, 308, 312, 323, 327, 331, 345
  procera, 347
Ulota, 54, 315, 326
  bruchii, 315, 318
  crispa, 315
  drummondii, 315
  hutchinsiae, 315
  phyllantha, 315, 318, 332
Umbilicaria, 32, 49, 97, 129, 134, 188, 372, 374, 381, see also *Lasallia*
  antarctica, 204, 206
  aprina, 201
  caroliniana, 191
  cristata, 196–197
  crustulosa, 374, 377
  cylindrica, 161, 373–374, 377
  decussata, 201
  deusta, 170, 374, 377, 382
  esculenta, 137
  hyperborea, 161, 170, 373–374, 377
  muehlenbergii, 136, 155
  polyphylla, 262, 371–374, 376–377
  polyrrhiza, 372, 374, 376–377, 381–382
  proboscidea, 161, 374, 377
  pustulata, 380, see also *Lasallia pustulata*
  torrefacta, 262, 372–374, 377
  vellea, 150
Uroplates finbriatus, 125
Usnea, 16, 83, 97, 129, 133–134, 218–219, 241, 260–261, 263, 305, 314, 326, 338, 340
  antarctica, 91–92, 186–187, 206
  articulata, 340
  "*barbata*", 91
  ceratina, 340–341
  "*dasypoga*", 234
  extensa, 342
  fasciata, 188, 192, 194, 196, 204
  fibrillosa, 340
  filipendula, 340
  flammea, 319, 338, 374
  florida, 340, 342

Usnea—continued
  fragilescens, 338
  fulvoreagens, 342
  hirta, 169, 260, 336, 340
  inflata, 260, 329, 338, 340
  intexta, 338, 340, see also *U. inflata*
  longissima, 151, 159
  subgen. *Neuropogon*, 201
  rubiginea, 340
  subfloridana, 260, 329, 336, 338, 340, 342, 369, 429–430
  sulphurea, 186, 191–192, 196–197, 200–201
Usneaceae, 22

V

Vaccinium
  myrtillus, 402
  uliginosum, 402, 406
  vitis-idaea, 407
Verrucaria, 11, 15–16, 36, 83–84, 103, 108, 218, 349–350, 353, 355, 389
  aethiobola, 390–391
  amphibia, 389
  aquatilis, 389, 393
  calciseda, 108
  coerulea, 351, 360, 363
  degelii, 389
  ditmarsica, 389
  dufourii, 351, 359–360
  elaeomelaena, 390, 393
  erichsenii, 389
  fusconigrescens, 378, 387
  glaucina, 351, 364
  hochstetteri, 351, 353, 359, 364, 395–396, 398, 401
  hydrela, 390–391
  internigrescens, 387
  kernstockii, 390, 393
  latebrosa, 393
  margacea, 390–393
  marmorata, 276
  marmorea, 32
  maura, 385, 388–389
  microspora, 201, 389

Verrucaria—continued
  mucosa, 389
  muralis, 258, 273, 351, 353, 358, 364, 395–398, 401
  mutabilis, 395–397
  nigrescens, 255–256, 279, 283, 351, 357–358, 394–396, 398
  papillosa, 284
  praetermissa, 391
  prominula, 387
  psammophila, 398
  sandstedei, 389
  silicea, 390, 393
  sphinctrina, 351, 353, 359, 364, 427
  striatula, 389
  viridula, 256, 351, 364, 395–397
Vezdaea
  aestivalis, 401
  leprosa, 401
Vireo
  atiloquus, 126
  flavifrons, 126
  griseus, 126
  philadelphicus, 126
  solitarius, 126

W

Wilsonema, 72

X

Xanthoria, 74, 87, 94, 97, 219, 260, 263, 305, 342–344, 361, 431
  aureola, 72–73, 256, 262, 279, 351, 354, 361, 397
  candelaria, 260, 344, 378, 381
  elegans, 37, 150, 163, 196–197, 275, 283, 355, 397
  fallax, 160, 237–238, 344
  isidioidea, 234
  lobulata, 55
  parietina, 15, 72–73, 92–93, 95, 102, 161, 163, 255–256, 258, 260, 262, 273–274, 277–279, 283–284, 287, 334, 339, 344, 347–348, 351, 361, 378, 385, 387, 397, 427
  polycarpa, 22, 55, 256, 260, 264, 284, 321, 344
  resendei, 355
Xylographa, 169
  abietina, 260, 264, 320
  vitiligo, 260

Z

Zercon, 82
Zoothamnium adamsi, 71
Zygodon
  baumgartneri, 325
  viridissimus, 332, 347

Subject Index

Note: Appendix A and Appendix B are not indexed.

A

Acarosporetum sinopicae, 377, 382–383
Acarosporion
　fuscatae, 378
　sinopicae, 371, 382
Aegricorpus, 11–12
Aesthetics, 276, 423
Agricultural chemicals, 138, 260, 302, 304, 331, 343, 390, 419, 426–428
Air pollution, 2–3, 35, 38, 55, 70, 100, 105–107, 111, 128, 136, 138, 160–162, 254, 269–272, 287, 302, 304–305, 308, 312, 314–315, 319–322, 331, 333–335, 338, 344, 355, 366, 372, 382–383, 417, 419–421
Alectorietum fremontii, 168
Aluminium, 282
Amaretum, 300, 316
Amphi-Atlantic, 150–151
Amphi-Pacific, 150–152
Anabiosis, 73
Antarctic, 84, 96–97, 99, 183 *et seq.*
Anthropochorous, 160
Apothecial production, 22
Arctic, 84, 96, 101, 111, 184 *et seq.*
Arctoeto-Callunetum, 405–406
Arid lands, 211–252
　impact of man, 243–245
　relationships of floras, 219–223
　sources of data, 215–217
Arthonietum
　impolitae, 307, 312

Arthonietum—continued
　luridae, 317
Arthopyrenietum
　conoideae, 357
　gemmatae, 344
　punctiformis, 315, 342
Asbestos-cement, 45, 271–273, 275, 277, 283, 287–288, 349, 361, 419
Ascospores, 73, 87, 205, 221
Asphalt, 275
Aspicilietum
　calcareae, 85, 352
　contortae, 353
　lacustris, 365, 391–392
　　subun. *Haplocarpon hydrophilum*, 391
　verrucosae, 362
Aspicilion calcareae, 349, 357, 390
Attraction of invertebrates, 85–87

B

Bacideetum chlorococcae, 316, 320–321
Barium, 420
Biatoreto-Chaenothecetum, 366
Biatoretum lucidae, 366
Biogeochemical erosion, 25–26, 32–33
Biomass, 45–46, 95, 132, 166, 169
Bipolar, 190, 201
Bird-perching stones, 287, 342, 360, 382, 391
Bone, 255

Boreal, 146 *et seq.*
  distribution types, 149–156
  ecoclimatic differentiation, 156–159
  edaphic-topographic differentiation, 159–160
  effect of man, 160–162
  exploration, 147–148
  origin, 148–149
  speciation, 162–164
Brick, 268–270, 274–275, 349, 366
Bryophilous, 397
Buellietum punctiformis, 331, 343, 345
Buellion canescentis, 342
Burning, 57, 129–130, 165, 402

C

Calicietum
  abietini, 307
  hyperelli, 305–306, 312, 320
Calicion
  hyperelli, 305–306, 312, 320
  viridis, 306
Callunetum, 402
Caloplacetum
  aurantiae, 353
  cirrochroae, 355–356
  citrinae, 353
  elegantis, 361
  heppianae, 85, 351–356, 358, 360
  marinae, 349, 384–385, 389
  murorum, 353
  phloginae, 320, 344
  variabilis, 353
Caloplacion
  decipientis, 349
  pyraceae, 350
Camouflage, 79, 102–104, 110, 122, 124–125, 128, *see also* Mimicry
Candelarielletum corallizae, 361, 378, 385
Cardboard, 266
Caribou, *see* Reindeer
Cement, 268, 273
Cephalodia, 10, 14
Chaenothecetum melanophaeae, 307

Chalk, 20, 389, 395, 397, 418
Chemotaxonomy, *see* Lichen products
Chomophytic communities, 170
Chorology, 3–4
Churchyards, 37, 271–273, 275–277, 285–286, 309, 355–356, 360–361, 386
Circumpolar, 149–150, 191, 196–197
Cladineto-Callunetum, 405
  -Vaccinetum, 405
Cladinetum, 171
Cladonieto-Usneetum tuberculatae, 314, 338
Cladonietum
  alcicornis, 394, 401
  alpestris, 166
  cenoteae, 313–314
  coniocraeae, 313–314, 338
    subass. *cladonietum digitatae*, 314
    var. *macrocladonietum digitatae*, 314
    var. *microcladonietum coniocraeae*, 314
    var. *sphaerophoretum globosae*, 314
Cladonion coniocraeae, 305–306, 313, 403
Coconut shell, 266
Coenogonio-Racodietum rupestris, 368
Cold desert, 184, *see also* Antarctic, Arctic *and* Desert
Collematetum
  crispi-Verrucarietum muralis, 357
  multipartitis, 357
  pulposi, 357
  tunaeformis, 350
Collemation tunaeformis, 350
Collemion rupestris, 350
Colonization, 2, 32–35, 78, 82, 85, 167–168, 263, 266–267, 273–275, 288–289, 320, 323, 394, 397, 420, 429, 431
Colour, 17, 226–227
Competition, 55–69, 108, 321
Competitive index, 57–58
Concealment, 102
Concrete, 271–273, 275, 349, 355, 427

Coniocybetum furfuraceae, 308, 312, 365–366, 370
Coniocybion gracilentae, 306
Conizaeoidion, 300, 319
Conservation, 2–3, 276–277, 415–416
  priorities, 421–423
  site evaluation, 433
  strategy, 432–434
Continental drift, 221
Copper, 283, 382, 420
Cork, 264–265
Corticolous, see Epiphytic communities
Cotton, 266
Cyphelietum
  inquinantis, 320
  tigillaris, 321
Cysts, 71, 73
Cystocoleion nigri, 365
Cystocoleo-Racodietum rupestris, 368

D

Decomposition, 101
Deserts, 25–26, 111, 211, 214 et seq.
  cold, 184, see also Antarctic and Arctic
  fog, 20
  hot, 3, 19–20, see also Arid lands
Digestion, 88, 101–111
Dirinetum stenhammariae, 351, 356
Disjunctions, 203, 220–223, 267, 324
Dispersal, 2, 34, 57, 71, 75, 79, 84, 87, 110–111, 160–161, 204–206, 211, 221
Dunes, 400–402, 425
Dust, 260, 263, 288, 345, 361, 421
Dyes, 1, 95, 137

E

Ecad, 19
Ecophene, 19
Ecophysiology, 3–4, 14, 232–239, 288–289
Ectocrine, 109

Endemism, 241, 391
Energetics, 95–99
Environmental modifications, 9–27, 231–232
Enzymes, 84, 89, 111
Ephebetum lanatae, 392
Epidiorite, 361, 400
Epiphytic communities, 167–169, 224, 240, 254, 372, 418, 421–422
  British Isles, in, 304–349
  factors affecting, 304
Epitaphs, 276–277
Epizoic symbiosis, 103
Erosion, 241, see also Biogeochemical erosion
Erratic, see Vagants
Establishment, see Colonization
Euphysodetum, 167–168

F

Feeding, 73–76, 80, 83, 88–89, 95–96, 100–101, 110–111
Fensterflechten, 227–229
Fertilizers, see Agricultural chemicals
Fibre-glass, 266
Flint, 267, 397
Fog oases, 240
Food
  for invertebrates, 73 et seq., 88–100, 243
  for man, 136–137
  for invertebrates, 122, 128, see also Reindeer
  webs, 99–101, 111
Freshwater, 389–393
Fulgensietum fulgentis, 398
Fur, see Hair
Fuscelletum, 312
Fuscideetum kochianae, 372–373

G

Galls, 16, 82–83, 88, 111
Germination, 2
Glaciation, 203, 205–206

Glass, 277–282, 383
Graphidetum scriptae, 316, 319
Graphidion, 314, 331
 scriptae, 305, 314, 326
Graphinetum platycaprae var. *graphinetum anguinae*, 316
Gravestones, *see* Churchyards
Grazing, 2, 75, 80, 83–85, 88–91, 94, 101, 108, 111, 131–132, 431
Growth, 2, 27, 35–46, 274, 323
Gyalectetum
 cupularis, 355
 jenensis, 351–352, 355–357, 390, 400
Gyalectinetum carneoluteae, 343
Gyalection cupularis, 349
Gyrophoretum cylindricae, 376

H

Hair, 257–258
Heathland, 95, 370, 390, 394, 402–407, 419, 422, 427
Heavy metals, *see* Metals
Hot deserts, *see* Arid lands *and* Deserts
Huilietum crustulatae, 370, 394, 403
Hygrochasy, 230
Hypertrophication, 254, 259–260, 265, 287, 342–343, 345
Hypogymnio physodis-Parmelietum saxatilis, 335

I

Impolitetum, 307
Index of Ecological Continuity, 51
Industrial melanism, *see* Melanism
Ionaspidetum
 odorae, 392
 suaveolentis, 392
Iron, 282–286, 383
Iron-rich rocks, 17, 286
Isidia, 16, 55, 58

L

Laminarinase, 84
Lasallietum pustulatae, 380

Lava, 275
Leaching, 287, 305
Lead, 282, 382–383, 420
Leather, 256–257
Lecanactidetum
 abietinae, 306, 308–309
 premneae, 309–311, 356
Lecanactinetum stenhammari, 352, 356
Lecanactinion stenhammari, 350
Lecanoretum allophanae, 318
 atrae, 386
 calcareae, 353
 campestris, 353
 carpineae subass. *atlanticum*, 318
  subass. *montanum*, 318
 coarctatae, 370
 dispersae, 353
 epanorae, 265, 383
 glabratae, 317
 laevis, 319
 orostheae, 366
 pityreae, 302, 316, 320–322
 rupicolae, 379
 sordidae, 371, 379–380, 382
 subfuscae, 316, 318–319, 342, 394
 variae, 320
Lecanorion
 calcareae, 349
 carpineae, 318
 dispersae, 350
 galactinae, 350
 subfuscae, 305, 318
 variae, 305–306, 319
Lecideeto parasemo-Phlyctideetum, 318
Lecideetum
 crustulatae, 370
 erraticae, 394–395
 juranae, 355
 kochianae, 372
 kochiano-aggegatilis, 372
 lithophilae, 371
 lucidae, 365–366
 orostheae, 366
 ostreatae, 322
 scalaris, 322
 watsoniae, 395–397

Lecideion
   ostreatae, 322
   tumidae, 370, 372
Leprarietum
   candelaris, 312
   chlorinae, 366
   incanae, 312, 365–366
Leprarion, 306
   chlorinae, 365
Leproplacetum chrysodetae, 351–352, 356–358
Letharietum divaricatae, 338, 340
Lichen products, 56, 86, 91–95, 105, 111, 136, 243, 394
Lichenase, 84
Lichenicolous
   fungi, 10, 16
   lichens, 10
Lichenin, 84, 89
Lichenometry, 41, 44, *see also* Growth
Lignicolous communities, 169–170, 224, 258–264, 268, 320–322, *see also* Epiphytic communities
Limestone, 267, 272–273, 288, 349–361, 382, 389–390, 393, 400, 402, 427
Linoleum, 265
Lobarietum pulmonariae, 322
Lobarion, 48–49, 54, 304, 323–327, 331
   pulmonariae, 304, 306, 315, 322–323, 325, 330, 343

M

Magnesium, 400
Man-made substrates, 2, 253–289, 382, *see also* named substrates
Manna, 19, 230
Marine and maritime, 3, 70, 83–84, 98–99, 103, 108–109, 112, 157, 267–268, 274, 287, 349, 365, 378, 380, 384–389, 418, 428
Medicines, 137
Melanism, 78, 102, 105–107, 110, 128

Memorials, 272–273, *see also* Churchyards *and* Monuments
Metals, 89, 100, 111, 131, 138, 282–286, 377, 382–383, *see also* named metals
Micareetum sylvicolae, 308, 366–367
Mimicry, 78, 85, 87, 102, 104–105, 122, *see also* Camouflage *and* Melanism
Mine waste tips, 420–421
Modifications, 4, 9–17, 231–232
Monuments, ancient, 267–268, 275–276, 386, 419, *see also* Churchyards
Morphotype, 19
Mortar, 268–270, 272–273, 349, 362, 401
Muscicolous, 397

N

Nephrometum
   laevigatae, 322
   lusitanicae, 324–325
Nests, 82, 126–128
Nitidetum, 314, 317
Nitrogen fixation, 4

O

Oil spills, 428
Opegraphetum
   fuscellae, 312–313
   herpeticae, 317
   horistico-gyrocarpae, 368, 390
   saxicolae, 356
   zonatae, 368
Oroboreal, 147
Over collecting, 433
Overgrowth, 58
Oxydated thalli, 17, 286, 383

P

Paint, 288
Paper, 266
Parasymbionts, 10, 53, *see also* Lichenicolous

Parmelietum
  acetabulae, 344–345
    var. *parmeliosum caperatae*, 334
  caperatae, 330–331, 334, 342
  carporrhizantis, 343–344
    subass. *parmelietosum endochlorae*, 343
  cervicornis, 331
  conspersae, 334, 381
  elegantulae, 344
  furfuraceae, 334–335
  furfuraceae-physodes, 335
  glomelliferae, 370, 379–382
  isidiotylae, 380
  laciniatulae, 344
  laevigatae, 309, 315, 327–330, 333
  olivaceae, 168
  omphalodis, 373–377, 382
  revolutae, 312, 318, 330–334, 344–345
    var. *caperatosum*, 331
    var. *parmeliosum laetevirentis*, 331
  saxatilis, 335
  subauriferae, 331
  sulcatae, 335
  trichotero-scorteae, 331
Parmelion
  caperatae, 330, 342
  conspersae, 378
  furfuraceae, 334
  laevigatae, 305–306, 326–327, 408
  omphalodis, 376
  perlatae, 305–306, 308, 330–331, 334–335, 344, 408
  saxatilis, 334, 378
Parmeliopsidetum, 337
  ambiguae, 168, 334, 337–338
    subass. *platismatietosum glaucae*, 335
Parmeliopsidion ambiguae, 334
Pebble communities, 365, 370, 373, 394–397, 403
Pedogenesis, 32–33, 50, see also Biogeochemical erosion
Perfumes, 137

Pertusarietum
  amarae, 316–317
  corallinae, 370–371
  hemisphaericae, 316
  wulfenii, 316
Pertusario-Racodietum rupestris, 368
Pesticides, 74, 100, see also Agricultural chemicals
Phenotypic modifications, see Modifications
Phyllidia, 58
Physcietum
  ascendentis, 320, 342, 344–349, 360–361
    subass. *physciosum leptaleae*, 348
  caesiae, 345, 350–352, 360–361, 378, 382, 391
  dubiae, 361
  elaeinae, 345
    var. *buelliosum canescentis*, 345
  teretiusculae, 361
Physciopsidetum elaeinae, 345
Physodeto-Sulcatetum, 168, 335
Physodetum, 335
Physodion, 167, 334
Phytogeography, 2–3, see named elements
Phytosociology, 2, 211, 225–226, 297–304
  British Isles, in, 304 et seq.
  nomenclature, 298–301
  taxonomy, 301–304
Placynthieto-Verrucarietum nigrescentis, 357
Placynthietum nigri, 351–353, 356–360
Plastic, 258, 266
Polygoneto-Rhacomitretum lanuginosi, 406
Porinetum carpineae, 317
Pottery, 274
Prairie, 3
Pruinosity, 17
Pseudevernietum furfuraceae, 168, 320, 322, 331, 333, 335–337, 372, 374, 382
  subass. *subfloridanae*, 342

Subject Index

Pseudevernion furfuraceae, 306, 331, 334–335, 371
Psoretum ostreatae, 319, 322
Ptilidietum, 168
Putty, 282
Pyrenuletum nitidae, 314, 316–317, 343

Q
Quarries, 267

R
Racodietum rupestris, 365, 368–369
Radionuclides, 100, 111, 132–133, 138
Ramalinetum
  duriae, 342
  fastigiatae, 319, 339, 342–343
  scopularis, 378, 380, 382, 384–387
  siliquosae, 386
Refuse, 275
Regeneration, 17
Reindeer, 51, 72, 89, 101, 129–133
Reproductive capacity, 57
Reservoirs, 273
Resistance to feeders, 90, 94
Rhacodietum, 368
Rhizocarpetum
  alpicolae, 373, 376
  concentricae, 364
  geographicae, 373
Rhizocarpion alpicolae, 372
Rubber, 266
Rufescentium, 317

S
Savanna, 3
Saxicolous communities, 170–171, 225, 232, 240, 264, 335, 431, see also named substrates *and* habitats
  British Isles, in, 349–397
Schizidia, 58
Sclerophytetum circumscriptae, 365, 384, 387–388
Sculptures, 275–276

Secondary plant substances, see Lichen products
Sewage farms, 273
Shingle, 394–397
Siliceous rocks, 364–393
  aquatic, 389–393
  exposed, 370–377
  marine and maritime, see Marine and maritime
  mineral rich, 382–383
  nutrient-rich, 378–382
  shaded, 365–370
Silk, 258
Silver, 382
Slate, 275
Soil formation, 2, 108, see also Pedogenesis
Soredia, 54–55, 58, 79, 87, 204, 365
Speciation, 162–164, 241–242
Steppes, 19
Subfossils, 258
Succession, 2, 46–55, 165, 171, 303, 305, 314, 355, 376
Symbiotes, 89
Synthetic organic materials, 266

T
Taiga, 147
Taxonomy, see Modifications
Teloschistetum
  chrysophthalmae, 342, 344, 349
  flavicantis, 340, 348
Teloschistidion chrysophthalmi, 342
Terricolous communities, 164–167, 223–224, 232, 240, 259, 264, 349, 418, 424, 428, 431
  British Isles, in, 313, 393–407
  Deserts, see Arid lands
Thatch, 265
Tombstones, see Churchyards
Trampling, 244, 302
Transplants, 4, 57, 428–432
Trichoterion, 330
Tropics, 3, 84

U

Umbilicarietum
　cylindricae, 370–373, 376–377
　deustae, 376–377
　pustulatae, 380–381
Umbilicarion
　cylindricae, 372, 374, 376
　hirsutae, 374
Usneetum
　articulato-floridae, 340, 342
　　var. *ceratinae*, 340–341
　barbatae, 338, 340, 342
　dasypogae, 338
　filipendulae, 340
　florido-neglectae, 340
　rubicundae, 340
　subfloridanae, 338, 340, 342
Usneion, 340, 342
　barbatae, 305–306, 318, 338
　dasypogae, 338
　florido-ceratinae, 338

V

Vagants, 19–23, 26–27, 220, 230–231
Verrucarietum
　laevato-denudatae, 393

Verrucarietum—continued
　maurae, 349, 384–385, 388–389
　siliceae, 391–393
Vicariants, 164

W

Wanderflechten, 20
Water relations, *see* Ecophysiology
Weathering, 257, 289, *see also* Biogeochemical erosion
Wood, *see* Lignicolous communities
Woodland management, 138
Wool, *see* Hair

X

Xanthorietum
　aureolae, 361
　candelariae, 344
Xanthorion, 168, 302, 320, 324, 334, 340
　parietinae, 305–306, 308, 342, 344, 360, 378, 382

Z

Zinc, 265, 282–283, 420

108357